ENGLISH PLACE-NAME SOCIETY. VOLUME LVI/LVII
FOR 1978-9 & 1979-80

GENERAL EDITOR
KENNETH CAMERON

CORNISH
PLACE-NAME
ELEMENTS

ENGLISH PLACE-NAME SOCIETY

The English Place-Name Society was founded in 1923 to carry out the survey of English place-names and to issue annual volumes to members who subscribe to the work of the Society. The Society has issued the following volumes:

    I. (Part 1) *Introduction to the Survey of English Place-Names.*
       (Part 2) *The Chief Elements used in English Place-Names.*
       (Reprinted as one volume).
    II. *The Place-Names of Buckinghamshire.*
    III. *The Place-Names of Bedfordshire and Huntingdonshire.*
    IV. *The Place-Names of Worcestershire.*
    V. *The Place-Names of the North Riding of Yorkshire.*
  VI, VII. *The Place-Names of Sussex,* Parts 1 and 2.
VIII, IX. *The Place-Names of Devonshire,* Parts 1 and 2.
    X. *The Place-Names of Northamptonshire.*
    XI. *The Place-Names of Surrey.*
    XII. *The Place-Names of Essex.*
    XIII. *The Place-Names of Warwickshire.*
    XIV. *The Place-Names of the East Riding of Yorkshire and York.*
    XV. *The Place-Names of Hertfordshire.*
    XVI. *The Place-Names of Wiltshire.*
    XVII. *The Place-Names of Nottinghamshire.*
    XVIII. *The Place-Names of Middlesex (apart from the City of London).*
    XIX. *The Place-Names of Cambridgeshire and the Isle of Ely.*
XX, XXI, XXII. *The Place-Names of Cumberland,* Parts 1, 2 and 3.
XXIII, XXIV. *The Place-Names of Oxfordshire,* Parts 1 and 2.
XXV, XXVI. *English Place-Name Elements,* Parts 1 and 2.
XXVII, XXVIII, XXIX. *The Place-Names of Derbyshire,* Parts 1, 2 and 3.
XXX, XXXI, XXXII, XXXIII, XXXIV, XXXV, XXXVI, XXXVII. *The Place-Names of the West Riding of Yorkshire,* Parts 1–8.
XXXVIII, XXXIX, XL, XLI. *The Place-Names of Gloucestershire,* Parts 1–4.
XLII, XLIII. *The Place-Names of Westmorland,* Parts 1 and 2.
XLIV, XLV, XLVI, XLVII, XLVIII. *The Place-Names of Cheshire,* Parts 1, 2, 3, 4 and 5, I: i.
XLIX, L, LI. *The Place-Names of Berkshire,* Parts 1, 2 and 3.
LII, LIII. *The Place-Names of Dorset,* Parts 1 and 2.
LIV. *The Place-Names of Cheshire,* Part 5, I: ii.
LV. *The Place-Names of Staffordshire,* Part 1.

The volumes for the following counties are in preparation: *Durham, Kent, Leicestershire & Rutland, Lincolnshire, the City of London, Norfolk, Northumberland, Shropshire, Staffordshire.*

All communications with regard to the Society and membership should be addressed to:

    THE HON. DIRECTOR, English Place-Name Society, School of English Studies, The University, Nottingham, NG7 2RD.

ENGLISH PLACE-NAME SOCIETY. VOLUME LVI/LVII

CORNISH PLACE-NAME ELEMENTS

By
O. J. PADEL

ENGLISH PLACE-NAME SOCIETY
1985

Published by the English Place-Name Society
Nottingham

© English Place-Name Society 1985

ISBN: 0 904889 11 4

Printed in Great Britain
by the University Press, Cambridge

The collection from unpublished documents of material for the Cornwall volumes has been greatly assisted by grants received from the British Academy

CONTENTS

	page
Acknowledgments	ix
Introduction	xi
Note 1: Classification of Usages	xiv
Note 2: Alphabetical Order	xvii
Alphabetical list of elements	xix
Abbreviations	
Abbreviated references	xxxi
County, etc., abbreviations	xxxix
Other abbreviations	xl
Dictionary of Cornish place-name elements . . .	1
Indexes	
Index of Cornish place-names cited	243
Index of non-Cornish place-names cited . . .	323
Index of rejected elements	333
Indexes of Welsh and Breton cognates . . .	337

MAPS

Cornwall: Hundreds	351
Distribution of **tre** as B2	352
Distribution of ***bod** as B2	353
Distribution of ***ker** as B2	354

ACKNOWLEDGMENTS

This dictionary has taken four years to compile. It is based on the collections of the Cornish Place-Name Survey at the Institute of Cornish Studies (University of Exeter); these collections are themselves based in part upon those of three men, J. E. B. Gover, Charles Henderson, and the Rev. Mr W. M. M. Picken: without their assistance it would have taken much longer. I am particularly grateful to Mr Picken, whose patient answers to endless inquiries have saved me from many mistakes and supplied much valuable material. In addition, many others have supplied information or made useful comments, either on individual points or on the typescript of the whole dictionary. They include: Gareth Bevan and the staff of Geiriadur Prifysgol Cymru; A. Buckley; Professor Kenneth Cameron; Miss C. R. E. Coutts; L. Douch and the staff of the Royal Institution of Cornwall; Professor Ellis Evans; John Field; Professor Léon Fleuriot; Dr Margaret Gelling; Dr K. George; Andrew Hawke; P. L. Hull and the staff of Cornwall County Record Office; Professor Kenneth Jackson; Professor Bedwyr Lewis Jones; E. Martin; Michael Polkinhorn; P. A. S. Pool; Miss A. Preston-Jones; Oliver Rackham; Tomos Roberts; Paul Russell; Richard Sharpe; Professor Charles Thomas; I. Thomas. For those errors and omissions which remain, I alone am responsible: additions, corrections and comments will be most welcome. If the book could have a dedication, it would be to the memory of the late J. E. B. Gover, whose unpublished work on Cornish place-names for the English Place-Name Society has been of such use to many people in Cornwall, and especially myself. The dictionary would not have been possible if it had not been for the far-sightedness of the University of Exeter and Cornwall County Council, which jointly set up the Institute of Cornish Studies in 1972. My greatest debt is to Mrs Margaret Bunt, secretary at the Institute: her scrupulous and patient help has made the preparation of the dictionary a much easier task than it would otherwise have been.

INTRODUCTION

THE present volume is necessarily provisional. It contains all the Cornish elements that I have found in Cornish place-names up to the end of 1983; as the place-name survey continues, no doubt further elements will be identified, and further instances of known elements. If an element does *not* appear in this dictionary, it can safely be concluded that no satisfactory evidence for it has yet been found. However, there may well be some respectable elements, especially among west-Cornish field-names, that do not appear here. See also the Index of Rejected Elements.

Since no comparable dictionary of place-name elements is available for any Celtic language, there is extensive reference to parallel place-name material in Welsh and Breton, particularly for the period before c. 1200, when the three languages formed almost a linguistic unity, with many place-names and usages common to all three. To some extent the citations are haphazard, consisting of parallels which I happen to have noticed (though the Place-Names Archive of Melville Richards in University College, Bangor, has been of great assistance in systematically following up particular points in Welsh); their importance lies in the fact that a particular derivation is greatly corroborated if it, or a similar usage, is found in more than one of the Brittonic languages; and in the fact that it is interesting in itself when parallels occur over an area reaching from North Wales (or even Southern Scotland) to Brittany.

Since there is no authoritative dictionary of Cornish, considerable attention has been given to citing the authority for a particular word. This is necessary, both for proving the occurrence of a word in the language, and also for citing examples of its forms and meaning: too much work on Cornish has been liable uncritically to cite from dictionaries words or forms which do not actually occur. It will be seen, too, that the context of a particular word in the texts is sometimes useful for suggesting more precisely its meaning in place-names. The work towards a dictionary of Cornish by Andrew Hawke at University College, Aberystwyth, has been of considerable use in this matter.

The selection of head-forms has been a very difficult problem. As the best of several possibilities, all with their own disadvantages, it has been decided to use attested Middle Cornish forms where available; where not available, either a hypothetical (starred) Middle Cornish form, or an Old or Modern Cornish form if suitable. In a few cases, where it seems unlikely that the element was in use later than the Old Cornish period, a starred Old Cornish form has been given instead. (Note that these are pronunciation-spellings, and not necessarily in accordance with Old Cornish orthography.) When a Middle Cornish form is cited without any source, it is so common in the texts as not to need specifying. The reasons for choosing Middle Cornish head-forms, rather than Old Cornish (as an English onomast might expect) are, first, that it is the period of the most extensive remains of the language, and thus requires the smallest number of starred forms; secondly, it coincides with the period of the greatest number of important place-name forms, and indeed the spellings of words in the texts often coincide to a surprising extent with those in documentary forms of place-names; and, thirdly, it is closer to the Revived Cornish spelling-system than Old Cornish is, and therefore more accessible to people within the county. The disadvantages are that Middle Cornish is further removed from Welsh and Breton than Old Cornish is (and for that reason, indexes of Welsh and Breton cognates, with their Cornish equivalents, are provided); that the spellings are sometimes ambiguous or imprecise as to the sounds intended (which have therefore been specified where appropriate); that in East Cornwall, the English often borrowed a Cornish name in the Old Cornish period, thus fossilising its form, so that many names, e.g. Trewint, Trenant, show an older form of an element than its Middle Cornish head-form (but, since it is, if anything, simpler to extrapolate the form of an element backwards than forwards, that is not a severe disadvantage); and, finally, that the alphabetical arrangement of such a hotch-potch of forms poses considerable difficulty: see below, note 2.

Since there is no treatment of Cornish phonology available (apart from the earlier period, in LHEB), a certain amount of detailed discussion is unavoidable in the case of certain elements. It goes without saying that an understanding of the development of the Cornish sound-system is essential before derivations can be suggested for place-names. Jackson's *Historical Phonology of Breton* has been invaluable here; but it should be noted that when parallels are cited from it, it is not necessarily being suggested that Cornish followed the Breton development in all respects: rather that the Breton changes cited (often irregular or sporadic ones) provide support for suggesting similar changes, often irregular or sporadic themselves (especially in place-names), within a closely similar phonological system. Alternatively, it may well be that one language influenced the other (at any time up to the 18th century) in respect of particular words: see, for example, *kemor, *pylas and *talar; this is a topic which needs further investigation.

There must have been a period when both Cornish and English were spoken by those who bestowed place-names, and that is sometimes reflected in the names themselves: sometimes an English word is used in a name with Cornish elements and Cornish syntax. For the most part these words are not included; a loan-word into Cornish is, in general, given as an element only if it occurs in more than one name (and thus seems likely to have been borrowed as a functioning element), or if it also occurs as a loan-word in the other remains of the language. In the case of some simplex names consisting of an English word borrowed into Cornish, or of a Cornish word which survived into English dialect, it is impossible to say whether the name was bestowed as a Cornish or English one.

The occurrences given of a particular element are not necessarily exhaustive; in the case of the very common elements, they are highly selective. However, when only a very few instances are given, or only one, it can usually be assumed that other definite examples are not available at the moment.

Citing genders of nouns is often a problem. The gender may be unclear in the remains of the language, or Welsh and Breton may give conflicting evidence, or the place-name evidence may disagree with the other evidence. For these reasons, and because genders in Brittonic are sometimes rather uncertain, these indications should be taken only as giving guidance, not as rigid classifications. Where not given at all, the gender is either unknown or irrelevant; in some cases, it is only deduced from the place-name evidence. Several common words seem to have been variable in gender (see, for instance, **rid**, ***ros** and **tal**). On Cornish genders, see Loth, RC 34 (1913), 159-75.

Brittonic personal names are a whole subject in themselves, very inadequately treated in print, and Cornish ones least of all. Many personal names seem to have been identical in form to individual common nouns, and in some cases the common nouns are here cited as possible elements, even though their use in place-names is much more likely to have been as personal names than as common nouns.

(See, for instance, *arwystel, *kenow and *kest.) This is a problem shared with many other languages, including the Germanic ones, and it is insoluble.

For reasons of space, it is out of the question to give the evidence for particular derivations; in the case of field and coastal names, there may not be any, since early forms for many of these names will probably never be available: in such cases, one can only suggest derivations on the basis of the available spellings (often only 19th century in date). Since these forms may well have been subject to corruption or folk-etymology, some such derivations may have to be revised in the unlikely event of earlier spellings being found. In fact, even a 10th-century form may have been corrupted in similar ways – a fact often overlooked – so that one is in effect applying the same technique in both cases: interpreting a name on the basis of the extant spellings. In general, it can be assumed that the more improbable a derivation looks on paper, the better the available evidence for it. Spellings or other evidence will be willingly provided to interested inquirers, and will be published in due course as part of a full treatment of Cornish place-names.

Note 1: Classification of Usages

Some system of classification for the different uses of elements, as in Hugh Smith's Elements, 1, liii, is essential. However, the different usages in the Celtic languages are rather more complex than in English, so that the simple classification employed by him, or the simple grouping of use as simplex, generic and qualifier, will not do. Unfortunately there is no systematic treatment or classification of Celtic place-name syntax, so a system has had to be invented. It is hoped that it will be found clear, and ideally one would like to see it used in treating place-names in the other Celtic languages: if it is found unsuitable, perhaps a better system will be evolved. Naturally there are some categories of name that do not fit readily into the system; but that would apply to any classification, and does not invalidate its general use.

The primary grouping is to use A to designate use of an element as a simplex name; B denotes use as the generic in a name with two elements; and C denotes use as qualifier in such names. Along with this, in names of two elements, type 1 indicates proper compounds (close or loose), where the qualifier precedes the generic; and type 2 denotes name-phrases (qualifier following the generic), which are the commonest type of name in all the Celtic countries in the historic period. (For examples, see below.) It is often said, and is undeniable as a general rule, that the Celtic languages originally formed place-names with the qualifier preceding the generic, forming a compound (just as normally in the Germanic languages, including English); and that, in about the 5th century, they changed and favoured instead the name-phrase type, familiar in any modern Celtic area. See, for instance, LHEB, pp. 225–7; Gelling, Signposts, pp. 51f.; and D. Mac Giolla Easpaig, 'Noun + noun compounds in Irish place-names', ÉC 18 (1981), 151–63. The difference and chronology of the two types are nicely illustrated by a contrasting pair such as Gaulish *Eburodunum*, a compound (4 exx., ACS, I, 1398–1400; DAG, pp. 45, 51 and 1215), compared with Welsh Din Efwr, Crm. (ÉC 13 (1972–73), 371), a name-phrase; and the pair Morchard, in Devon (PNDev., II, 380), a compound, 'great-wood', contrasted with Cutmere in East Cornwall, a name-phrase, 'great wood'.

However, while the chronological distinction is obviously valid as a general rule, there are several reasons why it should not be taken as a rigid one. Compounds are not necessarily earlier than the 5th century, since they can be formed at any time in any Celtic language (see, for instance, S. J. Williams, *A Welsh Grammar* (Cardiff, 1980), pp. 122–4, and Thurneysen, GOI, pp. 146 and 230); in Welsh they are particularly favoured in poetic diction, and thus have a slightly formal or archaic flavour which could well have been felt suitable for place-names as well. Thus Welsh *Sychnant* 'dry-valley', a compound, may simply feel more 'right', as a place-name, than *Nant Sych*, the same as a name-phrase, though both occur. Compound names, therefore, do not necessarily date from before the 5th century. In addition, the evidence that name-phrases were *not* used before the 5th century is inconclusive: for one thing, the attested Romano-British place-names may not be fully representative of British naming habits (indeed, the total lack of certain habitative elements common to all the Brittonic languages suggests that they are not entirely representative); and even they include one group, the *Duro-* names (PNRB, pp. 346–54; Gelling, Signposts, pp. 45 and 51f.), which have the qualifier following the generic, for whatever reason. Of names not attested until later, Penge in Surrey (PNSur., pp. 14f.) is more likely to be a name-phrase, 'top of the wood', like the common Welsh Pencoed, Cornish Penquite or Pencoose, and Breton Penhoat, than a compound, 'chief wood'; and, from its geographical location, it is unlikely to have been formed after the date when the change in naming habits took place. There are also names such as Pentyrch, Gla., Pentrich, Drb. (PNDrb., II, 490f.) and Pentridge, Dor. (PNDor., II, 235), and similar ones which probably contain vowel-affected genitives singular as the qualifying elements (*pace* Koch, BBCS 30 (1982–83), 208–10): if so, they clearly date from earlier than the 5th century, and are equally clearly name-phrases. These strictures do not, however, invalidate the general rule: a compound name is likely to be older than a name-phrase.

One thus has a classification of usages as follows.

A. The element is used on its own (or with the definite article or a preposition) to form a simplex place-name, e.g. **enys** 'island' in Enys and Ninnes (*an enys*); **tre** 'farmstead, village' in Drift (*an dref); ***devr-** pl. 'waters' in Ardevora (+ ***ar**, 'before-the-waters') and ***gothel** 'marsh' in *Arwothel* (+ ***ar**, 'beside-the-marsh').

B1. The element is used as generic in a compound name, e.g. **nans** 'valley' in *Hyrnans* 'long-valley', *Seghnans* 'dry-valley', and *Hethenaunt* 'easy-valley', or **men** in *Hyrven* 'long-stone' (+ **hyr**).

B2. The element is used as generic in a name-phrase, e.g. **nans** in Nanspian 'little valley' (+ **byghan**), Nansavallan 'apple-tree valley' (+ **auallen**), etc.; or **men** in Menhire 'long stone' (+ **hyr**).

C1. The element is used as qualifier in a compound name, e.g. **hyr** in *Hyrnans* 'long-valley', *Hyrven* 'long-stone', *Hyrveneth* 'long-mountain', etc.

C2. The element is used as qualifier in a name-phrase, e.g. **hyr** in *Doerhyr* 'long ground', Crowshire 'long cross', and Menhire 'long stone'; or **nans** in Trenance 'farm in a valley' and Chynance 'house in a valley'.

D. The element is used as a suffix, to distinguish different places of the same name, or different subdivisions of a single place: e.g. **tre**, **cres** and **pell** spv. in Kemyel Drea, Kemyel Crease and *Kemyel Pella*; **meur** and **byghan**, 'great' and 'small' in Ardevora Veor, Hellesveor and Hellesvean, Sellanvean, Trelanvean, etc.; **guartha** 'top' and **goles** 'bottom' in Grogoth Wartha and Wollas, Kemyel Wartha, Pennare Wartha and Wallas, etc.; and similar instances.

Thus each name containing two elements is either a compound, C1 + B1, or a name-phrase, B2 + C2, with the latter being much the commonest type of all. Note that in a compound the initial sound of the second element is lenited, if it is able to be.

A few examples of usages not easily classified by this system will show its shortcomings. Three-element names are difficult: in a name like Barlendew (**bar** + **lyn** + **du**) 'black pool summit', **bar** is obviously a generic (B2), and **du** is obviously a qualifier (C2); but **lyn** 'pool' is both qualifying **bar** and qualified by **du**, so cannot be put into a particular category. However, true three-element names are so rare that this is not a major problem. In other cases, it can be hard to decide whether a name is in fact a three-element one, or a straightforward two-element name with a suffix: Penscove (*Pencusgof* 1327: **pen** + **cos** + **gof**) could be 'the head of the smith's wood', or, more likely, 'Smith's *Pencos*'.

Another problem is to decide whether a particular compound existed as an element (or even as a common noun) in its own right, or was freshly formed in each instance. This is a problem that exists in other languages, including English, where some authorities will cite elements such as **fox-eorðe* 'fox-earth', while others prefer to cite it under its separate elements. Thus it is a moot point whether ***downans** 'deep-valley' or ***hyr-nans** 'long-valley' should be cited as elements in their own right (containing **down** and **hyr** as C1 and **nans** as B1), or under their constituent elements. The decision is a subjective one, and no attempt has been made to be consistent; where such a word is listed as a separate element, it is given a cross-reference under its constituent elements in the appropriate place. Similarly a few phrases, such as ***men hyr**, ***pen hal**, and one or two others, are also cited as elements in their own right.

Note 2: Alphabetical Order

Since the head-forms are such a varied assortment, and since Middle Cornish, which provides the majority, was itself not spelt consistently, the arrangement of elements in alphabetical order has itself been a problem. It has been decided to follow principles similar to those in the index of HPB, and arrange words by their intended sounds, rather than the literal spelling. The arrangement thereby itself provides a clue as to the sounds intended by a particular spelling. Thus *c* and *k* are treated as a single letter; vocalic and consonantal *i/y* are treated separately, with vocalic *i/y* coming after *h*, and consonantal *i/y* following it; *ch* = /tʃ/ is treated as a separate letter, following *c/k*; *wh/hw/w* = /χw/ as a separate letter, following *w*; and *gu/gw* is treated as a separate letter, following *g*.

It is appreciated that this arrangement may be confusing, or make it hard to find particular elements; but it has been settled on as the best compromise. An alphabetical list of elements, and cross-references within the body of the dictionary, are provided in an attempt to overcome these disadvantages.

ALPHABETICAL LIST OF ELEMENTS

A

*-a 'place of' (?)
agy 'nearer'
aidlen 'aspen-tree'
*alaw 'water-lily'
*aled (obscure)
als 'cliff'
alter 'altar'
*alun (obscure)
*amal 'edge'
amanen 'butter'
an 'the'
*-an² 'place of' (?)
anken 'grief'
anneth 'dwelling'
antell 'snare'
*ar 'facing'
ardar 'plough'
*arð 'height'
*argel 'retreat'
arghans 'silver'
*arghantell 'silvery stream'
arluth 'lord'
*arwystel 'pledge'
ascorn 'bone'
asow 'ribs'
*aswy 'gap'
atal 'rubbish'
aval 'apple'
auallen 'apple-tree'
aves 'yonder'
auon 'river'
awel 'breeze'
*aweth 'watercourse' (?)

B

bagh 'nook'
bagyl 'crozier'
bal 'mine'
ban 'peak'
banathel 'broom'
*bannek 'peaked'
bar 'top'
bara 'bread'
bargos 'buzzard'
barthesek 'wonderful'
*bas 'shallow'
bedewen 'birch-tree'
begel 'hillock'
beler 'water-cress'
ben 'base'
benyges 'blessed'
ber 'short'
*bery 'kite'
beth 'grave'
*bich 'small' (?)
byghan 'small'
*byly 'pebbles'
bynkiar 'cooper'
*byu 'cattle'
bleit 'wolf'
blyn 'tip'
blogh 'bald'
*bod 'dwelling'
*boðour 'filthy-water'
bogh 'billy-goat'
*bolgh 'gap'
bor 'protuberance' (?)
*both 'hump'

boudzhi 'cow-house'
bounder 'lane'
*****bounds** 'tin-mine'
bowyn 'beef'
*****bow-lann** 'cow-pen' (?)
bran 'crow'
bras 'great'
*****bre** 'hill'
bregh 'arm'
breilu 'primrose' (?)
*****breyn** 'putrid'
*****bren** 'hill'
brenigan 'limpet'
brew 'broken'
*****brygh** 'brindled'
*****bryth** 'variegated'
brytyll 'frail'
bro 'district'
brogh 'badger'
bron 'breast'
bronnen 'a rush'
bucka 'sprite'
budin 'meadow'
bugel 'shepherd'
bugh 'cow'
*****buorth** 'cow-yard'

C/K

cadar 'chair'
*****cad-lys** 'bailey'
*****cagh** 'dung'
kal 'penis'
kalx 'lime'
cales 'hard'
*****calon** (obscure)
cam 'crooked'
*****camas** 'bend'
can 'white'
*****canel** 'channel'

cans 'a hundred'
*****cant** 'district, border' (?)
capa 'cape'
*****car** 'cart'
*****car-bons** 'causeway'
*****car-jy** 'cart-house'
carn 'tor'
carow 'stag'
karrek 'rock'
cas 'battle'
*****casel** 'armpit'
cassec 'mare'
castell 'castle, etc.'
*****caswyth** 'thicket'
cath 'cat'
*****caubalhint** 'ferry'
*****cawns** 'causeway'
keber 'timber'
kee 'hedge'
*****kegen** 'kitchen' (?)
keyn 'back'
*****kel** 'cover'
kelin 'holly'
cellester 'pebble'
kelli 'grove'
kellys 'lost'
*****Kembro** 'Welshman'
*****kemer** 'confluence'
*****kemyn** 'a commoner' (?)
*****kenkith** 'imparked residence'
*****kendowrow** 'joint-waters' (?)
kenin 'ramsons'
*****kenow** 'puppy'
kenter 'spike'
*****ker** 'a round'
kerden 'rowan-trees'
cherhit 'heron'
*****kernan** 'mound' (?)
*****kernyk** 'corner'
*****kest** 'paunch'

*keun 'reeds'
*kevammok 'battle'
*kevar 'joint-tillage'
keverang '(administrative) hundred'
*kevyl 'horse(s)' (?)
*kew 'hollow, enclosure'
*kew-nans 'ravine'
*kew-rys 'hollow-ford'
ky 'dog'
*kyf 'stump'
*kyl 'nook'
*kylgh 'circle'
*kyniaf-vod 'autumn-dwelling'
kynin 'rabbit'
kio 'snipe'
clap 'clatter'
*clav-jy 'lazar-house'
cleath 'dyke'
*cleger 'cliff'
*cleys 'trench'
cloghprennyer 'gallows'
*clun 'meadow'
*clus 'heap'
*cnegh 'hillock'
*cnow 'nuts, hazel-trees'
*cok 'cuckoo'
kodna huilan 'lapwing'
*coffen 'lode-mine'
*coger 'winding' (?)
*coll 'hazel-trees'
*collwyth 'hazel-trees'
*colomyer 'dove-cote'
*comm 'small valley'
compes 'level'
conna 'neck'
*cor¹ 'tribe'
*cor² 'hedge'
*cores 'weir'
corf 'body'

*cor-lann 'enclosure'
corn 'horn'
*cors 'reeds'
cos 'wood'
coscor 'family'
coth 'old'
cough 'red'
*crak¹ 'sandstone'
crak² 'crack'
*crann 'scrub'
*cren 'trembling'
cres 'middle'
*creun 'dam' (?)
*crew 'weir'
krib 'crest'
*crygh 'wrinkled'
*cryn 'dry'
*crys 'fold' (?)
*croft 'rough ground'
crogen 'shell'
crom 'curved'
*cromlegh 'dolmen'
cronek 'toad'
crous 'cross'
crouspren 'crucifix'
krow 'hut'
*krow-jy 'hut'
cruc 'barrow'
cudin 'lock of hair'
cuic 'empty'
cul 'narrow'
kullyek 'cock'
cummyas 'leave'
kunys 'firewood'
*cun-jy 'kennel'
cuntell 'collect'

Ch (= /tʃ/)

chammbour 'chamber'

ALPHABETICAL LIST OF ELEMENTS

chapel 'chapel'
*chas 'chace'
cheer 'chair'
(**cherhit**: *see under* C/K)
chy 'house'

D

*daek (obscure)
dall 'blind'
dan 'under'
dar 'oak-tree'
daras 'door'
*darva 'oak-place' (?)
daves 'sheep'
deyl 'leaves'
den 'man'
denjack 'hake'
dentye 'dainty'
*derch 'bright' (?)
dev 'god'
*devr- 'water'
dew 'two'
*dewy (obscure)
dewthek 'twelve (men)'
dy- 'double' (?), etc.
diber 'saddle'
*dyjy 'small cottage'
*dy-les 'profitless'
*dyllo 'lively' (?)
*dyn 'fort'
*dynan 'fort'
*dynas 'fort'
dyner 'penny'
dyowl 'devil'
*dyppa 'small pit'
*dy-serth 'very steep'
*dywys 'burnt'
doferghi 'otter'
*dons 'dance'
dor 'ground'
dorn 'fist'
dour 'water'
*douran 'watering-place'
down 'deep'
*downans 'deep-valley'
dreyn 'thorns'
dreys 'brambles'
dres 'across'
*drum 'ridge'
du 'black'

E

ebel 'foal'
-ek aj. suffix
eglos 'church'
eys 'corn'
eithin 'furze'
(***-el**: *see* ***-yel**)
elaw 'elm-trees' (?)
*elen 'fawn'
elerhc 'swan(s)'
elester 'irises'
elin 'elbow'
-ell dimin. suffix
*emle (obscure)
-en sg. suffix
*enyal 'desert'
enys 'island'
epscop 'bishop'
er^1 'eagle'
êr^2 'fresh'
*erber 'garden'
erw 'acre'
*esker 'shank'
-esyk aj. suffix
ethen 'bird'
ethom 'need'
*evor (obscure)

ALPHABETICAL LIST OF ELEMENTS xxiii

*ewyk 'hind'
euiter 'uncle'

F

fav 'beans'
*faw 'beech-trees'
*fe 'fief'
*fenna 'overflow' (?)
fenten 'spring'
*feryl 'magician' (?)
fyn 'end'
*fynfos 'boundary-dyke'
fynweth 'end'
fok 'furnace'
fodic 'happy'
*fold 'pen'
forn 'kiln'
forth 'road'
fos 'dyke'
*fow 'cave'
Frank 'Frenchman, freeman'
*fry 'nose'
frot 'stream'

G

gahen 'henbane'
gajah 'daisy'
ganow 'mouth'
*gar 'leg'
garan 'crane'
garow 'rough'
*garth 'enclosure' or 'ridge'
gaver 'goat'
*gawl 'fork'
gawna 'calfless cow'
ghel 'leech'
geler 'bier'
geluin 'beak'

*genn 'wedge'
(gentyl: see under J)
gesys 'left'
glan 'bank'
glas 'green'
*glasen 'verdure'
*glasneth 'verdure'
glastan 'oak-trees'
*glynn 'large valley'
*glynn-wyth 'valley-trees' (?)
*glow^1 'clear'
glow^2 'coal'
*go- 'sub-'
*gobans 'little hollow'
*godegh 'lair' (?)
*godre 'homestead'
gof 'smith'
gofail 'smithy'
*gogleth 'north'
goyf 'winter'
gol 'feast'
goles 'lower'
golow 'light'
gols 'hair'
googoo 'cave'
goon 'downland'
*goon-dy 'moor-house'
*gor- 'over-'
*gor-dre 'over-farmstead' (?)
*gor-ge 'low-hedge'
gorhel 'boat'
*gorm 'dark'
*gortharap 'very pleasant'
gortos 'stay'
*gorweth 'wooded slope' (?)
*gosa 'bleed'
goth^1 'water-course'
goth^2 'goose'
*gothel 'watery ground'
gour 'man'

ALPHABETICAL LIST OF ELEMENTS

gover 'stream'
gras 'grace'
*gre 'flock'
*gre-dy 'cattle-shed' (?)
grelin 'cattle-pool'
gronen 'grain'
grow 'gravel'
(gruah: see under Gw)
*gruk 'heather'
gubman 'seaweed'
guillua 'look-out place'

Gw

gvak 'empty'
*gwailgi 'ocean' (?)
guaintoin 'spring'
*gwal 'wall'
*gwalader 'lord'
guan 'weak'
gwaneth 'wheat'
guary 'play'
guartha 'summit; upper'
guarthek 'cattle'
guas 'fellow'
*gwavos 'winter-dwelling'
guel 'open-field'
guely 'bed'
guella 'best'
gwels 'grass'
guennol 'swallow'
gweras 'ground'
guern 'alder-trees'
gvest 'lodging'
gwîa 'weave'
guibeden 'gnat'
*gwyk 'village' (?)
guicgur 'trader'
(guillua: see under G)
guyn 'white'

guyns 'wind'
guirt 'green'
guis 'sow, pig'
gwyth^1 'trees'
guyth^2 'tin-work'
*gwythel 'thicket' (?)
gulas 'land'
*gwleghe 'moisture'
*gwlesyk 'leader'
gruah 'hag'
gwrek 'woman'

H

ha(g) 'and'
haf 'summer'
hager 'foul'
*hay 'enclosure'
hal 'marsh'
hans 'yonder'
hanter 'half'
*havar 'summer-fallow'
*havos 'shieling'
*havrek 'arable land'
*heyl 'estuary'
hel 'hall'
*helgh 'a hunt'
heligen 'willow-tree'
hen 'ancient'
*henkyn 'iron peg' (?)
*hendre 'home farm'
*hen-forth 'old road'
*hen-lann 'old cemetery'
*hen-lys 'ruins'
hensy 'ancient house'
hep 'without'
heschen 'sedge'
*heth 'stag' (?)
hyly 'brine'
*hin 'border'

-hins- 'way'
hynse 'fellows'
hyr 'long'
*hyr-drum 'long-ridge'
*hyr-yarth 'long-ridge'
(**hiuin**: *see under* I/Y)
*hoby 'pony'
hoch 'pig'
*hogen 'pile'
horn 'iron'
horþ 'ram'
hos 'duck'
houl 'sun'
*huel-gos 'high-wood'
hueth 'happy'
(**hwilen**: *see under* Wh)

Vocalic I/Y

*-i name suffix
-yk aj. or dimin. suffix
idhio 'ivy'
yfarn 'hell'
*ygolen 'whetstone'
yn 'narrow'
ynter 'between'
*is¹ 'below'
*-ys² 'place of'
ysel 'low'
*yslonk 'abyss'
yst 'east'
hiuin 'yew-tree(s)'

Consonantal I/Y

yar 'hen'
*yarl 'earl'
yeyn 'cold'
*-yel aj. suffix
-yer pl. suffix

yet 'gate'
ieu 'yoke'
yorch 'roe-deer'
yow 'Thursday'
*yuf 'lord' (?)

J (= /dʒ/)

jarden 'garden'
gentyl 'noble'

K, *see* C/K

L

lakka 'stream'
lader 'thief'
*ladres 'sluice' (?)
lam 'leap'
lanherch 'clearing'
*lann 'cemetery'
le 'place'
lek 'lay'
ledan 'wide'
*lefant 'toad'
*legh 'slab'
*leyn 'strip of land'
les 'plant'
leskys 'burnt'
lester 'boat'
leth 'milk'
*lether 'cliff'
*lety 'dairy'
leven 'smooth'
leuerid 'sweet milk'
lêuiader 'pilot'
leuuit 'pilot'
lyen 'cloth'
lyn 'pool'

lynas 'nettles'
*****lys** 'court, ruin'
lyw 'colour'
*****lok** 'chapel'
*****loch** 'pool'
logaz 'mice'
*****lom** 'bare'
*****lon** 'grove'
*****lonk** 'gully'
los 'grey'
losc 'a burning'
lost 'tail'
lovan 'cord'
lowarn 'fox'
lowarth 'garden'
lowen 'happy'
lu 'army'
lugh 'calf'
*****luryk** 'breastplate'
*****lus** 'bilberries'
lusew 'ashes'

M

*****ma** 'place, plain'
*****mabyar** 'pullet'
maga 'feed'
*****magoer** 'wall'
mam 'mother'
manach 'monk'
manal 'sheaf'
margh 'horse'
marghas 'market'
marrek 'knight'
*****með** 'middle'
*****með-ros** 'middle of the hill'
*****meyn-dy** 'stone-house'
*****meynek** 'stony'
*****melek** 'honeyed' (?)
melhyonen 'clover'

melin[1] 'mill'
melyn[2] 'yellow'
*****melynder** 'miller'
*****melyn-jy** 'mill-house'
*****melyn wyns** 'windmill'
*****melwhes** 'slugs'
men 'stone'
*****menawes** 'awl'
*****meneghy** 'sanctuary' (?)
meneth 'hill'
*****men-gleth** 'quarry'
*****men hyr** 'long stone'
*****merther** 'saint's grave' (?)
*****merthyn** 'sea-fort'
mes 'open-field'
mesclen 'mussel'
methek 'doctor'
meur 'big'
midzhar 'reaper'
mẏdzhovan 'ridge'
mylgy 'greyhound'
myn 'edge'
*****mynster** 'endowed church'
moelh 'blackbird'
mogh 'pigs'
*****moyl** 'bare'
moyr- 'blackberries'
mols 'wether'
mon[1] 'slender'
*****mon**[2] 'mineral'
*****mon-dy** 'mineral-house'
*****mon-gleth** 'mine-working'
mor 'sea'
morhoch 'porpoise'
*****moryon** 'ants'
*****morrep** 'sea-shore'
*****morva** 'sea-marsh'
mosek 'stinking'
mowes 'girl'
munys 'little'

N

nans 'valley'
***nath** 'hewing' (?)
naw 'nine'
nerth 'strength'
nessa 'nearest'
***neth** (obscure)
***neved** 'sacred place'
nyth 'nest'
noth 'naked'
nowyth 'new'

O

***oden** 'kiln'
odion 'ox'
ogas 'near'
oghan 'oxen'
oye 'egg'
on 'lamb'
onnen 'ash-tree'
***op** 'alley'
***orguilus** 'proud'
ownek 'fearful'

P

padel 'pan'
***pans** 'hollow'
park 'field'
***pedreda** (obscure)
***pel** 'ball'
pell 'far'
pellen 'ball'
pen 'head'
***pen (an) dre** 'town's end'
***pen (an) gelli** 'grove's end'
***pen (an) pons** 'bridge's end'
***pen-arth** 'headland'
***pen cos** 'wood's end'
***pen-fenten** 'spring-head'
***pen hal** 'moor's head'
***pen-lyn** 'head-pool' (?)
***pen nans** 'valley's head'
***pennek** 'big-headed'
***pennyn** 'tadpole' (?)
***pen pol** 'creek's head'
***pen-ryn** 'promontory'
***pen ros** 'hill's end'
***pen-tyr** 'headland'
***perth** 'thicket'
perveth 'middle'
pêz 'peas'
peswar 'four'
***peulvan** 'pillar'
***peur** 'pasture'
***pever** 'bright'
***pybell** 'pipe'
***pybor** 'piper'
***pyk** 'point'
pyl 'pile'
***pylas** 'naked oats'
***pyll** 'creek'
pylles 'peeled'
pin- 'pine-trees'
***pystyll** 'waterfall'
pyt 'pit'
plas 'place'
plen 'arena'
plos 'filthy'
plu 'parish'
podar 'rotten'
pol 'pit, pool'
polgrean 'gravel-pit'
***pol ros** 'water-wheel pit'
pons 'bridge'
porhel 'piglet'
porth 'cove'
post 'post'

ALPHABETICAL LIST OF ELEMENTS

*poth 'burnt'
pow 'land'
pras 'meadow'
pren 'timber'
pry 'mud'
prif 'reptile'
pryns 'prince'
*prys(k) 'copse'
pronter 'priest'
pwl prî 'clay-pit'
pul stean 'tin-pit'
pur 'pure'
*puth 'well'

R

*radgel 'scree'
rag 'before'
*red 'watercourse'
reden 'bracken'
*rew 'slope'
rid 'ford'
*rynn 'point'
ryp 'beside'
*rodwyth 'ford'
*ros 'moor'
*rosan 'little promontory' (?)
*rud 'foul'
ruen 'seal'
*run 'hill'
ruth 'red'

S/Z

sans 'holy'
sawn 'cleft'
skath 'boat'
scawen 'elder-tree'
skyber 'barn'
scorren 'branch'

scoul 'kite'
scovern 'ear'
scrife 'write'
schus 'shade'
segh 'dry'
*seghan 'dry place'
*Seys 'Englishman'
serth 'steep'
seth 'arrow'
sevi 'strawberries'
sheft 'shaft'
*shoppa 'workshop'
sichor 'dryness'
sẏgal 'rye'
slynckya 'slither'
*soŏ 'depression'
sorn 'nook'
zowl 'stubble'
Zowzon 'Englishmen'
spern 'thorns'
spethas 'brambles'
*splat 'plot'
*stable 'stable'
*stak 'mud' (?)
stampes 'stamping-mill'
stean 'tin'
steren 'star'
steuel 'room'
stoc 'stump'
strail 'mat'
*stras 'flat valley'
streyth 'stream'
*stret 'street'
*stronk 'dung'
*stum 'bend'
*suant 'level'

T

*tagell 'constriction'

ALPHABETICAL LIST OF ELEMENTS

xxix

tal 'brow'
*talar 'auger' (?)
tam 'morsel'
tanow 'thin'
taran 'thunder'
tarow 'bull'
taves 'tongue'
*tawel 'quiet'
teg 'fair'
teil 'dung'
tenewen 'side'
ternoyth 'ill-clad'
*tescow 'sheaves' (?)
tevys 'grown'
*tewel 'pipe'
tewl 'dark'
ti (= chy)
tyn 'rump'
tyr 'land'
tyreth 'land'
*tnou 'valley'
*to- 'thy'
toll 'hole'
*toll-gos 'hole-wood'
*tol-ven 'holed stone' (?)
ton 'lea-land'
top 'top'
tor 'belly'
torch 'boar'
*torn 'district'
*torthell 'loaf'
towan 'sand-dune'
*towargh 'turf'
trait 'beach'
*trap 'stile'
tre 'estate'
trech 'cut wood'
*treth 'ferry'
*treu 'beyond' (?)
try¹ 'three'

*try-² 'triple; very'
trig 'ebb-tide'
trygva 'dwelling-place'
*tryonenn 'fallow-land' (?)
*trokya 'to full'
trogh 'broken'
tron 'nose'
*tron-gos 'nose-wood'
tros 'foot'
tu 'side'
tûban 'mound'
tum 'warm'
tus 'people'
*tusk 'moss'

U

ugens 'twenty'
ugh 'above'
ughel 'high'
ula 'owl'
usion 'chaff'
*ussa 'outermost' (?)

V

vooga 'cave'

W

war 'upon'
wast 'idle'
west 'west'
worth 'at'
(wuir: *see under* Wh)

Wh

whe 'six'

wheyl 'mine-working' **hwilen** 'beetle'
whelth 'story' **wuir** 'sister'
*wheth 'blow'
*whevrer 'lively' (?)   Z, *see* S/Z

ABBREVIATIONS

ABBREVIATED REFERENCES

ACL	*Archiv für celtische Lexikographie* (Halle, 1898–1907)
ACS	A. Holder, *Alt-celtischer Sprachschatz* (3 vols., Leipzig, 1896–1907)
An.Boll.	*Analecta Bollandiana*
Anc.Deeds	*Catalogue of Ancient Deeds in the Public Record Office* (London, 1890–)
Ann.Bret.	*Annales de Bretagne*
Ann.Camb.	Egerton Phillimore, 'The *Annales Cambriae* and Old-Welsh Genealogies from *Harleian MS. 3859*', *Y Cymmrodor* 9 (1888), 141–83 (cited by the year of the entry)
Antiquities	see Borlase
Arch.Camb.	*Archaeologia Cambrensis*
Armes Prydein	Sir Ifor Williams (ed.), *Armes Prydein: The Prophecy of Britain, from the Book of Taliesin*, English version by Rachel Bromwich (Mediaeval and Modern Welsh Series, vol. 6: Dublin, 1972)
Asser	W. H. Stevenson (ed.), *Asser's Life of King Alfred* (Oxford, 1904)
Ass.R.	Assize Roll in the Public Record Office: unprinted, except for 1201, Doris M. Stenton (ed.), *Pleas before the King or his Justices, 1198–1202*, part II, *Rolls or Fragments of Rolls from the Years 1198, 1201 and 1202* (Selden Society, vol. 68 for 1949: London, 1952)
Atlas Meirionydd	Geraint Bowen (ed.), *Atlas Meirionydd* (Y Bala, n.d. [1974])
Bannerman, Dalriada	John Bannerman, *Studies in the History of Dalriada* (Edinburgh and London, 1974)
BBC	J. Gwenogvryn Evans (ed.), *The Black Book of Carmarthen* (Pwllheli, 1906). See now also A. O. H. Jarman (ed.), *Llyfr Du Caerfyrddin* (Cardiff, 1982)
BBCS	*Bulletin of the Board of Celtic Studies*
BCS	W. de G. Birch, *Cartularium Saxonicum* (3 vols. and index, London, 1885–99)
Beauchamp Crtlry.	Emma Mason (ed.), *The Beauchamp Cartulary* (Pipe Roll Society, vol. 81, n.s. 43, for 1971–73: London, 1980)
Beauport	'Mots bretons dans les chartes de l'abbaye de Beauport', by H. d'Arbois de Jubainville (RC 3 (1876–78), 395–418) and G. Dottin (RC 7 (1886), 52–65 and 200–9, and RC 8 (1887), 65–75)
Bede	see Hist.Eccl.
Bégard	Claude Evans, 'Les noms bretons dans les chartes de l'abbaye de Bégard, 1156–1458', ÉC 15 (1976–78), 195–224
Beginnings	see Ifor Williams
Bezz.Beitr.	(Bezzenberger's) *Beiträge zur Kunde der indogermanischen Sprachen* (Göttingen, 1877–1907)

ABBREVIATED REFERENCES

Bieler	Ludwig Bieler (ed.), *The Patrician Texts in the Book of Armagh* (Dublin, 1979)
Bilbao MS	Tonkin's manuscript Cornish vocabulary in the Provincial Library of Bilbao, Spain (cited from a xerox in the Royal Institution of Cornwall, Truro, or from H. Jenner, JRIC 21 (1922–25), 421–37)
BM	Whitley Stokes (ed.), *Beunans Meriasek: The Life of Saint Meriasek* (London, 1872)
BNF	*Beiträge zur Namenforschung*
Bodm.	Max Förster, 'Die Freilassungsurkunden des Bodmin-Evangeliars', in *A Grammatical Miscellany offered to Otto Jespersen*, edited by N. Bogholm and others (London and Copenhagen, 1930), pp. 77–99 (cited by Förster's numbering)
Borlase, Antiquities	William Borlase, *Antiquities Historical and Monumental of the County of Cornwall* (2nd edn., London, 1769)
Borlase, Vocabulary	'A Cornish–English Vocabulary' in Borlase, Antiquities, pp. 413–64
Bosons	O. J. Padel (ed.), *The Cornish Writings of the Boson Family* (Redruth, 1975)
Bowen, Settlements	E. G. Bowen, *The Settlements of the Celtic Saints in Wales* (Cardiff, 1954)
BT	J. Gwenogvryn Evans (ed.), *The Book of Taliesin* (Llanbedrog, 1910)
CA	see Ifor Williams
Campanile	Enrico Campanile, *Profilo etimologico del cornico antico* (Pisa, 1974)
Carew	Richard Carew, *The Survey of Cornwall* (London, 1602)
Cart.Glam.	George Thomas Clark (edited by Godfrey L. Clark), *Cartae et Alia Munimenta quae ad Dominium de Glamorgancia pertinent* (6 vols., Cardiff, 1910)
Cart.Land.	R. Le Men and Émile Ernault (eds.), *Cartulaire de Landevennec*, in *Collection des documents inédits sur l'histoire de la France* (Mélanges historiques), 5 (1886), 535–600
Cart.Quimper	The Cartulary of Quimper (usually cited from Chrest.)
Cart.Quimperlé	The Cartulary of Quimperlé (usually cited from Chrest. or HPB)
Cart.Redon	A. de Courson (ed.), *Cartulaire de l'abbaye de Redon* (Paris, 1863) (or cited from Chrest. or HPB)
Cart.St M	P. L. Hull (ed.), *The Cartulary of St Michael's Mount* (Devon and Cornwall Record Society, n.s. 5: Torquay, 1962)
Catholicon	The *Catholicon* of Jean Lagadeuc (usually cited from GIB, GMB or HPB)
Celtic Folklore	John Rhys, *Celtic Folklore* (Oxford, 1901)
Chad	Old Welsh entries in the Book of St Chad (Lichfield Gospels) (cited from LLD, pp. xliii–xlviii)
Chrest.	J. Loth, *Chrestomathie bretonne* (Paris, 1890)
Chyanhor	'John of Chyanhor', folk-tale in Bosons, pp. 14–23
CIIC	R. A. S. Macalister, *Corpus Inscriptionum Insularum Celticarum* (2 vols., Dublin, 1945–49)
CLlH	Ifor Williams, *Canu Llywarch Hên* (3rd impression, Cardiff, 1970)
CMCS	*Cambridge Medieval Celtic Studies*
Co.Arch.	*Cornish Archaeology*

Colan	Dafydd Jenkins (ed.), *Llyfr Colan* (Cardiff, 1963)
Co.Studies	*Cornish Studies*
Courtney and Couch	M. A. Courtney and Thomas Q. Couch, *Glossary of Words in Use in Cornwall* (London, 1880)
CPNS	William J. Watson, *The History of the Celtic Place-Names of Scotland* (Edinburgh, 1926)
Cumbs. and Westms. Transactions	*Transactions of the Cumberland and Westmorland Antiquarian and Archaeological Society*
CW	Whitley Stokes (ed.), *Gwreans an Bys: The Creation of the World* (London and Edinburgh, 1864). See now also Paula Neuss (ed.), *The Creacion of the World* (New York, 1983)
Cymm.Trans.	*Transactions of the Honourable Society of Cymmrodorion*
DAG	Joshua Whatmough, *The Dialects of Ancient Gaul* (Cambridge, Mass., 1970)
Davey, Flora	F. Hamilton Davey, *Flora of Cornwall* (2nd edition, Ilkley, 1978). See now L. J. Margetts and R. W. David, *A Review of the Cornish Flora, 1980* (Redruth, 1981)
Davies, Charters	Wendy Davies, *The Llandaff Charters* (Aberystwyth, 1979)
Davies, Microcosm	Wendy Davies, *An Early Welsh Microcosm: Studies in the Llandaff Charters* (London, 1978)
DB	Domesday Book
DCNQ	*Devon and Cornwall Notes and Queries*
DEPN	Eilert Ekwall, *The Concise Oxford Dictionary of English Place-Names* (4th edn., Oxford, 1960)
Doble, Crowan	G. H. Doble, *A History of the Parish of Crowan* (Long Compton, 1939)
Doble, Lelant	G. H. Doble, *A History of the Church and Parish of St Euny-Lelant* (Long Compton, 1939)
Doble, Mewan and Austol	G. H. Doble, *Saint Mewan and Saint Austol* (2nd edn., Long Compton, 1939)
Doble, St Carantoc	G. H. Doble, *S. Carantoc, Abbot and Confessor* (2nd edn., Long Compton, 1932)
Doble, St Ewe	G. H. Doble, *A History of the Parish of St Ewe* (Long Compton, 1937)
Doble, St Ives	G. H. Doble, *St Ives: Its Patron Saint and its Church* (St Ives, 1939)
Doble, St Rumon	G. H. Doble, *Saint Rumon and Saint Ronan* (Long Compton, 1939)
Doble, Welsh Saints	G. H. Doble (edited by D. Simon Evans), *Lives of the Welsh Saints* (Cardiff, 1971)
Dottin	Georges Dottin, *La langue gauloise* (Paris, 1920)
Douch, Windmills	H. L. Douch, *Cornish Windmills* (Truro, n.d. [1963])
EANC	R. J. Thomas, *Enwau Afonydd a Nentydd Cymru* (Cardiff, 1938)
ÉC	*Études celtiques*
Eccl.Ant.	Charles Henderson, 'The Ecclesiastical Antiquities of the Four Western Hundreds of Cornwall', JRIC n.s. 2 (1953–56) and 3 (1957–60)
EEW	T. H. Parry-Williams, *The English Element in Welsh* (Cymmrodorion Record Series, no. 10: London, 1923)
EFN	John Field, *English Field-Names: A Dictionary* (Newton Abbot, 1972)
Elements	A. Hugh Smith, *English Place-Name Elements* (2 vols., English Place-Name Society, vols. 25 and 26: Cambridge, 1956)

ABBREVIATED REFERENCES

Elf.Ladin	*see* Lewis
EL1	Ifor Williams, *Enwau Lleoedd* (2nd edn., Liverpool, 1969)
ELlSG	John Lloyd-Jones, *Enwau Lleoedd Sir Gaernarfon* (Cardiff, 1928)
ERN	Eilert Ekwall, *English River-Names* (Oxford, 1928)
Eutych.	Old-Breton glosses on the Grammar of Eutychius (cited from VVB or FD)
EWGP	*see* Jackson, Gnomic Poems
EWGT	P. C. Bartrum, *Early Welsh Geneaological Tracts* (Cardiff, 1966)
FD	Léon Fleuriot, *Dictionnaire des gloses en vieux breton* (Paris, 1964)
FF	Joseph Hambley Rowe (ed.), *Cornwall Feet of Fines* (2 vols., Devon and Cornwall Record Society, Exeter, 1914 and 1950)
FG	Léon Fleuriot, *Le vieux breton: éléments d'une grammaire* (Paris, 1964)
Finberg, Lucerna	H. P. R. Finberg, *Lucerna* (London, 1964)
Fleuriot, Origines	Léon Fleuriot, *Les origines de la Bretagne* (Paris, 1980)
Foxall, Shropshire Field-Names	H. D. G. Foxall, *Shropshire Field-Names* (Shrewsbury, 1980)
GBGG	John Lloyd-Jones, *Geirfa Barddoniaeth Gynnar Gymraeg* (8 parts, Cardiff, 1931–63)
Geir.Mawr	H. Meurig Evans and others, *Y Geiriadur Mawr* (10th edn., Swansea and Llandysul, 1981)
Gelling, Signposts	Margaret Gelling, *Signposts to the Past* (London, 1978)
GIB	Roparz Hemon, *Geriadur Istorel ar Brezhoneg* (36 parts, Brest, Paris and Rennes, 1958–79)
GMB	Émile Ernault, *Glossaire moyen-breton* (Paris, 1895–96)
GMW	D. Simon Evans, *A Grammar of Middle Welsh* (Mediaeval and Modern Welsh Series, supplementary volume: Dublin, 1964)
GOI	Rudolf Thurneysen, *A Grammar of Old Irish*, translated by D. A. Binchy and Osborn Bergin (Dublin, 1946)
Gourvil	Francis Gourvil, *Noms de famille bretons d'origine toponymique* (Quimper, 1970)
Gover	J. E. B. Gover, *The Place-Names of Cornwall* (typescript, 1948, in the Royal Institution of Cornwall, Truro)
GPC	*Geiriadur Prifysgol Cymru: A Dictionary of the Welsh Language*, edited by R. J. Thomas and others (Cardiff, 1950–)
GPN	D. Ellis Evans, *Gaulish Personal Names* (Oxford, 1967)
Graves	Eugene van Tassel Graves, *The Old Cornish Vocabulary* (Ann Arbor, Michigan: Columbia University Ph.D., 1962)
Guyonvarc'h, Dictionnaire	Christian-J. Guyonvarc'h, *Dictionnaire étymologique du breton ancien, moyen et moderne* (*Celticum* 27, supplement to *Ogam* 24–25: 6 parts, Rennes, 1973–75)
HB	The *Historia Brittonum*, in John Morris (ed.), *Nennius: British History and the Welsh Annals* (Chichester, 1980)
Hemon, Grammar	Roparz Hemon, *A Historical Morphology and Syntax of Breton* (Mediaeval and Modern Breton Series, vol. 3: Dublin, 1975)
Henderson, Bridges	Charles Henderson and Henry Coates, *Old Cornish Bridges and Streams* (Exeter, 1928)
Henderson, Constantine	Charles Henderson (edited by G. H. Doble), *A History of the Parish of Constantine in Cornwall* (Truro, 1937)
Henderson, Essays	Charles Henderson (edited by A. L. Rowse and M. I. Henderson), *Essays in Cornish History* (Oxford, 1935)

ABBREVIATED REFERENCES

Henderson, Mabe	Charles Henderson, *Mabe Church and Parish, Cornwall* (Long Compton, n.d. [1930])
Hist.Eccl.	Bede, *Ecclesiastical History of the English People*, in Charles Plummer (ed.), *Venerabilis Baedae Opera Historica* (Oxford, 1896)
Hogan, Onomasticon	Edmund Hogan, *Onomasticon Goedelicum* (Dublin, 1910)
HPB	Kenneth Jackson, *A Historical Phonology of Breton* (Dublin, 1967)
Indo-Celtica	H. Pilch and J. Thurow (eds.), *Indo-Celtica: Gedächtnisschrift für Alf Sommerfelt* (Commentationes Societatis Linguisticae Europaeae, 2: Munich, 1972)
Iorwerth	Aled Rhys Wiliam (ed.), *Llyfr Iorwerth* (Cardiff, 1960)
Jackson, Gnomic Poems	Kenneth Jackson (ed.), *Early Welsh Gnomic Poems* (2nd impression, Cardiff, 1961)
JCS	*The Journal of Celtic Studies*
JEPNS	*Journal of the English Place-Name Society*
Jérusalem	Roparz Hemon and Gwennole Le Menn (eds.), *Les fragments de la 'Destruction de Jérusalem' et des 'Amours du Vieillard'* (Mediaeval and Modern Breton Series, vol. 2: Dublin, 1969)
JRIC	*Journal of the Royal Institution of Cornwall*
JRS	*Journal of Roman Studies*
JRSAI	*Journal of the Royal Society of Antiquaries of Ireland*
Juvenc.	Old Welsh glosses and poems in the Cambridge *Juvencus* manuscript (cited from VVB or Williams, *Beginnings*)
Lake	Joseph Polsue, (Lake's) *Parochial History of the County of Cornwall* (4 vols., Truro, 1867–72)
Largillière, Les saints	René Largillière, *Les saints et l'organisation chrétienne primitive dans l'Armorique bretonne* (Rennes, 1925)
Latin Texts	Hywel D. Emanuel (ed.), *The Latin Texts of the Welsh Laws* (Cardiff, 1967)
Le Gonidec	J.-F. Le Gonidec (edited by H. de la Villemarqué), *Dictionnaire français-breton* (Saint-Brieuc, 1847) and *Dictionnaire breton-français* (Saint-Brieuc, 1850)
LEIA	Joseph Vendryes and others, *Lexique étymologique de l'irlandais ancien* (Dublin and Paris, 1959–)
Leland, Itin.	*The Itinerary of John Leland in or about the Years 1535–1543*, edited by Lucy Toulmin Smith (5 vols., London, 1906–10)
Lewis, Elf.Ladin	Henry Lewis, *Yr Elfen Ladin yn yr Iaith Gymraeg* (Cardiff, 1943)
Lh.	Edward Lhuyd, *Archaeologia Britannica* (Oxford, 1707): note that Lhuyd's under-dotted *u* has been represented with *w* for typographical convenience
LHEB	Kenneth Jackson, *Language and History in Early Britain* (Edinburgh, 1953)
Lh.MS	Edward Lhuyd, Cornish vocabulary in National Library of Wales (Aberystwyth), MS Llanstephan 84 (usually cited from Nance, *Dictionary* or Nance, *Sea-Words*)
Lhuyd, Paroch.	Edward Lhuyd, *Parochialia* (supplements to *Arch.Camb.* 1909–11)
Litt.Wall.	J. Goronwy Edwards (ed.), *Littere Wallie preserved in 'Liber A' in the Public Record Office* (Cardiff, 1940)
Ll.Bleg.	Stephen J. Williams and J. Enoch Powell (eds.), *Cyfreithiau Hywel Dda yn ôl Llyfr Blegywryd* (Cardiff, 1942)
LlCC	Henry Lewis, *Llawlyfr Cernyweg Canol* (2nd edn., Cardiff, 1946)

LLD	J. Gwenogvryn Evans and John Rhys, *The Text of the Book of Llan Dâv* (Oxford, 1893)
Ll.Hendr.	John Morris-Jones and others, *Llawysgrif Hendregadredd* (2nd impression, Cardiff, 1971)
LP	Henry Lewis and Holger Pedersen, *A Concise Comparative Celtic Grammar* (2nd edn., Göttingen, 1961)
Mart.Cap.	Old Welsh glosses on Martianus Capella (cited from VVB)
MC ('Mount Calvary')	Whitley Stokes (ed.), 'The Passion: A Middle-Cornish Poem', TPhS 1860–61, Appendix, pp. 1–100, and separately (Berlin, n.d.)
Med.Arch.	*Mediaeval Archaeology*
Mél.Arb.	Joseph Loth (ed.), *Mélanges H. d'Arbois de Jubainville* (Paris, 1906)
MerSR	Keith Williams-Jones (ed.), *The Merioneth Lay Subsidy Roll, 1292–3* (Cardiff, 1976)
Monts.Collns.	*The Montgomeryshire Collections*
Morris-Jones	*see* WG
Myv.Arch.	Owen Jones and others (eds.), *The Myvyrian Archaiology of Wales* (2nd edn., Denbigh, 1870)
Nance, Dictionary	R. Morton Nance, *A New Cornish–English Dictionary* (St Ives, 1938)
Nance, Sea-Words	R. Morton Nance (edited by P. A. S. Pool), *A Glossary of Cornish Sea-Words* (Marazion, 1963)
NCPN	B. G. Charles, *Non-Celtic Place-Names in Wales* (London, 1938)
Neath and District	Elis Jenkins (ed.), *Neath and District: A Symposium* (Neath, 1974)
NED	*A New English Dictionary*, edited by J. A. H. Murray and others (Oxford, 1884–1933)
Nicolaisen	W. F. H. Nicolaisen, *Scottish Place-Names: Their Study and Significance* (London, 1976)
Norden	John Norden, *Speculi Britanniae Pars: A Topographical and Historical Description of Cornwall* (London, 1728)
Norris	Edwin Norris (ed.), *The Ancient Cornish Drama* (2 vols., Oxford, 1859)
NSB	Joseph Loth, *Les noms des saints bretons* (Paris, 1910: reprinted from RC 29–30, 1908–09)
OC	*Old Cornwall*
O'Donnell Lectures	J. R. R. Tolkien and others (edited by Henry Lewis), *Angles and Britons: O'Donnell Lectures* (Cardiff, 1963)
OM	*Ordinale de Origine Mundi* in Norris, I
Owen, Pembrokeshire or Pembs.	George Owen (edited by H. Owen), *Description of Penbrokshire* (Cymmrodorion Record Series, 1: London, 1892–1936)
Padel, Catalogue	O. J. Padel, (Catalogue of) *Exhibition of Manuscripts and Printed Books on the Cornish Language* (Redruth, 1975)
Payne, Roche	H. M. Creswell Payne, *The Story of the Parish of Roche* (Newquay, n.d.)
PB	Pierre Trépos, *Le pluriel breton* (Brest, 1957)
PC	*Passio Domini Nostri Ihesu Christi* in Norris, I
Penhallurick, Birds	R. D. Penhallurick, *Birds of the Cornish Coast* (Truro, 1969: = vol. I), and *The Birds of Cornwall and the Isles of Scilly* (Penzance, 1978: = vol. II)

Piette	J. R. F. Piette, *French Loanwords in Middle Breton* (Cardiff, 1973)
PKM	Ifor Williams (ed.), *Pedeir Keinc y Mabinogi* (2nd impression, Cardiff, 1951)
PNBrk.	Margaret Gelling, *The Place-Names of Berkshire* (English Place-Name Society, vols. 49–51, 1973–76)
PNChe.	John McNeal Dodgson, *The Place-Names of Cheshire* (English Place-Name Society, vols. 44–48 and 54, 1970–)
PNCmb.	A. M. Armstrong and others, *The Place-Names of Cumberland* (English Place-Name Society, vols. 20–22, 1950–52)
PNDev.	J. E. B. Gover and others, *The Place-Names of Devon* (English Place-Name Society, vols. 8–9, 1931–32)
PNDinas Powys	Gwynedd O. Pierce, *The Place-Names of Dinas Powys Hundred* (Cardiff, 1968)
PNDor.	A. David Mills, *The Place-Names of Dorset* (English Place-Name Society, vols. 52–, 1977–)
PNDrb.	Kenneth Cameron, *The Place-Names of Derbyshire* (English Place-Name Society, vols. 27–29, 1959)
PNFli.	Ellis Davies, *Flintshire Place-Names* (Cardiff, 1959)
PNGlo.	A. Hugh Smith, *The Place-Names of Gloucestershire* (English Place-Name Society, vols. 38–41, 1964–65)
PNHre.	A. T. Bannister, *The Place-Names of Herefordshire: Their Origin and Development* (Cambridge, 1916)
PNLnc.	Eilert Ekwall, *The Place-Names of Lancashire* (Manchester, 1922)
PNRB	A. L. F. Rivet and Colin Smith, *The Place-Names of Roman Britain* (London, 1979)
PNSur.	J. E. B. Gover and others, *The Place-Names of Surrey* (English Place-Name Society, vol. 11, 1934)
PNWLo.	Angus MacDonald, *The Place-Names of West Lothian* (Edinburgh and London, 1941)
PNWlt.	J. E. B. Gover and others, *The Place-Names of Wiltshire* (English Place-Name Society, vol. 16, 1939)
PNWml.	A. Hugh Smith, *The Place-Names of Westmorland* (English Place-Name Society, vols. 42–3, 1967)
PNWor.	Alan Mawer and F. M. Stenton, *The Place-Names of Worcestershire* (English Place-Name Society, vol. 4, 1927)
PNYoW.	A. Hugh Smith, *The Place-Names of the West Riding of Yorkshire* (English Place-Name Society, vols. 30–37, 1961–63)
Pokorny Festschrift	Wolfgang Meid (ed.), *Beiträge zur Indogermanistik und Keltologie (Julius Pokorny zum 80. Geburtstag gewidmet)* (Innsbrucker Beiträge zur Kulturwissenschaft, 13: Innsbruck, 1967)
Polwhele, Provincial Glossary	'A Provincial Glossary' in Richard Polwhele, *The History of Cornwall*, VI (Truro, 1808)
Pool, 'Tithings'	P. A. S. Pool, 'The Tithings of Cornwall', in JRIC n.s. 8 (1978–81), 275–337
Pryce	William Pryce, *Archaeologia Cornu-Britannica* (Sherborne, 1790)
PT	Ifor Williams (ed.), *The Poems of Taliesin*, translated by J. E. Caerwyn Williams (Mediaeval and Modern Welsh Series, vol. 3: Dublin, 1968)

PWCFC	*Proceedings of the West Cornwall Field Club*, n.s. (Camborne, 1953–61)
RBB	John Rhŷs and J. Gwenogvryn Evans (eds.), *The Texts of the Bruts from the Red Book of Hergest* (Oxford, 1890)
RC	*Revue celtique* (Paris, 1870–1934)
RCPS	*Annual Reports* of the *Royal Cornwall Polytechnic Society*
RD	*Ordinale de Resurrexione Domini Nostri Ihesu Christi* in Norris, II
Rev.Intnl.Onom.	*Revue internationale d'onomastique*
Rhestr	Elwyn Davies, *Rhestr o Enwau Lleoedd: A Gazetteer of Welsh Place-Names* (3rd edn., Cardiff, 1967)
Rhigyfarch	J. W. James (ed.), *Rhigyfarch's Life of St David* (Cardiff, 1967)
RIA	*Dictionary of the Irish Language, based mainly on Old and Middle Irish Materials* (Dublin: Royal Irish Academy, 1913–76)
Richards, Atlas Môn	M. Richards (ed.), *Atlas Môn* (Llangefni, 1972)
Richards, 'Tonfannau'	M. Richards, 'Tonfannau', *Journal of the Merioneth Historical and Record Society*, 4 (1963), 274–6
RP	J. Gwenogvryn Evans (ed.), *The Poetry in the Red Book of Hergest* (Llanbedrog, 1911)
RWM	J. Gwenogvryn Evans, *Report on Manuscripts in the Welsh Language* (2 vols., London, 1898–1910)
S., Sawyer	P. H. Sawyer, *Anglo-Saxon Charters: An Annotated List and Bibliography* (London, 1968)
SC	*Studia Celtica*
Smith, Top.Bret.	William B. S. Smith, *De la toponymie bretonne, Dictionnaire étymologique* (Supplement to *Language* 16, 1940)
Spurrell-Anwyl	J. Bodvan Anwyl (ed.), *Spurrell's Welsh-English Dictionary* (9th edn., Carmarthen, 1920)
SR	Subsidy Roll (unprinted) in the Public Record Office
Stoate, Survey	Thomas L. Stoate, *A Survey of West Country Manors, 1525* (Almondsbury, 1979)
Stokes, Glossary	Whitley Stokes, *A Cornish Glossary* (London, 1870, reprinted from TPhS 1868–69)
Surv.Denb.	Paul Vinogradoff and Frank Morgan, *Survey of the Honour of Denbigh, 1334* (British Academy, Records of the Social and Economic History of England and Wales, 1: London, 1914)
Symons, Gazetteer	R. Symons, *A Geographical Dictionary or Gazetteer of the County of Cornwall* (Penzance, 1884)
Tanguy	Bernard Tanguy, *Les noms de lieux bretons, I: Toponymie descriptive* (*Studi*, no. 3: Rennes, 1975)
Thomas, 'Dumnonia'	Charles Thomas, 'The Character and Origins of Roman Dumnonia', in Charles Thomas (ed.), *Rural Settlement in Roman Britain* (Council for British Archaeology, Research Report no. 7: London, 1966)
Thomas, 'Irish Settlements'	Charles Thomas, 'The Irish Settlements in Post-Roman Western Britain: A Survey of the Evidence', JRIC n.s. 6 (1969–72), 251–74
Thomas, Ling.Geog.Wales	Alan R. Thomas, *The Linguistic Geography of Wales* (Cardiff, 1973)
Thomas, North Britain	Charles Thomas, *The Early Christian Archaeology of North Britain* (Oxford, 1971)
Top.Bret.	see Smith

Top.Hib.	Breandán Ó Cíobháin, *Toponomia Hiberniae, I: Barúntacht Dhún Ciaráin Thuaidh* (Dublin, 1978)
Topon.Nautique	J. Cuillandre and others, *Toponymie nautique des côtes de Basse-Bretagne*, pamphlets reprinted from *Annales hydrographiques* (Paris, 1949–1973)
TPhS	*Transactions of the Philological Society*
Trg.	John Tregear, 'Homelyes XIII in Cornysche', British Library, Additional MS 46,397 (cited from a cyclostyled text published by C. Bice); cf. Nance, OC 4 (1943–51), 429–34, and 5 (1951–61), 21–27; JRIC n.s. 1 (1949–52), 119; ZCP 24 (1954), 1–5
Trois Poèmes	Roparz Hemon, *Trois poèmes en moyen-breton* (Mediaeval and Modern Breton Series, vol. 1: Dublin, 1962)
TYP	Rachel Bromwich, *Trioedd Ynys Prydein: The Welsh Triads* (2nd edn., Cardiff, 1979)
Units	Melville Richards, *Welsh Administrative and Territorial Units* (Cardiff, 1969)
VKG	Holger Pedersen, *Vergleichende Grammatik der keltischen Sprachen* (2 vols., Göttingen, 1909–13)
Voc.	The Old Cornish Vocabulary, in British Library, Cottonian MS Vespasian A.xiv (cited by the numbering in Graves' edition)
VSBG	A. W. Wade-Evans (ed.), *Vitae Sanctorum Britanniae et Genealogiae* (Cardiff, 1944)
Welsh Assize	James Conway Davies (ed.), *The Welsh Assize Roll, 1277–1284* (Cardiff, 1940)
WG	John Morris-Jones, *A Welsh Grammar, Historical and Comparative* (Oxford, 1913)
WHR	*Welsh History Review*
Williams, Beginnings	Sir Ifor Williams (edited by Rachel Bromwich), *The Beginnings of Welsh Poetry* (Cardiff, 1972)
Williams, CA	Ifor Williams, *Canu Aneirin* (3rd impression, Cardiff, 1970)
Williams, Lexicon	Robert Williams, *Lexicon Cornu-Britannicum* (Llandovery, 1865)
WM	J. Gwenogvryn Evans, *The White Book Mabinogion* (Pwllheli, 1907)
Wm.Worcester	John H. Harvey (ed.), *The Itineraries of William Worcestre* (Oxford, 1969)
WVBD	O. H. Fynes-Clinton, *The Welsh Vocabulary of the Bangor District* (Oxford, 1913)
ZCP	*Zeitschrift für celtische Philologie*

COUNTY, ETC., ABBREVIATIONS

Abd. Aberdeenshire
Agl. Anglesey
Ang. Angus
Arg. Argyllshire
Ayr. Ayrshire
Bdf. Bedfordshire
Bnf. Banffshire
Bre. Brecknockshire

Brk. Berkshire
Brt. Brittany
Bte. Bute
Buc. Buckinghamshire
Bwk. Berwickshire
Cai. Caithness
Cam. Cambridgeshire
Che. Cheshire

COUNTY, ETC., ABBREVIATIONS

Cla. Clackmannanshire
Cmb. Cumberland
Cnw. Cornwall
Crd. Cardiganshire
Crm. Carmarthenshire
Crn. Caernarvonshire
Den. Denbighshire
Dev. Devon
Dmf. Dumfriesshire
Dnb. Dunbartonshire
Dor. Dorset
Drb. Derbyshire
Drh. Durham
ELo. East Lothian
Eng. England
Esx. Essex
Fif. Fife
Fli. Flintshire
Gau. Gaul(ish)
Gla. Glamorgan
Glo. Gloucestershire
GtL. Greater London
Hmp. Hampshire
Hnt. Huntingdonshire
Hre. Herefordshire
Hrt. Hertfordshire
Inv. Inverness-shire
IoM. Isle of Man
IoW. Isle of Wight
Irl. Ireland
Kcb. Kirkcudbrightshire
Kcd. Kincardineshire
Knr. Kinross-shire
Knt. Kent
Lan. Lanarkshire
Lin. Lincolnshire
Lnc. Lancashire
Mdx. Middlesex
Mer. Merionethshire
MLo. Midlothian

Mon. Monmouthshire
Mor. Morayshire
Mtg. Montgomeryshire
Nai. Nairnshire
Nfk. Norfolk
Ntb. Northumberland
Ntp. Northamptonshire
Ntt. Nottinghamshire
Ork. Orkney
Oxf. Oxfordshire
Peb. Peebles-shire
Pem. Pembrokeshire
Per. Perthshire
Rad. Radnorshire
Rnf. Renfrewshire
Ros. Ross and Cromarty
Rox. Roxburghshire
Rut. Rutland
Sco. Scotland
Sfk. Suffolk
She. Shetland
Shr. Shropshire
Slk. Selkirkshire
Som. Somerset
Ssx. Sussex
Stf. Staffordshire
Stl. Stirlingshire
Sur. Surrey
Sut. Sutherland
Wal. Wales
War. Warwickshire
Wig. Wigtownshire
WLo. West Lothian
Wlt. Wiltshire
Wml. Westmorland
Wor. Worcestershire
YoE. Yorkshire (East Riding)
YoN. Yorkshire (North Riding)
YoW. Yorkshire (West Riding)

OTHER ABBREVIATIONS

aj. adjective
Br. Breton
Co. Cornish
coll. collective
cpd. compound
deriv. derivative
dimin. diminutive
Eng. English, England
ex(x). example(s)

f., fem. feminine
fld. field(-name)
indic. indicative
ipv. imperative
len. lenition
m., masc. masculine
MlBr. Middle Breton (12th–16th cent.)
MlCo. Middle Cornish (12th–16th cent.)

MlWe. Middle Welsh (12th–15th cent.)
MnBr. Modern Breton (17th cent. onwards)
MnCo. Modern Cornish (17th–18th cent.)
MnWe. Modern Welsh (15th cent. onwards)
MS manuscript
n.d. no date
n.s. new series
OBr. Old Breton (9th–12th cent.)
OCo. Old Cornish (9th–12th cent.)
OE Old English
OWe. Old Welsh (9th–12th cent.)
pers. personal name(s)
pl. plural

p.n. place-name
ppp. past participle passive
pres. present
pret. preterite
RB Romano-British
r.n. river name
s.a. *sub anno*
sb. substantive
sg. singular
sgv. singulative
spir. spirant mutation
subj. subjunctive
s.v. *sub verbo*
vb. verb
v.n. verbal noun
We. Welsh

* denotes a hypothetical form.

Place-names in *italics* are obsolete, i.e. not found in standard 19th–20th century sources.

DICTIONARY
OF
CORNISH
PLACE-NAME
ELEMENTS

A

*-a 'place of' (?): this name-forming suffix seems to occur, though its derivation is obscure, and Welsh or Breton parallels are lacking. It can be suggested in the following names: Leah, Leha (both + ***legh**); Alsia (+**als**); **whyla* (+**hwilen** pl.), used as C2 in Trewhela and Trewhella (both +**tre**); and one or two other, more doubtful names (see e.g. ***tusk**).

agy 'hither, within, nearer', OM 953 (contrasted with **aves** 'outside'), etc.; a preposition and adverb formed from the preposition *a* 'from' and **chy** 'house'.

    D. *Gweale Gollas a Choy*, fld. (contrasted with *Gweale Gollas a Vese*)

aidlen (f.; OCo., *d* = /ð/), 'aspen', only at Voc. 700, *abies* (OE *æps*) glossed *aidlen vel sibuit*. The first word, *aidlen* (plural presumably **aiðel*) is evidently 'aspen', judging by both Middle Breton *ezlen* (GMB, p. 230), '[peuplier] tremble' (MnBr. *eflenn*) and Welsh *aethnen*, but there is no need, despite Williams (BBCS 11 (1941–44), 10f.) and Graves (Voc. s.v.), to emend the Cornish form in order to bring it in line with the Welsh: the Middle Breton spelling proves that it is a good form. It does not, of course, correctly translate Latin *abies* 'fir-tree': apparently the Cornish glossator first translated OE *æps* 'aspen', and then, realising that the OE gloss on *abies* was incorrect, inserted *vel sibuit*; the latter is obscure, but is presumed to mean 'fir, pine' (Graves ad loc.). On aspens in Cornwall, see Davey, Flora, pp. 417f.; 'rather rare', but recorded for most parts, including those where the three names are. The names all seem to show -ðl- > -dl- in their modern forms.

    A. ?Idless, ?Eddless (both + ***-ys²**)
    C2. aj. ?Trevadlock (+**tre**)

*alaw 'water-lily', equivalent to Welsh *alaw*. See Richards, Atlas Môn, p. 156, river-name Alaw, Agl., whence Aberalaw (Units, p. 1; PKM, p. 216); Ekwall (ERN, p. 5) disbelieves the derivation of the Welsh and Cornish names from *alaw* 'water-lily'; and Nicolaisen, Nomina 6 (1982), 39, has it derived from **el-/*ol-* 'flow' (?); but in view of Richards' satisfaction with the derivation, it is accepted here, with reservations (e.g. that one might expect the

*aled

plural, or a suffix, to occur in the river-names, rather than the singular: but cf. s.v. **heligen**).

C2. Porthallow (+**porth**): more probably the qualifier in this name is not *****alaw** as a common noun, but a simplex r.n., *****Alaw**. If so, the whole name is very similar to Welsh Aberalaw.

*****aled** (OCo.). This element of obscure meaning and derivation occurs as a river-name in Alt (PNLnc., p. 95; ERN, pp. 9f.), Welsh Aled (Den., Rhestr, p. 3; ELl, p. 47) and as a town-name in Breton Aleth (Smith, Top.Bret., p. 9). M. Richards (Atlas Meirionydd, p. 200) derives the Welsh river-name Aled from the root *****al-** 'grow, nourish' (cf. LEIA, p. A57), with a suffix *-ed*; if the Cornish ones could be river-names as well, the same derivation would apply. Note also Gaulish *Aletanus pagus*, DAG, p. 176; *Aletum*, ibid., p. 595; and in personal names: Untermann, BNF n.s. 5 (1970), 89.

    A. Allet (= former r.n.?)

    C2. *****Lannaled* (+ *****lann**; = St Germans)

als 'shore, cliff, slope', Voc. 733 glossing Latin *litus*, OE *sæstrand*; Welsh *allt*, Breton *aod* (cf. Tanguy, p. 61). Both Welsh and Breton share with Cornish the variation of sense from 'cliff, slope' (which can occur inland) to 'shore'. Note the early Welsh gloss *Altgundliu* = *Collis Gundleii*, VSBG, p. 28. See also Fleuriot, ÉC 11 (1964–67), 152–5. In the list which follows, 'sea-cliff', the commonest sense, is not specified, but where the element appears to illustrate a different sense, it is given.

    A. derivative (+ *****-a**?) ?Alsia

    B1. Camels (+**cam**: cf. Welsh Gamallt, Rad.)

    B2. Halsferran (+**yfarn**); *Als Roshusyon* (+p.n., 'Rosudgeon Cliff'); Holseer Cove (+**hyr**)

    C2. Penhalt (+**pen**); Trenault (+**tre, an**: '(inland) slope'); Trenalls (+**tre, an**); Chynhalls, Chinalls (both '(sea) slope'), and Chenhalls ('shore') (all +**chy, an**); Parc-an-Als Cliff, *Park an Aule Cliff*, fld. (both +**park, an**: '(sea) slope'); Dornolds, fld. (+**dor, an**); Wheal Owles, mine (+**wheyl**); Porth-en-Alls (+**porth, an**)

alter 'altar', OM 1170, etc.; *altor*, Voc. 747 glossing Latin *altare*: a loan from Latin.

    B2. Altarnun (+St)

*****alun** (u = /y/), obscure; see Elements, 1, 8; JEPNS 1, 43; LHEB, p. 306; ERN, pp. 6–8. This common Brittonic river-name, of unknown meaning, may occur in the River Allen (though the second vowel is unsuitable in many early forms), and in St Allen (which may well be a river-name taken over as a parish name and turned into a saint); also perhaps as an earlier stream-name in the following, where, however, a personal name is also possible.

C2. Tregarland (+**cruc**); Redallen (+**rid**); Penhallam (+**pen**) – on this last name, cf. Penally/Penalun (Pem.), and Padel, Med.Arch. 18 (1974), 127f.; Penally is *aluni capitis*, LLD, pp. 149 and 151.

***amal**, equivalent to Welsh *ymyl* 'edge'. The vowel of the second syllable is unexpected: **amel* would be preferred, but the forms indicate **amal*. Compare the OE stream-name *Amal-burnan* (Suffolk, BCS, no. 1289; Sawyer, no. 1486). Cf. also ***emla**, which has been supposed to be a plural, **emlow*, of ***amal**; but early spellings insist upon **emla* or **emle* as the form of that element.

A. Amble (= r.n.)

B2. Amal-ebra /-veor/-whidden (+ ***bre**?/**meur**/**guyn**)

amanen 'butter', Voc. 849 glossing *butirum*; Boorde (JRIC 22 (1926–28), 369) gives *hag a manyn* 'and butter'; Lh. 45a *manyn* glossing *butyrum*. Welsh *ymenyn*, Breton *amanenn*. For place-names containing the Welsh word, see Richards, Arch. Camb. 113 (1964), 176f. (Carn Ymenyn); Indo-Celtica, p. 189 (Dryll yr Ymenyn); Studia Celtica 2 (1967), 77 (Ffrith y Menyn); also Pantymenyn, Crm. In the case of Carn Ymenyn, Richards suggests that the shape of the tor was responsible for the name; but in several of the Cornish names, it is likely that the element (especially after **hal**, **goon**) is connected with the custom of making butter on the summer pastures when transhumance was a common form of husbandry. Alternatively, and especially after **park**, it may be related to (or derived from) the English use of 'butter' in field-names, presumably to denote lush ground (e.g. Field, EFN, pp. 34f.; PNBrk., 1, 283 and 11, 541). For coastal names in Brittany containing *amanenn*, see Guyonvarc'h, Dictionnaire 6, 426f. Note that three of the Cornish names parallel Lhuyd's form *manyn*, with loss of the first syllable.

C2. Halamanning (+**hal**); Gunnamanning (+**goon**); Cannamanning (+**carn** aj.); Enysmannen (+**enys**); *The Hellamannin*, fld. (+**hal**?); Park-Mannin, Park-Manin, flds. (+**park**)

an 'the', definite article; Welsh *y(r)*, Breton *an, ar*. Old Cornish *en* (in Voc. 37 and 39; MC 182a, etc.) is also the normal form in the 12th-

and 13th-century forms of place-names, and is still common in the 14th century. The definite article in Cornish place-names, as in the other Celtic languages, is used much more frequently than in English. It can also come and go quite freely, so that for many names it is impossible to say whether they 'originally' contained it or not: for example, Chycoose (St Clement) was *Chincois* 1337, *Chicoys* 1556 (**chy + cos**); Trekillick was *Treenculyek* 1327, *Trekelyek* c. 1500 (**tre + kullyek**); places called Ninnis were usually *Enys* in the middle ages; Trengove in Constantine was *Tregoyf* 1327 and *Trengoff* 1369 (**tre + gof**); Trengilly was *Trekelly(-byan)* 1287 and *Trengelly(-byghan)* 1304 (**tre + kelli**). Note also instances, such as Pendrift, etc. (see ***pen (an) dre**), Tolgarrick (**tal + karrek**), and the many examples of Pengelly (see ***pen (an) gelli**), where the lenition of a feminine second element after a masculine word must be due to a lost definite article, even though no form of those names shows it (whereas Old Welsh *penn i celli*, LLD, p. 272, does). (The same explanation is therefore possible, *pace* HPB, p. 320, n. 5, for Breton Penhoat, where the spirant mutation could be due to a definite article understood.) It is therefore not possible to say under what circumstances the definite article was used or not. On the whole it is the minor names that tend to show it, and the major names that do not: it may well be that a name was felt to be more formal, more 'name-like', without it. But some major names (e.g. Tywarnhayle, **ti + war + an + *heyl**) do show it, though its early forms are variable: *Ti Wærnhel*, 960, shows it, but *Tiwarhel*, mid-11th century, does not. In fact the 960 spelling is the only instance of the definite article from any Anglo-Saxon document, which means that the many minor names in the boundaries are entirely lacking in it. This lack is in keeping with the conclusions of D. Flanagan concerning the use of the article in similar phrases in Old Irish place-names: 'Place-names in early Irish documentation: structure and composition' (Nomina 4 (1980), 41–5). She finds that the structure, 'noun + article + noun' is very rare in the oldest stratum of Irish place-names, and must be presumed to be a later type, starting to appear in the 9th century and becoming popular in the 11th and 12th centuries. For lack of evidence, this pattern cannot be confirmed or denied in Cornish, but it is possible a similar development occurred. More closely related, however, are the early Welsh and Breton charters and their boundaries: Fleuriot (FG, p. 271) notes the rarity of the definite article in the 9th-century material in the Cartulary of Redon (and also in the glosses: he compares its rarity in early Welsh poetry as well); but the boundaries in the Llandaf charters provide plenty of examples, for instance: *lost ir inis* and *bronn ir alt* (LLD, p. 73); *tal ir cecyn* and *tal ir fos* (LLD, p. 122); *guern i drution* and *luch i crecion* (LLD, p. 123). However, the early date of these examples is in each case very uncertain, as they occur in later, partly-modernised, copies.

An incorrect definite article is very common in late minor names and field-names: it can occur either before an adjective, as in Street-an-Garrow (***stret + garow***) and Parkanhere, fld. (**park + hyr**), or before a proper name, as in *Chyambarret* (= Chybarrett), *Whele an Phelp*, mine, and *Park an Rogers*, fld., or before an English phrase, as in Park-an-starve-us, fld. These are obviously formed by analogy with the many names starting correctly with *Park an*, *Street an*, *Wheal an*, etc., and the same phenomenon is found in the other Celtic languages: Welsh Rhyd Fraith 'speckled ford' (Mon.) appears incorrectly as *(tire) rid y frayth*, 1630 (ÉC 10, 212); and Pant y Meriog is for Pant Meriog (BBCS 16, 28); Irish Faill an Bhric is probably 'speckled cliff' (Top.Hib., 1, 143); and see Nicolaisen, p. 125, for useful discussion of Scottish place-names showing it.

*-an², a possible suffix for forming place-names; compare the British suffix *-ana* (PNRB, pp. 327b and 338a), and Gaulish *-an-* in names (Holder, ACS, 1, 134). The early Welsh gloss *Nant Caruguan, id est, Uallis Ceruorum* (VSBG, p. 54) seems also to presuppose that such a suffix was acceptable to the glossator. It probably occurs in the following elements or names: ***seghan** (from **segh**); ?Manaccan (+ **manach**?); Fraddon (+ **frot**?); ?Ludgvan (+ **lusew**?), ***douran** (+ **dour**) and perhaps one or two others.

anken 'grief, sorrow, pain'. Welsh *angen* 'need, necessity'; Breton *anken* 'chagrin, peine, douleur'. The meaning in place-names is unclear, but for a late place-name rhyme, apparently bewailing the poverty of Crankan farm, and presumably based on the derivation of its name, see Nance, JRIC 21 (1922–25), 146–53 (cf. Loth, RC 41 (1924), 275f.). The best sense would come from taking **anken** in the sense of the Welsh cognate, 'hardship', though that is not the attested meaning in Cornish: compare similar names in English (Field, EFN, p. 276).

C2. Grankim, Crankan (both + ***ker**)

anneth 'dwelling', PC 705; Welsh *annedd*, Breton *annez*. Confusable with *aneth* 'a marvel' (RD 1302), but that is unlikely to occur in place-names.
    C2. ?Bosanath, ?Bessy Benath (both + ***bod**: meaning?)
antell 'snare', MC 19a. Breton *antell* 'trap, snare'; Welsh *annel* 'snare, deceit'.
    C2. ?Carnantel, fld. (+**carn**)
***ar** 'before, facing, beside', acting as a leniting prefix. Gaulish *are* 'ante', *Aremorici* 'antemarini' (DAG, p. 547; PNRB, p. 254). Welsh Arfon (Units, p. 7) '(the district) facing Môn'; *arfordir* '(land) facing the sea' (PKM, p. 278); Argoed '(land) facing the forest' (several, Units, p. 7), etc.; Old Welsh *Lisarcors* 'Court beside a swamp' (VSBG, p. 190); Orchard, Dor. (DEPN, s.v.); Breton Argoad, etc. (Smith, Top.Bret., p. 11). See discussions by Caerwyn Williams of Old Welsh *ar*, *ir*, Middle Cornish *er*, Celtica 2 (1954), 306-8 and 318-24; and by D. S. Evans of Welsh *ar*, BBCS 15 (1952-54), 1-12 and 169-83; also K. Meyer, *Zur keltischen Wortkunde* (Berlin, 1912-19), no. 190.

    The preposition ***ar** is not active in Middle Cornish, which uses **rag** for 'before' and **war** for 'upon'; so it can be assumed that names containing ***ar** are early in date: from the Old Cornish period, at any rate. Similarly in Welsh, such names as Arfon and Argoed are likely to be old, since they show an earlier sense of *ar* than the usual Middle Welsh one 'upon' (which is due to crossing with early Welsh *guar*); note, however, *y tu ar Gernyw* 'the side facing Cornwall' (PKM 45.8), and other instances (GMW, p. 184). The prefix is also suggested in the Cumbrian Arlosh (PNCmb., II, 291, where, however, I. Williams is quoted as suggesting that **ar-* is the intensive prefix, as at GPC *ar*^{2}), and Dollerline (**dol* + **ar* + r.n., PNCmb., I, 55). In one or two of the names below, where the second part is obscure, it may be that we are dealing with the intensive prefix (not otherwise attested in Cornish); but in all those where the meaning is clear, it has the sense 'before, facing'.

    Ardevora (+ ***devr-** pl.); Perranarworthal (+**St**), Northwethel (both + ***gothel**); Arwenack (+ ?); Harvose (+**fos**); Harlyn (+**lyn**); ?Harcourt (+ ***crak**[1]?); ?Arrowan (+**auon**?); ?Relowas (+**lyw** aj.?) – this last only if the variant **er-* is allowed for Cornish, as for Welsh (GPC *er-*).
    If, in *hryt ar þugan* (Sawyer, no. 1019; DCNQ 29 (1962-64), 27;

= Retew?), the last word is a stream-name (*Dugan?*), then 'ford upon the D.' would be a usage similar to Welsh Pont-ar-Ddulais (Gla.), and would be a unique Cornish instance of the sense 'upon'. For the phrase *hryt ar*, compare the Welsh construction, *Rhyd ar* + river-name: R. J. Thomas, BBCS 7 (1933–35), 127, and Richards, ÉC 10 (1962–63), 225–6.

ardar 'plough', Lh. 43b, Old Cornish *aradar*, Voc. 342, both glossing *aratrum*; *arder*, Pryce, p. Ff2r. Welsh *aradr*, Breton *arar*. The Cornish plural would have been **erder*, older **ereder*, judging by Welsh (pl. *erydr*, *ereidr*) and Breton (pl. *erer*), and it may occur in one name, as suggested by Picken, DCNQ 30 (1965–67), 38f. (*pace* Quentel, ZCP 39 (1982), 195–200).

C2. pl. ?Powder (+**pow**)

*arð (OCo.) 'height'. Welsh *ardd* 'a height' is rare and early; the best examples are in LLD, p. 174.5 *di'r ard*; *ar hit ir ard* 'to the height; along the height' (also at 184.20f.) and 261.1 *di ard ir allt* 'to the height of the slope'; otherwise it is restricted to a few occurrences in early poetry and to place-names. In Cornish and Breton the word does not occur independently at all. Furthermore, I. Williams makes the point (PKM, p. 293; ELl, pp. 22f.) that Welsh *ardd* in place-names is hopelessly confused with *gardd* 'garden' and *garth* (1) 'ridge, hill', (2) 'enclosure'; the same applies to Cornish ***arð** and ***garth**, except that an equivalent of Welsh *gardd* is not to be expected, since it is a Norse loan in Welsh. Note that a document quoted by I. Williams (PKM, p. 293) also appears to have the Welsh pl. *arddau* translated as *terra arabilis*, if the *terra arabilis monachorum* of one document is to be equated with the *Arddau'r Myneich* of another. In Breton, **arð* > **arz* is unattested in the language and very rare in place-names: Tanguy, p. 62, cites two, both of which may be compounds. Cornish ***arð**, as well as being confusable with ***garth**, would be indistinguishable in place-names from ***arth** 'a bear' (Breton *arzh*, Tanguy, p. 117), if that occurred in place-names (or in the language), but it can be presumed not to do so, except, of course, in compound personal names. (Breton personal names containing both words are discussed by Fleuriot, Ann.Bret. 64 (1957), 529–32; and Gaulish ones by H. Birkhan, *Germanen und Kelten bis zum Ausgang der Römerzeit* (Wien, 1970), pp. 435–40.) Despite all this confusion, a Cornish ***arð**, equivalent to Welsh *ardd*, can safely be postulated, and it is reasonable to assume its presence in a few place-names;

*arð

note also Old Irish *Ardd Machae* glossed *Altum Machae* (Bieler, p. 190, etc.).

A. Ayr (formerly *Arth*); Aire Point: cf. Breton l'Ile d'Arz (Quentel, Ogam 7, 388).

B1. Several names appear to contain, as second element (C2), some word which is itself a compound ending in *-ard*. This is best explained by supposing that the compounds contain ***arð** as B1, and that in a compound, after the stressed syllable, ***arð** became **-ard*. Approximate parallels for this occur in Breton, though the phenomenon is generally late and sporadic: HPB, p. 533, final *-θ* > *-d* (cf. RC 17, 60–3); HPB, pp. 665–7 (medial *-ð-* > *-d-*, and very occasionally in final position); HPB, p. 700. There is also a similar change in the spoken forms of some Welsh place-names ending in *-rth* (giving *-rt*): EANC, p. 5. In addition, the two Breton names Rozart and Crénard (Tanguy, p. 62; earlier *Crenarth*, Chrest., p. 121) appear to show the same change, and so does Dinard (formerly *Dinart*: Quentel, Ogam 7 (1955), 387–9; it is not there considered that the name might be a compound, 'fort-height'). Examples of this are as follows; all are tentative. In each case, the stress falls on the middle syllable (i.e. the first one of the disyllabic second element), so that if the names contain ***arð** > *-ard*, a compound word used as C2 and itself containing ***arð** as B1 is indicated.

*goon-arð 'down-height': Trewonnard (+**tre, goon**)
*moyl-arð 'bald-height': Penmillard (+**pen, *moyl**)
*gwyn-arð 'white-height': Trewinnard (+**tre, guyn**)
*gol-arð 'watch-height': Trewellard (+**tre, gol**)
?**scoul-arð* 'kite-height': Treskillard (+**tre, scoul**)

In addition, the name Lizard, if the stress, as pronounced, is an accurate guide, ought to be a compound, **lys-arð* 'court-height'. If the stress is misleading and the name is not a compound, it contains ***arð** as C2, **lys arð* 'height's court'. It must be emphasised that these compounds are not well established, there being only one instance of each; but the three Breton names (Crénard, Dinard, Rozart) look as though they might well be comparable in structure, as in phonology. Compare ***garth** as B1.

B2. *Arthia* (+St)

C2. pl. Nanjarrow (+**nans**); see also ***pen-arth**, which may contain ***arð**; an aj. **arðek* is possible as C2 in Trevarrick (+**tre**),

and, very doubtfully, in Carharrack (+*ker); a diminutive
*arðin perhaps as C2 in Carharthen (+*ker).
argel 'secluded place, retreat', Welsh *argel*. The Breton *argil* 'recul'
and Cornish *argila* 'to recoil', Lh. 245a, are a different word, from
*ar+*kyl 'back', whereas *argel is from *ar+*kel 'cover'.
   A. Argal
arghans 'silver', also *arhuns*; *argans*, Voc. 225. Welsh *arian(t)*,
Breton *arc'hant* (cf. Tanguy, p. 58). There is possibly some slight
(but inconclusive) evidence for a variant Cornish form *argans*, like
the southern Breton *argant* (HPB, pp. 716–17); the similarity of the
19th-century variant *Ariance* to Middle Welsh *aryant* may also be
mere coincidence. On 'silver' in place-names, see PNDor., II, 34n.
   C1. Ardensaweth (+*aweth?); Arrallas (+*lys) (a compound
very much akin to the Gaulish *Argentorate* (DAG, p. 1088) and
Argantomagus (DAG, p. 360): the sense is obscure; compare early
Irish compounds in *Airget-*, Hogan, Onomasticon, p. 23)
   C2. Venton Ariance (+**fenten**): cf. *fenten bryght avel arhans* 'a
spring bright as silver', OM 771; *Wheal Arrans*, mine (+**wheyl**);
Pol Lawrance, coast (+**pol**)
***arghantell**, a stream-name corresponding to Welsh *Ariannell* and
variants, 'silvery one', a derivative of the preceding. Fourteen
examples of the Welsh name are given (EANC, pp. 92–6), including
one correctly spelt in Old Welsh as *arganhell* (LLD, pp. 75 and
173). A river in the Vosges, now called L'Arentelle, was formerly
Argentilla (EANC, p. 95; DAG, p. 1088); see also Nicolaisen,
BNF 8 (1957), 231–3. Note the Breton coastal name *Kleier
Arc'hantell* (= Roches d'Argenton), Ann.Bret. 69 (1962), 594. At
least two streams in Cornwall must formerly have borne this name,
which now survives, as often, only as that of farms lying on their
courses.
   A. Tregantle (2 exx.)
   C2. *Whele Arantall*, mine (+**wheyl**)
arluth 'lord', fem. *arlothes*; *arluit*, Voc. 188 glossing *dominus vel
herus*; fem. *arludes*, Voc. 189 glossing *domina*. Welsh *arglwydd* (the
word does not occur in Breton) has a variant *arlwydd*, first appearing in the 13th century (GPC); the fem. *arlwyddes* also occurs;
note Waunarlwydd 'the lord's moor', Gla. (Units, p. 221).
Cornish universally shows forms without the *-g-*, except for one
remarkable spelling of the first name, *Trenargluth* 1327: this form
proves that the variant with *-g-* must have existed in Cornish too.

11

*arwystel

C2. Trenarlett (+**tre, an**); (fem.) *Park-an-Arlothas, Parke an arlothas*, flds. (both +**park, an**)

*arwystel 'pledge'. Welsh *arwystl*; Old Breton *aruuistl*, FD, pp. 74f. Since the same word as a personal name is quite common (OWe. *Arguistil*, LLD; Br. *St Arewestl*, Smith, Top.Bret., p. 11; Loth, NSB, pp. 11 and 147; Doble, Welsh Saints, pp. 70f.), it is much more likely to be a personal name (as suggested by Picken, JRIC n.s. 7 (1973-77), 223) than the common noun in the three place-names. Note other personal names in Bodm. containing *guistel* 'hostage, pledge' (Voc. 174 glossing *obses*; OBr. *guuistl*, FD, p. 204, and Welsh *gwystl*): *Tancwuestel*, xlvi; *Medwuistel*, xlvii; and *Anaguistl*, xxv.

C2. ?Trussall, ?Trusell, ?Treroosel (all +**tre**)

ascorn 'bone', Voc. 44 glossing *oss*; pl. *escarn, yscarn*; Welsh *asgwrn*, Breton *askorn*. Cf. Breton Coatascorn, Smith, Top.Bret., p. 34. In the field-names there may have been some confusion with *askellen* (sg. f.) 'thistle' (Voc. 660 glossing *cardus*; Welsh pl. *ysgall*, Breton pl. *askol*), though that is not definitely attested in place-names.

C2. sg. Crasken, Creskin (+***ker**); *Park-an-Askerne*, fld. (+**park, an**); dimin. *Croft-an-Askernel*, fld. (+***croft, an**)

asow 'ribs', OM 99, MC 218d; sg. *asen*, Voc. 83 glossing *costa*, CW 385. But the name shows a new singular, *asowan*, formed from the plural, and found also in Trg. 2a (sic MS: Nance, ZCP 24 (1954), 2; OC 4 (1943-51), 431). Welsh sg. *asen*.

C2. sg. *Denys Azawan* (+***dynas**)

*aswy (f.) 'gap'; Welsh *adwy*, Breton *ode* (HPB, p. 226, LEIA, p. A99, and GMB, p. 448). The Breton word means particularly 'gap in a hedge'; 'ouverture dans un fossé pour le passage d'une seule bête à la fois' (Le Pelletier, in GIB; cf. Ann.Bret. 51 (1944), 136), or (in coastal names) 'passage' (Topon. Nautique 1419, p. 517); and in Welsh the word can mean not only this, but also 'movable hurdle for blocking a gap', while a 'mountain pass' is rather the commoner *bwlch* (cf. ***bolgh**): WVBD, pp. 4 and 60f. (But note Adwy-wynt, 2 exx.: PNFli., p. 2). Since most of the Cornish names are obsolete field-names, it is probable that 'gap in a hedge' was the usual meaning in place-names; but some examples could have rather the sense 'pass, notch in the skyline'. OE *geat* shows the same extension from 'gap in a hedge' to 'mountain pass' (as in Yetts o' Muckart, Per.): Elements, I, 198, and note especially *wind-geat*, Elements, II, 268; PNDrb., III, 700 and 754; PNCmb., II, 388; PNYoW., VII, 268.

B2. Adjawinjack (+**guyns** aj.); Adgewella (+**ughel** spv.); the remaining names are all those of fields: Nangidnall (+(**goon**), ***enyal**); *Adjaporth* (+**porth**); *Assawine* (+**guyn**); *Assa Govranckowe* (+**keverang** pl.); Aja-Bullocke, *Adga Bullocke Bounds*, mine (both +English?); *Adgyan Frank* (+**an**, **Frank**); *Ajareeth* (+**ruth**); Aja-Gai (+**kee**); Agahave (+**haf**)

C2. ?Clydia (+**guel**)

atal 'rubbish', OM 427; *attle* 'mining waste', Pryce (s.v. *atal*); dialect *attle* 'rubbish containing little or no ore' (Symons, Gazetteer, p. 139). Borrowed from Old English *adela* 'liquid filth'.

C2. *While an Attol*, mine (+**wheyl, an**)

aval 'apple', OM 823, etc. (pl. *avalow*, OM 176); Welsh *afal*, Breton *aval*. See Hamp, 'The North European word for "apple"', ZCP 37 (1979), 158–66, and next entry.

C1. ?Levalsa (+**ti**; meaning?)

auallen (f., *u* = /v/) 'apple-tree', Voc. 679 glossing *malus*; pl. unknown. Welsh *afallen* (Rhydafallen or Rhydyfallen, 6 exx., ÉC 10, 222); Breton *avalenn*, Old Breton *aballen* glossing *malus* 'apple-tree', FD, p. 51 (also *terra Ennavallen/terra Pomerii*, 13th-cent., Tanguy, p. 103). It is curious that all the names listed below appear to contain the singular, 'one apple-tree'. It suggests that (wild) apple-trees were sufficiently unusual for a single one to be a noteworthy feature in the landscape. Or, if the singular could stand for the plural (cf. **heligen**, etc.), Nansavallan could be translated 'orchard valley'. See Davey, Flora, pp. 183–5 on crab-apple trees in Cornwall: 'rather common'.

C2. Redevallen (+**rid**); Nansavallan, Landevale Wood (both +**nans**); Rosevallon in Cuby, Rosevallen, Rose in the valley (all +**rid**/***ros**, (**an**)); ?Worthyvale (+**guartha**?)

aves 'yonder, further, outside', OM 953 (contrasted with **agy** 'within'), etc.; a preposition and adverb formed from the preposition *a* 'from' and **mes** 'open-field'. Middle Welsh *o vaes* (GMW, p. 201), Breton *a-vaez*.

C2. ?Gullaveis (+**guel**?)

D. *Te Wyn Ha Vaes*; *Gweale Gollas a Vese*, fld. (contrasted with *Gweale Gollas a Choy*); *Dorheere a Veese*, fld.

auon (OCo., *u* = /v/) 'river', Voc. 731 glossing *flumen vel fluvium*;

awel

Lhuyd's Cornish instances of the word (3b, etc.) may be derived from that text, though such a phrase as *tornewan an awan*, 141a, may be genuine. Welsh *afon*, Middle Breton *auo(u)n*. The word is virtually unknown in Cornish place-names, as well as in the language, but the one certain example forbids the idea that it was entirely lost; other, commoner, words for 'river, stream' were **dour, gover, frot, streyth** and **goth**[1].

A. ?Arrowan (+ ***ar**): 'facing the river' does not suit the site, but is suggested by early forms.

B2. *Awen-Tregare* (+ place-name, = *Tregeare Water*)

C2. ?Trevone in Padstow (+ **tre**), if an unusual stress-shift is allowed as possible.

awel 'breeze', OM 1147 and PC 1209; Voc. 446 *auhel* glossing Latin *aura*, OE *hwiða oððe weder*. Welsh *awel*, Breton *avel*. Pryce's *Hagar awell, ha auel teag* 'Bad, or foul weather, and fair weather' (p. Ff1v) is the only instance of the sense 'weather'.

C1. Haggerowel Croft, fld., Haggarowel (both + **hager**)

C2. *Chiawel* (+ **chy**)

*****aweth** 'watercourse' (?): a possible equivalent of Breton *naoz* 'lit (de rivière), canal' (*aos* and *naöz*, 18th-cent.: derivation?); or compare Welsh *aweddwr* 'fresh water'. Both form and meaning are uncertain: the form depends upon early spellings of the first name, and the sense upon the comparison with Breton *naoz* (assuming that to have gained *n-* through misdivision of *an aos*). Compare possibly Breton Karreg an Awezed, rock (Topon. Nautique 1419, no. 9646).

B1. ?Ardensaweth (+ **arghans**)

C2. ?House Oath, fld. (+ ***havos**?)

B

bagh 'hook, fetter', BM 3562 (so Parry, BBCS 8, 132f., *pace* Cuillandre, RC 49, 117f.); pl. *bahaw an darraz*, Lh. 46b glossing Latin *cardo*. Welsh *bach*, Breton *bac'h*; Old Breton *bach* 'hook', FD, p. 77. In Welsh place-names the meaning 'nook, corner' occurs, e.g. *bachlatron*, LLD, p. 78; Bach(y)sylw, Crm. (+ pers.: BBCS 23, 323); etc. This sense is not found in the sparse Cornish occurrences of the word, but it is very likely to be the one present in place-names, as with OE *hōc* (Elements, I, 255).

A. pl. Bahow

B1. Corva (+ *cor²)
B2. Bahavella (+**auallen** ?); Bohago (+pers.); Barteliver (+ ?)
C2. Trembath (+**tre, an**)

bagyl 'crozier', BM 3007. Welsh *bagl* 'staff, crozier, crook', aj. *baglog* 'crooked'; *hamella de Brenbagle* 1334, Surv.Denb., p. 21. (The specialised Breton *beleg* 'croziered one, priest', HPB, p. 163, is not found in Cornish.) The Cornish aj. was **baglek*.

C2. aj. Carbaglet (+ ***kyl**)

bal 'mine, area of tin-working'. The origin of this word is difficult. It might be from *pal* 'spade', OM 392 and 396 (*palas* 'dig', OM 681 and 865), via some such form as **wheyl bal*, but that phrase is very rare, and the word seems in fact to be contrasted with, rather than a subdivision of, **wheyl**. Loth (RC 39 (1922), 47–58, at 51f.) compares Gaulish *balma* 'peak, grotto' (cf. DAG, pp. 32f. and GPN, pp. 147f.). Pokorny, 'Cornisch *bal* m., "Mine, Erzgrube, Zinngrube"' (*Indogermanische Forschungen* 65 (1960), 164–7) equates the word with Welsh *bal* 'peak, summit' (comparing, for the sense, Welsh *bol(y)*, meaning both 'cavity' and 'bulge'), but Welsh *bal* is itself not attested until 1788. In *Vox Romanica* 10 (1948–49), 226f., he had suggested that **bal** was perhaps not Indo-European at all. As for the meaning, there are three 18th-century witnesses, possibly interdependent. T. Tonkin (MS 'A', 1710×1733: JRIC n.s. 7 (1973–77), 200), says 'A Ball (the usuall name they give to a large parcell of tin works)', although he was referring to a working called *Wheal* Dreath; Borlase, Antiquities, translates it 'a place of digging' (he was evidently thinking of *palas* 'dig'); and Pryce s.v. translates as 'a parcel of tin works together'. Both *bal* and *wheal* continued in dialect use through the 19th century (e.g. as heard by Loth in 1911, RC 39, 51), and the usage may be of significance: Symons, p. 139, *bal* 'the miners' term for a mine, a very old word'. The general impression is that **bal** is an area where surface working was carried out, perhaps a group of workings, whereas **wheyl** is a specific tin-work. This still does not explain the name *Wheal Bal*, but otherwise is consistent with the evidence. On present information, **bal** as a Cornish generic is much rarer (with only about a dozen instances) than **wheyl** or **guyth²**, the normal terms for individual mines; and it is first found slightly later than those elements, not until the second half of the 16th century: *Bal dew* 1593, *Balrosa* 1571 and *Hemball* 1569 are the earliest examples. This means that

ban

bal as an English dialect word (see below) is attested as early (in Norden) as those names where it occurs as a Cornish element.

B1. Hembal (+**hen**)

B2. Balwest (+**west**); Baldhu (3 exx.), *Bal dew* (all +**du**); Balrose (+ ***ros**?); Balleswidden (+p.n.); Balnoon (+**an**, **goon**); *Ballamoone* (+**an**, ***mon**²?); Balmynheer (+***men hyr**); Bal Dees, *Ballandees*, mines (both +(**an**), **tus**)

C2. Zawn a Bal, coast (+**sawn, an**); Wheal Bal Hill (+**wheyl**) This word, taken over into dialect, was also used as an English element in place-names, e.g. Trevenen Bal 'T. mine'; also *Carmeall-Ball* and others; however, confusion with English *ball* 'hill' (Elements, I, 18f.) is sometimes possible.

ban 'peak' occurs in the language only in such phrases as *warvan* 'up', BM 1450 and 3671; *avadn* 'atop', CW 1809, *in badn* 2202, *in ban* 1826, 1847, 1900, etc.; also *ban* glossing *altus*, Lh. 42b. OWe. *bann* 'height' in *bannguolou*, Ann.Camb. s.a. 873, Welsh *ban* 'top, tip', cf. Williams, Beginnings, p. 188; OBr. *bann-* 'high' in *banncepr*, FD, p. 78; *bann* 'horn', ibid.; Tanguy, p. 62. For the adjective of this word, see ***bannek**.

A. The Vans, coast

B1. Callevan (+**kelli**); ?Trevan Point (+ ***try-**²: or **men**)

C2. Trevan, Tolvaddon, Tolvadden (all +**tal**); Pednvadan, coast (+**pen, tal**); Venton Vadan (+**fenten**)

banathel 'broom-plants', Voc. 694 glossing *genesta*: the Cornish word is plural, for sg. ***banathlen**; Lh. 240c *bannolan* 'a broom [sc. bush]'; dialect *bannal* 'broom' (e.g. Davey, Flora, p. 111). Since, in OCo. *banathel*, the *-e-* is an epenthetic vowel (LHEB, pp. 337f.), it will not have counted as a syllable, and the stress in Cornish (as in Welsh *banadl, banal*, Breton *banal, balan*) was on the first syllable; the development, with loss of internal *-th-* before *-l*, is parallel to the Welsh form *banal* (15th-cent., GPC) and Breton *banal* (HPB, p. 488: the forms lacking d/z can be traced as early as the 12th century, *-Benalec*, and 13th, *-Banalec*, in place-names; cf. Tanguy, p. 103; RC 3, 400). When lenited, in the form *-vannel* or similar, this word would be indistinguishable from a lenited form of **manal** 'sheaf', but **banathel** is much the more probable in all such names, and they are therefore entered here rather than under that word. Davey, Flora, p. 111, broom is common throughout most of Cornwall. It was of economic value as a fuel, and *Skebervannel* 'broom barn' would be one where the plants were stored.

A. pl. Bannel, fld. (several); aj. Bonallack, Benallack (2 exx.), Menallack, Penadlick, Penhallick

C1. Bonyalva (+ *ma)

C2. pl. Carvannel in Gwennap (+ *ker); Rôspannel (+ *ros); Skebervannel (+ skyber); Park Bannel, fld. (+ park); sg. ?Trenadlyn (+ tre); cf. Breton Kerbannalen, etc. (Tanguy, p. 103)

*bannek 'peaked, prominent'; Welsh *bannog* (Jackson, Gnomic Poems, p. 59, and Antiquity 29 (1955), 81); Breton *Bannec* (= island), Tanguy, p. 62. Being an aj. formed from **ban**, this word should mean simply 'peaked, possessing a peak or peaks', but for the Welsh word the further sense 'prominent, conspicuous' is generally given as well; either sense would do in the first of the four names, St Agnes Beacon being both pointed and prominent.

C2. *Bryanick* (= St Agnes), Brevadnack (both + *bre); Colvannick, Calvadnack (both + kal) (= Welsh Calfannog, Bre., Units, p. 27)

bar 'top, summit' (f.?, but m. in Welsh and Breton). Welsh *bar*; Breton *barr*; Old Breton *barr* (FD, p. 80); in Cornish the word occurs only in *bar an pedn* (Lh. 172b), glossing *vertex* 'the top or crown of the head'. An aj. *barrǫg occurs in Barrock Fell (PNCmb., I, 201) and in *Berk-* of Berkshire (PNBrk., I, 1f.), etc. See Elements, I, 19f.; JEPNS 1, 43f.

A. ?The Var, coast.

B2. Barbican (+ byghan); Barlendew (+ lyn, du); *Bargudul* (+ ?); Bervanjack (+ ?); Burgotha (+ goth^1 pl.); Barnoon, Barnewoon (both + an, goon); Barbolingey (+ *melyn-jy); Bargus, ?Burgois (both + cos)

C1. A compound *bar-drogh (+ trogh), 'summit-broken; having a broken summit' (cf. Welsh *barlwm* 'summit-bare', in Bryn Barlwm, Gla., quoted by Williams, Lexicon, p. 18a, and Twmbarlwm, Mon., ST 2492: correctly Twyn Barlwm) may appear (as C2) in Levardro (+ *lann) and Trevaddra (+ tre).

bara (m.) 'bread', Voc. 851 glossing *panis*; etc. Welsh and Breton *bara*. Cf. Welsh *Pant y bara* 1610 (Cart.Glam., no. 1495) and Breton *Karreg ar Bara* (Topon.Nautique 1419, no. 8705). Baragwaneth Field is more likely to contain the surname of that form (= **bara + gwaneth**), than a place-name in its own right.

C2. *Gwyth an bara* (+ guyth^2, an); Nampara (+ nans); Carn Barra, coast (+ carn)

bargos (m.) 'buzzard', OM 133. Welsh *barcud* means primarily 'kite', though also 'buzzard', but Breton *barged* is solely 'buzzard' (GIB; Tanguy, p. 125; ÉC 20,108-9). The Cornish word is most likely to be 'buzzard'; for 'kite' see ***bery, scoul**. In addition, most of the place-names (+ **cruc, carn**) suggest the behaviour of buzzards rather than that of kites. However, in one name, Rosebargus (+ **lost**, 'tail'), where the land has an indented shape, **bargos** signifies 'kite'. Cf. NED s.vv. *glede* and *puttock* for similar vagueness in such words in English; and also Penhallurick, Birds, II, 84f. and 97–100, on kites and buzzards in Cornwall. The *c* in the Welsh form is explained by Williams, BBCS 1, 20, as being due to preservation by a lost lenited *g* (*barg- or *berg-) before -*cud*, but since his arguments would apply equally to Cornish and Breton, which have -*g*-, the anomaly of the Welsh form is unexplained. In theory, **bargos** might be indistinguishable from a word *barges equivalent to Welsh *bargod* 'eaves, edge, border'; however, there is no evidence for the existence of such an element.

C2. Carn Barges (3 exx.), Carn Bargus, all coastal (all + **carn**); Rosebargus (+ **lost**); *Gweal-Bargus*, fld. (+ **guel**); ?Perbargus Point (+ **porth** ?); ?*Crukbargos* (+ **cruc**; = Bargoes); ?Busvargus (+ ***bod**)

barthesek 'wondrous, wonderful', RD 109, *barthusek* [sic MS], RD 1177 (an aj. of *marthus* 'a wonder').

C2. *Wheal Varizick*, mine (+ **wheyl**)

*bas aj. 'shallow', as a noun, 'shallows': only in *basdhowr* glossing *vadum*, 'a ford', Lh. 169a; the verb occurs, ppp. *basseys* 'abated', OM 1098 and 1127. Welsh and Breton *bas*. (A borrowing from Late Latin *bassus*, according to GPC; not in Elf. Ladin or Piette.)

C2. Carn Base, coast (+ **carn**); ?Park an Bays, fld. (+ **park, an**)

bedewen (f.; OCo.) 'birch tree', Voc. 693 incorrectly glossing *populus* 'poplar', following the mistake of Ælfric who glossed *populus* with *byrc*. The stress was on the first syllable, /ˈbedəwen/, since the second -*e*- is an epenthetic vowel. Plural *bedew > *bezow. The word does not occur elsewhere, since Lhuyd's *bedewen* 'a birch' (241c) is probably derived from Voc., and Pryce's *bedho*, *bezo* 'a birch tree' [sic, for pl.] may be derived from Lh. 44c, *betula* glossed Breton *bezo*, *bedho*. Welsh *bedwen*, pl. *bedw*; Breton pl. *bezv*, *bezo* (cf. Tanguy, p. 104); Old Breton dimin. *beduan* 'birchlet' (FD, p. 81). The Breton word shows a sound-change -*dw*- > -*ðw*- (> -*zw*-), also found elsewhere (HPB, pp. 501f.). This same sound-change, *-dw-* > *-ðw-*, appears in the Cornish place-name Penburthen, implying that Old Cornish contained *beðwen as a variant of *bedwen. (Note that the variant would have to have arisen before the date of epenthesis, probably not before the 9th century, LHEB, pp. 337f., in order to have the same phonetic context as in Breton.) The early forms of Penburthen are virtually the only evidence for such a variant, but they do require its existence; if the variant is accepted, then some names listed as containing **beth** 'grave' pl. may belong here instead. For a similar sound-change, compare the parish-name Petherick < *Pedrek. According to Davey, Flora, p. 405, the birch is 'rather common' in Cornwall, occurring in all areas except for one; but he does not mention the earliest attestation of them, twelve trees growing at Goran churchtown in A.D. 1271 (JRIC 6 (no. 21, 1879), 238).

A. pl. Bissoe. For the plural used as A, cf. the North Welsh *boscus...qui vocatur le Bedowe* 1334, Surv.Denb., p. 24: the Cornish Bissoe occurs as 'wood of *Bydo*' 1261 (FF, no. 176).

C2. pl. Trevedda, Treveddoe, ?Treviddo (all +**tre**); Lambessow (+***lann**); sg. Penbetha in Creed, Penburthen (both +**pen**), Laveddon (+***lann/nans**). Under **beth**, note particularly Lampetho and Penbetha in Probus, and compare the fairly common Welsh Penbedw.

begel (m.) 'navel; hillock', Lh. 17a; *bigel*, Pryce, p. Ff4r. Breton *begel*, Old Breton *becel* glossing *bulla* (FD, p. 80): cf. Welsh *bogail*, *bogel*, supposed to be a borrowing from Old French *bocle* (PKM, pp. 110f.); but, as Fleuriot says, it is hard to see how the Cornish and Breton words (especially Old Breton *becel*) could be so derived, and in any case their first vowel is unexplained if they are to be cognate with the Welsh word. As qualifier, **begel** is indistinguishable from **bugel** 'shepherd' unless there are good early forms.

A. The Beggal, fld.

B2. *Begeledniall*, fld. (+***enyal**); Beagletodn (+**tum** ?)

C2. *Croft en begel* (+***croft, an**); Penbeagle, *Penbegle* (both +**pen**); *Ponsbeggal Bridge* (+**pons**); *Bounds Begall*, mine (+***bounds**)

beler 'water-cress', Voc. 656 glossing *carista vel kerso*; dialect 'belercresses', Davey, Flora, p. 30; Breton *beler* (cf. GMB, p. 40 and Tanguy, p. 111), Welsh *berwr* (for the variation *l/r*, cf. Gaulish

berula (DAG, p. 441), and Middle Irish *birar*, *bilar* 'cress': HPB, p. 813; VKG, I, 491).
   A. sgv.pl. 'cress-plants' may occur in Polurrian.
   C2. aj. 'cress-bed', Polbelorrack (+**pol**)
ben (m.) 'foot, base, stump', OM 779 and 788. Loth (ACL 3, 258) assumed this word to be the same as Welsh *bôn* 'base, root, stump' (Irish *bun* 'base, stump'; also 'river-mouth, estuary'), but if so the word has behaved irregularly in Cornish, possibly by crossing with **pen** 'head'; most of the place-names containing **ben** have subsequently been changed to *Pen-* by folk-etymology. It is used mainly with words for rivers or streams, and the places bearing the names lie at the foot of their respective streams, not at the head (as with true names in *Pen-*): compare the similar usage of Scottish Gaelic *bun* (Watson, CPNS, pp. 47, 116 and 477), and of Irish *bun* (Hogan, Onomasticon, pp. 133f.); also Breton *Ben-* 'mouth of a river' in Bénodet (Pinault, Ogam 13 (1961), 617–18), and note particularly Banastère, from OBr. *Bonester*, MlBr. *Benester* (Bernier, Ann.Bret. 76 (1969), 650).
   B2. Pendower in Philleigh (+**dour**); Pendavey in Egloshayle, ?Pendewey, ?Pendavy (all +r.n. ***Dewy***); Penhale in Davidstow (+**hal**); Poldowrian (+***douran***); Penberth (+***perth***)
   C1. ?Bonnal (+***gwal***?)
benyges 'blessed', OM 819, and frequently in various spellings (ppp. of *benyga* 'bless': Welsh *bendigo*; Breton *binnigañ*).
   C2. ?Crowsmeneggus (+**crous**/***ros***)
ber 'short', Voc. 949 glossing *brevis*; etc. Welsh *byr*, fem. *ber*; Breton *berr*.
   C1. ?Burras (+**rid** or ***red***)
   C2. *maenber* (+**men**: Sawyer, no. 770)
***bery** 'kite', equivalent to Welsh *bery(f)* 'kite'. On kites in Cornwall, see **bargos**.
   C2. ?Reperry (+**rid**)
beth (m.) 'grave'; pl. *bethow* PC 2999, *bethov* BM 1333. Elements, I, 32; JEPNS 1, 44. Welsh *bedd* (used in place-names such as Beddgelert and Bedd y Ci Du, Units, p. 10; Bedd Captain Morgan, PNFli., p. 8), Breton *bez*. As generic, **beth** is in one case replaced by ***bod**; this confusion may also occur in Wales (ELlSG, p. 49). As qualifier, the four names containing the plural could contain ***beðow** 'birch-trees', if that is allowed as a variant of the regular ***besow** (see **bedewen**).

B1. ?Morvah (+**mor**?)

B2. Boscathnoe (+pers.); Benbole in St Kew (+***bolgh**); *beð cywrc* (+ ?; Sawyer, no. 684); *Bethkile* (+***kyl**?)

C2. Trembleath in St Mawgan in Pydar (+**tre, an**); pl. Penbetha in Probus (= *pen beðow*, Sawyer, no. 770) (+**pen**), Lampetho (+**nans**), Trembethow (+**tre, an**), Trevethoe (+**tre**), *Tonenbethow* (+**ton, an**)

***bich** (OCo.) 'small' (?). This element would be the equivalent of postulated Welsh *bych (m.), *bech (f.), attested only in place-names such as Dinbych, Dwyfech (GPC s.v. *bych*[1]; ELl, p. 46), supposedly equivalent to Old Irish *becc*. (But this must have contained *-gg-*, not *-kk-*, in Primitive Irish: Thurneysen, GOI, p. 93.) The normal word and element for 'small' was **byghan**. In theory the word might be confusable with Middle Cornish *be* 'load' (OM 1057 and 1299 [sic MS]), Welsh *baich*, Breton *bec'h*, but there is no evidence for that word in place-names.

C2. Denby, Dunveth (+***dyn**; the lenition in the latter is inexplicable. The former is in contrast with nearby Dunmere); Carveth in Cuby (+***ker**); Rospeath (+***ros**); Trebick (PNDev., p. 211) may contain this word.

byghan 'small'; Welsh *bychan*, Breton *bihan*. The Old Cornish form *boghan* glossing *parvus* (Voc. 951) shows a different vowel (VKG, 1, 282 and 385), which survived in the noun *bohes* 'a little', but not otherwise. Compare ***bich**. Though the usual modern form is *Vean*, etc., there are some names, not confined to East Cornwall, where the voiceless velar spirant has become a stop, voiced or voiceless. This is more likely to be an English treatment of the sound than a variation within Cornish; or else possibly a spelling-pronunciation. Some of the mediaeval forms in *-bigan*, etc., probably contain *g* for *gh*, but they can often be produced for those names which now contain a stop, and may thus represent the stop as being already in existence. This word is extremely common as an affix, to denote the smaller part of a subdivided tenement; usually it is contrasted with *vear* (= **meur**), but there can be other contrasted words, instead or in addition.

C2. Barbican (+**bar**); Nanpean (4 exx.), Nanspian (all +**nans**); Colbiggan, Collibeacon (both +**kelli**); Halviggan (+**hal**); Lanvean, Laddenvean (both +***lann**); Trebeigh, Trebyan, Trevean (10 exx.), Trevine, Treveans, Trebighan, Trebiffin (all +**tre**); *Porthpyghan* (= West Looe), Porth Pean, Porthbean Beach,

*byly

Perbean Beach, Perprean Cove (all +**porth**); Goonvean (2 exx.), Gunvean (all +**goon**); Bosvean (+***bod**); Coosebean (+**cos**); Carvean (+***ker**); Fentervean (+**fenten**); Tirbean (+**tyr**); Carbean, Carn Bean, coast (both +**carn**); Croftvean (+***croft**); Park Bean, etc., flds.; Enys Vean, rock (+**enys**); Marazion (+**marghas**)

D. Upwards of a hundred names (many, but not all, obsolete forms) of subdivided tenements or other places, such as Trelanvean, Hellesvean/-veor, Sellanvean (= Little Sellan), Truro Vean. Note *Trescavyghan* = *Parva Trescau* (= Trescowe); *Roscarrec bian* = *Roscarec minor* (= Scarrabine); sometimes one cannot tell if various subdivisions with different affixes are separate tenements, or simply the same ones differently named: e.g. it is very likely that *Trevajegmur/Trevagecbyghan* 1300 and *Trevasekwartha/-woles* 1392 represent only two, not four, subdivisions of Trevassack.

*byly 'pebbles' (?), equivalent to Breton *bili* 'cailloux' (cf. GMB, p. 68). This word would be indistinguishable from an Old Cornish personal name ***Bili**, equivalent to the common Breton *Bili* (Chrest., p. 110), or (if it is not the same), Welsh and Old Cornish *Beli*; any of the names below could contain either.

C2. Trevilley (2 exx.), Trevilly (all +**tre**); Pedn Billy, coast (+**pen**); Chybilly (+**chy**); ?Menabilly (+**meneth**; or **mcn**+**ebel** pl.)

býnkiar (MnCo., $\acute{y}$ = /ʌ/) 'cooper', Lh. 174a glossing *vietor*. The word may be derived from Lhuyd's *bynk* 'a blow', 67a glossing *ictus*, though the semantics would be unclear.

C2. ?Park Banker, ?*Park-an-Munkyer*, flds. (+**park, (an)**)

*byu 'cattle'; Middle Welsh *biw*, Breton *bioù* (cf. Tanguy, p. 117).

C2. Lisbue (+***lys**); *mayn biw* (+**men**: Sawyer, no. 832; so Loth, RC 34, 144, who makes it singular however). Other names that appear to contain this word are listed under **bugh**: the two would have fallen together, in sound, by the Middle Cornish period.

bleit (m.; OCo., *t* = /ð/) 'wolf', Voc. 559 glossing *lupus*; *ramping blythes* 'ravenous wolves', Trg. 19a; *blygh*, CW 1149; Welsh *blaidd*; Breton *bleiz* (Tanguy, p. 117). Cf. the early Welsh 'lapides ...qui Brittannico sermone Cunbleid vocantur, id est, lupina saxa' VSBG, p. 92; Middle Breton *Pol-bleiz* in A.D. 1242 (RC 7, 209). This last name is the equivalent of Cornish Blable, Welsh Blaiddbwll (Owen, Pembrokeshire, I, 265n.) and of Old English *wulf-pytt* (Woolpit, Sfk., Sur., etc.: DEPN). In theory the element could be a personal name every time it occurs: compare, for instance, 'Walter called *Bleyt*' in A.D. 1285 (Anc.Deeds, IV, A.9489); but in practice all names probably contain the animal, either from the sense (found with **cruc**, **pol**), or because of the presence of the definite article. However, the one instance of the adjective may be a personal name instead. Other place-names occur which incorporate a compound personal name of which one element is **bleit**.

C1. Blable (+**pol**)

C2. *Cruplegh*, Carplight (both +**cruc**: cf. Welsh *Cruckebleith* in 1316–17, Owen, Pembrokeshire, IV, 366n.); Carblake (+ ***ker**); Trembleath in St Ervan (+**tre**, **an**); aj. Treblethick (+**tre**)

blyn (m.) 'tip', OM 779; also 'top, summit'. Welsh *blaen*, Elements, I, 38; JEPNS 1, 44; Breton *blein* (GMB, p. 70; Tanguy, p. 76); Old Breton *blein* (RC 9, 419; FD, p. 85; in place-names, HPB, p. 157; also *Blen*-, Chrest., p. 110). For further instances, and the sense 'edge, frontier', see Quentel, Ogam 8 (1956), 435f. The vowel of Breton *blein*, as compared with Welsh *blaen*, is unexplained (HPB, p. 157); the Cornish hapax **blyn** is compatible with the Breton form, since it could represent a Primitive Cornish **blein*. VKG, I, 125 and LP, p. 43 accept the equation of the forms without offering a solution. There are six names containing the Cumbric reflex, usually as *Blen*-: see PNCmb., III, 462; but the forms are too late to be a reliable guide as to what that Cumbric reflex would have been. Alternatively, a Welsh *blen* 'hollow' (Old Irish *mlén* 'groin') is possible, and could have a Cornish reflex: see CA, p. 386; LEIA, pp. M56f.; and Hogan, Onomasticon, p. 117.

B2. ?Blankednick (+ ?)

blogh 'bald, bare' (*pen blogh*, BM 3828). Middle Breton *blouch* (GMB, pp. 71f.); for Welsh *blwch* 'bald' see Padel, BBCS 29 (1980–82), 523–6; also *Aidan Bloch*, VSBG, p. 120. Lamplough (PNCmb., II, 405; cf. Quentel, Ogam 7 (1955), 81–3), and Welsh Cwmblwch, Crm., are equivalents of Nanplough: a 'bare' valley would be one devoid of (tall?) vegetation; compare the common Welsh name Nantmoel, Moelnant.

A. ?Blouth Point: cf. Breton Ar Bloc'h, rock, 'le dénudé' (ÉC 10 (1962–63), 286)

C2. Nanplough (+**nans**); as a noun ('bald/beardless man'), *Chymblo* (+**chy**, **an**)

*bod (f.; OCo.) 'dwelling'. Welsh *bod*, Old and Middle Breton *bot*

*bod

(Chrest., pp. 110 and 192). There are virtually no instances of the word used as a common noun in any Brittonic language; 'dwelling, residence' is agreed to be the general sense. Ultimately it is identical with the verbal noun *bos*, etc., 'to be' (LEIA, p. B74). The element is translated *vilar* in Old Breton (FD, p. 88), and *mansio* in '*Bothmenaa*, id est mansio monachorum', Gotha *Vita Petroci* (An.Boll. 74, p. 163), though the gloss inspires the less confidence inasmuch as the author has got the second part of the name slightly wrong. In early Irish, by contrast, *both* 'hut, bothy' is used both as a common noun and as generic in place-names (Hogan, Onomasticon, p. 120). From all this, it seems likely that ***bod** was in use early as an element, and had probably gone out of use before the period when the remains of the languages become common; that is, it seems likely to belong more to the 5th–11th centuries, rather than any later period, though it would still doubtless have been understood in place-names. Some other evidence also favours the theory of early date. The word seems to be approximately consistent in its occurrences over the three Brittonic countries, which again favours its early use; Wales and Cornwall are even similar in having **bod* distributed unevenly over their respective areas: in Wales, it is mainly found in the north-west (information from Tomos Roberts, Bangor), and in Cornwall it is found especially in the area south of Bodmin Moor (= Westwivelshire), and the Land's End peninsula (= West Penwith). In most other areas it is rare, though it does occur even in strongly-anglicised areas right up to the Tamar (Bohetherick, Bamham, and note also Bodgate, PNDev., 1, 158). There seems to be no geographical reason for this variation, in either Wales or Cornwall, nor is the distribution in any way complementary to that of **tre** (which, on the contrary, is evenly distributed). Information on the distribution of **bod* in Brittany does not seem to be available, but note examples in the cartularies of Redon (Chrest., p. 110), Landevennec (*Bot Frisunin, Bot Tahauc*, Chrest., p. 165), Beauport (RC 3 (1876–78), 401–2) and Quimper (Chrest., p. 192); in Port-Louis (Ann.Bret. 59 (1952), 328), the Rhuis peninsula (ibid. 74 (1967), 502–3), and elsewhere (Top.Bret., p. 19).

The element must have been contrasted in some way with **tre**, since they were in use at the same time; they both occur quite frequently in the names of Domesday manors, with about the same proportion of ***bod** and **tre** names as the overall numbers of those elements (and thus, in absolute terms, many more *Tre*-manors than *Bod*- ones). The most likely contrast is a legal one: a **tre** probably had some particular legal status (if only in being, nominally at least, a hundredth of a *cantref* in Wales), whereas a ***bod** was humbler, being simply somebody's 'dwelling'. This is also in keeping with the fact that ***bod** is found particularly with personal names as qualifiers, and several times with words referring to women: **tre** often has a personal name, of course, but ***bod** does so in a greater proportion of cases; and the use of ***bod** in terms for temporary, seasonal dwellings (see B1 below) is also in keeping with the idea of its humbler status.

It is interesting to note that *Bod*- manors very rarely became mediaeval tithings, and, curiously, not at all in the hundred of Penwith, where ***bod** is commonest: the few such tithings are elsewhere in the county. Of the total of about 230 *Bod*- names in Cornwall, over 60 (or more than a quarter) are in West Penwith: this strange distribution can only be explained as local fashion. It cannot be explained by reference to the Cornish language having continued in use later there, for virtually all such names in the district are attested before the end of the fourteenth century (when Cornish was still widespread elsewhere), and five of them by A.D. 939 (if Sawyer, no. 450, represents an original charter of Athelstan) – further support for the early use of ***bod** as a formative element.

A. Vose, Foage, Forge, Anvoaze (+**an**); Vorvas, ?Worvas (both + ***gor-**); pl. Bussow, Bawdoe (compare Romano-British *Botis*, PNRB, p. 273; Irish pl. Botha, Hogan, Onomasticon, p. 119)

B1. See the compounds ***kyniaf-vod** 'autumn-dwelling', ***gwavos** 'winter-dwelling' and ***havos** 'shieling'.

B2. Only a selection is given. Qualified by a word for a natural feature: Bodinnick in St Stephen in Brannel (+**eithin** aj.); Bodulgate (+ ***huel-gos**); Bokelly (+**kelli**); Boscarne, Boscarn (both +**carn**); Boscawen (+**scawen**)

Qualified by a word for a man-made feature: Boscreeg, Boscrege (both +**cruc**); Boskear (+ ***ker**); Bodmin (+ ***meneghy**); Bosneives (+ ***kyniaf-vod**)

Qualified by a word denoting a person (not a personal name): Bamham (+**mam**); Bosoar, Besore (both +**wuir**); Borah (+**gruah**); Bosfranken (+**Frank** pl.); Boswase (+**guas**); Bodieve, Bogee, Bojea (all + ***yuf**)

*boðour

Qualified by a personal name: Boswinger (+*Wengor*, Bodm.); Bosworgey in St Columb Major (+*Wurci*, Bodm.); Boswellick (+*Mæiloc*, Bodm.); Bodiggo (cf. Old Welsh *Guidci*, LLD); Bephillick (+*Felec*, ÉC 3, 149); Bohetherick (cf. Old Breton *Hedroc*, Chrest., p. 137); Biscovellet (cf. Old Welsh *Cimelliauc*, etc., LLD); etc.

Qualified by the saint of the same or an adjacent parish: Bosulval (+Gulval); Bosliven (+St Levan); Boderlogan (+Illogan)

Qualified by a place-name: there are several cases where a name in ***bod** appears to have as its qualifier another place-name; such cases are unexplained, and in most cases the place-names that form the qualifiers are not otherwise known nearby: Bosporthennis (+**porth, enys**); Bosigran (+**Chygaran*, **chy**+**garan**?); Bossava (+**Tysava*); Bosullow (+**Tywoelou*, **ti**+**golou**); Burghgear (+**Baghkeyr*, **bagh**+***ker** pl.); Butter Villa (+**Trevely*?); Boderwennack (+**Trevenek*, **tre**+***meynek**?)

*boðour (OCo., *bo-* = /bœ/) 'filthy-water'; Old Welsh r.n. *baudur* LLD, p. 74, from Welsh *baw* 'filth, filthy' (Old Cornish **bo*); a compound of **dour**.

C2. Polborder, ?*Polmorder* (both +**pol**)

bogh (m.) 'billy-goat', BM 1625; *boch*, Voc. 588 glossing *caper vel hyrcus*. Welsh *bwch* 'billy-goat, roebuck'; Breton *bouc'h* (cf. Tanguy, p. 117), 'billy-goat'. The Cornish word probably refers to the domestic animal each time. The female, 'nanny-goat', is **gaver**. Compare Drumburgh (PNCmb., I, 124; +**drum*); Breton Penboc'h, etc. (Tanguy, p. 118); Welsh *nant i buch* (LLD, p. 228); *llam yr bwch* (CLlH, p. 7); Carreg y Bwch, Crn. (ELlSG, p. 15); Glyn Bwch (Myv.Arch., p. 357b); Alt Bough, Hre. (cited PNCmb., I, 124). Rather surprisingly, the word is, like **gaver**, rare in Cornish place-names, and only two are known, with a third possible: as generic, in the name of a rock, **bogh** would be comparable with other animal-words used for rocks: see **ebel**, etc.

B2. ?Bo Cowloe, rock (+p.n.)

C2. Penbough, Pemboa (both +**pen**)

*bolgh 'gap, pass'; Welsh *bwlch*, Breton *boulc'h*. Diminutive **bolghan* (?). This element was superseded by ***aswy**.

C2. Benbole in St Kew (+**beth**); Treboul (+**tre**); Carvolth (+***ker**; or **moelh**); dimin. ?*Trevolen*, ?*Trevolland* (both +**tre**)

bor 'fat', Voc. 942 glossing *pinguis*. Welsh *bwr*, Breton *bour-* in *bourbell* 'someone with big eyes', and in place-names; Old Breton *borr* glossing *corpolentus*, ÉC 16 (1979), 199 and 207. See also Williams, BBCS 7 (1933-35), 35-6; Vendryes, ÉC 4 (1948), 310-13; Quentel, Ogam 6, 23f.; Evans, GPN, pp. 154-6. Smith's Cornish **bor*² 'rampart, fort' (Elements, 1, 42) should be deleted (so JEPNS 1, 44); the meaning of this element as a noun is rather 'swelling, protuberance' (cf. Quentel's note). Used as a noun, it seems to be feminine.

B2. Borlase (+**glas**); Burlerrow (+ ?)

***both** 'boss, hump'; Welsh *both*. (Breton *bos* is borrowed from French: Piette, p. 82; see Tanguy, p. 63 for its use in place-names.) The aj. occurs, Laurence *Bothek* (1327 SR, Paul), 'the humped'.

B2. Bolster (+**lester**): 'boat(-like) hump' describes part of a nearby earthwork.

C1. ?*Bothle* (+**le** ?)

C2. ?Lambo (+**nans** ?)

boudzhi (m.; MnCo.) 'a cow-house', Lh. 10b; also *bowdzhe devaz* glossing *ovile* 'a sheep-coat, a fold, a sheep-house', 110c. Welsh *beudy*; Old Breton *boutig* glossing *stabulum*, FD, p. 89. A compound of **chy** 'house', and the root seen in **bowyn**.

A. Bowgie; Boughie, The Bowgey, etc., flds.

B2. Bowgyheere (+**hyr**; 2 exx.)

C2. *Park-an-Bowgy*, flds. (+**park**, **an**)

bounder (f.) 'lane; pasture'. In Voc. 724 this word glosses Latin *pascua* (Old English *læswe* 'leasow, pasture-land'), but its meaning in place-names was widely recognised to be 'lane', and there is no other trace of the sense 'pasture'. (E.g. Tonkin, JRIC n.s. 7, 203: 'Chyvounder, i.e. the House in the Lane'; Borlase, Antiquities, p. 461, *Vounder* 'a lane'. Lhuyd's 'common', 113c, is derived from Voc.) Thomas, 'Irish Settlements', pp. 270-2, notes the similarity of the word to Welsh (Pembrokeshire) *meid(i)r*, *moydir*, Irish *bóthar*, both also meaning 'lane'; on these, see O'Rahilly, Celtica 1, 160; M. Richards, Lochlann 2 (1962), 128-34; A. W. Thomas, Ling.Geog.Wales, fig. 172; RIA Dictionary, s.v. *bóthar*. These words are so similar in both meaning and usage to the Cornish one (and the Pembrokeshire word also close geographically), that there must be some connection, even though there are problems in a phonological equation, primarily the *-n-* in the Cornish word.

Richards suggests that the first part of the Welsh and Irish words

*bounds

is the root *bow- 'cattle', and leaves the second part unexplained, but shows that the -*i*- in -*dir* is epenthetic. Thomas accepts Richards' derivation for the first part of the Cornish word, but suggests that the second part is **tyr** 'land'. However, that explanation will not do for the Welsh word if the -*i*- is epenthetic, and Cornish -*der* for -*dir* would not be very good, either. It is better to assume that the Cornish *e* is also epenthetic, and that the element has the same derivation as the Welsh and Irish words, leaving the -*n*- as intrusive. It was already present in the twelfth century, and might be due to influence from Mediaeval Latin *bunda* 'boundary', or, more likely, a vernacular equivalent: the lanes often serve as territorial bounds. (Graves' suggestion, Voc. s.v., of influence by English 'boundary' itself is unlikely, since that word is not attested until 1626 in English. Campanile, p. 16, suggests a derivation from Late Latin *bunnarium*, etc., via an English or French intermediary form; but that ignores the Welsh and Irish words.)

The distribution of the Cornish element, as shown by Thomas, is entirely typical of an ordinary Cornish word appearing predominantly in field-names. The main sense is 'lane leading from the farmstead to pasture-land'; whether Voc.'s use of it to gloss *pascua* is simply wrong, or whether it is the original meaning of **bounder** (with 'lane' being an extension), or whether 'lane' is the original sense, and 'pasture' an early extension, will remain unsolved until a satisfactory derivation can be produced. But the sense 'lane' in Welsh and Irish encourages the idea that that is the original meaning.

Some examples of parallel usages in Welsh and Cornish (the Welsh ones taken from Richards' article): Co. Pednanvounder, We. Penyfeidr (also Irish Ceann an Bhóthair, Top.Hib., 1, 35); Co. Bounder Vean, Welsh Feidr Fach; Co. Vounder, We. Foidir; Co. Chyanvounder, We. Tŷ'rfeidr.

A. Vounder (2 exx.).

B2. All but the first are names of fields or lanes: Bounder Vean (+ **byghan**); *Bounder-en-parson* (+ **an**, English); *Vounder Gogglas* (+ ***gogleth** ?); *Bounder Hebwotham* (+ **hep**, **ethom**); Vounderlidden (+ **ledan** ?); *Vounder Bullock* (+ English ?); Voundervour Lane (+ **meur**); *Vounder Feene* (+ **fyn**)

C2. Pednavounder (2 exx.), Pednanvounder, Pednvounder, Penny Vounder, etc. (all + **pen**, (**an**)); Chyanvounder (+ **chy**, **an**)

*bounds 'an area of tin-working; a tin-mine', borrowed from English dialect *bounds* 'the limits of a given area for mining' (Symons, Gazetteer, p. 140); as a Cornish word it is only found in a few 17th- and 18th-century mine-names, and even then the names may be translated from English.

B2. *Bounds Coath* (+**coth**); *the Great Bounds* als. *Bounds Brose* (+**bras**); *Bownds an Coskar* (+**an, coscor**); *Bounds Begall* (+**begel**): all mines.

bowyn 'beef', BM 3224; *bowin* glossing *bovina*, Lh. 33a; *boen, bowen* 'an ox; beef', Pryce. Breton *bevin*, Middle Breton *beuyn*, 'viande de bœuf'; compare Welsh *beu-*, as in *beudy*, etc.

C2. Park (an) Bowen, flds. (+**park, (an)**): compare English Beef Close, etc., flds.

*bow-lann 'cow-pen' (?); a compound of *bow-* (as in the previous entry and **boudzhi**) and ***lann** in the sense 'enclosure'.

C2. Carbouling (+**corn**); Trebowland (+**tre, (an)**)

bran (f.) 'crow', OM 1099; *an vrane vras* 'the raven', CW 2464; pl. *bryny*, OM 133. Welsh *brân*, pl. *brain*; Breton *bran*, pl. *brini* (PB, pp. 46, etc.; Hemon, Grammar, p. 32). Some names listed as containing the singular probably contain instead the common personal name equivalent to Old Welsh *Bran* (LLD), Old Breton *Bran* (Chrest., p. 111); however, the definite article occasionally appears, indicating that a particular name contains the common noun, or 'the crow' as a nickname. Note Welsh Dinas Brân, Dinbrain (ÉC 13 (1972–73), 367–8 and 383): it is likely that Caer Brân in particular contains the name of the mythological figure known in early Welsh literature. Compare *Branodunum* (PNRB, pp. 274–5) and Irish *dún mbrain, ráith brain*, etc. (Hogan, Onomasticon, pp. 378 and 568).

A. ?The Brawn (rock)

C2. Brane, Bodbrane (both +***bod**); Caer Brân, Crane (both +***ker**); Hensafraen (+**hensy**); Cutbrawn (+**cos**); Mellanvrane (+**melin**[1]; 2 exx.); Partonvrane (+**fenten**); Penfrane (+**pen**); Chyverans (+**chy, an**); Neithvrane (+**(carn), nyth**); *Errow-Brane*, fld. (+**erw**); pl. Zawn Brinny, coast (+**sawn**), ?Respryn (+**rid**)

bras 'great', Voc. 946 glossing *grossus*; etc. Welsh, Breton *bras*. The normal contrast with **byghan** was **meur** (but cf. 'Crofts called Owen Vean and Owen Brose', A.D. 1665); **bras** seems to have been more specialised, in the sense 'fat, well-built' (cf. William *Bras*,

*bre

1327 SR, Kea); as a noun, *an bras* is evidently 'the fat man'. The vowel of **bras** behaved irregularly in Modern Cornish, giving /brɔːz/, with rounding.

C2. Creegbrawse (+**cruc**); Zawn Bros, coast (+**sawn**); *Fosebrase* (+**fos**); Park Braws, Park Brase, fld. (+**park**); Carn Brâs, rock (+**carn**); *an Cogh'n bras, Coffin broaz*, mines (both + ***coffen**); Owan Vrose, Owen Brose, fld. (both +**goon**); Guthen Brose, coast (+ ?); *Menvrause*, fld. (+**men**); as noun, Trembraze (+**tre, an**; 2 exx.)

D. *Tregyffian Vrose* (= Tregiffian Veor)

*bre (f.) 'hill'; Welsh, Breton *bre*. The fact that this word is not attested in Cornish is not surprising: the Breton word seems to be very rare, apart from place-names (Tanguy, p. 63); but for a modern usage in Cornouailles, see GMB, p. 80. In Welsh, too, the word is restricted to place-names and early poetry. For this reason, and because it occurs quite often in compounds, names containing it are likely to be early ones – probably from the Old Cornish period or earlier. What the difference was, if any, between ***bre** and the many other words for 'hill' is unknown. Perhaps a ***bre** was always the most prominent hill in a district. The element may, as B2, have been liable to confusion with ***bren** and **bron** (as in Breton: Tanguy, loc. cit). See also Elements, I, 50f.; JEPNS 1, 44; and, in compounds, Richards, ÉC 13 (1972–73), 366; and with Mulfra, etc., compare the common Welsh Moelfre (Units, p. 158; etc.), and Mellor (PNLnc., p. 73; PNDrb., I, 144). For *-vre > -ver* in Dinnever, compare Kinver, Stf. (DEPN), and both instances of Mellor.

A. Bray (2 exx.), Brea (2 exx.), Bray Hill; a pl. ***bre-yer** probably in Bryher (Scilly): see Quentel, Ogam 7 (1955), 239–241.

B1. Mulfra, Mulvra, Mulberry (all + ***moyl**); Torfrey (+**tor**); Numphra (+ ?); ?Dinnever Hill (+ ***dyn** ?).

B2. Brill (+ ***helgh**); *Bryanick*, Brevadnack (both + ***bannek**); Bartinney (+ ?); Brewinney (+ ?); *Breglos* (+**eglos**)

C2. Goonvrea (+**goon**); Carn Brea (+**carn**); Lennabray (+**melin¹** pl.)

It is just possible that Amalebra contains a place-name ***Amal** and an equivalent of Welsh *obry* 'down, below', which is from ***bry** = *bre*, according to GPC and GMW, p. 223; note also Tŷ-Fry, Agl. (SH 5176).

bregh 'arm'; *brech*, Voc. 73 glossing *brachium*. Welsh *braich*, Breton *brec'h*. An aj. *breghyek 'armed; with inlets, convoluted' may occur as A in *Brechiek* (= St Martin, Scilly).

breilu (OCo.) 'rose; primrose (?)', Voc. 663 glossing *rosa*. Welsh has *briallu* 'primroses', and if the Cornish form were emended to *brielu the words would be fairly similar (though the variation -*e*-/-*a*- would still be unexplained); Breton *brulu* 'digitales' (= foxgloves) (Tanguy, p. 111) has evidently been changed by vowel harmony, as well as undergoing semantic change. There is a Richard *Brelu* (1327 SR, Redruth): the Welsh word is used for yellow hair-colour. Note also OBr. *briblu*, a plant, FD, p. 89.

    C2. ?Menaburle (+**meneth**): the equivalent of English Primrose Hill, if so.

***breyn** 'putrid', equivalent to Welsh *braen*, Breton *brein*. With the first name compare Breton Poull Brein (Top.Naut. 1419, no. 9091).

    C2. Polbreen (+**pol**); ?Rosevine (+***ros**); ?Pol Bream, ?Polbream Cove, ?Polbream Point, all coastal (all +**pol**)

***bren** 'hill'; Welsh *bryn*; Old Welsh *tal ir brinn* 'the brow of the hill', LLD, p. 173; *brinn i cassec* 'the mare's hill', LLD, p. 262; in Breton *Bren*- occurs only in a few early place-names: Chrest., pp. 111 and 193; Tanguy, pp. 63f. The element is certainly confused with **bron** 'breast, hill', and probably with ***bre** as well. Some names cited under those elements may belong here instead.

    A. Brynn, Burn; pl. *brennion (?), Brunnion; pl. *brenniou, Breney

    B2. Burniere (+**er**[1] or **hyr**); Brownsue (+**du**)

    C2. *Goen Bren* (+**goon**: = Bodmin Moor; Fleuriot, ÉC 14, 52); pl. Lambrenny (+**nans**)

brenigan (MnCo.) 'limpet', Lh. 241c; *bernîgan*, pl. *brennik*, glossing *patella*, Lh. 114a. Welsh *brennig* 'limpets', Breton *brennig* 'patelles' (cf. GMB, p. 81).

    C2. *Porthe Brenegan* (+**porth**): the use of the sg. is curious.

brew 'broken, wounded; fragment': see Padel, Co.Studies 7 (1979), 41. Welsh *briw*.

    A. Brew: the sense may be 'fragment (of land)' or perhaps 'broken (land)'.

***brygh** 'variegated, speckled, brindled'. Welsh *brych*, fem. *brech*, 'brindled'; Breton noun *brec'h* 'pox'. A separate word from ***bryth** 'variegated', though the two may not be distinguishable in place-names.

bryth

C1. Frightons (+ ***dyn** or **tyn**); Brightor, ?Brighter (both + **tyr**); compare Welsh Brechfa, Crm., 'speckled-place'.

C2. Coldvreath, Killifreth (both + **kelli**); a derivative r.n. *Bryghy (+ *-**i**) may appear in Suffree (+ **rid**: formerly *Resvreghy*)
***bryth** 'variegated', equivalent to Welsh *brith*, fem. *braith*; Breton *brizh* 'tacheté, bariolé'. The element is distinct from ***brygh** 'speckled'. Old Welsh *i main brith* 'the speckled stone', LLD, p. 191; *lapis in i guoun breith* 'a stone in the mottled marsh', LLD, p. 196. The Breton derivative *brezhek* 'pommelé' justifies a Cornish *brethek 'dappled', occurring as C2 in Namprathick (+ **nans**). Another derivative, *brythey, with an obscure suffix, is seen as A in Burthy. Another aj., (?) *bryth-yel (+ *-**yel**), may occur as C2 in Trythall (+ **tre**).

brytyll 'brittle, frail', Trg. 9; *brvttall*, *brotall*, CW 452, 614 and 2223; *brettal* 'brittle', Lh. 33c. For the name, cf. English Quaking Bridge (PNBrk., 1, 256; etc.).

C2. Ponsbrital (+ **pons**)

bro 'country, district', MC 250b; Welsh and Breton *bro*.

C2. Penbro (+ **pen** 'end'): cf. Welsh Penfro (Pembroke), a compound, and Breton Pembro (Paimbœuf: Top. Bret., p. 93)

brogh (m.) 'badger', PC 2926; BM 1280; *broch* glossing *taxo vel melus*, Voc. 564. Welsh *broch*, Breton *broc'h*.

A. ?The Vro, rock

C2. Polbrock, Polbrough (both + **pol**: cf. Breton Poulbroc'h, Tanguy, p. 118)

bron (f.) 'breast, hill', OM 1755, etc.; *duivron*, Voc. 55. Welsh *bron*, Breton *bronn* (cf. FD, p. 90), both also used topographically: Tanguy, p. 64; Old Welsh *bronn ir alt* 'the breast of the slope', LLD, p. 73; *hyd ym bronn* 'stag on hill', Jackson, Gnomic Poems, p. 25. This element is certainly confused with ***bren**, and some names listed here may well belong under that element. Where the initial is lenited, however, and the generic is masculine, as in Penvearn, Tolverne, one can assume the word is **bron** (f.), not ***bren** (m.), the lenition being caused by a lost definite article; so also when, as a generic, it lenites the next element, as in Brown Willy, Berrangoose (but the converse, absence of lenition, does not prove ***bren** rather than **bron**).

A. pl. Burnow

B1. Camborne (+ **cam**)

B2. Brown Willy (+ **guennol** pl.); Burngullow (+ **golow**: cf.

Breton Brengoulou, etc. (with *_bren_): Gourvil, Ogam 6, 87–90 and 136; Quentel, Ogam 6, 233–36); Bozion (+ ***seghan**); Burnuick (+ ***ewyk**); Brimboyte (+ ***poth**); Burncoose (2 exx.), Barncoose, Berrangoose, Pencoose in Stithians (all + **cos**: cf. early Welsh _bronn y coet_ 'the breast of the wood', CLlH, p. 39); Burnwithen (+ **gwyth**[1] sg.); Barlewanna, Barlowennath (both + **lowen** deriv.); Burnoon (+ **goon**); _Bronhiriard_ (+ ***hyr-yarth**; = Herodshead)

C2. Trebrown (+ **tre**, (**an**); 2 exx.); Tolverne (+ **tal**); Penvearn (+ **pen**); Lanthorne (+ ***lann/nans**); aj. (= Welsh _bronnog_ 'hilly, undulating') Trefronick (+ **tre**)

bronnen (f.) 'a rush', RD 2096 (N. Williams, BBCS 23, 320f.); _brunnen_, Voc. 668 glossing _iuncus_; _brwdnan_, Lh. 146b glossing _scirpus_ 'a rush without a knot, a bulrush'. Welsh _brwyn_ (sg. -_en_), aj. Brwynog, Agl. (SH 3586); Breton _broenn_ (sg. -_enn_) 'joncs', _broenneg_ 'jonchère' (cf. Tanguy, p. 111). The original diphthong had already disappeared in Cornish by the time of Voc., but appears in an earlier charter boundary-point. Which of the many types of rushes was meant is obscure; see also ***keun**, ***cors**, **elester**, and **heschen**.

C2. pl. _le_(_i_)_n broinn_, _lenbrun_ (both + **lyn**); Lambourne (+ **lyn** ?; 2 exx.); aj. Hendraburnick (+ ***hendre**)

bucka 'sprite, hobgoblin', CW 1196 and 1589 (also in English dialect: Courtney and Couch, p. 7). Compare Welsh _bwci_, and M. Richards, 'The Supernatural in Welsh Place-names', in _Studies in Folk Life: Essays in Honour of Iorwerth C. Peate_ (edited by J. Geraint Jenkins: London, 1969), 304–13, at p. 306: note especially Tŷ'r Bwci (3 exx.), equivalent to the Cornish Chybucca; also P. L. Henry, 'The goblin group', ÉC 8 (1958–59), 404–16.

C2. Chybucca (+ **chy**); Polpuckey (+ **pol**); ?Izzacampucca, coast (+ ***yslonk** ?, **an** ?); _Woon Bucka_, ?Ony Pokis, flds. (both + **goon**)

The word is also used as qualifier in English – e.g. Bucca's Lane.

budin (f.; OCo., -_d_- = /ð/) 'meadow', Voc. 727 glossing _pratum_; Modern Cornish _bidhen_, Lh. 33c and 127c; _vethan_ and _vythan_, Pryce. Pryce's forms are genuine: the place-names show the same fossilised lenition in many cases (cf. **bounder**). Welsh _buddyn_; Old Welsh _budinn_, LLD, p. 257 (cf. CA, p. 105: from *_bu_ 'cattle' + *_dynn_ 'place', as in _tyddyn_ 'house' ?).

B2. Bethanel (+ **an**, **hel**); all the following are fields: _Buthan_

bugel

Green Grass, Buthan Glebe, Buthan Ten Acre (all +Eng.); *Buthen en Hesk* (+**an, heschen** pl.); *Vithen Vorne* (+**forn**); *Vithin Thomas* (+pers.); *Vithen Cornett* (+ ?); *Bithen Lidgeo* (+**les** pl. ?)

C2. *Gwyth en futhen*, mine (+**guyth**[2]**, an**); *Park-an-Vethen*, fld. (+**park, an**); Polveithan (+**pen**); pl. Penbothidnow (+**pen**)

bugel (m.) 'shepherd', PC 893, BM 2979, Voc. 197 glossing *pastor*; *begel*, MC 48c; John *Bugel*, Ralph *Bugel*, 1327 SR. Welsh *bugail* 'shepherd', Breton *bugel* 'shepherd, child'. Cumbric Barnbougle (PNWLo., pp. 4f.; CPNS, p. 351) is agreed to contain this word as qualifier, but the first element is uncertain. In view of the Cornish names, one might suggest Cumbric **brinn* 'hill'. The word is found as qualifier in various compounds (GPC s.vv. *bugeil-*; and Bugeildy, Rad., Units, p. 23). Other place-names: Old Welsh *brinn bucelid* 'shepherds' hill', LLD, p. 155; Irish Lios na mBuachaillí, Top.Hib., I, 66; Welsh Maen y Bugail (= West Mouse rock, Agl., SH 3094) shows a usage similar to the Cornish rocks called Beagle, etc. As stated under **begel**, that element and this are formally indistinguishable without early spellings.

A. Beagle, Beagles Point, coast; this rock-name is very common in Scilly, with e.g. Biggal of Gorregan, Biggal of Mincarlo, etc. (named from neighbouring islets): the sense is probably 'watchman', or possibly the Breton sense of 'child', not otherwise known in Cornish, except for dialect *beagle* 'child' (Courtney and Couch, p. 3). *Bogullas* (a manor) looks like the feminine, 'shepherdess' (Welsh *bugeiles*), but the meaning would be odd (or **bugel**+ ***lys**?). Compare early Welsh *in predio Buceles* and *villa Nant-bucelis*, VSBG, pp. 76 and 120; and Breton Buguélès (near Plougrescant).

C2. Penbugle (+**pen**; 2 exx.); Rosenbeagle, fld. (+***ros, an**); Hallenbeagle (+**hal, an**): the last two names could equally well contain **begel**.

bugh (f.) 'cow', OM 123 and 1185; *buch*, Voc. 599 glossing *vacca vel buccula*; *bewgh*, CW 403; *biwh*, Lh. 60c. Welsh *bu(w)ch*, Breton *buoc'h* (GMB, pp. 86f.). The normal word for 'cow'. Its plural was originally ***byu**, but since the two words must have fallen together in sound by the Middle Cornish period, one would expect to find a fresh plural formed, as in Breton (PB, p. 74: plural *biou, buohed*, and also *saout*, etc.). However, there does not seem to be any trace of such a new plural, and it may be that some of the field-names listed here really contain the plural, ***byu**; see also

bowyn 'beef', and its derivatives **boudzhi**, ***bowlann**, and perhaps **bounder**.

C2. Penvith (+**pen**); flds. Park (an) Bew, *Park-an-Beau*, *Parkenbewgh*, etc. (+**park**, (**an**))

***buorth** 'cow-yard'. Welsh *buarth*; Old Breton *buorth* glossing *bovello* and *buuoort* [sic] glossing *bobello*, FD, pp. 92 and 331; *Buorht*, CL. From **hu-* 'cattle' and ***garth**. The difference in vowel between Welsh -*arth* and Cornish/Breton -*orth* appears in other compounds, too (see Loth, RC 36, 174); it is ascribed to an Indo-European alternation by Pedersen, VKG, 1, 180.

A. pl. Bohortha

C1. An Old Cornish compound, **buorth-del*, is the equivalent of Welsh *buarthdail* 'land manured by enclosing animals upon it': see Latin Texts, p. 548; Colan, p. 157; Loth, RC 34, 143. For the second element, see **teil** 'manure'. The compound is used as C2 in *nantbuorðtel* (+**nans**: Sawyer, nos. 832/1027).

C2. Cusveorth (+**cos**)

C

cadar (f.; MnCo.) 'chair, seat', Pryce; Welsh *cadair* (ELl, p. 25; ELlSG, pp. 84 and 109), Breton *kador* (cf. Tanguy, p. 64). The Cornish word is virtually confined to rock-names, especially on cliffs (see Bosons, p. 9, for a seventeenth-century folk-tale of a cliff-seat, not involving this element). Cf. Elements, 1, 75; JEPNS 1, 44. See also **cheer** 'chair'.

A. The Gaider (cf. The Chair, also a coastal name)

B2. *Garder-Wartha/-Wollas* (+**guartha/goles**); ?Cathebedron (+ ?)

C2. Chapel Engarder, coast (+**chapel, an**); *Meane Cadwarth*, rock (+**men**); Pen a Gader, coast (+**pen**, (**an**)); *Pengadoer* (+**pen**: note that this spelling, dated 1302, is exactly like the correct Middle Breton form of the word, *cadoer*: HPB, p. 223. The Middle Cornish form would evidently have been **cador*). For the last two names cf. Welsh Pen y Gadair (2 exx., Rhestr, pp. 96f.) and Breton Penhador (Tanguy, loc. cit.); aj. (+ ***-yel**?) ?Tregatherall (+**tre**)

***cad-lys** (f.; OCo.) 'bailey, courtyard, camp', a compound of **cas** 'battle' (OCo. **cad*)+ ***lys**; Welsh *cadlys*. The Welsh variant *cadlas* appears in Pen y Gadlas (PNFli., p. 130). Note the three *catlys* 'baileys' that had to be crossed to enter Wrnach's fort in the

*cagh

story of Culhwch and Olwen, WM 488.11-13.
  A. Gadles
*cagh 'dung'; *caugh*, PC 2715; *kâwh*, Lh. 154c glossing *stercus*.
Welsh *cach*, Breton *kaoc'h*. The early forms of the place-names show *-k*, not *-gh*: that is probably an English sound-substitution, though it could also represent a genuine variant *cak: Borlase has *cagal* 'rubbish, rubble, dirt, sheep-dung', which could be a derivative of *cak; cf. Welsh *cagl*, Breton *kagal*.
  C2. Polcocks, Polcatt (both +**pol**): cf. Irish Loch an Chaca, Top.Hib., I, 39.
  It is likely that Caca-Stull Zawn, coast, contains this element in some form, though whether as B2 or C1, or what the second element was, is obscure.
kal (f.) 'penis', Bosons, p. 58; Lh. 116c. Welsh *cal(y)*, Breton *kalc'h*. The meaning in place-names is probably 'pointed rock', 'prominent hill' or the like.
  B2. Colvannick, Calvadnack (both + ***bannek**: cf. Welsh Calfannog, Bre., Units, p. 27); Colvennor (+**meneth**; 2 exx.: cf. Welsh *Calchuynid*, EWGT, p. 16? but see Antiquity 29, 83)
kalx (MnCo., $x = /\chi/$) 'lime', Lh. 45c; Welsh *calch*. With the name compare Welsh *odyn galch* 'lime kiln': Jackson, BBCS 23 (1968–70), 116f.; Padel, Co.Studies 1 (1973), 58f.; WVBD, p. 403.
  C2. *odencolc*, Dev. (+ ***oden**: Sawyer, no. 298)
cales 'hard', BM 671; *calys*, Trg. 53 ('difficult'). Cf. Elements, I, 78; JEPNS 1, 44.
  A. Gallas, fld.
  C2. Noongallas (+(**an**), **goon**); *Wheal Callice*, mine (+**wheyl**); Carrick Calys, rock (+**karrek**)
***calon**, river-name of unknown meaning: ERN, pp. 90f.; Elements, I, 76; JEPNS 1, 44. A Cornish instance may appear as C2 in Lancallen (+**nans**), though early forms agree in giving *-an*, not *-on*.
cam 'crooked, curved, bent'; Voc. 381 glossing *strabo* 'squint-eyed'; *cam-* 'crooked, wrong', Voc. 306, 403 and 436. Welsh *cam*, Breton *kamm*. The element is quite often used as C1; in this use compare early Welsh *campull*, LLD, p. 73; *rit i cambren, ryt y cambrenn* ('the ford of the crooked tree'), LLD, pp. 42 and 134; *cam lann* 'crooked bank', Ann.Camb. s.a. 537; and Camros, Pem.
  A. Gam (as a noun, 'bend in a river')

36

C1. *Candern Water*, Candor (both +**dour**: Welsh *cam dubr*, Chad 4; Camddwr, EANC, p. 46); Camels (+**als**: Welsh Gamallt, Rad.); Camborne (+**bron**); Cambrose (+ *****prys** ?); Camerance (+**gweras**)

C2. Lo Cabm, coast (+ *****loch**); Gilly Gabben (+**kelli**); *Park-Cabben*, fld. (+**park**); *Gweal Cabben*, fld. (+**guel**); *Monhek-cam*, mine (+ *****mon**^{2} aj.); ?*Frogabbin*, fld. (+**forth** ?)

*****camas** (f.) 'bend, bay', corresponding to Welsh *camas* (in place-names and early poetry; ELl, p. 7).

A. Gamas Point; ?The Gabmas, fld.

can 'white', in *bara can*, Voc. 852, glossing *panis albus*. Welsh *can*; Old Breton *-cann*, FD, p. 95. Cf. Old Welsh *Candubr*, Ann.Camb. s.a. 1073; Middle Breton *Rethcand*, *Redgand*, Cart.Quimper (Chrest., p. 227). Note the adjacent pair Polcan and Poldew (with **du**).

A. ?Cansford (= r.n., +English)

C1. ?Candra (+**tre** ?)

C2. Polcan (+**pol**); Trecan (+**rid**); Fentrigan (+**fenten**); ?Cascadden (+**cassec** ?)

*****canel** (f.) 'channel' (from English). Cornish dialect and NED *cannel*, *kennel* 'roadside gutter; channel'. Cf. Pryce, *shanol* 'a channel, the gutter, the kennel'. Welsh *canel* 'gutter, drain'; Breton *kanol* 'canal, chenal', 'désigne le plus souvent un chenal de navigation', Toponymie Nautique 1399, p. 348.

A. Gannel Rock, coast; The Gannel, river

B2. Cannalidgey (+saint: this name is attested from 1201, so that the loan must have occurred very early)

C2. Venton Gannal (+**fenten**)

cans 'a hundred'. Welsh *cant*, Breton *kant*.

C1. Kelsters (+**cellester**), formerly *Cansclester* c. 1270; ?Cans Loggers, fld. (+**logaz** ?)

*****cant** (OCo.), obscure, perhaps 'district, region' or 'edge, border', or possibly 'host, throng, troop'. This root is discussed by Jackson, JRS 38 (1948), 55; by E. Evans, 'Some Celtic forms in *cant-*', BBCS 27 (1976–78), 235–45; by Rivet and Smith, PNRB, pp. 297–9; and, very usefully, by Quentel, 'Le nom celtique du canton en Gaule et en Grande-Bretagne', Rev.Intnl.Onom. 25 (1973), 197–223. Among other place-names thought to contain it, the Quantock Hills (Somerset) show an adjectival form (Turner, BBCS 15 (1952–54), 18: formerly *Cantuc*, etc.); cf. Elements, 1, 80.

The idea of a pre-Indo-European root *kant- meaning 'stone, rock', still occasionally cited, is no longer considered valid: E. Evans, op. cit., p. 241 and references.
 A. Cant
 B1. Hingham (+ **hin**); Loggans (+ ?)
 C2. aj. ?Hantergantick (+ ***hendre***); aj. (+ ***-yel**) ?Cargentle (+ ***ker**? or **cuntell**)

capa 'cap, cape, cope', Voc. 811 glossing *cappa*; *cappioz* 'caps', Lh. 243a; borrowed from English. Welsh *câp*; Breton *kab* 'cap, promontoire'. Breton *Capquern* (17th century), etc. (Tanguy, p. 64). The topographical meaning of the Cornish word is evidently 'headland', as in Welsh and Breton.
 C2. Pencabe, coast (+**pen**)

***car** 'cart' occurs only in compounds: see the next two entries. Welsh *car*, Breton *karr*, Old Breton *carr* glossing *vehiculis*, FD, p. 97.

***car-bons** (m.) 'paved road, causeway' (a compound of ***car**+**pons**); Breton *karrbont* 'voie pavée, chausée'. Gourvil (no. 249, *Carpont*) notes that, as a Breton place-name, the word can be applied to places far from streams, so that the meaning 'bridge' is inappropriate. He also notes the Middle Breton equation (in the *Catholicon*, 1463) of this word with *strehet* 'street'; cf. OE *brycg* as 'causeway', Elements, I, 54. The phonology of the Cornish word is interesting: early forms usually show *-bous*, *-bows*, from as early as the 14th century, with a further change to *-bis* from the 16th century. It is clear that the *-n-* was lost, as elsewhere in this context: Middle Cornish *kemmys* < **kemmyns* (= Welsh *cymaint*, Middle Breton *quement*) shows the same loss, and so does **nans** in compounds (*-nas*): see remarks under that entry.
 A. Carbis (5 exx.).

***car-jy** (m.) 'cart-house' (a compound of ***car**+**chy**); Breton *karrdi* 'remise, hangar, garage'.
 A. Cargey Gate (+Eng.)
 C2. Park-an-Cady (+**park, an**)

carn (m.) 'rock-pile, tor'; RD 2297, 2333 and 2335 (all meaning 'a rock in the sea'). Welsh *carn*, Breton *karn* both mean 'cairn' or 'hoof'; the latter meaning occurs in Old Cornish, Voc. 98 *ewincarn* glossing *ungula*, but there is no reason to suppose that it occurs in place-names. (The Old Welsh place-name *carn cabal* (HB, §73) appears to be a pun, with *carn* bearing both its Welsh senses.)

carn

Fleuriot (FD, p. 97) suggests that *carn* 'cairn' and *carn* 'hoof' are actually two unrelated words. Further early Welsh place-names: as B2 'cairn, rock', 'ille lapidum cumulus vel congeries...Carn, id est rupes, Tylyuguay, vocaretur', VSBG, p. 90; B2 'hill', [*Carningli*] glossed *Mons Anglorum*, ibid., p. 10; as C2 'stony', *Bochriucarn*, glossed *maxilla lapidee vie*, ibid., p. 26. There is little evidence for **carn** referring to an artificial structure in Cornwall: usually it denotes the natural tors distinctive of SW England. But there is one clear instance of **carn** 'barrow' (not 'cairn'), at Carne in Veryan, and there may be others. In field and coastal names the element is much commoner in the west than in the east, where English 'Tor' serves the same purpose. Further senses that occur are 'point, headland' (e.g. Carn Boscawen = Boscawen Point); especially in the Camborne–Redruth area, 'hill' (cf. the Old Welsh gloss *mons*); 'isolated rock in the sea': Carn Brâs (Longships); Carn Du (Manacles); this is the sense of the occurrences of the word in RD. Occasionally **carn** and **karrek** are confused, owing to the similarity of sense: e.g. Carn Gloose, formerly *Carrick gloose* 'grey rock'. See also Elements, 1, 81; JEPNS 1, 44.

A. Carne (several); Carnanton (+Eng.)

B1. Dry Carn, rock (+**try**[1])

B2. A small selection is given, classified according to the sense of the qualifier; coastal, rock and hill names are starred.

Qualified by pers., or words denoting people: Carn Peran (+St); Carne Biskey (+OE *Beorhtsige*); Carnarthen (+Arthur); Carnstabba; *Carn Epscoppe* (+**epscop**)

Qualified by animals or birds: Carnyorth (+**yorch**); Carmouth, Carnemough (+**mogh**); ?Carrancarrow (+**carow**?); Carnkie (+**ky**; 2 exx.); *Carnwhyl* (+**hwilen** pl.); Carn Bargus*, Carn Barges* (3 exx.) (all +**bargos**)

Qualified by plants: Carneadon (+**eithin**); perhaps also Carn Sperm* (+**spern**?); Carn Lodgia* (+**les** pl. ?). Plant-names are rare as qualifiers of **carn**, though some may remain unrecognised.

Qualified by colour: Carn Glaze (+**glas**); Carn Du*, Carn-du*, Carn-du Rocks*, Carnsew in St Erth (all +**du**); Carn Mellyn* (+**melyn**[2]; 2 exx.); perhaps Carnwidden (+**guyn**?)

Qualified by other physical description, or analogy from shape: Carn Bean, Carbean (both +**byghan**); Carnowall (+**ughel**); Carncrees, Carn Creis* (2 exx.) (all +**cres**); Carn Brâs* (+**bras**);

Carn Greeb* (+**krib**); Carn Leskys* (+**leskys**); Carn Scathe* (+**skath**); Carn Cheer* (+**cheer**); Carnmenellis, Carn-Mên-Ellas* (both +**manal** vb. ppp.); Carn Barra* (+**bara**)

Qualified by archaeological or other human works: Carneglos (+**eglos**); Carn Galver*, Carn Olva* (both +**guillua**); perhaps Carn Creagle* (+**cruc** dimin.)

Qualified by place-name: Carn Camborne, Carn Boscawen*, Carn Gwavas*, Carn Venton Lês*, Carn Fran-Kas*, *Carnetrembethow*, and many others

C1. No examples in Cornwall, but Charles, Dev., is supposed, not very plausibly, to be from **Carn-lys* 'tor-court' (?), PNDev., 1, 61.

C2. Boscarn, Boscarne (both +***bod**); Chycarne (+**chy**; 2 exx.); Tregarne (2 exx.), Tregarden in Luxulyan (all +**tre**: cf. *Tref carn*, LLD, p. 125); Gwealcarn (+**guel**); Tolcarne (several); *Talcarn* (= Minster), Trecarne (2 exx.) (all +**tal**)

D. Clahar Garden

The word was borrowed into the English dialect of Cornwall, and used as generic in a number of English place-names; e.g. Yellow Carn, Silver Carn. In this usage it is especially common to find a place-name as qualifier; e.g. Trengwainton Carn, Carwynnen Carn, *Chegwin Carn*, fld., *Bolinnow Cairn*, and many others.

Derivatives of **carn**: Welsh has *carnen* 'small cairn or tumulus' (from 1893). A word **carnan* appears as C2 in Roscarnon (+***ros**), but it would be confusable with a word equivalent to Welsh *carnen* 'wild sow', if that existed in Cornish, and also with **kernan*. The plural of a possible diminutive **carnel* may occur as A in Carnelloe (2 exx.: or **corn** deriv.). Tanguy (p. 65) recognises an aj. **karnek*, *karniek* in Breton place-names; most Cornish place-names that might contain this are listed instead under **kernyk*. However, **carnek* certainly occurs once, as B2: Cannamanning (+**amanen**), and **kernek* probably as C2 in Leskernick (+**lys**; 2 exx.). Finally, there is a word **carneth*, equivalent to Welsh *carnedd* and Breton *karnez* (Tanguy, p. 65). Its adjective (Welsh *carneddog* 'rocky; abounding in cairns') occurs as A in Carnethick.

carow (m.) 'stag', MC 2b; *karow*, OM 126; *caruu*, Voc. 582 glossing *cervus*. Welsh *carw* (cf. ELISG, p. 15), Breton *karv* (cf. Tanguy, p. 118). The Welsh pl. is *ceirw*, but *kerwyt* may be another pl. (Jackson, Gnomic Poems, p. 56; taken as sg. by GPC, s.v. *cerwyd*);

Breton pl. *kirvi* (PB, p. 53). The Cornish pl. is unknown; if early Welsh *kerwyt* is a pl., we could postulate **kerwys*, which may then occur in two place-names, one of which, if so, shows -*w*- > -*v*- as in Breton from the Middle Breton period onwards (HPB, pp. 470-3); Lifris of Llancarfan, in his Life of St Cadog (c. A.D. 1100), implies another plural (or derivative: see *-**an**²) in the following gloss: 'Nant Caruguan, id est, Uallis Ceruorum, inde Nancarbania, ex ualle scilicet et ceruo', VSBG, p. 54. (But his explanation may be fanciful.) In some names, especially after **nans, carow** is indistinguishable from **garow** 'rough', of which the **g**- would be unvoiced after *Nans*-; such names (Nancarrow, Lancarrow) always contain -*c*- or -*k*- in their forms, but may be the equivalent either of Welsh Nantgarw, Gla. (PNDinas Powys, p. 58), or of Old Welsh (*pen*-)*nant ir caru* (Chad 6).

C2. ?Carrancarrow (+**carn**?); *Polcarowe* (+**pol**); Pencarrow in St Austell and Eglosbayle, Pencarrow Head (all +**pen**); ?Nancarrow (2 exx.), ?Lancarrow (all +**nans**; or **garow**); pl. (see above) ?Liskeard (+***lys**); ?Nankervis (+**nans**)

karrek (f.) 'rock', BM 1016 and 1024; *carek*, BM 1072; *garrak*, OM 2464; pl. *karrygy*, OM 478. Welsh *carreg*, Breton *karreg*. Old Welsh shows various plurals: *cerricc* (Mart.Cap.), *carrecou* (Juvenc.), *carreic* (FD, pp. 97f.), while Breton has generally *kerreg*, but shows others in place-names, such as Carrégui, Kerregou (PB, pp. 150f.; Tanguy, p. 65).

A. Angarrack (+**an**); Carracks, The Carracks, both coastal; pl. Creggo

B1. ?Padgagarrack Cove (+**peswar**?)

B2. *carrec wynn* (+**guyn**: Sawyer, nos. 832/1027); Cargreen (= *carrecron*, Sawyer, no. 951), *Caregroyne*, coast (= Black Rock) (both +**ruen**); *Careck-an-googe* alias the Cuckoe Rock (+**an**, ***cok**); *Carrack an deeber*, rock (+**an**, **diber**); *Carek Veryasek*, rock (+St: BM 1072); Caragloose, Carrag-Luz, Carrick Lûz, coast, Carn Gloose, coast, Cataclews Point (all +**los**); Carrick Du, coast (+**du**); Carrack Gladden, coast (+**glan**?); Carrickowel Point (+**ughel**); Carricknath Point (+***nath**)

C2. Carcarick (+**carn** or ***ker**); Tolgarrick (+**tal**; 3 exx.); Pengarrock (+**pen**); Roscarrock in St Endellion (+***ros**); Tregarrick (+**tre**; 2 exx.); Halgarrack (+**hal**); flds. Park-Garrack, Park-an-Garratt (+**park**, **an**), Croft-an-Garratt (+ ***croft**, **an**); pl. *Park Caregy*, fld. (+**park**)

cas (f.) 'battle', MC 64b and RD 2517; Old Cornish *cad-* in *cadwur*, Voc. 179 glossing *miles vel adletha*. Welsh *cad*. In place-names this word occurs only in the Old Cornish compound ***cad-lys**; personal names containing the element (cf. Evans, GPN, pp. 171-5; Chrest., pp. 115 and 195; GBGG, pp. 89-93) are quite common in place-names, e.g. *hryt catwallon* (Sawyer, no. 755), Tregajorran (**Cadworon*), Bocaddon (**Cadwen* = Old Welsh *Catguen*, LLD, p. 186), etc.; note also *Caduualant* holding Nancekuke in 1066, DB.

***casel** (f.) 'armpit', BM 1419 (*indan the gasel*); *kazal*, Lh. 44b glossing *axilla*. Welsh *cesail* 'armpit; nook, recess', Breton *kazel* (cf. Tanguy, p. 66). The Welsh word can mean 'a small inlet' (WVBD, p. 254), which is probably relevant for the three coastal names; it may also apply to some of the Breton ones cited by Tanguy, such as Corn ar Gazel.

A. Gazzle, Gazell, Gazells, all coastal

cassec (f.) 'mare', Voc. 566 glossing *equa*; *cazock* 'a mare', Carew, p. 56; *casak*, CW 406; *kazak dhàl* 'a blind mare', Lh. 243c. Welsh *caseg*, Breton *kazeg* (cf. Tanguy, p. 118); Old Welsh *porthcassec*, LLD, p. 150; *brinn i cassec* 'the mare's hill', LLD, p. 262.

A. Gazick, coast: cf. Breton Ar Gazek (rock, off Ile de Sein); ?Cassock Hill, ?*Cassick-well* (both +Eng.)

B2. ?Cascadden, ?*Kasekcan* (both + **can**?)

C2. Nancassick (+**nans**); Polgazick, Polgassick Cove (both +**pol**: cf. MlBr. *Polcasec*, RC 3, 404); Scarsick (+ ***ros**?); Goengasek (+**goon**)

castell (m.) 'castle, village, tor', BM 245; also *castel*; pl. *castilly* PC 133, *castylly* BM 305, *kastilli* Lh. 242c. Welsh *castell* 'castle, town, village', Breton *kastell* 'château'. The sense 'village' is present at OM 1709, PC 174 and RD 1295, and Lewis (LlCC, Glossary, s.v.) compares similar Welsh usages. The word should not, therefore, be thought necessarily to imply any archaeological feature when it occurs in Cornish place-names: it could mean simply 'settlement'. Moreover, in Breton place-names the word can mean nothing more than a 'steep rock': Le Berre, Topon. Nautique, 1971, p. 643: '*kastell*, traduit par château, désigne un énorme rocher escarpé, à terre ou en mer'; cf. Topon. Nautique 1399, p. 348: '*kastell*, château, désigne deux gros rochers de 15m. de hauteur'; GPC, *castell* 1(a), cites similar cases in Wales and this sense could well be present in some of the Cornish place-names.

See also VSBG, p. 46: 'tumulum in modum urbis ['fort', Wade-Evans] rotundum de limo terre exaggerari, ac in tumulum erigi fecit quod Brittonum idiomate Kastil Cadoci nuncupatur': that is, Cadoc's *Kastil* was the structure that he had built on top of the mound – a palisade, or similar?

Although the normal Breton plural is *kestell* (but also *kistilli, kastilli*: PB, pp. 55 and 70), and the Welsh plural is *cestyll* (but also *castelli*), yet the Cornish form *kestell*, common in place-names, appears to denote the singular; either that, or plural and singular were interchangeable for place-name purposes, for several names show forms in both -*a*- and -*e*- at an early date: Castle Wary is spelt *Castel-* 1337, *Kestel-* 1347; Castle Horneck is *Castel-* 1335, *Kestel-* 1395; Castle Gotha is *Kestel-* 1297, *Castel-* 1304; Rôskestal is -*kestel* 1302, -*kastel* 1369. This variation (or interchange) must have occurred in the Old Cornish period, for two Anglo-Saxon charters show forms with -*e*-: *Kestelcromlegh* [sic leg.] (Sawyer, no. 450), and *Cestel(l) merit* (Sawyer, nos. 755 and 832/1027). On the other hand, the frequent simplex place-name *Kestle* invariably shows -*e*- in the early forms, and thus probably does represent the plural in some specialised sense: cf. Middle Breton *decimagium de Kestel* (Beauport: RC 7, 59).

A. Kestle (7 exx.), Kestal; fld. Kestles; pl. Castella (= rocks, Scilly); aj. Castallack

B1. *Hengastel* (+**hen**)

B2. *Castel Uchel Coed* (+ ***huel-gos**; = Ashbury, ÉC 14, 53); Castle Goff (+**gof**); Castlezens (+**sans** pl.?); Castle Horneck (+**horn** aj.); Castle Wary (+**guary**); *Kestelcromlegh* (+ ***cromlegh**); Castledewey (+ ***dewy**); Castlemawgan (+pers.); Castle-an-Dinas (+**an**, ***dynas**; 2 exx.); Castle Gotha (+**goth**[1] pl.)

C2. Penkestle (+**pen**; 2 exx.); Rose an Castle (+ ***ros, an**); Rôskestal (+ ***ros**)

***caswyth** 'thicket, bramble-brake'; Welsh *cadwydd*. This element is a good instance of a word attested only late in Welsh (earliest instance 1632), and then mainly as a dictionary-word, yet present in Cornish place-names. As C2, it might be confusable with a possible pers. **Cadwyth*, though that name does not seem actually to occur in Brittonic.

A. Cadgwith

C2. Tregaswith, Tregadgwith (both +**tre**); Rosecassa (+ ***ros**)

cath (f.) 'cat', BM 3413; *cathe*, CW 407, 2301 and 2378; *kat*

*caubalhint

(-*t* = /θ/), Voc. 578 glossing *cattus vel murilegus*. Welsh *cath*, Breton *kazh*.

C2. Killigarth (+ ***kyl**); Langarth (+ **lyn** ?, **an**); Longsongarth (+ **lost, an**); Ponsongath (+ **pons, an**); *Polgath* (+ **pol**)

*caubalhint (OCo.) 'ferry-crossing': Old Breton (*camp*) *Caubal hint, Caupalhint*, Cart.Redon (FD, p. 99; Chrest., p. 115n., from *caubal* 'boat' and *hint* 'way'). The diphthong of the first syllable is uncertain, for Breton always shows -*au*- in this word, while Old Welsh shows -*ou*- in *Coupalua* (+ ***-ma** 'place'), LLD, p. 151: see LHEB, pp. 321–3, for the varying treatment of Latin -*au*-, and cf. HPB, p. 261 n. 3. In any case, in the Cornish compound the first syllable, being unstressed, would have been reduced at least by the Middle Cornish period. There was probably another word for 'ferry', ***treth**, also unattested in the language. The re-appearance of ***caubalhint**, found only in Old Breton (and only once, in the 9th century), in 16th–19th century Cornwall is remarkable.

A. *Kebellans* (= King Harry Ferry)

*cawns 'causeway'; from English. As simplex (e.g. Caunse, Conce) it is impossible to tell whether the element occurs as a Cornish or English word. Cf. Welsh *cawsai* (Parry-Williams, EEW, pp. 190 and 196) and Breton *chaoser* (Piette, p. 89).

C2. Chycoanse (+ **chy**); Pedn y coanse (+ **pen**)

keber 'timber', Voc. 836 glossing *tignum* 'rafter'. Welsh *ceibr*, Breton *kebr*; Old Breton -*cepr* in *banncepr*, FD, p. 78, and pl. *cepriou*, FD, p. 103. Breton aj. *kebrek* 'contenant des chevrons, des solives [joists]'. A Cornish aj. may appear as A in Kiberick Cove (+ Eng.).

kee (m.) 'hedge, bank', BM 1253 and 1896, and Trg. 40a; *in neb toll kea* 'in some hole of a hedge', CW 1128; Modern Cornish pl. *keaw*, J. Jenkins (Pryce, p. Ff3r), *kēou*, Bosons, p. 17, §17. Welsh *cae*, Breton *kae*; Old Breton *caiou* glossing *munimenta*, FD, p. 94; see also DAG, p. 554. Note also the Middle Cornish gloss *fossatum vel kaeum*, Cart. St M., no. 35. This word never developed the meaning 'field', as in Welsh, and is not nearly as common as the Welsh word. However, ***kew** may (especially in place-names) have fallen together with ***keow**, the plural of **kee**, so that some instances of ***kew** may belong here instead. The normal word for 'enclosed field' was **park**. See also ***gor-ge**.

B2. Keigwin (+ **guyn**); Kergilliack (+ **kullyek**: the lenition is odd); ?Kedue, fld. (+**du**?); Kenython (+**an, eithin**); Causilgey (+ ?)

C2. *Mankependoim* (+ ?), Menkee, Mankea (all +**men**); Kearn Kee, Corgee, *Chycornekye* (all +(**chy**), **corn**); Tregea (+**tre**); ?Ventin Gay, fld. (+**fenten**); ?Noon Gay (+(**an**), **goon**); an aj. *ke-ek (cf. Welsh *caeog* 'wearing a torque') probably appears in Tregeage (+**tre**).

***kegen** (f.) 'kitchen', *gegen* BM 3928, CW 2012, and Bosons, p. 17, §§ 22 and 23; *keghin*, Voc. 882 glossing *coquina*. Welsh *cegin*, Breton *kegin*. Note the approximate early Welsh gloss *atrium Coquine...id est, Cayr i coc* (VSBG, p. 120). In theory this element is indistinguishable from two others: from an equivalent of Welsh *cegin*, Breton *kegin* (f.) 'jay', for which there is no evidence in Cornish, though it could well have existed; and from a Cornish equivalent of early Welsh *cegin* (m.) 'ridge' (LLD, passim, e.g. *tal ir cecyn* 'the brow of the ridge', p. 122; *cecin i minid 'n i hit* 'the ridge of the mountain in its length', p. 155; cf. CA, p. 362 and ELISG, p. 96), for which there is no evidence in either Cornish or Breton. In the first name, ***kegen** 'kitchen' is the most likely (cf. English field-names containing 'kitchen', Field, EFN, p. 118); in the second, *kegyn 'ridge' would make the better sense, though it is the least well-attested word, and one would have to suppose the Cornish word was feminine, unlike the Welsh one (?), to account for the lenition.

C2. *Park an Gegen*, fld. (+**park, an**); Pengegon (+**pen**); ?*Creggan-Geggan*, fld. (+**cruc, an**)

keyn (m.) 'back, ridge'; *war geyn margh* 'on a horse's back', BM 3411; *chein*, Voc. 71 glossing *dorsum*. Breton *kein*; Welsh *cain* 'ridge' is rare, and *cefn* is much commoner (Old Welsh *cemn*, LLD, p. 156): see HPB, pp. 160 and 631 for the variants, and Sims-Williams, BBCS 29 (1980–82), 205 and n. 2. The phrase **keyn march/me(i)rch* 'horse's back' is common in Welsh and Breton place-names (Ceinmeirch, Den., Units, p. 39; *Quenmarch*, etc., Tanguy, p. 66; Smith, Top.Bret., p. 33; Loth, RC 22, 98), often showing -*e*- or -*ei*- in the second word: rather than the plural, 'horses' ridge', it is likely that the name preserves a trace of an old vowel-affected genitive singular **me(i)rch* of **margh**, as does Pentyrch, Gla., 'boar's head', of *twrch*: cf. Quentel, Ogam 7 (1955), 79–81 and Rev.Intnl.Onom. 8 (1956), 302–3. The phrase would then be 'horse's back', a good name for a ridge. (Compare the

*kel

English topographic term *hogback, hog's back*: PNSur., pp. 8 and 380; NED.) It is likely that Kilmar Tor and Kilmarth are further instances of this name: both are found only with *Kyl-, Kil-* forms (from the 16th and 14th centuries respectively); but since Carmar (*Chenmerch* 1086) had itself become *Kilmerigh* by the 14th century, it is evident that **keyn** was rather prone to become corrupted into *Kil-*. Alternatively, ***kyl** might itself mean 'back'.

B2. Carmar, *Carrack Kine marh*, rock (both +**margh**); ?Kilmar Tor, ?Kilmarth (both +**margh**; or ***kyl**); Tencreek in Talland (+**cruc**); *Carrack Kine hoh*, rock (+**karrek**, **hoch**)

C2. aj. ***keynek** (cf. Welsh *cefnog* 'ridged'): Castle Canyke (+**castell**), Trekennick (+**cruc**), Treskinnick (+***ros**?)

*kel 'hiding, cover, concealment', Welsh *cêl*; the word occurs only in *nynsyv grefons in dan geyl*, BM 1438 (sic MS: Nance, OC 3 (1937–42), 426), 'it is not a complaint under concealment': cf. *ydan gel* 'concealed', BBC 52.1. Cf. the prefixed form ***argel** 'secluded place'. The element is confusable with ***kyl**. In place-names the meaning may be 'shelter'.

C1. A compound ***kel-lom** 'shelter-bare', i.e. 'devoid of shelter' (+***lom**) may appear (as C2) in Porthcollumb (+**pol**), and possibly in *Goengellom* (+**goon**; = Gyllingdown)

C2. ?Coskeyle (+**cos**); Crill (+***ker**); Bojil, Boskell (both +***bod**); ?Coombekeale (+***comm**?)

kelin (f. pl.) 'holly', Voc. 692 glossing *ulcia*; Pryce's *kelynnek* 'a place where holly trees grow', though correct, is probably borrowed from Lhuyd's Breton *kelennek*, 42a, with identical definition. Welsh *celyn, celynnog* (cf. Clynnog, Crn., etc.: ELlSG, pp. 23 and 29; ELl, pp. 67f.; PNFli., p. 33; r.n. Celynen, EANC, p. 107: for the sg. as a r.n., cf. ***alaw**, **heligen**); Breton *kelenn, kelenneg* (cf. Tanguy, p. 104; Gourvil, nos. 929, 1812–14 and 1817–18); Old Breton *colænn* (FD, p. 113b; HPB, pp. 91 and 297); Middle Breton *Kelennec* in 1268 (RC 7, 58).

A. aj. Kelynack. With loss of the first vowel at an early date, Clinnick, Clennick and Linnick may also belong here. A sgv. pl. ***kelennow** 'holly trees' is seen in Carlidna: cf. Breton *quelennenneu*, GIB, s.v.

C1. Calendra (+**tre**)

C2. pl. Tregallon (+**tre**); *Park en Gellyn* (+**park**, **an**); Treskilling (+***ros**); aj. Pencalenick (+**pen**), Treglennick (+**tre**), *Venton kelinack* (+**fenten**)

cellester 'a pebble, or small stone', Pryce. Welsh *callestr*, Breton *kailhastr* 'a flint' (cf. GMB, p. 91).
  B1. Kelsters (+**cans**)
  C2. sg. or dimin. ?Rosecliston, ?Relistien (both +**rid**)
kelli (f.) 'grove, small wood', Voc. 709 glossing *nemus* (OE *holt*). Welsh *celli*, Breton **killi* (Tanguy, p. 100; Chrest., pp. 115 and 198 n. 5). Elements, I, 88; JEPNS 1, 45; PNRB, pp. 219f. Found all over Cornwall: see Henderson, Essays, pp. 146f. The difference between a **kelli** and a **cos** was presumably one of size.
  A. Kelly (2 exx.); Gilly (3 exx.); Gelly; Engelley (+**an**); Kelley; pl. Kellow (2 exx.), Killiow (2 exx.); dimin. (cf. Breton **killian*: Chrest., p. 198; Tanguy, p. 100), *Killian*
  B2. Qualified according to size, shape or position: *Gullivere* (+**meur**); Colbiggan, Collibeacon (both +**byghan**); *Killigam*, Gilly Gabben (both +**cam**); Callabarrett (+**perveth**); Killiserth (+**serth** ?); Killiwerris (+**gweras**); *Callyvais*, fld. (+**mes**)
  By nearby features: Clann (+ **lann*); Killianker (+**an, cruc**); Kehelland (+ **hen-lann*); *Killivorne* (+**forn** ?); Killivose (+**fos**)
  By colour: Coldvreath, Killifreth (both + **brygh*); Killicoff (+**cough**)
  By animals or birds: Colligeen (+**ky** pl. ?); Kilgogue (+ **cok*)
  By dominant tree(s): Callenowth (+ **cnow* pl.); Killeganogue (+ **cnow* aj.); Killiganoon (+ **cnow* sg.); perhaps Kilhallen (+**onnen** pl. ?) and Kellygreen (+ **cren*): one would expect this type to be frequent, but it is not. Similarly, there is no clear instance so far of **kelli**+pers., though note *Kellyengof* (+**an, gof**). The lost *Kellewik* is not a diminutive as the two Breton instances of *Quillivic*, etc., and the Welsh Gelliwig, Crn., are assumed to be (see Loth, RC 41, 390–3, and GPC, s.v. *celliwig*), but a name-phrase, **kelli**+ **gwyk* 'forest grove' (?), as is shown by the spelling as two words (*kelli wic*) in early Welsh texts (Padel, Co.Arch. 16 (1977), 115–18 and references); cf. Welsh Gelli-wig, Mon. (Richards, Units, p. 74).
  C1. Callevan (+**ban**)
  C2. Lankelly in Lanreath, Nankilly (2 exx.), Nankelly (all +**nans**); Roskilly (+ **ros*); Voskelly (+**fos**); Stephen Gelly, Stephengelly (both + **stum*); Cargelly, Cargelley (both + **ker*); Bokelly (+ **bod*); Trengilly (+**tre, an**); Treskelly (+**tre**, **is*[1]); pl. Tregilliowe (+**tre**). See also the phrase **pen (an) gelli*.
  D. *Preisk Gelley alias Priesk meor* (= Priske)

kellys

kellys 'lost', e.g. PC 2465 and RD 187; ppp. of *kelly* 'lose', CW 840 and 2029; Welsh *colli*, Breton *koll* (ppp. *kollet*, but *quellet*, Jérusalem, no. 168: on the fluctuation of the vowel, see HPB, p. 295). The sense in place-names must be 'hidden': cf. Lost Valley (Glencoe, Scotland), a hidden valley, and OE *derne* as an element.

C2. Porkellis (+**porth**); Zawn Kellys, coast (+**sawn**)

***Kembro** (m.) 'a Welshman', plural *Kembrîon* 'the Welsh', Lh. 242c. Welsh *Cymro*. Walter *Kembro* 1302 Ass.R. An epithet *Kembre* 'of Wales' also appears, e.g. Richard *Kembre*, John *Kembre* (1327 SR): Breton *Kembre* 'Pays de Galles'.

C2. Chykembro (+**chy**); Hayle Kimbro Pool (+**hal**?)

***kemer** 'confluence', also *camper (see below). Welsh *cymer* (Old Welsh *cimer*, Chad 6); Breton *kember*, common in place-names (and Quimper is called *Confluencia* in the Life of St Rumon: Doble, St Rumon, p. 13; cf. Tanguy, p. 89). On the survival of British *-mb-* in Breton (or > *-mp-*), see HPB, pp. 787–9. For discussion of the element in the Scillies see Charles Thomas, 'An archaic place-name element from the Isles of Scilly', BBCS 28 (1978–80), 229–33. The *-mp-* in West Cornwall (and one instance of *-mb-*) is very odd, especially when the regular *-m-* appears further east. It is possible that the word was re-borrowed at a late stage from Breton (and the spellings with *am*, /æm/, in the coastal examples would support this, as representing sound-substitution for a French-type nasal /ɛ̃/, or the like: cf. Quimper, with *-im-*, /ɛ̃/): this would most naturally have been by fishermen, and it is in sea-names that it occurs, except for Tregembo, which cannot be thus explained. The meaning in sea-names must be 'the meeting of two currents'. At Roskymmer there is a stream-crossing (now a bridge) at a point where two streams join. Tregembo (*Trethigember* 1311) is at a point where two small streams enter the River Hayle.

A. Gamper, coast (3 exx.)

B2. Camperdenny, Scilly (+ ?)

C2. Roskymmer (+**rid**); ?Tregembo (+**tre**, **dy-**?); Zawn Gamper, coast (+**sawn**)

***kemyn**, obscure. This word occurs four times in the plural, always qualifying **tre**. Three interpretations are possible. (1) An equivalent of Welsh *cymyn* 'bequest', Breton *kemenn* 'instruction' (cf. GMB, p. 533); a corresponding verb is well attested in Cornish: *kemynna* BM 503; 3 sg. pres. 'commends', *a gymmyn* OM 2363, *a gemmyn*

PC 2986, etc.; pret. *kemynnys* OM 2394: thus 'farmstead of bequests'. (2) An equivalent of Welsh *cymyn* 'a felling', Middle Breton *quemenas* 'cut' (GIB, s.v. *kemenañ*; cf. GMB, p. 533), Old Irish *con-ben* 'cuts away': thus 'farmstead of fellings' (= clearings?). (3) The plural (used substantivally) of *kemyn* aj. 'common', BM 3215, cf. Breton *kumun*: thus 'farmstead of commoners' (i.e. of non-nobles, or people holding land in common?). Formally there is nothing to choose between the three. Semantically, the first is least likely (though the word is the best-attested of the three in Cornish); the second is possible, although the word is not actually attested in Cornish; the third is perhaps the most likely, though the precise meaning would be uncertain.

C2. pl. Tregaminion (+**tre**; 4 exx.)

***kenkith** 'imparked residence', equivalent of Breton *kenkiz*, for which GIB quotes two meanings from earlier dictionaries: 'maison de plaisance, qui a ordinairement un Bois, ou même un Parc, qui sert de décoration à la maison', and 'plessis, bosquet, petit bois'; cf. GMB, p. 548, Chrest., p. 197, FD, p. 284, and Flatrès, 'Breton settlement names: a geographical view' (Word 28 (1972), 63–77), p. 72: it translates French *plessis* 'imparked residence'. The sense in the one Cornish instance is clear: the house is set in Pinsla Park, which was disparked in the 18th century (Henderson, 'Cornish Deer Parks', Essays, at p. 162). The name presumably represents a borrowing from Breton, at some date before c. 1250 when it first appears.

A. Kenketh

***kendowrow** 'joint-waters' (?). The form is necessitated by three places called Condurrow (2 exx.) and Condurra, with early forms such as *Kendorou, Condorou*. All are at or near places where streams meet, and a derivation from **ken-* 'together, joint' (LHEB, p. 659) and a plural *dowrow of **dour** 'water' is possible. The matter is complicated, however, by the fact that one of the places called Condurrow is on a stream which, in the 10th and 11th centuries, is called *Cendefrion* (Sawyer, no. 755) and *Cendeurion* (Sawyer, nos. 832/1027). The *-ion* ending of this name is at variance with the *-ow* of the three later names; it is difficult to see how they can be reconciled, unless the earlier *-ion* plural was replaced by the commoner *-ow*. Alternatively some relationship with Ekwall's Candover is possible (ERN, pp. 68f.: Welsh *cain*, Middle Breton *quen* 'fair, beautiful'?).

kenin (f. pl.) 'ramsons, wild garlic', in *kenin euynoc* ('clawed garlic'), Voc. 629 glossing *algium* [sic], and *goitkenin* ('wild-garlic'), Voc. 661 glossing *hermodactula vel tilodosa* ('dogsbane, saffron'). Welsh *cennin* 'leeks' (Old Welsh *cennin* glossing *cipus* 'food', VVB, p. 69), Breton *kignen* 'garlic'. For the adjectival form, cf. the Old Welsh r.n. *ceninuc* (LLD, p. 188), and Breton Quignénec (Tanguy, p. 112). The plant is native, 'locally abundant' (Davey, Flora, p. 438). For the Scilly instance cf. the comment by Leland, Itin., I, 191: 'Diverse of these islettes berith wyld garlyk'.
    A. aj. Ganinick
    C2. *lyncenin* (+**lyn**: Sawyer, nos. 832/1027)

*****kenow** 'puppy'; Welsh *cenau*, Old Breton *-ceneu* in personal names, dimin. *ceneuan* glossing *catulaster*, FD, p. 101. The normal word for 'puppy' in Cornish is *caloin*, Voc. 610 glossing *catulus*, and it is likely that all the following names contain rather *Kenow as a personal name. Cf. Old Welsh *nant cenou*, LLD, p. 155.
    C2. Boskenna (+ *****bod**); ?Tregenna in Blisland and St Ewe (both +**tre**); dimin. Treganoon (+**tre**)

kenter (f.) 'nail, spike', PC 2746 and 2766. Welsh *cethr* (see LHEB, p. 498), Breton *kentr*. The Cornish gender is shown by Lhuyd, *an genter-ma* 230c.
    C2. *Chingenter* (+**chy, an**)

*****ker** (f.) 'fort, a round', Welsh *caer*, Breton *kêr* (cf. GMB, p. 345). In Old Welsh *Cayr* is glossed *atrium* in *atrium Coquine...id est, Cayr i coc*, VSBG, p. 120; and glossed *civitas* in *Cair (i. ciuitas) Trigguid*, ibid., p. 76; Old Welsh *cair* and *civitas* are also evidently synonymous in the 9th-century list of the twenty-eight *civitates* of Britain in HB, §66a, the names of which all begin with *Cair*: Jackson, 'Nennius and the Twenty-Eight Cities of Britain', Antiquity 12 (1938), 44–55. It interchanges with *din* in *castellum dinduici l, id est cairduicil*, LLD, p. 226. In Old Breton it is glossed *villa* in *villa Bannhedos* (= *Caer Banhed*), RC 5 (1881–83), 432. The derivation is difficult; the best suggestions are a British Latin **quadra*, or some derivative of *cae* 'hedge': see Loth, RC 24, 298f.; Lloyd-Jones, BBCS 2, 292; Pokorny, JCS 1, 135; Jackson, LHEB, p. 252n. and HPB, p. 162.

    In Welsh the word is very varied in meaning, ranging from 'Roman camp', through '(non-Roman) fort' and 'castle' to 'city'; in Breton it is the normal word for 'homestead' and can also mean 'town, village' (and similarly perhaps in Cumbric:

Jackson, O'Donnell Lectures, pp. 8of.). In Cornwall the general assumption is that in the great majority of cases the word refers to the field monuments called 'rounds', univallate curvilinear hill-slope enclosures. This raises several problems. Archaeologists accept that the heyday of the round as a dwelling-type was the immediate pre-Roman centuries, carrying on into the Roman era. This is not altogether supported by the very small number of rounds excavated, since at least two of recent years have produced imported Mediterranean pottery: Trethurgy (Co.Arch. 12 (1973), 25–9), and Grambla (Co.Arch. 11 (1972), 50–2); but it does apply to a number of others: they show occupation in the first or second centuries B.C. or A.D., with no Dark-Age habitation of any significance: Threemilestone (Co.Arch. 15 (1976), 51–67); Shortlanesend (Co.Arch. 19 (1980), 63–75); Carlidnack (Co.Arch. 15 (1976), 73–6); also Castle Dore (Co.Arch. 10 (1971), 49–54 and references) and Carvossa (Co.Arch. 9 (1970), 93–7), though not 'rounds', fall into this category and both have adjacent names in *Car-*. But two non-archaeological factors suggest that the round should have continued as one (or even perhaps the predominant) type of settlement into the post-Roman era. One is the point, already mentioned, that it is the standard word for a habitation in Breton: this should indicate that the word represented the standard type of habitation at the period when the settlement of Brittany from Celtic Britain took place (probably 4th to 6th centuries: LHEB, pp. 12–30; Fleuriot, Origines, pp. 110–18 and references). The area from which the settlers went is not known, but may have been South West Britain, at least in part. The other factor suggesting a continued use of the round as a settlement-type is the Cornish place-names themselves: nearly all are of the name-phrase type, with the qualifier following the generic (Carloggas, Carvedras, etc.), not compounds. According to the accepted canons of Brittonic place-name formation, these names must date from the 6th century or later, and should not date from as early as the supposed period of the rounds: see the Introduction. To explain this anomaly, one could suppose that the names in *Car-* were all given after the monuments to which they referred had become obsolete – in other words, that in place-names the element always denoted an antiquity, not a living settlement: that is good for some of the names (e.g. those qualified according to flora or fauna), but not for others, particularly those qualified by words indicating

*ker

people. Another solution might be to suppose that ***ker** in Cornish, as in Breton, could denote things other than rounds – could, in fact, mean simply 'a settlement'; but that, too, is unsatisfactory when there is no hint of ***ker** having that meaning in the remains of the language, and when Cornwall has at least three other words for settlements or dwellings (***bod, tre, chy**) – words which are commoner than in Brittany and which serve the function that *kêr* serves there. In any case, a fair number of names in *Car-*, etc., do refer to known rounds or fortifications. Or, finally, we could suppose that the round as a habitation-type (not necessarily the dominant type) lasted into the 6th and 7th centuries. That is the most satisfactory answer, and it is also in keeping with an insular origin for the Breton usage; evidence from the small number of excavated rounds does not altogether contradict this suggestion.

There are about 130 names containing ***ker** as B2 in Cornwall (and one in Devon, near the border: Carley, PNDev., 1, 189); they are much commoner in the western hundreds of Penwith, Kerrier and Powder (each with about 30 or more) than in the remaining hundreds (each with 5–10, except for Stratton, with none). The element has frequently been confused with others, especially **cruc** and **carn**, and a particular name cannot be ascribed to ***ker** with certainty unless there are spellings in *Caer-* (normally in the 14th century or earlier), though reduction of the diphthong, giving *Car-*, may appear as early as 1086. Occasionally ***ker** refers, not to a 'round', but to a larger fort, e.g. at Tregeare, SX 0379, and Carvossa, SW 9148; but the usual term for that is ***dynas**.

A. Cair; Gare; Gear (8 exx., and field-names); Angear (+**an**); note the occasional disyllabic forms *Cayer* 1327 (= Gare); *Cay(e)r* 1444 (lost); and *Gaier* c. 1570 (= Castle Gayer): a similar phenomenon appears for **men** as a simplex name; pl. Kerrow (5 exx.), Cairo, Keiro (cf. Welsh Caerau, Rhestr, p. 20); pl. in **-yer**, ?Kerrier

B1. Hengar (+**hen**)

B2. A selection only is given. Qualified by a description of the fort: *caer lydan* (+**ledan**: Sawyer, no. 770); Carvean (+**byghan**); Carvear (+**meur**); Carlyon (2 exx.), Carleon in Morval (all +***legh** pl.); Carvinack (2 exx.), Carwinnick in Goran (all +***meynek**); perhaps Crasken and Creskin (both +**ascorn**) come into this category.

Qualified by a colour-word: Carwen (+**guyn**; 2 exx.); Cardew (2 exx.), Carthew (4 exx.) (all +**du**: cf. Cardew, PNCmb., 1, 131f.); ?Carloose (+**los**?)

Qualified by a word denoting the atmosphere of the place: Carrine (+**yeyn**); Keskeys, *Caerskes* (both +**schus**); Crankan, Grankim (both +**anken**)

Qualified by words for animals or birds: Carloggas in St Columb Major and Constantine (+**logaz**); Carvath (+**margh**), Carveth in Mabe (+**margh** pl.); Crohans (+**oghan**); in this category may come Crawle (if +**yorch** dimin.), Carblake (if it is *ker+**bleit**), Carvolth (if +**moelh**), and Crane (if +**bran**)

Qualified by a word denoting vegetation: Cargelley, Cargelly (both +**kelli**); Caruggatt (+***huel-gos**); Carnkief (+***kyf**); Carvannel in Gwennap (+**banathel**); Carwalsick (+**gwels** aj.)

Qualified by a personal name: Carvedras in Kenwyn (+*Modred*, Bodm.); Carwither (+*Guaedret*, Bodm.); Carvallack (+*Mæiloc*, Bodm.); Carworgie (+*Wurci*, Bodm.); Carbilly (cf. OBr. *Bili*, Chrest., p. 110); Carwarthen (cf. OBr. *Uuenran*, Chrest., p. 175); Crelly (cf. OWe. *Ili/Eli*, LLD); Crenver (cf. early Welsh *Kynvawr*, etc.)

Qualified by a word denoting a person (not a personal name): *caer uureh* (+**gruah**: Sawyer, no. 770); Carzise (+***Seys**)

Qualified by a word relating to a structure at the camp: Cardinham (+***dynan**; 2 exx.); Tregorland (+***cor-lann**); *Carfos* (+**fos**)

Qualified by a place-name: *Kaerclewent* (+Clowance)

There seems to be a variant of *ker as B2, which appears as *Care-*, *Cari-*, etc., in the mediaeval forms of some names: Carworgie (*Carewrge* 1086, *Karewurge* 1244); Carvossa (*Carawoda* 1302: +***gosa**?); Cargentle (*Carhegintell* c. 1150, but *Karkentel* 1284); Crinnis (*Caryhunes* 1354; *Caryones* 1477); and possibly Caerhays (*Karihaes* 1259; *Kerihayes* 1313) and Carracawn (*apud Cariconam* 1291). It is uncertain what these forms represent: since the second elements in the names are mostly obscure, it is not always certain where the forms should be divided, and thus not definite that they contain *ker at all; but the probability is that they do. (With Caerhays compare Breton Carhaix, and see Loth, RC 24 (1903), 288-98, and Ernault, RC 35 (1914), 491f.; but Loth's strictures against Breton Carhaix, etc., containing *ker* can be countered: HPB, pp. 97f. and 163, shows that it can appear, and be pronounced, as *Car-*, and the further Cornish evidence cited here may also help with the problem.) Assuming that the names do contain *ker, it could be simply a bad spelling for *Caer-*, but the cumulative evidence makes that unlikely. Such a variant could have arisen as a result of metathesis of the *i* and *r* in Old Cornish *cair, but that would be very odd. Perhaps one should compare the variant **penna-* of **pen** as a possible parallel. In Breton, note Kerivaladre, described as a 'forme archaïque' (Chrest., p. 171, n. 5), and perhaps early forms of the various Carhaix, such as *Caerahes* 11th-13th cent., *Kerahes* 1348, *Carahais* 1533 (Chrest., pp. 186f.).

C2. Boskear (+ ***bod**); Tregear (6 exx.), Tregeare (2 exx.), Tregair, Trengayor (all +**tre**, (**an**)); Roskear in St Breock, Rosecare (both + ***ros**); Roskear in Camborne, Rezare (both + **rid**); Parkengear (+ **park, an**); Pencair (+ **pen**); Kilgeare (+ ***kyl**); Polgear (+ **pol**); Messenger, fld. (+ **mes, an**); pl. *keyr, Burghgear (+(***bod**), **bagh**); pl. *kerow, Pencarrow in Advent (+ **pen**), Cutcare (+ **cos**)

The ppp. of a vb. equivalent to Welsh *caeru* 'fortify', or else a noun, equivalent to Welsh *caered* 'wall', may occur as A in Kirriers, Kerris; and as C2 in Tregays (+ **tre**).

kerden (f. pl.) 'rowans', Lh. 109c glossing *ornus* 'the quicken tree'. Welsh *cerddin, cerdin*; Breton *kerzhinenn* 'alisier' (= service-tree; cf. GMB, p. 346). The word *kerdinen* in the margin of Voc., p. 9a, is probably Welsh (so Graves, p. 300). On the Welsh word (and a similar personal name), see Foster, 'The Irish influence on some Welsh personal names', in *Féil-Sgríbhinn Eóin Mhic Néill* (edited by J. Ryan, Dublin, 1940), 28-36, at pp. 30f. All the Cornish place-names show forms with /ð/, not Lhuyd's *-d-*. Rowans are 'common' in most of Cornwall, less frequent in the far west (Davey, Flora, pp. 181f.). Dialect *care* 'rowan' (Courtney and Couch, p. 9) may be derived from *kerthen.

A. pl. Kerthen

C2. pl. ?Tregerthen (+ **tre**); Parc an gerthen, fld. (+ **park, an**: RC 41, 274)

cherhit (OCo., *ch* = /k/, *t* = /ð/) 'heron', Voc. 501 glossing *ardea*; Welsh *crychydd*, Middle Breton *quercheiz* (GMB, p. 550); Old Breton *cherched* glossing *gallina*, and *corcid* glossing *ardea* (FD, pp. 105 and 118). Two of the names, if they contain the element, show a sound-change /ð/ > /d/ (> *s*), which would be unusual;

however, it is supported by Pryce's Modern Cornish *kerhez* (s.v. *cherhit*: but that could be a semi-learned modernisation of the Voc. form). Compare /ð/ > *s* also perhaps in ***gogleth**. Polkerth in St Keverne cannot contain this element, because of the form *pollicerr/pollcerr* (Sawyer, nos. 832/1027: cf. *ryt y cerr*, LLD, p. 247?), though it might contain it by folk-etymology. For lack of early forms, its presence in all three names is somewhat uncertain. Penhallurick, Birds, I, 57-9, discusses the incidence of the bird in Cornwall. The word **garan**, 'crane', may sometimes mean 'heron' in place-names.

C2. ?Polkirt, ?Polkerris, ?Polkerth in Gerrans (all + **pol**)

***kernan** 'mound, hill' (?), either corresponding to Welsh *curnen*, *cyrnen* 'cone, rick' (WVBD, p. 321; ELl, p. 15), or alternatively a diminutive of a word corresponding to Welsh *cern*, Breton *kern* (GMB, pp. 550f.), 'top or side of the head; hillside'. Confusable with *carnan, diminutive of **carn**.

C2. Boscarnon (+ ***bod**)

***kernyk** (f.) 'little corner, little horn', a dimin. of **corn** 'horn, corner': *gernygou* 'little horns', BM 3396. Cf. Welsh *cyrnig* 'horned'. Some of the following names might theoretically contain *carnek or *kernek, adjectives of **carn**, instead.

A. Kernick (8 exx.), Kernock, Gurnick; pl. Carnegga

B2. Carnaquidden (+ **guyn**)

C2. Porth Cornick (+ **porth**); Park Gernick, fld. (+ **park**); Street an Garnick, street (+ ***stret, an**)

***kest** 'paunch, receptacle', equivalent to Welsh *cest* 'belly, basket', Breton *kest* 'basket, beehive' (cf. GMB, p. 747), Old Breton *cest*, FD, p. 104 (probably borrowed from Latin *cista*, 'chest', but cf. Gallo-Latin *cisium*, RC 48 (1931), 398, and DAG, p. 559). The Cornish word occurs as a personal epithet ('paunch') in *Gurcant Cest*, Bodm. xlv (cf. Förster ad loc.), the modern surname Keast, like Welsh *Einion Kest* (BBCS 3 (1926-27), 46). The meaning in place-names would be 'cavity'; compare Welsh Moel-y-Gest, ELl, p. 64.

C2. ?Lankeast (+ **nans** ?)

***keun** (pl.; *eu* = /œ/) 'reeds, rushes', Welsh *cawn* (from Late Latin *cāna*?).

C2. Penquean (+ **pen**); ?Halankene, ?Helencane Cove (both + **hal, an**); Trecaine (+ **tre, an**); ?Trequean (+ **rid** ?); ?Brown-queen (+ **bron**/***bren** ?)

*kevammok

An aj. **keunek* (f.) 'reed-bed' (*kenegan* 'reed-bed, bog', Trg. 8a), pl. **kenegy*, occurs as follows. (Alternatively, for these names one could compare Jackson's unknown Primitive Welsh **cönǫg*, Elements, I, 120; JEPNS 1, 46; PNDor., II, 71. Tanguy's Breton *kenek*, pl. *kenegi* 'hillock' (p. 69), said to be a form of *kreh*, *cnech*, is presumably irrelevant, despite the similarity of forms.)
    A. aj. Kennack; aj. pl. Kenneggy, Kenegie, Carnegie
    C2. aj. Leskinnick (+ ***lys**); aj. pl. Tregoniggie (+ **tre**)

*kevammok 'fight, battle'; Welsh *cyfamwg* 'defends, fights', 3 sg. pres. indic. of *cyfamwyn*. Welsh *cyfamwg* also occurs as a (verbal?) noun, *ae ketwyr yn kyuamwc* 'and its warriors as a defence' (CLlH, pp. 45 and 235), and in Breton the uncompounded form *amoug* 'retard' occurs (for the verb, cf. Lindeman, BBCS 28 (1978–80), 603–5). For the sense of the two names, compare Welsh *Brynn Kyuergyr* 'conflict hill' (PKM, p. 289).
    C2. *Meskevammok* (+ **mes**); *Gwael Kephamoc* (+ **guel**)

*kevar 'joint-tillage'; Welsh *cyfar* 'joint ploughing; land ploughed jointly', Breton *keñver* 'arpent' (cf. GMB, p. 553); early Middle Breton *(c)emer*, *c(h)emer* glossing *(are)pennum* (FD, p. 101). The third name contains the word with a suffix *-as* or *-es*, as in Welsh *cyfarad* 'partnership'. Co-tillage is treated in the Welsh Laws: see Colan, §§147ff. and pp. 67f. and 71; Latin Texts, p. 151, etc.
    A. ?*Kevar*, fld.; aj. ?Keveral (+ ***-yel**)
    C2. Tregavarras (+ **tre**)

keverang 'hundred, cantref', BM 2217 (in the phrase *keverang penweth* 'the hundred of Penwith'); Welsh *cyfranc* 'meeting, encounter, battle', Old Welsh *cibracma* 'battle-field' or 'meeting-place', Chad 4; Middle Breton *cuuranc* 'military assembly' and *coufranc* 'dispute, disagreement' (see Loth, Mél.Arb., p. 225, and Pinault, Ogam 15 (1963), 241–3; also HPB, p. 592 n. 2); Old Irish *comrac* 'battle' (the verbal noun of *con-ricc* 'meets, arrives'). The semantic development, which is most suggestive, must have been from 'meeting, encounter, battle' to 'military assembly', and thus to 'district operating as a unit for military service' (like the origin of the English hundred) – confirmation of the ancient function of the Cornish hundreds: compare the remarks s.v. ***cor**[1]. The two places named are at each end of the boundary between the hundreds of Penwith and Kerrier.
    A. pl. *Kyvur Ankou*

C2. pl. *Meenkeverango* (+**men**), *Assa Govranckowe* (= *Kyvur Ankou*; +***aswy**)

kevyl 'horse(s)' (?). This word could either be an equivalent of Welsh *ceffyl* 'horse' (but we would expect **kefyl*), or, less likely, a plural **kevyll* of an equivalent of Welsh *cafall* 'horse' (but that is rare and early), both perhaps ultimately related to each other, and to Latin *caballus* (cf. DAG, p. 554; Evans, GPN, p. 318 and n. 2; FD, pp. 99 and 331). Compare Welsh Bryn Ceffyl, Mer.

C2. Penkevil (+**pen**); ?Carkeval (+***ker**?). (But Nanskeval contains rather an unknown element **cuvel*, perhaps a personal name.)

kew (f.) 'a hollow, enclosure'; Welsh *cau* (m.) 'hollow', Breton *kev* (m.) 'creux' (*keo*, Tanguy, p. 67; cf. HPB, p. 242). The meaning is not always clear in place-names; no doubt 'hollow' is the usual sense, as both generic and qualifier, but Whitaker in 1791 knew the sense 'enclosure' (JRIC n.s. 7 (1973–77), 140), and Pryce, s.v. *gew*, says 'On many estates, especially in the West, one of the best fields is called *the gews*'. It is likely that this sense is restricted to the simplex occurrences, and even among those, the coastal examples are more likely to mean 'hollow, cavity'. The gender is odd, differing from Welsh and Breton (but cf. Coat-ar-Guéo, Tanguy, p. 67). It is possible that the sense 'enclosure' appeared partly under the influence of **keow* (the Middle Cornish plural of **kee**), which may have fallen together with ***kew**.

A. Gew (3 exx.), Angew, The Gew (2 exx.), and fields with these names; Kew Croft, fld., Gue Hole, coast (both +Eng.); aj. with **-esyk**, Cowyjack (= Welsh *ceuedig*)

B1. The element **vooga** 'cave' was originally apparently a compound of ***kew** in the sense 'cavity'.

B2. Gew-Graze, coast (+**cres**); *Gewan Gampe* (+**an**, ?)

C1. Keyrse, Case-(hill) (both +***ros**); see also the compounds ***kew-nans**, ***kew-rys**.

C2. Tregew (2 exx.), Tregue (3 exx.), Treguth (all +**tre**); Park Gue, fld. (+**park**); Forgue (+**forth**); ?Porthcew (+**porth**). A derivation with aj. suffix *-el* (= Welsh *ceuol*: see *-**yel**) may appear as C2 in Cargoll (+***ker**) and Tregole (+**tre**).

kew-nans (m.) 'ravine' (a compound of ***kew**+**nans**), Welsh *ceunant*.

A. Kynance (2 exx.), Cowlands, Trekelland in Lezant

kew-rys 'hollow-ford' (a compound of ***kew**+**rid**): cf. Welsh

Rhyd Gau (3 exx.), Richards, ÉC 10 (1962), 211. Alternatively, a personal name or *kew+ *red is possible in both cases.

   C2. Boskerris (+ *bod); Tregowris (+tre)

ky (m.) 'dog', PC 2242 and RD 2026; *ki*, Voc. 608 glossing *canis*. Welsh *ci*, pl. *cŵn*; Breton *ki*, pl. rarely *co(u)n* (Hemon, p. 38; PB, pp. 132–7). The Cornish plural is difficult: one would expect *cun /ku:n/, but the only instance is *kuen*, RD 172 (rhyming with *buen* /by:n/) and BM 3223; the spelling probably means /kœ:n/: cf. Lhuyd's *kên* (46a) and the place-name Colligeen, below; see also *cun-jy 'kennel'.

   C2. Carnkie (+ carn; 2 exx.); Tregye (+tre); ?Pedn Kei, rock (+pen); pl. ?Colligeen (+kelli)

*kyf 'stump'; Welsh *cyff*, Breton *kef* (in place-names: GMB, p. 525; Tanguy, p. 99). The name Cutkive is an exact equivalent of Breton Coetqueff, cited by Tanguy. Note also Ffynnon-cyff, Crd.; and, with Cutkive, compare English Stubwood, PNBrk., II, 306; Stubbing(s) Wood, PNWml., I, 86, PNYoW., VI, 153 and 208 and PNDrb., II, 304; Stockwood, Stock Wood, PNGlo., I, 156, III, 239, etc.

   C2. Cutkive Wood (+cos); Carnkief (+ *ker); Roskief (+rid or *ros); Trekeive (+tre); ?Millen Keeve (+melin¹; = Kieve Mills)

*kyl (m.) 'nook, back'; *chil* (*ch* = /k/), Voc. 26 glossing *cervix* (OE *hnecca*); *kylban* 'nape of the neck', CW 1114 (with which Loth, RC 23, 252, compares Breton *kilpenn* 'nuque'). Welsh *cil* 'corner, back, nape, nook' (cf. ELlSG, p. 110), Breton *kil*, meaning 'dos, revers' in place-names (Tanguy, p. 67; Chrest., p. 198). See also Elements, I, 93 and II, 3; JEPNS I, 45 and 48. The choice of meanings is thus between 'back, ridge' and 'nook, back part': the latter is best in Colquite, etc., 'nook of a wood', and Coldrinnick 'thorny corner', and perhaps with the animal-names as well. There are no instances where it clearly means 'ridge', except for those where it may have replaced **keyn**. Particularly as B2, *kyl readily takes other forms, and cannot be easily distinguished from **kelli**, or perhaps **cul**; note also Carglonnon and Carbaglet, where Kyl- > Car-. The element is common as B2, but very rare as C2 except perhaps in the possible plural *kylyer (v. -yer).

   A. pl. ?Kellyers, flds.

   B2. Colquite (4 exx.), Kilquite, *Kilcoys* (all +cos; some might contain **cul** instead); Coldrinnick (2 exx.), Coldrenick (all + **dreyn** aj.); Kilmanant, Kilminorth (both + *menawes); Killaworgey (+pers.); Carbaglet (+bagyl aj.); Killigarth (+cath); ?Kilmar Tor, ?Kilmarth (both +margh: or keyn); Calamansack, ?Kilmansag (+ ?)

C2. ?Coombekeale (+ *comm ?), Carkeel (+ *ker), and *Bethkile* (+beth) are all uncertain: confusion with *kel is possible; pl. ?*Park an Kellier*, fld. (+park, an), ?Dorkillier, fld. (+dor), ?Porth Killier (+porth)

*kylgh 'circle'. Welsh *cylch*, Breton *kelc'h*. What this would mean in place-names is unclear: possibly a stone circle or hill-fort, or compare Welsh Cylchau, Crm., SN 7520.

A. Kilkhampton (+Eng. *-tūn*)
B2. *Kylkethewe* (+du)

*kyniaf-vod (OCo.) 'autumn-dwelling', from *kyniaf* 'autumn' (glossing *autumpnus*, Voc. 463: Welsh *cynhaeaf*) + *bod. Not in Welsh or Breton, but cf. Welsh *cynaeafdy, cynhaefdy* (*kynhayafty*, Iorwerth, §139), and *cynhaefdre* (Richards, Monts.Collns. 56 (1959–60), 180), and compounds such as *meifod and *hafod*, also denoting seasonal dwellings (ibid., 13–20 and 177).

A. Kernewas; ?Canavas
C2. Bosneives (+ *bod)

kynin (MnCo., *y* = /ʌ/) 'rabbit', Lh. 53a glossing *cuniculus*; Welsh *cwning* (pl.), Breton *koulin*.

C2. ?*Hellan Conen* (+hal, an?); ?*Goudekining* (+ *godegh*?); Park Connin, fld.

kio (MnCo.) 'snipe', Lh. 146b glossing *scolopax* 'a snipe or snite'; *kîo*, Lh. 286a; Welsh *gïach*, Breton *gioc'h*. Penhallurick, Birds, I, 114–17.

C2. Trekee (+tre, an); Hendragoth (+ *hendre: cf. S. Welsh Lluestygïach 'the snipe's shieling', Monts.Collns. 56 (1959–60), 179)

clap 'chatter, clatter', RD 1113 (cf. *clapier* 'speak', Pryce, p. Ffiv). Welsh *clep* (and *melin (y) glep* 'clap mill') and *clap* (of which GPC quotes, of a mill, *a'i chlap megis hwch lipa* 'and its clap like a flaccid sow'). Borrowed from English (cf. Parry-Williams, EEW, pp. 57 and 70). See the remarks under Clap Mill, PNDev., II, 579; and ELlSG, p. 39.

C2. *Melynclap* (+melin[1]); perhaps *Mellenlappa* (= Lappa Mill)

*clav-jy (m.) 'lazar-house', from *claf* 'sick'+chy 'house'; Welsh

cleath

---

clafdy, Breton *klañvdi* (see Gourvil, Ogam 7 (1955), 149 for Breton place-names containing the word). Henderson, 'Clodgy', OC 2, iv (1932), 37f.; M. I. Somerscales, 'Lazar houses in Cornwall', JRIC n.s. 5 (1965-68), 60-99. The element is confusable with dialect *clidgy* 'sticky' (Courtney and Couch, pp. 12 and 81). Names such as Clodgy Moor, Clodgy Lane are perhaps more likely to contain the latter unless associated with a known lazar-house, as are the following examples.

A. Clodgy Point, coast (St Ives); Clodgy Moor (Budock and Paul); Clugea Lane (St Keverne); pl. *Clausiow* (Truro)

C2. *Nansclegy* (+**nans**)

cleath (m.; MnCo.) 'dyke, ditch, bank', CW 1140 (= 'a dyche', 1134 stage direction); *kledh*, Lh. 61b and 244c; Welsh *clawdd*, Breton *kleuz* (cf. Chrest., p. 198; GMB, p. 106; Ernault, RC 27 (1906), 50-7; Quentel, Ann.Bret. 59 (1952), 342f.; and HPB, p. 683 and n. 1). This element is liable to confusion with ***cleys** and ***clus**, both of closely similar meaning, and the three may have fallen entirely together after c. 1600. See also the compounds ***mengleth** and ***mon-gleth**, where it has the sense 'excavation, quarry'.

C2. *Wheal an Clay*, mine (+**wheyl, an**); dimin. ?Cargloth (+***ker**)

***cleger** (m.) 'rock(s), cliff': *cleghar*, *cleggo* 'a rock, cliff or downfall', Pryce; Welsh *cleg*(*y*)*r*, Breton *kleger* (Tanguy, p. 67). Note the place-names, OWe. *Pennclecir* (LLD, pp. 127 and 255); Welsh Clegyr, Mer.; OBr. *Cleker, Clecher* (Chrest., p. 116; FG, pp. 229 and 350).

A. Cligga Head, coast, Clicker Tor (both +Eng.)

***cleys** (m.) 'trench, groove'; Welsh *clais*, Old Irish *class, clais* 'trench'. This element is confusable with **cleath** and ***clus**, both of closely similar form and sense. In particular, it is hard to believe that the instances of the plural as simplex (Clijah, etc.) are different from the Breton occurrences of *Cleuziou*, etc. (Chrest., p. 198; HPB, p. 683 and n. 1) containing *kleuz* (= Cornish **cleath**), but that appears to be the case. There may have been confusion between the three within the language.

A. Cleese, Clies; *Clise Craft*, Clys Croft, *Clise Close*, flds. (all +Eng.); pl. Clijah, Clidga

B2. *Clise Nawiddon which is an old pitt or quarry* (+p.n.); *Clise Gwenna* (+p.n.)

C2. *Parken Clyes*, fld. (+**park, an**); *Croft-an-Clyes*, fld.
cloghprennyer 'bell-beams, gallows', BM 923 and 1241; *kloxprednier*, Chyanhor 29 (see Bosons, p. 23).
A. Comprigney
***clun** (*u* = /y/) 'meadow', equivalent to Welsh *clun* 'meadow, moor, brake' (cf. Rhys, Celtic Folklore, II, 513 n.); Irish *clúain* 'meadow, pasture': Hogan, Onomasticon, p. 253. Old Welsh *in guoilaut clun* 'at the bottom of the meadow', Chad 6. There may also have been an adjective *clunyek, 'marshy place' (?).
A. aj. ?Calenick
C2. ?Treglyn (+***dyn**?)
***clus** (*u* = /y/) 'heap', equivalent to Welsh *clud* 'load, bundle', Breton *klud* 'tas' (Tanguy, p. 67; GIB gives only *klud* 'perchoir, juchoir [perch]'). Compare the closely similar **cleath** and ***cleys**.
C2. pl. Menaclidgey (+**meneth**), identical to Breton Ménez-Klujeau (Tanguy, loc. cit.).
***cnegh** or ***cnogh** (m.) 'hillock'; Welch *cnwch* (also *cnwc* and *clwch* in p.ns.: Richards, JRSAI 90 (1960), 147–61); Breton *krec'h* (cf. GMB, p. 347; Tanguy, p. 72), Old Breton *cnoch* glossing *tumulus*, FD, p. 110; *Cnoch villa*, Chrest., p. 118.
C2. Penknight (+**pen**)
***cnow** (f. pl.) 'nuts, nut-trees'. The singular *cnowen 'nut-tree' occurs in Lhuyd, 51c, *gwedhan knyfan* glossing *corylus* 'an hasle-tree', and 74a, *kẏnẏphan frenk* glossing *juglans* 'a wall-nut'; the aj. *cnowek would mean 'a nut-grove, a hazlett' (cf. ***coll**). Welsh *cnau, cneuog*; Breton *kraoñ, kraoñeg*. Lhuyd's *kẏnẏphan* and two of the place-names show epenthesis between *k* and *n*, as in Vannetais *keneuen*, etc. One of the names, Callenowth, shows in its early forms variation between the plural, ***cnow**, and the adjective, *cnowek.
C1. Crockett, Knagat (both +**cos**)
C2. pl. Callenowth (+**kelli**); sg. Killiganoon (+**kelli**); aj. Killeganogue (+**kelli**)
***cok** (f.) 'cuckoo', perhaps in *arluth an gok*, PC 2890; otherwise only in *an gôg* glossing *cuculus*, Lh. 52c. Welsh *cog* 'cuckoo' (cf. ElSG, p. 15), but Breton *kog* is 'cock' (so PB, p. 140: *kogo* in place-names means 'cocks'). Cf. Blencogo (PNCmb., I, 122). Cuckoos occur in all parts of Cornwall: Penhallurick, Birds, II, 137–40 and 459. The element may be confusable with **cuic** 'empty, blind'.

kodna huilan

C1. Rooke is apparently *Ruth-gok* 'red-cuckoo' (+**ruth**: meaning? Cf. *rud cogeu* 'red are (the) cuckoos', CLlH, p. 9)

C2. Kilgogue (+**kelli**); Fentengo (+**fenten**); *Careck-an-googe alias the Cuckoe Rock* (+**karrek, an**); *Park an goag Cuckow*, fld. (+**park, an**; cf. English field-names such as Cuckoo Pen, etc., Field, EFN, p. 56); *Wheal an Gogue*, mine (+**wheyl, an**); pl. Trecoogo (+**tal**)

kodna huilan (MnCo.) 'lapwing', Lh.241b; cf. Welsh *cornchwiglen*; also *cornicyll*, and Breton *kernigell* (cf. GMB, p. 551, and Old Breton *cornigl* glossing *cornix*, FD, p. 119). The meaning in place-names may have been extended to 'desolate, remote spot', as was certainly the case with dialect *horniwink* 'lapwing; tumbledown place' (Courtney and Couch, pp. 29 and 88): that is especially likely in the case of the simplex examples. Cf. South Welsh *cornicyll* used of a shieling: Lluestcornicyll (Monts.Collns. 56 (1959–60), 179); and, in English field-names, The Lapwing (Baschurch, Shr.; from 1609) and Lapwing (*Lapwingeflight* 1673: PNDrb., III, 643); and cf. Foxall, Shropshire Field-Names, p. 45.

A. Codnawillan, Codna Willy, flds.

C2. pl. *Chapelkernewyl* (+**chapel**); *Pehel-Carnawhenis alias the Hornawink House* (+ ?)

***coffen** 'a dug mine on a lode' (as opposed to a tin-streaming work, or to a digging for alluvial tin); dialect *coffin* 'old workings open to the surface', Symons, p. 141. The earliest spelling known at present is *an Cogh'n bras* 1502, and the word may be a derivative of **cough** 'red', because of the red iron-ore often found on the upper surfaces of lodes.

A. Koffan, fld.

B2. *an Cogh'n bras, Coffin broaz*, mines (both +**bras**); *Coffenoola*, mine (+**ula**); *Coffin Garrow*, mine (+**garow**)

C2. *Huel Goffen*, mine (+**wheyl**)

***coger** (?) 'winding (stream)', from British **kukrā*, like Coker (Som.), Cocker (Cmb.), etc.: ERN, pp. 83f.; LHEB, p. 578.

A. ?Kuggar; dimin. (+**-yk**), ?Corgerrick: the two places are adjacent.

***coll** (pl.) 'hazel-trees', sg. **collen*; Welsh *coll* (sg. *-en*); Old Welsh *coll* glossing *corilis* (Mart.Cap.: VVB, p. 78), *finnaun he collenn* 'the spring of the hazel' (LLD, p. 247); Breton *kel-* in *kelvez*, Old Breton *coll* (FD, p. 114) and *limn collin* 'supple hazel' glossing *tilia* 'lime' (FD, p. 243). There was a derivative **collas* 'hazlett'

(apparently not the expected *colles, as in Breton Nancollet, Tanguy, p. 78). See also the other words for 'hazels', ***collwyth** and ***cnow**. Hazel in Cornwall is 'Very common in all the districts' (Davey, Flora, pp. 406-8).
 A. sg. ?Collon; ?Gollen Orchard, fld.
 B1. sg. Engollan (+**hen**?)
 C2. Presingol (+***prys, an**); Trescoll (+***ros**); sg. Fentongollan (+**fenten**), Park an Gollen, fld. (+**park, an**); deriv. *collas*, Roscollas (+**rid**), Tregolls (3 exx.), Tregolds in St Merryn, Tregullas (all +**tre**)

*collwyth (f. pl.) 'hazel-trees'; sg. *colwiden* (-*d*- = /ð/), Voc. 677 glossing *corillus*: a compound of the preceding with **gwyth**[1] 'trees'. Welsh *collwydd* (sg. -*en*); Breton *kelvez* (sg. -*enn*), Old Breton *collguid* (FD, p. 114). The aj. **collwythek* corresponds to Breton *kelvezeg* 'coudraie' (cf. GMB, p. 533). Some of the names seem to show internal *i*-affection, as in the Breton forms (HPB, p. 297).
 A. pl. Colwith; aj. Colvithick Wood, *Kylwethek*

*colomyer 'dove-cote'; Breton *koulomer*. A derivative of *colom* 'dove', Voc. 504 glossing *columba*, OM 132, etc.; *kylobman*, Lh. 241b. The Cornish form ***colomyer** is determined by the place-names: it is probably a borrowing from the Breton word, or from French *colombier*. See further Henderson, 'Cornish Culverhouses', Essays, pp. 211-14; R. Robertson & G. Gilbert, *Some Aspects of the Domestic Archaeology of Cornwall* (Redruth, 1979), pp. 8-10, 38-42 and Map 4.
 A. Clumyer, Clumyers, etc., flds.
 C2. *Parke an Clibmier*, fld. (+**park, an**)

*comm (m.) 'small valley', Welsh *cwm*, Breton *komm* (Tanguy, p. 68); Gaulish *cumba* 'valley', DAG, p. 455. (Cf. Elements, I, 119; JEPNS 1, 46.) A contrast with ***glynn** is indicated (in Welsh) by the following sentence: *y cwm mawr a welwch glynn coet oed* 'the great *cwm* that you see was a wooded *glynn*', WM 491.10f. The word was borrowed into English, where it was very productive as a place-name element, especially as a simplex name and in South-West England (Elements, I, 119). Since Breton *komm* is also used as a simplex name (Tanguy, loc. cit.), the numerous Cornish instances of Combe and Coombe could theoretically be either Cornish or English; however, they are much more likely to be English, because of the great frequency of the name further east

compes

(82 exx. in PNDev.). The element is in fact rare in Cornish, with only one certain, and a few doubtful, instances. For what it is worth, the topography of the sites suggests that the meaning in Cornish might have been specialised to 'tributary valley', leading off a main valley. See also RC 29 (1908), 69, showing the Breton word used as A (Le Coum, Co(u)mou), B2 (Coum bras/bihan) and C2 (Prat ar c'houm); and Welsh Llangwm, Mon. (*Lann Cum*, LLD, p. 173), also as C2.

    A. Gomm, *Gome*; *Gumm Park*, fld.; pl. ?Gummow (cf. Breton Coumou, Tanguy, loc. cit.)

    B2. ?Coombekeale (+ ***kel** or ***kyl**)

    C2. Pencobben (+ **pen**); ?Kilcobben Cove (+ ***kyl**?)

compes 'level, even', OM 2472 and CW 492; *compos*, OM 2442 and 2485; *compys*, PC 1206. Welsh *cymwys*, Breton *kompez*.

    C2. Goon Gumpas, Woon Gumpus Common (both + **goon**); Ventongimps (+ **fenten**): presumably 'evenly-flowing spring'.

conna (f.) 'neck', Voc. 27 glossing *collum*, OM 2813, etc.; *kodna*, Lh. 46c, 61c. The word is of unknown derivation (Campanile, p. 30).

    A. Cudno; Godna, fld.

    B2. Cudna Reeth (+ **ruth**); Codnagooth (+ **goth**[2]); *Codna-coos* (+ **cos**); *Codnidne* (+ **yn**)

    C2. ?Vinegonner, fld. (+ **fyn**?); see also **crak an gonna*, s.v. **crak**[2].

*cor[1] or **corð* (OCo.) 'clan, tribe, family, army'; Welsh *cordd* (only in early poetry and compounds such as *gosgordd* and *gwelygordd*), Old Breton **cor* (FD, p. 118; cf. FG, p. 89); early Irish *cuire* 'host, troop'. The word appears in **coscor**, perhaps in ***cor-lann**, and also in one early district name: Trigg (+ ***try-**[2]), called *pagum quem Tricurium vocant* (= 'Greater Triggshire') in the Life of St Samson, is of the same derivation as Breton Tréguier (Smith, Top.Bret., pp. 72f. and references) and Gaulish *Tricorii*, *Petrucorii*, etc. (DAG, pp. 186 and 402; PNRB, pp. 317–19); *Vertamocori pagus* (ACS, III, 243). The meaning of *Tricorii* and *Tricurium* (and of *Petrucorii*) would be 'district or tribe supplying three (four) armies', a confirmation of the ancient status of the north Cornish hundreds as a single unit (the Domesday hundred of Stratton, 'Greater Triggshire') subdivided into three parts (the later hundreds of Trigg, Lesnewth or 'Middle Triggshire', and

Stratton): see Charles Thomas, 'Settlement-History in Early Cornwall: I, The Antiquity of the Hundreds' (Co.Arch. 3 (1964), 70–9); Picken, 'The Names of the Hundreds of Cornwall' (DCNQ 30 (1965–67), 36–40). Compare the remarks s.v. **keverang** 'hundred'. A pl. *corðou might possibly appear as C2 in *Langorthou* (+ *****lann**; = Fowey).

*****cor**² 'hedge, boundary' (?); equivalent to Welsh *côr* 'a plaiting, a bound', as seen in Ban-gor 'plaited hedge', *amgor* 'limit'. This word might be responsible for Lhuyd's *kwr* 'limit', glossing *ora* (p. 108c), but note also Welsh *cwr* 'corner, border'. This element, rather than *****cor**¹, may be the first part of the compound *****corlann**, and appears also in *****cores** 'weir'.

B1. Spargo (+**spern**)
B2. Garlenick (+ ?)
C1. Corva (+**bagh**)

*****cores** (f.) 'weir, fish-pond', Welsh *cored*, Breton *gored* (cf. GMB, p. 280; the word is derived from *****cor**² in the sense 'plaiting, weaving'). Latinised as *coretis*, the word appears in a 10th-century Cornish charter (Sawyer, no. 1207; Padel, Co.Studies 7 (1979), 43f.); the Welsh word is common, both in the Llandaf charters (LLD, index s.v. *coretibus*; Davies, Microcosm, p. 36, n. 3) and in the Laws. The usual Latin word was *piscina* (*id est*, *coret*: Latin Texts, p. 289.29) or *piscarium* (e.g. *coretibus*, LLD, p. 69.13, glossed *piscariis*, p. 358), and in Cornwall *piscaria* (sg.) is used to refer to *Coresturnan* in the 12th century; see also Padel, Co.Studies 6 (1978), 24 n. 6 and references. Welsh names containing *cored* are given by Richards, 'Some fishing terms in Welsh place-names', Folk Life 12 (1974), 9–19, at pp. 10–15.

A. *Gorres* (= an inland earthwork)
B2. *Coresturnan* (+ ?), now half-translated as Turnaware (= '-weir')
C2. Nancorras, Nancollas (both +**nans**); Gorrangorras (+**goon, an**)

corf (m.) 'body (living or dead)'; Welsh *corff*, Breton *korf*.
C2. Nancor (+**nans**); ?Zawn Carve, coast (+**sawn**)

*****cor-lann** (f.) 'fold, enclosure', a compound of *****cor**¹ 'clan' or *****cor**² 'hedge, bound' + *****lann** 'enclosure'; Welsh *corlan, corddlan*. The element is confusable with *gorlan* 'a church-yard', Lh. 48c glossing *cœmiterium*, *korhlan* [for *korphlan*?], Lh. 149a glossing *sepulchretum*, Welsh *corfflan* (confused with *corddlan* in Welsh:

65

corn

see GPC and Colan, pp. 155f.; Iorwerth, p. 120); ***cor-lann** is more likely in the place-names.

A. Carland; Gurland; Garland, fld.

C2. Rescorla in St Ewe (+**rid** or ***ros**); Lancorla (+**nans**); Tregorland (+***ker**); *Park Gorland*, fld.

corn (m.) 'horn, corner', Voc. 259 glossing *cornu*; in the sense 'corner', RD 2163 and 2185; Welsh *corn*, Breton *korn*. For the topographical use, compare Early Welsh pl. *brinn cornou*, LLD, p. 172. See also the diminutive ***kernyk**.

A. dimin. ***cornell**, pl. ?Carnelloe (2 exx.: or **carn** deriv.)

B2. Carno (+**hoch**); Corgee, Kearn Kee (both +**kee**); Carbouling (+***bow-lann**)

C2. *Chycornekye* (+**chy**, **kee**); pl. Tregurno, Trecorner (both +**tre**), *Chapel Curnow* (+**chapel**), Porth Curno (+**porth**); dimin. ?Trekernell (+**tre**)

***cors** (f. pl.) 'reeds, fen', sg. ***corsen** (Voc. 645, *koisen* glossing *calamus*); aj. ***corsek** (f.) 'reed-bed'. Welsh *cors* 'marsh', aj. *corsiog*, Breton *korz* 'roseaux', *korzeg* 'roselière' (e.g. *corsec* glossing *cannetum*, Catholicon). With Pencorse compare Welsh Pen y Gors (Pem., Units, p. 178) and Breton Penhors (Tanguy, p. 113). Landcross (PNDev., 1, 94) and Llangors, Bre., contain this element with ***lann**. Cf. Gauze Brook (PNWlt., p. 7; ERN, p. 95); Corsley (PNWlt., p. 152); *curiam Lisarcors* 'the court beside the fen', VSBG, p. 190.

B2. aj. Gersick-an-Awn, coast (+**an**, ?)

C1. *Corslen* (+**lyn**)

C2. Pencorse (+**pen**); Tregoss (+**tre**, whence Goss Moor); *Halgrosse Moore* (+**hal**); Park en Course, fld. (+**park**, **an**); *Gwythancorse*, mine (+**guyth^2**, **an**); aj. Pengersick (+**pen**; 2 exx.); deriv. ***kersys** (+***-ys^2**; = Breton Gorzit, Tanguy, p. 113), ?Porthkerris (+**porth**)

cos (m.) 'wood', OM 364; *coys*, OM 2589; Old Cornish *cuit*, Voc. 705 glossing *silva*; Modern Cornish *kûz*, Lh. 244c, *cooz*, Pryce, p. Ff3r. Welsh *coed*, Breton *koad*. For the former distribution of woodland in Cornwall, see Henderson, 'An Historical Survey of Cornish Woodlands' (Essays, pp. 135-51); and Ravenhill, in *The Domesday Geography of South-West England* (edited by H. C. Darby and R. Welldon Finn: Cambridge, 1967), pp. 325-8: in the middle ages woodland was widespread over most of Cornwall except in the farthest west, but individual woods tended to be larger in the east of the county. The difference between **cos** 'wood' and **kelli** 'grove' is presumed to be one simply of size. For the frequent and early reduction of the diphthong when the element occurred as B2, compare RC 3, 405; HPB, p. 195; and Welsh *Cod-* in *boscus...Codragheyn*, 1334 (Surv.Denb., p. 27), and *Codanew*, 1292 (BBCS 9 (1937-39), 68).

A. Engoyse (+**an**; the lenition is curious); see also **rag-gos* (?), s.v. **rag**.

B1. Lidcott, Ludcott, Lydcott, Lidcutt (2 exx.), Liskis, Liskey Plantation (all +**los**); Crockett, Knagat (both + ***cnow**); Dowgas (+**dew**); Devichoys Wood (+ ***dywys**); Tagus (+**tal**); Morchard (PNDev., II, 380) belongs here (+**meur**); see also the compounds **tron-gos* and **toll-gos*.

B2. With a qualifier denoting position or size: Cutparrett (+**perveth**); Coosewartha (+**guartha**); Coosebean (+**byghan**); Cutmere (+**meur**). Denoting appearance: *Coyseglase* (+**glas**); *Coeswyn* (+**guyn**); Cuttivett (+***dywys** or **tevys**); Cutlinwith (+***glynn-wyth**?); Polskeys (+**pol**, **schus**). Denoting birds: Cutbrawn (+**bran** sg. or pl.); Cusgarne (+**garan**). Denoting owners: Cutmadoc (+pers.); *Cosesawsyn* (+**Zowzon**); Coosehecca (+pers.). Denoting nearby structures: Cutcare (+***ker** pl.); Cusveorth (+***buorth**); Cutcrew (+***crew**); Crosspost (+**post**). Denoting other features: *Coysynchase* (+**an**, ***chas**); Cotehele (+***heyl**); Cusvey (+***fe**?); *Coyskentueles* (+**cuntell** deriv.). With a place-name as qualifier: *Coyspenhilek*, *Coyse Laydocke*, *Coys Penryn*. With a river-name: *Coysfala*, *Coysbesek*, *Codde Fowey*. With obscure qualifier: Cogegoes, Carpuan, Cosawes, *Cossabnack*, Coskallow Wood, etc. With a type of tree: Kus Skewes (+**scawen** deriv.). It is remarkable that there is no other clear instance of **cos** as B2 qualified by a word denoting a type of tree.

C2. Qualifying a word for 'dwelling': Trequite (2 exx.), Tregoose (8 exx.), Trequites (all +**tre**); Trevisquite, Lesquite in Pelynt (both +**tre**, ***is**[1]); Chycoose (+**chy**; 4 exx.); Melancoose, Mellangoose (2 exx.), Mellingoose (all +**melin**[1])

Qualifying a word for 'hill, top': Burncoose (2 exx.), Barncoose, Pencoose in Stithians, Berrangoose (all +**bron**); Goongoose (+**goon**); Burgois, Bargus (both +**bar**); Pendoggett (+**pen**, **dew**); see also ***pen cos**.

Qualifying a word for 'side, edge': Tucoyse (2 exx.), Trecoose (both +**tu**); Transingove (+**tenewen**); Mingoose (+**myn**)

Qualifying other words: Colquite (4 exx.), Kilquite (all + *kyl: or possibly as B1 + cul?); Stencoose (2 exx.), Stencooce, Tamsquite (all + *stum); Lesquite in Lanivet (+lost); Ventongoose (+fenten)

D. Withielgoose

coscor (OCo.) 'family, retinue', in Voc. 191, *den coscor* glossing *cliens*, and in Voc. 137, *goscor pi teilu* glossing *familia*. Welsh *cosgor(dd)* and *gosgor(dd)* (CA, p. 380; Old Welsh *casgoord*, VVB, p. 65); Breton *koskor*. The second part of the word is *cor¹* 'troop, clan'.

C2. *Bownds an Coskar*, mine (+ *bounds, an)

coth 'old', Voc. 208 glossing *senex*, etc.; Breton *kozh*. Note the superlative as a personal epithet in *Methuthala Coththa*, A.D. 1660. Compare **hen**.

C2. *Dolcoath* (2 exx.), *Dorcoath*, flds. (all +dor); *Todden Coath* (+ton); *Croft Coothe*, fld. (+ *croft); *Bounds Coath*, mine (+ *bounds); *Gone Goth*, fld. (+goon)

D. *Helliscoth* (= Helston); *Beaulugooth*; *Poundergourth*, fld. (+*Penvounder*); *Merrose Coath* (contrasted with *Merrose Noweth*); *Rosculiangoth* (= Roscullion)

cough 'blood-red', PC 2326; Welsh *coch*. See *coffen for a possible derivative; but 'red' was normally **ruth**.

C2. *Killicoff* (+kelli)

***crak¹** 'sandstone'; Middle Breton *cragg* 'grès' (cf. GMB, p. 131 and Tanguy, p. 59); Welsh *craig* 'rock' (Elements, I, 112; JEPNS 1, 45; PNCmb., II, 266). The Breton word has a diminutive **kragen* (Tanguy, ibid.), and Welsh has *creigen* 'rock': the diminutive **cragen* would be confusable with **crogen** 'shell, skull'.

A. *Crackington* (+English *-tūn*); ?*Harcourt* (+ *ar?)

C2. pl. ?*Pennycrocker* (+pen); dimin. *Tregragon, Tolgroggan* (both +tal)

crak² 'crack!' RD 397; *my a der crak ov conne*, OM 2184, 'I will break my neck, crack!' (cf. Cuillandre, RC 50, 49). Welsh *crac* 'crack'; possibly Breton *krak* 'short'. The exact construction of the names is obscure; **crak an gonna* (+ **an, conna**) might be 'the break of the neck' (cf. *crak taran* 'a clap of thunder', RD 294); or, more likely, an (exclamatory) imperative of the verb *crakye* 'break' (MC 139a and 164b; BM 1582), as in *Break My Neck Farm*. Compare the common English field-name *Break Back*, etc. (Field, EFN, p. 28). Examples of **crak an gonna*, all fields, are: Crackagodna, Crockagodna, Crack-an-Godna (several), *Croc-an-codna*, Cargodna. According to Henderson, Constantine, p. 240, it generally applies to steeply-sloping fields, but hardness of tilling the soil, as in the English names, is another possible factor.

*crann 'bracken, scrub'; Breton *krann*, 'endroit où il y a des restes de fougère' (GIB), 'racines, broussailles' (Tanguy, p. 100); cf. Chrest., p. 121 and n. 1; Loth, RC 29, 70; *kranneg*, 'endroit où il y a des restes de racines d'ajoncs, etc.' (GIB). All the Cornish instances are doubtful. The normal word for 'bracken' was **reden**.
 A. aj. ?Crannow; ?Crannis (+ ***-ys²** ?)
 C2. ?Goodagrane (+ **goth¹** pl. ?); pl. ?Pencrennow (+ **pen**)

*cren 'trembling' (noun and aj.) ,Welsh *crŷn*, Breton *kren*; cf. the verb *gren* 'shakes', BM 3450 and 3529, *kerna* 2257, Breton *krenañ*, Old Breton *arm-criniat*, FD, p. 73. Indistinguishable from *kren* 'round' (Lh. 141c), if that occurs at all in place-names, and liable also to be confused with ***cryn** 'dry'. Breton *kren* also means the trees 'aspens' (GIB; Tanguy, p. 105; Evans, ÉC 15 (1976–78), 211f.), and that sense may occur in Cornish place-names.
 A. ?Creens
 C1. Greenwith (+ **gwyth¹**; or **crynwyth* 'dry-trees', like Welsh *crinwydd* ?); ?Kirland (+ ***lyn** ?)
 C2. Kellygreen (+ **kelli**)

cres 'middle', OM 2481, etc., always as a noun; as aj. *bêz krêz* 'the middle finger', Lh. 172a. Welsh *craidd*, Breton *kreiz* (on the difference of the Welsh form from the Cornish and Breton, see LP, p. 15 and HPB, p. 94). The Cornish word is used mainly as a suffix, where there are three or more subdivisions of an original single settlement. Most of these have been lost with the death of the language. The older word for 'middle' was **perveth** (and, earlier still, probably ***međ**), and **cres** evidently occurs only in later names: for example, **tre** is always found with **perveth**, never with **cres** as **Tregres*.
 C2. Screws (+ ***ros/rid**); Carncrees, Carn Creis, coast (2 exx.) (all + **carn**); Gew-Graze, coast (+ ***kew**); Park Crees, *Park Crease*, Park Crazie, flds. (all + **park**)
 D. *Treludrou Gres* (= Treludderow, also *T. Barton*, *T. Masek*); *Bodriancres* (= Bodrean, also *B. pella*); Colgrease (*Crowgres*); *Ammalgres* (= Amble, also *Ammalmur*, *Ammaleglos*); *Porthylly Gres* (= Porthilly, also *P. eglos*); Meaver-crease (also *M. vean*,

*creun

M.-wartha); Carnon Crease (also Higher, Lower C.); *Scoryacres* (= Scorrier, also *S. wartha, S. woles*); *Polcrybowe Grees* (= Polcrebo, also *P. woles*); Kemyel Crease (also K. Drea, K. wartha, K. Mear, K. nessa, K. pella); Vorvas Crease (also V. Vean, Higher V., Lower V.)

*creun (*eu* = /œ/) 'dam' (?); Welsh *crawn* 'hoard', *crawnbwll, cronbwll* 'reservoir'; compare Breton *kreun* 'croûte'? The stem is seen in a verb, 3 sg. pret. *crunys* 'accumulated', MC 224b. The status of this element is dubious: the common Polgrean, etc., contains rather *gron: see **gronen** and **polgrean**. However, *Park Crane*, fld., adjoined a mill-pool and may have contained the word.

*crew 'weir', Welsh *cryw* 'creel, weir', also pl. 'stepping stones' in place-names (GPC); Elements, 1, 118; JEPNS 1, 46, and 2, 73. The element is almost indistinguishable from **krow** 'hut'.

C1. ?Scraesdon (+Eng. *dūn*), ?Crowlas (both +**rid**; or **krow**+*lys)

C2. Lescrow (+**rid**; Welsh Rhyd Cryw, ÉC 10, 217); Cutcrew (+**cos**)

krib (f.) 'comb, crest', Lh. 53a *krib an tshÿi* 'the ridge of an house', glossing *culmen*; 115b *krîb* glossing *pecten*. Welsh *crib*, Breton *krib* (cf. Tanguy, p. 73); a diminutive *kriban* 'a birds crest', Lh. 240c, also occurs, like Welsh *cribin*, Breton *kribenn, kribin*. All the names are coastal rock-names, unless noted.

A. Greeb (farm), The Greeb, Greeb Point (3 exx.), Grebe Rock, Creeb (Scilly); pl. ?Cribba Head, ?Cribbar Rocks; dimin. Gribbin Head (cf. Welsh Gribin, Den., sJ 2146)

B2. *Crubzu* (+**du**; inland?); *Cribbensaune* (+**an, sawn**)

C2. Mên-y-grib Point (+**men**; **an**?); Carn Greeb (+**carn**); Pengreep (+**pen**; inland, 'ridge's end'); pl. Polcrebo (+**pol**; inland: meaning?)

*crygh 'wrinkled', Welsh *crych*, Breton *krec'h*. The Welsh word occurs as qualifier in several compounds.

C1. Knightor (+**tyr**); *crygh-don (+**ton**) 'wrinkled lea' possibly (as C2) in Pedn Crifton, coast (+**pen**)

C2. pl. (as noun) ?Tregrehan (+**tre**)

*cryn 'dry, withered', Welsh *crin* (Old Welsh *crin* glossing *ar[i]dum*, VVB, p. 88), Breton *krin*. Liable to confusion with *cren 'trembling; aspens'.

C1. ?Crylla (+**le**?; the same compound may appear as C2 in Tregrill, +**tre**)

crom

C2. ?Menacrin Downs (+**meneth**; or ***cren**? but cf. Welsh *Crynfenyht*, Welsh Assize, p. 301)

A possible derivative of ***cryn**, East Cornwall dialect *crinnicks* 'dry sticks, kindling-wood', may appear (probably as the dialect word) as the East Cornwall field-name Crinnicks, Crennick, etc.

***crys** 'fold, wrinkle', equivalent to Breton *kriz* 'pli, ride' (GMB, p. 134); or to Welsh *cryd* 'a trembling', Old Breton *crit* (FD, p. 123), which appears in Cornish *dorgrys* 'earthquake', PC 3086. The former is the more likely meaning in place-names.

C2. pl. Tregidgio (+**tre**)

***croft** 'uncultivated enclosed land, rough ground', borrowed from English, like Welsh *crofft*, *grofft* 'cultivated field' (cf. PKM, p. 241; ELlSG, p. 41). According to Henderson (Constantine, p. 229), Cornish 'croft', though uncultivated, is enclosed land, containing furze grown high for fuel (whereas 'down' is unenclosed and used for pasture). Cf. PNCmb., II, 420 (*Croftbladen*), 421 (*Croftmorris*), and 444 (*Croftbathoc*). Cornish field-names in *Croft-* are mainly, if not entirely, restricted to the western half of the county. The word was already naturalised in Cornish in the 13th century (*Croftengrous*, 1265).

A. pl. Crofthow, Croftoe

B2. Croftnoweth (+**nowyth**); Croft Pascoe, Croft Michell (both +pers.); Croft-an-Creeg (+**an**, **cruc**); Crofthandy (+**hensy**?); Croft West (+**west**); Croftvean (+**byghan**); Croftengweeth (+**an**, **gwyth**[1]); the following are flds. Croft Coothe (+**coth**); Croft Horspole (+p.n.); Croftengrous (+**an**, **crous**); Croft en begel (+**an**, **begel**); Croft en Ferngy (+**an**, *forn-jy? see **forn**); Croftenorgellous (+**an**, ***orguilus**)

crogen (f.) 'shell, skull', Voc. 557 glossing *concha*; PC 2141 and RD 2558. Welsh *crogen*, Breton *krogen*. The plural would have been *cregyn, equivalent to Welsh *cregyn*, Breton *kregin*, but liable to be confused with **cruc** dimin. See also *cragen 'rock' (?), s.v. ***crak**[1].

C2. Roscroggan (+***ros**); ?Scrawsdon (+***ros**? and Eng. *dūn*); pl. ?Boscregan (+***bod**: or **cruc** dimin.)

crom 'curved', OM 2443, *krwm*, Lh. 53b; Welsh *crwm*, Breton *kromm* (Old Breton *crum*, FD, p. 124). The first, and perhaps the second, of the names contains the word as a noun: the meaning is probably 'the hunched', like the Middle Breton personal name *le Croum* (Beauport: RC 3, 407); but in the case of the first name,

*cromlegh

it might conceivably mean 'the arc', referring to an adjacent hill-fort (see the next element).

C1. See ***cromlegh**.

C2. Trencrom (+**tre, an**); Trecombe (+**tre, an**; or **tyr**); ?Mennabroom (+**meneth**); Marcradden (+**men**)

*cromlegh (f.) 'dolmen, quoit', Welsh *cromlech* (cf. Richards, BBCS 25, 269), Breton *krommlec'h* (cf. GMB, p. 360). The word is a compound, 'curved-slab', originally (like English 'quoit') referring only to the capstone of a dolmen, then by extension to the whole structure. The Breton word has been said to mean 'stone circle' (e.g. le Gonidec, s.v.), i.e. 'arc of slab(s)' or 'curved place'? If that is so, it is a curious divergence. *Kestelcromlegh* in Sawyer, no. 450, is much the earliest instance of the word in any Brittonic language, if the charter is genuinely of Athelstan's reign (925-39).

A. Grumbla; Grambler (2 exx.); Grambla

C2. *Kestelcromlegh* (+**castell**)

cronek 'toad'; *croinoc*, Voc. 614 glossing *rubeta* 'toad'. Welsh *croeniog* 'made of skin'; cf. Breton *kroc'henek* 'thick-skinned' (see VKG, I, 125).

C2. pl. Polkanuggo, Polkernogo (both +**pol**)

crous (f.) 'cross'; Old Cornish *crois*, Voc. 755 glossing *crux vel staurus*. Welsh *crwys*, *croes*, Breton *kroaz* (also Middle Breton *croes* in place-names: HPB, p. 189). The Old and Middle Cornish forms are at variance: **crous** is not the natural descendant of *crois*. Lewis, LlCC, p. 7, says that *crois* became **crōs*, which then became *crous*; but that ignores the fact that **crous** appears as early as the 10th century (see below), some 200 years before the *crois* of Voc. It seems rather that there were two forms side by side, the regular *crois* and a variant *crous* (compare *Zowz*, Lh. 42c, and ***Seys** 'Englishman', and LHEB, p. 535; cf. HPB, p. 189). See the useful collection of names containing **crous**: Henderson, 'Missing Cornish Crosses', OC 1, xii (1930), 6-12. Welsh *croes*, as well as meaning 'a cross', can have other meanings such as 'thwart' (e.g. Tŷ-croes) and 'crossroads', but there is no evidence whether these occur in Cornish.

A. Angrouse (+**an**)

B2. Crosswyn, *Crouse-widen* alias White Cross (both +**guyn**); Crouse-Harvey (+pers.); Crowsmeneggus (+**benyges** or ***menawes**; **crous** and ***ros** alternate in this name); Crousa (*crouswra(c)h*,

10th century: Sawyer, nos. 755 and 832/1027), Crows-an-Wray (both +(**an**), **gruah**); Crowshire (+**hyr**); Persquiddle, fld. (+**park**, ***gwythel**?); *Crousbronnou* (+p.n. Burnow); *Croustroch* (+**trogh**); *Meene-Crouse-an-Especk* (+**men, an, epscop**); *Goen an Groushire*, fld. (+**goon, an, hyr**)

C2. Rose-an-Grouse (+**rid** or ***ros, an**); Trengrouse (+**tre, an**); *Porthencrous* (+**porth, an**); Messengrose (+**mes, an**); *Croftengrous*, fld.; Parcan Growes (+**park, an**); *Hewas-an-grouse*, fld. (+ ***havos, an**); *Gwythengrouse*, mine (+**guyth²**, **an**)

crouspren 'cross-beam, crucifix' (written as one or two words; cf. Stokes, ACL 1 (1898–1900), 169 and 172f.). A compound (with unvoicing of *-b-* to *-p-*) of the preceding element with **pren** 'timber'; Welsh *croesbren*, Middle Breton *croaz pren* (GMB, p. 134).

A. *Crouspren*; *Crowse predon bounds*, mine

krow (m.) 'hut', Lh. 4c and 47a (also *krow môh* 'a hog sty', 15a), dialect *crow* (cf. M. Wakelin, '*Crew, cree* and *crow*: Celtic words in English dialect', Anglia 87 (1969), 273–81); Welsh *crau* (cf. PKM, p. 260; TYP, p. 534), Breton *kraou*. Liable to confusion with ***crew** 'weir'; and theoretically indistinguishable from *crow* 'blood, gore' (MC 74c and 131d; Welsh *crau*), but that is unlikely in place-names. Cf. PNCmb., III, 467; PNWml., II, 244, for names containing an English derivative of the Brittonic word.

A. *Crou* (now subdivided as Carevick, Carines, Colgrease); dimin. Croan (= Welsh *crowyn*, etc.)

C1. ?Scraesdon (+Eng. *dūn*), ?Crowlas (both + ***lys**; or ***crew**+**rid**); see also ***krow-jy**.

C2. Roskrow (+ ***ros**); Lancrow (+**nans**/***lann**); Kilgrew (+ ***kyl**); Park-and-Crow, Parken-Crow, *Croft-an-Crow*, flds.

***krow-jy** (m.) 'hut, cottage', a compound of **krow**+**chy**.

A. Crowgey (4 exx.), Crowdy Marsh

C2. Roscrowgey (+ ***ros**); Pencrowd (+**pen**)

cruc (m., u = /y/) 'barrow, hillock', Voc. 718 glossing *collis*; Welsh *crug*, Breton *krug*; Old Welsh *cruc*, glossed *cumulum* (VSBG, p. 118), Old Breton *cruc*, glossing *acceruum* and *gibbus* (Chrest., p. 122 and FD, p. 124). In Cornwall it is often assumed that the word always refers to an archaeological feature, a 'tumulus', but that may not necessarily be true; the assumption in England generally tends to be that it refers to a natural hill: Elements, I, 115 (cf. JEPNS 1, 46); BBCS 15, 15f.; PNDor., II, 268 and 275; PNDev., II, 638.

cudin

That is in agreement with the Old Cornish use to gloss *collis*, and also perhaps with the original sense (cf. Irish *crúach* 'stack, hill'); so some of the Cornish examples may refer to natural features, though in minor names 'barrow' is more likely. See Gourvil, Ogam 7 (1955), 219-23, and Tanguy, p. 73, for the use of the word in Breton place-names. There are two diminutives: *crugell (cf. Welsh *Crugyll*, Agl., and perhaps *Pencryc(h)gel*, VSBG, p. 120), Breton *krugell* (cf. PB, p. 271; Tanguy, loc. cit.; FD, p. 110); and (theoretically, but not certainly found) *crugyn, Welsh *crugyn*, *crugen* (*agro Crucin*, VSBG, p. 128), Breton *krugenn (Tanguy, loc. cit.). Note that *Cruc-* as B2 is often corrupted (or changed by folk-etymology) into *Car-*.

A. pl. Crugoes, Cregoe (cf. Crugiau, Pem., Units, p. 51; *villa crucou*, LLD, p. 262); dimin. Criggan

B2. Qualified by shape or size: *crucmur* (Sawyer, nos. 832/1027), Crugmeer (both +**meur**); Carluddon (+**ledan**); Creegbrawse (+**bras**); *Curkheir* (+**hyr**)

By colour: Carclaze (2 exx.), Creeglase, Cruglaze (all +**glas**; cf. *cruc glas*, VSBG, p. 136, and LLD, p. 258); Creegdue (+**du**)

By animals, etc.: Carloggas (in St Mawgan in Pydar and St Stephen in Brannel), Creeglogas (all +**logaz**); Carplight, *Cruplegh* (both +**bleit**); Crigmurrian (+***moryon**); *Crukbargos* (+**bargos**?; = Bargoes)

By other features: Cartole, Creeg Tol (both +**toll**); Trekennick (+**keyn** aj.); *Krucgruageh* (+**gwrek** pl.); Crugsillick (+pers.); Tregarland (+***alun**); Cargurrel (+**geler**); *cruc drænoc* (+**dreyn** aj.); pl. *crucou* mereðen (+***merthyn**?)

C2. Trencreek (5 exx.), Tencreek in Menheniot, Tencreeks, Trecreege, *Trencruk* (all +**tre**, (**an**)); Boscrege, Boscreeg (both +***bod**); Egloscrow (+**eglos**; = St Issey); Langreek (+***lann/nans**); Roskruge (+***ros**); Polcreek (+**pol**); Tencreek in Talland (+**keyn**); *Goyncrukke* (+**goon**); Killianker (+**kelli, an**); Croft-an-Creeg (+***croft, an**); Screek Wood (+**lost**?); dimin. Carn Creagle, rock (+**carn**); ?Boscregan (+***bod**: or **crogen** pl.); pl. Trecrogo (+**tre**)

cudin 'lock of hair', Voc. 33 glossing *coma*; Welsh *cudyn* (Old Welsh pl. *cutinniou*, VVB, p. 92), Breton *kudenn* (cf. GMB, p. 136). In coastal names the sense is presumably 'tresses' of seaweed. Some

of the names might instead contain *cudon* 'pigeon' (Voc. 505 glossing *palumba*; Breton *kudon*, cf. Chrest., p. 122, and Tanguy, p. 126).
 A. pl. Cadedno (= reef, Scilly)
 C2. Boscudden, ?Tregidden (+ **bod, tre**: as pers. ?); pl. ?Croft Codanna, fld., ?Pargodonnel Rocks (+ **porth** ?)
cuic (OCo.) 'empty, blind', Voc. 380 glossing *luscus vel monoatalmus*, 'blind in one eye'; also *coic-*, Voc. 650 translating OE *blind-*. Welsh *coeg*: note the usage as C1 in Coegnant (several), considered to mean 'hollow-valley', and in other Welsh compounds with the meaning 'pseudo-, vain' (GPC); in the Cornish compound the meaning might rather be 'worthless' (cf. *kuk*, BM 3481). Note also *pen cok* 'block-head' ('empty head'?), OM 1529, etc.; the element may be liable to confusion with ***cok** 'cuckoo'.
 C1. Cogveneth (+ **meneth**)
 C2. Nancekuke (+ **nans** ? *non cuic*, Sawyer, no. 684)
cul (*u* = /y/) 'thin, narrow', Voc. 945 glossing *macer vel macilentus*; Welsh *cul*; Old Breton *cul* glossing *macer*, *culed* glossing *macies*, and *culion* glossing *macilentos* (FD, p. 124); cf. GMB, pp. 137-8; Elements, 1, 118; JEPNS 1, 46.
 C1. Some examples of Kilquite, Colquite, etc., may be compounds of **cul** + **cos**, but a phrase ***kyl** + **cos** is more probable. (Cf., however, Culgaith and Culcheth: PNCmb., 1, 184f.; PNLnc., p. 97; LHEB, p. 320.)
 C2. Porthcuel (+ **porth**)
kullyek (m.) 'cock', PC 903; *colyek*, MC 49b and 86a; *chelioc*, Voc. 517 glossing *gallus*; *kwlliag*, Lh. 240c. Welsh *ceiliog*; Breton *kilhog* (on which see HPB, p. 301 for the *-i-*, and pp. 136-8 for the *-o-*). The first vowel of the Middle Cornish form is unexpected: one would expect **kelyek*. However, it is confirmed by 14th-century forms of the first four names. It must stand for /u/, not /y/.
 C2. Trekillick, Treculliacks (both + **tre**, **an**); Portkillock (+ **porth**); Kergilliack (+ **kee**: the lenition is odd); *Wheal an Cullieck*, mine (+ **wheyl, an**)
cummyas 'leave, farewell', OM 750, etc.; Modern Cornish *kibmiaz*, *kibnias*, Bosons, p. 16, etc. Note particularly *gays the cumyys* 'take thy leave' (BM 2969; cf. *gase farwel*, BM 1286). Middle Breton *quemiada* 'prendre congé' (GMB, p. 534).
 C2. Porth Kidney Sands (+ **pol**)

kunys

kunys (*u* = /œ/) 'firewood, fuel', OM 1296, etc.; *kinnis*, Pryce, p. Ff3r. Welsh *cynnud*; Breton *keuneud* 'bois de chauffage', aj. *keuneudek* 'abondant en bois à brûler'.
  A. aj. Kenidjack
***cun-jy** (m.) 'kennel', a compound of **ky** pl. and **chy**; Welsh *cyndy*, *cwndy*; cf. Breton Guern-Condy (Ogam 7 (1955), 153).
  A. (sg. or pl.) Kingey
cuntell 'collect, gather' (2 sg. ipv.), CW 1293; 3 sg. pres. indic. *kuntel*, BM 1877, *guntell*, CW 1091; Modern Cornish v.n. *kentel*, *centle*, Bosons, pp. 52 and 38; pl. (translating English 'communion' in the Creed) *cuntillian*, Bosons, p. 41, and *kontiliow*, ibid., p. 56; ppp. *cuntullys*, MC 88d and 92b; *kontlez*, Bosons, p. 52. Welsh *cynnull*, Breton *kutuilh*. Nouns derived from this verb include *contulva*, Trg. 31, 'assembly-place' (+ ***ma**; Welsh *cynullfa*), and Old Cornish *cuntellet* glossing *congregatio vel concio* (Voc. 185), Old Breton *contulet*, etc., (FD, pp. 117 and 125): cf. Welsh *cynullawd* 'assembly'. The meaning of the v.n. *cuntell in place-names was presumably 'assembly'.
  C2. Cargentle (+ ***ker** ?; cf. Middle Breton *Kerguntuill*, etc., GMB, p. 138; or see ***cant**); pl. Tregatillian (+ **tre**); in the name Coyskentueles 1337, Coiscuntell 1556 (= Engoyse), the noun, OCo. *cuntellet*, or adjective, ppp. *cuntullys*, was later replaced by the vn. *cuntell, then dropped altogether.

Ch (= /tʃ/)

chammbour 'chamber', OM 2110; *tshombar*, Lh. 52c glossing *cubiculum*. Borrowed from English, like Welsh *siambr* (Parry-Williams, EEW, p. 226); cf. Breton *kambr* (Piette, pp. 42 and 86).
  B2. Chamber Byan (+ **byghan**); Chamber an Tresousse (+ **an**, ?)
  C2. Park an Chamber, fld.; Gulchamber, fld. (+ **guel**)
chapel 'chapel', BM 642, 644, etc.; *tshappal* glossing *sacellum*, Lh. 143a; borrowed from English; cf. Welsh *capel* (a learned borrowing), Breton *chapel*, Middle Breton *capel* (Piette, p. 86).
  B2. Chapel Jane in Zennor (+ ***enyal**); Chapel Ainger (+ St); Chapelkernewyl (+ **kodna huilan** pl.); Chapel-an-Grouse (+ **an**, **crous**); Chapel Maria (+ St); Chapel Curnow (+ **corn** pl.)
  C2. Porth Chapel, *Porthchaple* (both + **porth**); Parken Chapel, Park Chaple, *Park Cheple*, etc., flds.
  D. *Prospidnack-an-Chaple* (= Prospidnick)

chas 'hunting-ground'; Middle Breton *chacc* 'pack of hounds', etc., Piette, p. 88. (Borrowed from English or French. Cf. *ch(e)as* 'chases', etc.: Stokes, Glossary, s.v. *chasye*.) The Cornish element appears only in a series of names around Chacewater, lying in the upland manor of Blanchelande, named in Eilhart of Oberg's 12th-century *Tristrant* as King Mark's hunting-ground in Cornwall, and implied to be King Theodoric's centre for hunting in the Breton *Life of St Ke* (17th-century): G. H. Doble, *Four Saints of the Fal* (Exeter, 1929), pp. 10 and 16.

    A. Chacewater (+Eng.)

    C2. *Penanchase* (+**pen, an**); *Coysynchase* (+**cos, an**); *Goenchase* (+**goon**)

cheer 'chair', BM 3002; *chear*, Trg. 49 (borrowed from English). The native word was **cadar**.

    B2. Chair Ladder, coast (+ *****lether** or **lader**)

    C2. Carn Cheer, coast (+**carn**)

cherhit, see under C/K.

chy (m.) 'house, cottage'; Old Cornish *ti* glossing *domus*, Voc. 744; Welsh *tŷ*, Breton *ti*. The older form appears: (1) in early (12th–13th cent.) forms of some minor names, now *Chy*-; (2) in the names of various manors, given below, where the written tradition has preserved the old form and has influenced the pronunciation; (3) in some compounds where *-dy* (> *-sy*) has not become *-jy* (see below). The palatalisation of *t-* to [tʃ] is unique to this word in Cornish; judging by place-name forms, it occurred in the 13th century: forms in *Ty-* are quite common in that century, but almost unknown thereafter (except in the few manorial names), while forms in *Chy-* first appear in that century. In initial position, **ti** became **chy** directly; but medially, in compounds, the progression was *-dy* > *-sy* (= /ʒi/?) > /dʒi/.

The element as now represented is a late one: though extremely common in the four western hundreds, it is unknown in the five eastern ones (where there is a single instance of *Ti-*, now Trethevey in St Mabyn). Moreover, even within the western half the density of *Chy-* names increases westwards, being higher in the hundreds of Penwith and Kerrier. Approximate proportions are of the order: Pydar, 7; Powder, 18; Kerrier, 33; Penwith, 41. (These figures are all minimal, based on names surviving today on the map: full numbers will eventually be much higher.) Such a distribution is typical of a late, minor-name element, and it is clear that that is

chy

what **chy** represents, denoting as it does a humble dwelling, 'cottage', very much the equivalent of English *cot, cote*. In fact, the parallel extends further, for, like *cot*, **chy** is frequently used (in various compounds, listed below) for buildings which were not dwellings, but were used for some agricultural or industrial purpose; also like *cot*, Old Cornish **ti** appears in the names of a few Domesday manors.

Chy- names are for the most part readily comprehensible: those where the qualifier is obscure are much rarer than with *Tre*- or other elements, and that is in keeping with its use as a late, minor-name, element. The vast majority of *Chy*- names are qualified by a word (or, often, a prepositional phrase) denoting the location of the dwelling, and that, too, is in keeping with its use for minor names. Personal names are not very common as qualifiers.

The near-total absence of *Ti*- or *Chy*- names from the eastern half of Cornwall is surprising, even so. It is unlikely that **ti** was not used during the period when Cornish was spoken there (say, up to the 12th century over most of the eastern half): Asser's *tig-guocobauc* 'speluncarum domus' (= Nottingham) shows the element in use in the early 10th century, and there is no other common Cornish element for 'cot' in either east or west Cornwall. The implication must be that the names of such minor dwellings and tenements tended to be rather short-lived, especially if dependent purely on oral, not written, preservation, and that they were liable not to survive the change to English-speaking; the *Chy*-names in west Cornwall, on the other hand, were current in spoken Cornish at a time when written records ensured their continuance after the death of the language. Nonetheless, even that does not fully explain the lack in east Cornwall, for the occasional Cornish field-name does occur there, and one would expect minor tenements to have a survival rate at least equal to that of field-names. Nor is there any evidence for widespread translation of such names, which might have rendered them unrecognisable. (Some Old Cornish *Ti*- names may have become *Tre*-, like Trethevey in St Mabyn, before the earliest written records, which would partially explain the lack in east Cornwall; but it is insufficient to explain it entirely.)

Chy is frequently followed by the definite article before the qualifying noun, expecially in the mediaeval period (13th–14th centuries, often spelt *en* at that date), and the definite article often comes and goes within a single name over the centuries, so that it is impossible to say whether a name 'really' contained it or not: see further remarks s.v. **an**.

As a separate word, **chy** was pronounced /tʃʌi/ in the Modern Cornish period (e.g. *tshyi*, Lh. 55c; *tshei*, Lh. 251a; *choy*, Bosons, p. 27; *chuy*, ibid., p. 42); however, when unstressed, as generic in the normal (unemphatic) pronunciation of place-names, correct usage is normally 'Che-' or 'Sha-', or the element may even be omitted altogether, as in 'Zoister' (= Chysauster): cf. Loth, RC 35 (1914), 151f.

Compounds containing **chy** as second word are mainly listed as separate elements. In certain compounds, notably some of those where the first element ended in *-n*, the palatalisation never occurred, and the words evidently contained *-d-* in the spoken language. (The intermediate stage, spelt with *-s-*, also survives in some compound names.) The following have been noted: **boudzhi** 'byre'; ***clav-jy** 'sick-house, lazar-house'; ***krow-jy** 'hut'; ***cun-jy** 'kennel'; ***dyjy** 'cottage' (?); ***goon-dy** 'moor-house'; ***gre-dy** 'herd-house' (?); **gwesty* 'inn' (s.v. **gvest**); **hensy** 'ruin'; ***lety** 'dairy'; ***meyn-dy** 'stone-house'; ***melyn-jy** 'mill-house'; ***mon-dy** 'ore-house'; see also **agy** 'hither, within'.

B1. Levalsa (+**aval**?); Rinsey (+***rynn**); ?Erisey (+?)

B2. In manorial names. Tywardreath (+**war, trait**); Tywarnhayle (+**war, an, *heyl**); Tybesta (+?); Degembris (+pers.); Trethevey in St Mabyn (+**war, *dewy**).

In other names. Qualified by location: Chytane, Chyton, Chytan (2 exx.), Chytodden (4 exx.), Chiverton (3 exx.), Chyverton, Chyvarton (all +(**war, an**), **ton**); Chywoon (4 exx.), Choone (2 exx.), Chyoone, Chûn (all +(**an**), **goon**); Chyreen (2 exx.), Chirwyn (all +(**war, an**), ***run**); Chycoose (+(**an**), **cos**; 4 exx.); Chenhall (3 exx.), Chynhale (2 exx.), Chyenhâl, Chenhale (all +**an, hal**); Chinalls, Chenhalls, Chynhalls (all +**an, als**); Chycarne (+(**an**), **carn**; 2 exx.); Chynance (+**nans**; 2 exx.); Chyangwens (+**an, guyns**); Cheesewarne, *Cheiensorn* (both +(**an**), **sorn**); Chingweal, Chy-an-Gweal, Chyangweale (all +**an, guel**); Chyrose (+***ros**); Chyvarloe (+**war, *loch**); Chypraze (+**pras**; 2 exx.); *Chyporth* (+**porth**); Chylason (+(**an**), ***glasen**); Chyandour, Chyandower (both +**an, dour**); *Chywarmeneth* (+**war, meneth**); *Chiawel* (+**awel**); Chyrase (+**rid**).

By a neighbouring structure: Chitol (+**toll**); Chyvogue (+**fok**);

*daek

Chypons (+**pons**; 6 exx.); Chypit (+**pyt** ?); Chyanvounder (+**an, bounder**); Chyandaunce (+**an,** ***dons**); Swingey (+ ***goon-dy**)

By other quality: Chynoweth (+**nowyth**; 7 exx.); Chygwyne, Chegwidden in Constantine (both +**guyn**)

By a personal name, or other word denoting a person or animal, etc.: Chybarrett, Chycowling, Chyweeda, Chepye (all +pers.); Chysauster (+*Sylvester*); Chykembro (+***Kembro**); *Chestewer* (+ *Gwas Dewy: v. **guas**); Chirgwidden (+**gour, guyn**); *Chymblo* (+**an, blogh**); Chymder (+**midzhar** ?); Chybucca (+**bucka**); Chyverans (+**an, bran**); Chyanhor (+**an, horþ**)

C2. *Hwelan Tshei*, mine (+**wheyl, an**)

D

***daek** (disyllabic), element of unknown meaning; it might be an adjectival derivative of *da* 'good, goods', perhaps meaning 'full of riches'. Some of the names listed under **teg** 'beautiful' might also belong here instead. Alternatively, it might be a personal name: compare Gaulish personal names in *Dago-*, etc. (Evans, GPN, pp. 188f.).

C2. Trethake (2 exx.), Tredeague, Tretheague, Tretheake (all +**tre**); ?Rosteague (+ ***ros**)

dall 'blind, unseen', MC 192b and 220c; *dal*, Voc. 373 glossing *cecus*. Welsh, Breton *dall*.

B2. (as a noun) Dal Jo (+**du**), a hidden rock.

C1. ?*Dal-thour (+**dour**), used as C2 perhaps in Bydalder (+**myn**)

dan, preposition, 'under' (usually *yndan* in the texts); Welsh, Breton *dan*. It occurs in such field-names as *Dannandre*, *Park dannandre* (+(**park**), **an, tre**).

dar 'oak-tree', Voc. 675 (*glastannen vel dar* glossing *quercus vel illex*), pl. *dery. Welsh *dâr*, pl. *de(i)ri* (*deri emreis* 'Ambrosius' oaks', LLD, p. 42); Old Breton *dar* (FD, p. 129); Breton pl. *diri in place-names (Tanguy, p. 105). There was also a set of forms with the stem **derw-* (see BBCS 28 (1978–80), 551, n. 5): pl. *derow*, OM 1010; Welsh *derw*, sg. *derwen* (Derwgoed 'oak-wood', Mer.; *pul ir deruen* 'the pool of the oak', Chad 6), Breton *derv*, sg. *dervenn* (cf. GMB, p. 152). The Breton derivative *dervoed (Tanguy, ibid.) is paralleled by a Cornish name which appears to contain *derves (+ *-*es* < *-*et*). Oak-names are surprisingly rare in

Cornwall (even when those containing **glastan** are counted as well), considering that oak is 'very common' throughout the county (Davey, Flora, p. 408).
    C1. See ***darva** (+ ***ma**).
    C2. pl. *dery, Eglosderry (+**eglos**), ?Tendera (+ ***dyn**); pl. *derow*, Nanterrow (+**nans**); deriv. *derves, ?Pendarves (+**pen**)

daras 'door'; *darat*, Voc. 763 glossing *hostium*. The Cornish word (VKG, II, 36; RC 50 (1933), 257) is without Brittonic cognates, except for Breton pl. *dorojoù*, etc. (PB, p. 90). The use, restricted to field-names, is probably borrowed from English Foredoor Field, etc.
    C2. Park Darras, etc. (+**park**); Gweal Darras, *Gweal Darros*, ?Gold Arish (or Eng.) (all +**guel**)

darva** 'oak-place' (?), from **dar**+ma**.
    D. ?Penhaldarva

daves (f.) 'sheep', OM 127; *dauas*, OM 2230; pl. *deues*, PC 894, *dewysyov*, BM 2981. Old Cornish *dauat*, Voc. 602 glossing *ovis*. Welsh *dafad*, Breton *dañvad*. For *an navas cf. Middle Breton *han affuat* 'and the sheep (sg.)' (Trois Poèmes, p. 155), and Cornish *an nor* 'the earth', OM 272; cf. ÉC 10, 281-2.
    A. Davas, rock
    C2. *Parken Davers, Park-an-Davers*, ?Park-Nevas, flds.; Porth Navas (+**porth, an**); pl. ?Toldavas (+**tre**, ?), Park an Devers, fld., Trethevas (+**tre**)

deyl (pl.) 'leaves', OM 254; *deel*, CW 858; sg. *delen*, Voc. 672 glossing *folium*; double plural *delyow*, OM 30, *dylyow*, OM 777. Welsh *dail* (double plural *deiliau*), Breton *deil*. An adjective *delyek, corresponding to Breton *delieg* 'leafy place' (Tanguy, p. 101), might theoretically exist, but in practice the names are more likely to contain **teil** aj.
    A. ?Deli (+ ***-i** ?)
    B2. ?Delawhidden (+ ***gwythel** ?)
    C2. ?Trendeal (+ ***dyn**); double pl. Pennatillie (+ ***pen nans**), Nantillio (+**nans**)

den (m.) 'man', Voc. 18 glossing *homo*; etc. Welsh *dyn*, Breton *den*. See also **tus**, the plural of **den**.
    A. The Dean, rock
    C2. Ponsandane (+**pons, an**)

denjack (MnCo.) 'a hake fish' (Borlase); *denshoc dour*, Voc. 556 glossing *luceus* (OE *hacod*), dialect *tinsack* (Nance, Sea-Words,

dentye

p. 72). Literally 'toothy' (cf. Nicholas *Densc.*, 1327 SR); Welsh *deintiog*, Breton *dantek*. The word also meant 'pike' (fish), and 'pike's head' could be the meaning of the field-name.

C2. Pedn Tenjack, fld. (+**pen**)

dentye 'dainty', CW 1456 (borrowed from English).

C2. *Taban Denty*, fld. (+**tam**)

*derch (OCo.) 'bright, clear', equivalent to Old Breton *derch* 'face' (FD, p. 135; cf. pp. 162f.), *-derch* in personal names (Chrest., pp. 123 and 201); Middle Breton *derch* 'pure' (GIB); cf. Welsh *ardderchog* 'excellent'. If present, it could be a stream-name, like the English Dork (Ekwall, ERN, pp. 128f.); but the name could alternatively contain **torch**.

C2. ?Reterth (+**rid**)

dev ($v = u$) 'god', PC 3, etc.; *duy*, Voc. 1 glossing *deus*; plural *dewow*, OM 178, etc.; *dewyow*, CW 812, Pryce, p. Ee3r. Welsh *duw* (pl. *duwiau*, etc.), Breton *doue* (pl. *doueou*, etc., PB, p. 81). In the place-name the pl. should perhaps be understood as 'fairies'.

C2. pl. Carnsew in Mabe (+**carn**)

*devr- 'water': this stem is presumably a derivative of **dour**, but the vowel is unexplained. It appears in Lhuyd's *devrak* in *tîr devrak*, 112a glossing *palus* 'marsh, bog'. The following words containing it appear to exist; the two plural forms probably have the singular meaning 'water-course', as the Breton plural does (PB, p. 267). Neither **devryon* nor **devren* is likely to be an equivalent of Welsh *dyffryn*, which one would expect to have kept its original *-nt* (> *-ns*) in Cornish.

A. aj. (+ **-yel*) Deveral, Derval (Welsh *dyfr(i)ol*; cf. Deverill, PNWlt., p. 6); pl. **devrow*, Ardevora (+***ar**); pl. **devryon*, Devoran, ?*Cendefrion* (v. ***kendowrow**): cf. Lindifferon, Fife (CPNS, p. 383)

C2. deriv. **devren* (?), Pennydevern (+**pen**); ppp. **devrys* 'watered' (cf. Breton *dourañ* 'to water'), Treheveras (+**tre**)

dew 'two', fem. *dyw*; Welsh *dau*, fem. *dwy*; Breton *daou*, fem. *div*. The form *deu* occurs in Old Cornish, *deumaen* (Sawyer, nos. 832/1027). Cf. Old Welsh *dy'r dou pull* 'to the two pools' (LLD, p. 143); Old Breton *pagus daudour/doudur* 'two-waters' (FG, p. 258). Welsh (Aber) Daugleddyf; (Aber) Deunant (ELlSG, p. 69); Dwyryd (ÉC 10, 215). In Irish *dá* 'two' is common in place-names: D. Flanagan, 'A reappraisal of *da* in Irish place-names, I', *Bulletin of the Ulster Place-Name Society*, Series 2,

dy-

Vol. 3 (1980-81), 71-3. In some cases, especially before **nans** 'valley', this element is hard to distinguish from **down** 'deep', but most instances of **Downans* probably contain the latter. See also **dy-**.

C1. Dowgas (+**cos**); Pendoggett (+**pen, cos**); Duloe (+ ***loch**); Duporth (+**porth**); Tretherras, Tretherres (both +**tre, rid**); *deumaen coruan* (+**men**, ?)

*dewy, river-name of unknown meaning; cf. Ekwall, ERN, pp. 6n. and 125; EANC, p. 139; Loth, RC 33, 271n.; compare Vendryes, 'Saint David l'Aquatique', ÉC 7 (1955-56), 340-7. Two rivers, called Dewey and *Deuy* (= Allen) bore this name, which also appears in farms along their courses: on the Dewey, Lantewey (+**nans**) and Castledewey (+**castell**); on the *Deuy*, Pendavey in Egloshayle (+**ben**), *Nansdeuy* (+**nans**), Trethevey in St Mabyn (+**ti, war**) and Pendavey in Minster (+**pen**). Two other farms called Pendavy and Pendewey (both +**ben**?), both at the foot of smaller streams, appear to show further instances of the streamname.

dewthek 'twelve men', PC 228; *dewʒek* 'twelve', MC 72c, *dowʒek* 'twelve men', MC 61a. Welsh *deuddeg*, Breton *daouzek*. Cf. Twelve Men's Moor, and also Twelmin (PNWml., II, 145).

C2. *Wheale an Dowthick*, mine (+**wheyl, an**)

dy-, prefix, either intensive or negative: Welsh, Breton *di-* (cf. FD, p. 136). The privative sense is seen in ***dy-les** and in *dicreft*, Voc. 244, and *dyflas*, PC 2604; and the intensive sense in ***dyserth**. What the sense is in the following three instances is obscure: *difrod* (Sawyer, no. 951) (+**frot**); ?Tregembo (*Trethigember* 1311: +**tre, *kemer**?); Trengwainton (*Trethigwaynton* 1319: +**tre, guaintoin**). However, in the first two it appears to have the meaning 'double': cf. Old Irish *dé-*, Thurneysen, GOI, pp. 242 and 246; VKG, II, 127. Pokorny, Vox Romanica 10 (1948-49), 253f., doubts the existence of such a prefix, and points out that Old Cornish *difrod* is the only powerful evidence for it in Celtic; he suggests that *difrod* could contain privative *di-* instead. However, the boundary-point appears to be at the confluence of two streams. The Welsh and Breton prefixes *di-* discussed by Caerwyn Williams, BBCS 16 (1954-56), 105-8, do not help with the Cornish place-names, though one might note Morris-Jones' *dy-* 'to, together' (WG, p. 266) and Ernault's *di-* 'pour *de-*, "à, vers"' cited by Caerwyn Williams, p. 107. Note also Gaulish personal

diber

names beginning with *Di-* (GPN, p. 193, and references there cited).

diber 'saddle', Voc. 961 glossing *sella*; *deeber* 'saddle', Symonds, 17th century (OC 4, ii (1943), 87). Breton *dibr*.

C2. *Carrack an deeber*, rock (+**karrek**, **an**)

***dyjy** 'small cot or farmstead', dialect *dijey* 'small farm' (Courtney and Couch, p. 17). Perhaps a compound of **chy**. Two names could represent the dialect word instead.

A. The Digey; *The Degey or Diggey*

***dy-les** 'profitless', from **dy-**, negative prefix, and *les* 'benefit' (Voc. 323, glossing *commodum*; etc.); Welsh *diles*.

C2. Trelease in St Hilary (+**tre**)

***dyllo** 'lively, active' (?), equivalent to Breton *dillo* (but that is not attested before the 19th century). Presumably as a personal name or nickname.

C2. ?Trethella (+**tre**)

***dyn** (m.) 'fort'; Welsh *din*, Old Breton *din* (FD, p. 143); Elements, I, 133, 139; II, 180; JEPNS 1, 46. The variant **tyn*, as in Welsh Tenby, Tintern, etc. (Lh. 5c, 16c; cf. VKG, I, 493–5, and M. Richards, ÉC 13 (1972–73), 367–77) occurs in several Cornish names: theoretically it is confusable with **tyn** 'rump', if that was used as generic in place-names; but in practice most names can be shown to contain **dyn*, either because they have forms in both *D-* and *T-* (as do Tintagel, Treglyn, and Trendeal) or because they are situated beside forts (as are Trenarren and Tendera, both attested with *T-* only). The meaning was usually either 'hill-fort' or 'cliff-castle'; it should normally mean something more substantial than a 'round' (v. ***ker**), but in several cases there is nothing now to be seen. For the sense 'refuge' see Jackson, CMCS 3 (1982), 33f. Those sites where a fort is visible are followed by an asterisk.

B1. Meudon (+ ?); see also ***merthyn** and s.v. **horþ**.

B2. Dunmere* (+**meur**); Dunveth, Denby (+***bich**: the lenition in the first name is odd, but cf. possibly Welsh Dindryfol (+*tryfal*?), ÉC 13, 370); Demelza*, Domellick* (both +pers.); Tintagel* (+***tagell**); Tinten (+ ?; cf. Tintinhull, Som., BBCS 15, 19); ?Treglyn (+***clun**?); Trenarren* (+**garan**); Trendeal (+**deyl**?); Tendera* (+**dar** pl. ?)

C2. Pendeen* (2 exx.), Pendine (all +**pen**); Treen* (+**tre**; 2 exx.); ?Retyn (+**rid**)

An aj. *_dynek_ occurs, but after **tre** it is hard to distinguish from **eithin** aj. and **reden** aj., both of which are liable to give -_dinnick_: even when no forms such as _Treveythynek_ or _Treredenek_ appear, the extra syllable may have been lost before the date of the earliest forms. But ***dyn** aj. is likely in the following names.

C2. Bodinnick in St Tudy (?) and Lanteglos by Fowey (both + ***bod**); Pordenack Point (+**pen**); ?Mendennick (+**myn**?); Tredinick in St Breock (+**tre**)

***dynan** 'fort', a diminutive of ***dyn**. Welsh Dinan (ELl, p. 79; M. Richards, ÉC 13 (1972–73), 378f.), Breton Dinan (Top.Bret., p. 43).

A. Dinham

C2. Cardinham (+ ***ker**; 2 exx.); Tredenham (+**tre**; 2 exx.)

***dynas** 'fort', a derivative of ***dyn**; Welsh _dinas_.

A. Dennis; Dinness; Denas; Andennis (+**an**); St Dennis (see Thomas, Co.Arch. 4 (1965), 31–5); Dennis Head; Dennis Point; Dinas Cove; Dinas Head

B2. _Dynas Ia_ (+St: = St Ives Head); _Denys Azawan_ (+**asow** sg.); _Dennis Cockers_ (+ ?)

C2. _Pendinas_ (= St Ives Head), Pendennis Castle (both +**pen**); Hall Dinas, coast (+**hal**)

dyner 'penny', PC 505; _dinair_, Voc. 915 glossing _nummus_ (see Jackson, JCS 1, 74; Campanile, ZCP 33, 23). Breton _diner_.

B1. _Trydinner_, mine (+**try**[1]); Padge Dinner, fld. (+**peswar**): cf. Threepenny Close, Sixpenny Close, etc. (Field, EFN)

C2. _Wheal an Dinner_, mine; pl. ?Polandanarrow Lode (+**pol, an**)

dyowl 'devil', OM 301, etc.; Old Cornish _diauol_, Voc. 388. Welsh _diawl_, Breton _diaoul_.

C2. Stamps and Jowl Zawn, coast (+**stampes, an**)

***dyppa** 'small pit', borrowed from English dialect _dippa_ 'a small pit', Symons, p. 141. As an English word, it occurs in e.g. Dippa Meadow, fld.

C2. Maze Dippa, fld. (+**mes**)

***dy-serth** 'very steep', from **dy-**, intensive prefix, and **serth**. The element is theoretically confusable with a word equivalent to Welsh _diserth_ 'wilderness, hermitage', from Latin _desertum_: compare Richards, SC 3 (1968), 13, and Breton place-names _Désert_, etc. (Gourvil, Ogam 7 (1955), 152); also the common Irish element _diseart_ (Hogan, Onomasticon, pp. 345–7). But in practice the

*dywys

topography dictates the sense, and there is no evidence for a Cornish **dyserth* 'wilderness, hermitage'. In Wales, note Diserth, Fli., at the foot of a steep hill (SJ 0579).

A. Dizzard

*dywys 'burnt', ppp. of *dywy* 'burn', OM 1397 (cf. PC 693 and 1221), Welsh *deifio*, Breton *deviñ* (ppp. *devet*). Although all three occurrences of the word in the plays show -*w*-, the place-names show -*v*-. For the discrepancy cf. Hamp, ÉC 14 (1974-75), 465, and Sims-Williams, BBCS 29 (1980-82), 217 and n. 11. As A, it is a noun, 'burnt (place)'. Some of the names might contain **tevys** instead.

A. Devis (+**an**); Davis Farm
C1. Devichoys Wood (+**cos**)
C2. ?Cuttivett (+**cos**: or **tevys**); Tredivett (+**tre**)

doferghi (OCo., *f* = /v/, *gh* = /g/) 'otter', Voc. 573 glossing *lutrius*, a compound of **dour** and **ky**; Welsh *dyfrgi, dwrgi*, Breton *dourgi* (Tanguy, p. 119). In the place-name this is almost certainly used as a pers., like the frequent Old Irish pers. *Dobarchú* (cf. Loth, RC 37, 205n.).

C2. Trethurgy (+**tre**)

*dons 'dance', borrowed from English; Welsh *dawns*, Breton *dañs* (Piette, p. 101). The corresponding verb occurs: *downssya*, CW 2547; *thonssye*, RD 2646. In the first two names 'dance' refers to a stone circle: see Bosons, p. 10, and Padel, Co.Studies 3 (1975), 23.

B2. Dawns Men (+**men** pl.)
C2. *Myne an Downze*, mine (+**men** pl., **an**); Chyandaunce (+**chy, an**); Park an daunce, fld. (+**park, an**)

dor (m.) 'ground', OM 64, etc.; *doer*, Voc. 12 glossing *terram*. Welsh *daear* (South Welsh *dâr*, WG, p. 100), Breton *douar* (see HPB, p. 231); Old Breton *doiar* (FD, p. 148). Note that the disyllabic spelling of Voc. appears in the 14th century in place-names.

A. Doar
B2. *Dormayne* (+**men**); Dorminack (+***meynek**); Dorheere, *Doerhyr, Doreheer*, flds. (all +**hyr**); Dor-se, fld. (+**segh**?); *Doerpoys*, fld. (+***poth**?); *Dorcoath*, flds., Dolcoath (2 exx.) (all +**coth**); Derese, fld. (+**eys**)

dorn 'fist, hand', PC 657, etc. Welsh *dwrn*, Breton *dorn*. See Jackson, JEPNS 1, 47, and Rivet and Smith, PNRB, p. 345 for its occurrence and meaning in Brittonic place-names in England; also

Dornock, Dornoch, etc., in Scotland (Watson, CPNS, pp. 182 and 488). The derivative *dornel (+ *-**yel**) used as C2 could be a nickname or a true personal name, similar to Gaulish *Durnacus* (ACS, 1, 1382).

C2. ?Tredorn (+**tre**?); aj. Botternell (+***bod**)

dour (m.) 'water', Voc. 728 glossing *aqua vel amnis*; note *dour tyber* 'River Tiber', RD 2136, and *thour cedron* 'River Cedron', OM 2804; *dowr*, MC 211a; *douer vel dur*, Voc. 855 glossing *aquam*. Welsh *dŵr*, Breton *dour*. The normal pl. was presumably *dowrow, *dowryow, as in Breton (PB, p. 159), but that occurs only in a late translation of Genesis, Book 1 (D. Gilbert, *The Creation of the World* (London, 1827), p. 189), and possibly in one place-name (Poldower). But there seem to have been other plurals, using a stem ***devr-**, which is unexplained.

A. aj. (Welsh *dyfr(i)og*, Breton *dourek*; but cf. *devrak* s.v. ***devr-**) Dowrack, fld., pl. ?Dovrigger, fld.

B1. Ruthdower (+**ruth**); *Candern Water*, Candor (both +**cam**; cf. Welsh *Camddwr*, EANC, p. 46); Gwarder (+ ?); Bydalder (+**myn, dall**? Formerly *Mindaldur*, Ekwall, ERN, p. 292); see also ***boðour**.

B2. Dowreth (+**ruth**); *Endourbyhan* (+**an, byghan**); *Dower Meor* (+**meur**); *Dower Ithy* (+ ?); *Dour Conor*, *Dour-Tregoose*, *Dourragenys* (all +p.n.).

C1. Durloe; Durla, Durlah, flds. (all +**le**); ?Durva, fld. (+***ma**)

C2. Chyandour, Chyandower (both +**chy, an**); Maen Dower, coast (+**men**); Ogo-Dour Cove (+**googoo**); Pendower in Philleigh (+**ben**); *Parkandower*, Park-a-dour, flds.; sg. or pl. Poldower (+**pol**); aj. Restowrack (+***ros**)

***douran** 'watering-place', a derivative of **dour**+*-**an**².

A. Dowran

C2. Poldowrian (+**ben**)

down 'deep'; Welsh *dwfn*, Breton *don*. Both the Welsh and Breton words are used also as nouns, 'depth', and the one or two Cornish instances of **down** as a generic must also represent such a usage. Note the epenthetic vowel in *Dofen soðo* A.D. 960 (Sawyer, no. 684) and *Duuenant*, etc. 1086 (LHEB, pp. 337f.); but then the *v* became *w* and the vowel was lost again, giving **down**, by the 14th century.

B2. Downathan (+**eithin**?); ?Downinney (+ ?)

C1. Dansotha (+***soð** pl. ?); see also ***downans**.

*downans

C2. Pooldown, Pulldown (both +**pol**); Tretawn, Rosedown (both +**rid**: cf. Welsh Rhyd ddofn, ÉC 10, 212); ?Kildown Cove (+***kyl**?)

*downans 'deep-valley', Welsh *dyfnant*; Old Welsh *dubnnant du* 'black deep-valley', LLD, p. 172 (but also *nant duuin* 'deep valley', ibid., p. 78). Not easily distinguished from **Dew nans* 'two valleys' (v. **dew**), but in practice most names probably contain **down**. Most names show loss of the *-n-* in *-nt/-ns*, even as early as 1086 (though in that instance the loss could be scribal), but the *-n-* can reappear at a later date; cf. **car-bons*, etc.

A. Dannett, Dawna, Downas Valley, *Dounans*; Dannon(-chapel, etc.)

dreyn (pl.) 'thorns, thorn-bushes', PC 2119, etc.; *drein*, Voc. 695 glossing *sentes*; *dreyne ha spearn*, CW 1091. Welsh pl. *drain*, sg. *draenen*, aj. *dreiniog*; Breton pl. *drein*, sg. *draenenn*, aj. *draenek*, *dreinek* (cf. Tanguy, p. 113). The Cornish singular would have been **dr(a)enen* 'thorn-bush', and the adjective **dreynek*. (But 'thorn' singular is *drain*, Voc. 698 glossing *spina*, Welsh and Breton *draen*.) There are two words for 'thorn-bushes', **dreyn** and **spern**; and two types of thorn-bushes, sloes (or blackthorns, *Prunus spinosa*) and hawthorns (or whitethorns, *Crataegus monogyna*). Both species are 'very common' throughout Cornwall (Davey, Flora, pp. 139 and 185). In the case of *cruc drænoc*, **dreyn** almost certainly indicates sloes, while **spern** can mean either: it is possible (though no more than that) that **dreyn** generally means 'sloes', and **spern** generally means 'hawthorns'.

A. aj., meaning 'spinney', Drinnick (2 exx.), Drannack, *Draennek*; deriv. **dreynes* (cf. **-ys²*), Draynes

C2. pl. Trendrine, Trendrean (both +**tre**, **an**); Landreyne, Landrine, Landrends (all +***lann/nans**); Hendragreen (+ **hendre*); *Tereandreane* (+**tyr**, **an**); ?Halldrine Cove (+**hal**?); Park-an-Drean, fld.; sg. Trendrennen (+**tre**, **an**); aj. Coldrinnick (2 exx.), Coldrenick (all +***kyl**), *Gueldrenek* (+**guel**), *cruc drænoc* (+**cruc**: Sawyer, nos. 832/1027); deriv. **dreynes* (cf. **-ys²*), ?Lestrainess (+***lys**?)

dreys (pl.) 'brambles', Trg. 9; *dreis*, Voc. 699 glossing *vepres*; MnCo. *drize*, Pryce, p. Ff3r. Welsh *drys*, *drysi*, Breton *drez* (cf. Tanguy, p. 113). The diphthong in the Cornish forms is unexplained, but is confirmed by the early forms of several of the place-names; cf. HPB, pp. 93–4. Other Cornish words for 'brambles' are moyr- and **spethas**. Many types of brambles are found all over Cornwall (Davey, Flora, pp. 142-66, Suppl. pp. 49-52).

A. aj. *Drisack*, fld.

C2. Lantreise (+**nans**); *Park-an-Drise*, fld.; aj. (Welsh *drysog*, Breton *drezek*) Nantrisack, Nancetrisack (both +**nans**), Tredrizzick, Treisaac (both +**tre**), Striddicks (+**rid**?), *Gwaeldreysec*, *Gweal Drisack*, flds. (both +**gucl**), Pendrissick, fld. (+**pen**), Ventontrissick (+**fenten**), Park Drysack, fld.

dres 'across, beyond'. Welsh *dros, traws*; Breton *dreist*. Compare the Welsh place-names Trawsfynydd, Trawsgoed, and Y Drostre, Bre., Trostre, Mon. (LLD, p. 321); also Troustrie, Fif. (CPNS, p. 350); see PKM, pp. 239f. So far found only with **cos**, 'beyond the wood', which occurs three times: Pendriscott (+**pen**); Caduscott (+ ***kyl**?); Truscott. The last name might theoretically be English, *trūs*+*cot* 'brushwood cottage' (so Elements, II, 188); but as English *trūs* is rather rare in place-names (Trusley, PNDrb., III, 613; Trewsbury, PNGlo., I, 68; Trussenhayes, PNWlt., p. 153; *Trussmore*, fld., PNGlo., IV, 179), and the farm is 'beyond the wood' from Launceston, the Cornish derivation is more likely.

***drum** 'back, ridge', Welsh *drum, trum* (ELl, p. 29; ELlSG, p. 88; PNFli., p. 168; PNCmb., I, 124 and 139f.); Old Irish *druimm*. The Cornish names imply ***drum** rather than *trum, except for ***hyr-drum**, which presumably has lenited *tr-*.

A. Drym; ?Drum Head

B1. See ***hyr-drum** 'long-ridge'.

C2. Pendrim (+**pen**); ?Menadrum, fld. (+**meneth**); Trethem (+**tre**)

du 'black'; *duw*, Voc. 483 glossing *niger*; Welsh, Breton *du* (FD, p. 153; VSBG, p. 176n.). Note the remarkably archaic form *Carduf* 1327 (= Carthew in St Austell): cf. LHEB, pp. 415-16, and HPB, pp. 609-10.

C2. Qualifying hill and rock words: Menadue (5 exx.), Menerdue, Menadews (all +**meneth**); Pen Diu, coast (+**pen**); Toldhu, coast (+**toll**); Tater-du, rock (+ ***torthell**); Carn Du, Carn-du, Carn-du Rocks (all coastal), Carnsew in St Erth (all +**carn**); Carrick Du, coast (+**karrek**); *Crubzu* (+**krib**); Maen-du Point (+**men**)

Qualifying words for water: Lanjew (+**lyn**); Barlandew (+**bar, lyn**); Poldue (2 exx.), Poldew (2 exx.), Poldhu Cove, Polsue in St Erme, St Ewe and Goran (all +**pol**)

Qualifying **nans** 'valley': Lanyew, Landue (or ***lann**?)

Qualifying words for archaeological features: Cardew (2 exx.), Carthew (4 exx.) (all + ***ker**; cf. Cardew, PNCmb., I, 131f.); Creegdue (+**cruc**); *Kylkethewe* (+ ***kylgh**)

Qualifying words for dwellings (where **du** may refer to a person): Trethew, Trew in Breage (both +**tre**); Polsue in Philleigh (+ ***bod**)

Qualifying other words: Baldhu (+**bal**; 3 exx.); Gwealdues (+**guel**); ?Kedue, fld. (+**kee**?); Haldu (+**hal**); *Opetjew*, street (+ ***op**)

D. *Argeldu* (= Argal; also *Argelwen*, +**guyn**); *Hay thu* (= Blackhay; also Whitehay)

E

ebel (m.) 'colt, foal', PC 177, etc.; *ebol*, Voc. 520 glossing OE *fola*; pl. *ebilli*, Pryce. Welsh *ebol*; Breton *ebeul*. For the simplex used as a rock-name, compare other animal-words so used: **mols, bogh, horþ**, and perhaps **brogh**.

A. Ebal Rocks

C2. *Park an Eball*, fld.; pl. *Poole-an-abelly* (+**pol, an**; 'alias the Poole where the colts doe drink'), ?Menabilly (+**men**: or **meneth**+ ***byly**?)

-ek, adjectival ending; in place-names it can form a feminine noun meaning 'place of'; nearly always spelt *-oc* (= /œg/) in Old Cornish, e.g. *galluidoc*, Voc. 245 glossing *potens*, Middle Cornish *gallosek*, and *drænoc* 'thorny', Sawyer, nos. 832/1027; but note *morgeonec*, ibid. (11th century). Welsh *-og* (feminine when used as a noun: I. Williams, BBCS 16 (1954–56), 28), Old Welsh *-auc*, Middle Welsh *-awc*. Breton *-ek/-eg* (with plural by adding *-ier*, giving *-eier*: Hemon, Grammar, p. 33), Old Breton *-oc* (FG, pp. 244 and 342). See further PNDinas Powys, pp. 41f.; Williams, ELl, p. 67; EANC, p. 1; PNRB, p. 357a; VKG, II, 30; Hubschmied, RC 50 (1933), 254–60; Quentel, Onoma 24 (1980), 45–50. The adjectives, once formed, may act as nouns and be used to form simplex place-names, as well as in combination with other elements. The following words cited as elements are originally adjectives containing this suffix: ***bannek, barthesek, cronek, denjack, marrek, *meynek, *melek, mosek, ownek**, and ***pennek**; and the following elements also have adjectival forms using this suffix: **aidlen** (?), ***arð** (?), **bagyl, banathel, beler, bregh** (?), **bron, bronnen, *cant** (?), **carn, castell, keber** (?), **kee** (?), **keyn, kelin, kenin, *keun, *clun** (?), ***cnow, *collwyth, *cors, *dyn, dour, dreyn, eithin, elaw** (?), **elester, *fry** (?), **goth¹, gwels, guern, guibeden, guyn** (?), **guyns, gwyth¹, haf, hal, horn, les, leth** (?), **lyw, logaz, lost** (?), **lowarn** (?), ***lus** (?), **manach, melhyonen, meneth, mes** (?), ***mon², *moryon, *neved, onnen, pin-, pry, reden, scawen, schus** (?), **sevi, spern, stean, tal** (?).

eglos (f.) 'church', Voc. 745 glossing *ecclesia*, etc. Welsh *eglwys*; Breton *iliz*, rare in place-names, and occurring mainly as C2: Flatrès, Word 28 (1972), 67. For a list of examples, and discussion, see Padel, 'Cornish names of parish churches', Co.Studies 4/5 (1976–77), 15–27. As generic, **eglos** is usually followed by the name of the patron saint of the church; such phrases did not usually constitute true place-names, but rather phrases in the language, referring to the church site. They die out from east to west along with the language, and only one (Egloskerry) has survived as a place-name today. These phrases could apply to churches that had names in *Lan-*. Alternatively, **eglos** could be used with a descriptive second element to make a true place-name; this usage probably goes back to an early period, since Egloshayle (= Maker) is unlikely to have been formed later than the 10th century.

B2. *Eglosmadern, Eggloscraweyn, Egglostetha*, Egloskerry, etc. (all + St); Egloshayle (2 exx.), *Egloshayle* (= Phillack) (all + ***heyl**); Eglarooze, Eglosrose (both + ***ros**); Egloscrow (= St Issey) (+ **cruc**); *Eglos-Withiel* (= Withiel), *Eglospenbro* (= Breage) (both + p.n.)

C2. Treviglas (2 exx.), Treveglos (4 exx.), Treneglos, *Treneglos* (all + **tre, (an)**); *Foregles*, Wriggles, etc., flds. (+ **forth**); Lanteglos (2 exx.), Nanzeglos (all + **nans**); Carneglos (+ **carn**); Roseglos (+ ***ros**); *Goeneglos* (+ **goon**); *Park Eglos* (+ **park**); *Park an Eglos*, fld.

D. *Withiel-eglos* (= Withiel; also W.-*mur*, W.-*goose*); *Lusuoneglos* (= Ludgvan; also L. Lease: see ***lys**); *Ammaleglos* (= Chapel Amble; also *A.-gres, A.-mur*); Burlorne Eglos (also B. Pillow, B. Tregoose); *Porthillieglos* (= Porthilly; also *P.-gres*)

eys 'corn', OM 1058, 1559; CW 1089; *ys*, PC 881; *yees*, CW 1189. Welsh *ŷd*, Breton *ed*.

C2. Derese, ?Druse, flds. (both + **dor**)

eithin

eithin (pl.) 'furze, gorse', in *bagas eithin* 'a bush of furz', Lh. 56a; *tha trehe ithen* 'to cut furze', Pryce, p. Ff2r; sg. *eythinen* glossing *ramnus*, Voc. 697; *eithinan* 'a furz-bush', Lh. 240c. Welsh *eithin*, Old Breton *ethin* glossing *rusci .i. inculti agri*, FD, p. 168. Without good early forms the plural, preceded by the definite article (*'n eithin*) is confusable with the personal name *Neythan (< *Nechtan); but in practice most of the following names can safely be assigned to **eithin**, because of forms in *-in* or *-yn*. Gorse would have been of use as fodder for animals.

C2. Rosenithon (+ ***ros, an**: cf. *ros ir eithin*, LLD, p. 221); Trenithan (+ **tre, an**; 2 exx.); Carneadon (+ **carn**); Kenython (+ **kee, an**); ?Downathan (+ **down**); Park-an-Nithen, Park an ithan, flds.; aj. (= Welsh *eithinog*), Tredinnick in Morval and Probus, Tretrinnick (all + **tre**: Old Welsh *tref eithinauc*, LLD, p. 126; contrast **reden** aj. and ***dyn** aj.)

***-el**, adjectival suffix: see ***-yel**.

elaw 'elm-trees' (?), Lh. 13c and 175c glossing *ulmus* 'the elm tree' (but **elaw** should be plural); Pryce has *ula* 'an elm', pl. *ulowe* (but that was probably invented to explain a place-name). Both the form and the meaning of this element are thus uncertain: the form, because of the variance, and the lateness, of the words; the meaning, because a closely similar word in Breton means 'poplars': Tanguy, p. 106, has *evleh* 'ormes [elms]' (*evlec'h*, GIB), but also **el(a)ouet* 'poplar-grove', from *efl*, *elv* (*eflenn*, GIB; but cf. Cornish **aidlen**). The confusion is too great to sort out here; moreover, there appears to be some doubt as to whether elms are native in Cornwall, or, if introduced, when (though Davey, Flora, p. 401, says that they are native and common): the earliest record currently known is 'the Ealme garden in Winsor' (Cubert, 1613). All the following names are therefore to be taken as tentative, both as to whether they contain the word, and, if they do, as to whether it necessarily means 'elms', and not (say) 'aspens'. But it is at least possible that most of the following names contain some tree-name, based on a stem **el-*.

C2. pl. *elow (?), ?Crellow (+ ***ker**), ?Trevelloe (+ **tre**); pl. *el-wyth (+ **gwyth¹**), ?Trevella (+ **tre**), ?Bodelva (+ ***bod**); sg. *elowen (?), ?Burlorne (+ ***bod**), ?Prynullin (+ **pren** ?)

***elen** 'fawn' (?), equivalent to Welsh *elain, alan*. Confusable with **elin** 'elbow, corner'.

C2. ?Trezelland (+ ***ros**); ?Bodellan (+ ***bod**)

elerhc (OCo.) 'swans', Voc. 509 glossing *olor vel cignus*. The Cornish word is properly plural; Welsh *alarch*, Breton *alarc'h*. 'Swan' is *swan* in Middle Cornish, OM 133. The Welsh plural is used as C1 in *eleirch vre*, CA, lines 288f.
    A. Elerkey (+*-i?; cf. the Welsh place-name Elerch, Crd., EANC, p. 165)

clester (f. pl.) 'yellow flags, irises', in Voc. 842; sg. *elestren*, Voc. 667 glossing *carex*. Welsh and Breton *elestr*; Old Breton *elestr* glossing *hibiscum* (FD, p. 156). Though originally pronounced /ˈelestər/, the stress in the Cornish plural must have moved to the second syllable, as shown by dialect *laister* 'yellow water iris' (Courtney and Couch, p. 33). The adjectival form, 'iris-bed', was regularly **elestrek* (f.) (cf. Breton *elestrek*, Tanguy, p. 113).
    C2. aj. *Park an Lastrack*, fld.

elin 'elbow, nook', Voc. 74 glossing *ulna*, 746 glossing *angulus*. Welsh *elin* (OWe. *elin*, VVB, p. 116); Breton *ilin* (HPB, p. 301); Old Breton *-olin-*, pl. *elinou* (FD, pp. 276 and 157; HPB, p. 296 n. 1).
    C2. Nanjulian (+**nans**)

-ell, diminutive suffix, appears in such elements as **cornell*, **crugell* (see **corn**, **cruc**); in **arghantell* it is perhaps not a diminutive but a stream-name suffix.

***emle** (?), meaning unknown; perhaps a compound of **le**. It can hardly be a plural of **amal*.
    A. Embla
    C2. ?Penimble (+**pen**)

-en, feminine singular ending for plant-names; Welsh *-en*, Breton *-enn*. As well as constituting the singular, this suffix was used to form river-names and place-names (including ones where the plants must have been found in the plural): see **heschen**, **heligen**, **glastan**, and cf. the Welsh river-name Celynen (EANC, p. 107); see also **alaw* and **brenigan**.

***enyal** 'desert, wild', equivalent to Welsh *anial*, *ynial* 'desolate', from **ande-gal-*: Lloyd-Jones, BBCS 1 (1921–23), 3; cf. Williams, CA, pp. 241f.; PT, pp. 103f.; WVBD, p. 13; also Lloyd-Jones, 'The Compounds of *Gal*', in *Féilscríbhinn Torna* (edited by S. Pender, Cork, 1947), pp. 83–9, at p. 84.
    C2. *Begeledniall*, fld. (+**begel**); Chapel Jane in Zennor (+**chapel**); Nangidnall, fld. (+(**goon**), **aswy**)

enys (f.) 'island', Lh. 19a; Welsh *ynys*, Breton *enez*. The meaning of this word when it occurs inland is difficult: it is to be compared with similar usages, not only in Welsh and Breton, but also in Irish (and Scottish Gaelic), and in Old English and Old Norse (Elements, I, s.vv. *ēg*, *holmr*). In general the meaning is 'land beside a river, river-meadow', but sometimes in Cornwall the meaning appears to be 'isolated, remote spot': this occurs also in Breton (Gourvil, p. 68). For the meaning 'river-meadow, land partly cut off by water', cf. Owen, Pembrokeshire, II, 423; Loth, RC 35, 289f.; RC 46, 161f.; ELISG, p. 117; ELl, pp. 36f.; Bowen, Settlements, p. 115; and, in Breton, ÉC 18 (1981), 371f. In the following list, inland examples are starred; they are all farms, and all are either between or beside streams.

A. Ennis* (4 exx.), Innis*, Enys*, Ennys*; Ninnes* (3 exx.), Ninnis* (4 exx.), Ninniss Farm* (all +**an**); Ince, Inswork (+Eng.); The Enys (2 exx., islands); Enys Head; Raginnis (+**rag**, = 'opposite the island', not 'offshore island' as in Welsh; cf. Gourvil, no. 1889, Raguénès); St Agnes, island (Scilly: + ?)

B1. Molinnis* (+ ***moyl***)

B2. Enniscaven* (+**scawen**); *Ennisveor** (+**meur**); Ennisworgey* (+pers.); Enysmannen* (+**amanen**); Enestreven* (+ ?); *Enyshall** (+**hal**); Savath* (+**margh** pl.); Enys Dodnan, rock (+**ton**); Enys Vean, rock (+**byghan**); Illiswilgig, rock (+**gwels** aj.); Innis Pruen (= Mullion Island; +**prif** deriv.); *Inisschawe* (= Tresco; +**scawen** pl.)

C2. Gooninis* (+**goon**); *Porthennis*, Bosporthennis*(+(***bod**), **porth**); Pen Enys Point (+**pen**); Carn Enys, coast (+**carn**)

D. Carines* (+p.n. *Crou*, **worth**?)

epscop (m.) 'bishop', OM 2601, etc. Welsh *esgob*, Breton *eskob*. A form without the first *p* is also found: *escop*, Voc. 103 glossing *episcopus*; *ispak*, Lh. 57a. The place-names show both forms.

C2. Carn Epscoppe (+**carn**); Meene-Crouse-an-Especk (+**men**, **crous**, **an**); Maenenescop (+**men**, **an**; = Bishop Rock: Loth, RC 32, 444f.)

er[1] 'eagle', Voc. 496 glossing *aquila*; OM 133. Welsh *eryr*, Breton *er*, *erer*.

C2. ?Burniere (+***bren**; cf. Welsh Bryn-yr-Eryr, Mer., if so; or **hyr**)

êr[2] 'fresh, green', Lh. 136c (Welsh *ir*, Old Irish *úr*, LEIA, pp. U26-7) may occur in one or two names, but it is confusable with **hyr**.

C2. ?Lannear (+**lyn**); ?*Ventonear* (+**fenten**)

*erber 'garden', pl. *erberow*, OM 32 (borrowed from Middle English). Only in field-names.
    B2. *Erbyer Gwarra* (+**guartha**)
    C2. *Park-an-Erbyer*, *Park an Herbio*, *Parke-an-Erbbear*, etc.; *Wheal an Harbier*, mine (+**wheyl, an**)

erw (OCo.) 'acre, field', Voc. 721 glossing *ager*; *ereu*, Voc. 338. Welsh *erw* 'acre, measure of land', Breton *erv* 'sillon' (cf. GMB, p. 221). Note the early Welsh glosses, *acre legalis, id est, eru* (Latin Texts, p. 230); *Eruguenn .i. candidus ager* (VSBG, p. 46). See also PNDinas Powys, pp. 340f.
    A. Erra, fld.
    B1. ?Eggens-warra, fld. (+**ugens**?); Padzhuera, fld. (+**peswar**)
    B2. All of the following are fields: *Eru Marut* (+ ?); *Errow-Porth* (+**porth**); *Arrowe Jellard* (+ ?); *Erov Porthm'* (+p.n.); Arra Venton (+**fenten**); *Erwereden* (+**reden**); Ara Gayan (+**gahen**?); *Errow-Brane* (+**bran**); Harry Mussy (+**mowes** pl.)
    C2. Trevarra (+**tre**); ?Roserrow (+***ros**); *Park-an-Errow*, fld.

*esker 'shank, leg', appears only in Voc. 93, *elescher* glossing *tibia*; it probably stands for ...*vel escher* (following I. Williams, BBCS 11, 92). Welsh *esgair* (also 'ridge' topographically, GPC; cf. PKM, pp. 39, lines 24f., and 40, lines 12f.); Breton *esker* 'genou'. The meaning in the place-name would be 'spur' of a hill.
    C2. ?Penisker (+**pen**)

-esyk, aj. ending; like Breton *-idik*, Old Breton *-etic* (FG, pp. 314f.; Hemon, Grammar, p. 202 n. 1) and Welsh *-edig* (Evans, GMW, pp. 165f.) it can be either a ppp. or a derivative aj. with almost the force of a present participle (active). As ppp.: *genesyk, genesek, genygyk* 'born' (BM 3211, 2287; RD 2186); as aj.: Old Cornish *treuedic* glossing *rusticus*, Voc. 227. The Breton variant with *-id-* appears also in Cornish (*genygyk*) and in the place-name. It occurs in one place-name, Cowyjack, apparently 'hollowed (place)' (+***kew**).

ethen 'bird', OM 223 and 1111; pl. *ethyn*, OM 43, etc. Old Cornish *hethen*, Voc. 495 glossing *avis vel volatile*; Welsh *edn*, Breton *evn* (see HPB, p. 487).
    C2. (pl. ?) ?Carn Pednathan, coast (+**carn, pen**)

ethom 'need, necessity', PC 182, etc.; also (with metathesis) *othem*, BM 356, etc., *otham*, CW 1132, etc., and *othom*, OM 967, etc. Breton *ezhomm* (cf. HPB, pp. 655f. and FD, p. 155).

*evor

C2. *Bounder Hebwotham* (+**bounder**, **hep**)

*evor, some kind of plant: Welsh *efwr* 'cow-parsley, hogweed'; Breton *evor* 'bourdaine, ellébore' (cf. Tanguy, p. 106); Old Irish *ibar* 'yew-tree'. The word probably occurs in Romano-British *Eburacum* (PNRB, pp. 355–7) and perhaps in Evercreech (Som., BBCS 15, 15f.); also in Gaulish *Eburodunum* (ACS, I, 1398–1400: 3 exx.) and Welsh Din Efwr.

C2. ?Lantivers (+**nans**)

*ewyk 'hind, doe', Voc. 583 *euhic* glossing *cerva*; *yweges*, OM 126, is the expected plural form, but appears rather to contain a (redundant) feminine singular ending. Welsh *ewig*, common in place-names: Old Welsh *nant yr eguic* 'the hind's valley' (LLD, p. 72), *rit ir euic* 'the hind's ford', (LLD, p. 229); Middle Welsh *parcus qui vocatur Moillewyk* (Surv. Denb., p. 51).

C2. Rosuic (+ ***ros**); Nanjewick (+**nans**); Burnuick (+**bron**); Pinnick (+**pen**)

euiter (m.; OCo., t = /θ/) 'uncle', Voc. 149 glossing *patruus*; Modern Cornish *ountr*, *ownter*, Lh. 114c, 44b glossing *patruus*, *avunculus*. Welsh *ewythr*, Breton *eontr*. The Modern Cornish form is the regular and expected one; the Old Cornish form is irregular in agreeing with Welsh against Breton (it disobeys LHEB, p. 498: *-ntr-* > *-thr-* in Welsh, but remains in Cornish and Breton); it could be dismissed as Welsh, but the place-name supports it (cf. **arluth** for a similar instance).

C2. Henderweather (+ ***hendre**)

F

fav (pl.) 'beans' in *kwthw fav* 'bean-cods', Lh. 13b (cf. 150b); *fa*, BM 2616; sg. *faven*, BM 3481. Welsh *ffa*, Breton *fav* (cf. GMB, p. 231).

C2. Park Fave, *Park an Vave*, *Gwell fave* (+**guel**), all flds.; sg. ?Park Favin, fld.

*faw (pl.) 'beech-trees'; Welsh *ffawydd* (+*gwŷdd*), Breton *faou*; Old Breton *Fau*, Chrest., p. 129. See LHEB, pp. 373 and 443; Davey, Flora, pp. 410f.

A. Fowey, river (+ *-**i** suffix)

C2. ?Penfound (+**pen**): cf. Breton Penfao: Loth, RC 28 (1907), 393.

*fe 'fief, feudal estate', borrowed from English *fee*. It is uncertain whether this was ever a fully-naturalised Cornish word; but if it was the third place-name cited here may contain it.

B2. *Fe Kenel, Fe Mareschal* (both +p.n.: see Nance, OC 3 (1937-42), 426; Doble, RCPS n.s. 9 (1937-41), 62)

C2. ?Cusvey (+**cos**)

*****fenna**, vb. 'spread, overflow' (?); Breton *fennañ*, 'éparpiller, déborder'.

C2. ?Ventonvedna (+**fenten**)

fenten (f.) 'spring, well', OM 836, 1845, etc. Old Cornish *funten*, Voc. 737 glossing *fons*; *fonton*, Charters (see below), *funttun* glossed *fontis*, An.Boll. 74, 154. Welsh *ffynnon*, Old Welsh *finnaun* (LLD), *fennun(n)* (VSBG, pp. 62 and 90), *finnun* (ibid., p. 72). Breton *feunteun* (cf. Tanguy, p. 94), Old Breton *funton* (FD, p. 172; HPB, p. 135). The word denotes a natural spring, not a dug well (see *****puth**), though it can include a built superstructure, as at many of the holy wells. As Nance says (Dictionary, p. 56), as B2 the commonest form in East Cornwall is *Fenter-*, and in West Cornwall *Venton-*.

B1. Suffenton (+**segh**?)

B2. Qualified according to the quality (colour, flow, etc.) of the water: Venton Ends, *fonton gén* (Sawyer, no. 755), Ventonjean (all +**yeyn**); Ventongimps (+**compes**); Ventonvedna (+*****fenna**?); Venton Ariance (+**arghans**); Ventonzeth, Park Venton Sah, fld. (+**segh**); *Ventonear* (+**êr²**?); Ventonwyn (+**guyn**); Fentafriddle (+**frot** aj.); Fentrigan (+**can**); Carn Venton Lês, coast (+**carn**, **glas**?)

Qualified by words for birds or animals: Partonvrane (+**bran**); Fentengo (+*****cok**); *Fentonscroll* (+**scoul**); Ventonveth (+**margh** pl.); *fonton morgeonec* (+*****moryon** aj.; Sawyer, nos. 832/1027)

Qualified by words for plants: Ventontrissick (+**dreys** aj.); *Venton kelinack* (+**kelin** aj.); Fentongollan (+*****coll** sg.: cf. *finnaun he collenn* 'the spring of the hazel', LLD, p. 247)

Qualified by location: Ventongoose (+**cos**); Venton Vadan (+**ban**); Venton Vaise (+**mes**); Fentonadle (+Eng. p.n.?)

Qualified by saints' names: Ventongassick (+Cadoc); Fentonladock (+Ladock); Ventonberron (+Piran); Ventonglidder (+Clether); Ventontinny (+Entenin); *Fentyn Carensek* (+Carantoc); perhaps Venton Veor (+*****Mayr** = Mary?). No definite instances are known of **fenten**+pers. *not* that of a known saint, but note *Vinten Wicorrian* (+**guicgur** pl.).

*feryl

Qualified by other words: Fentervean (+**byghan**); Ventonraze (+**gras**: cf. *fenten ras*, OM 836); Ventonleague, Vent-an-League, fld. (both +**lek**?); Venton Gannal (+***canel**)

C2. *Nansfonteyn* (= Lt. Petherick; +**nans**); *Placea Enfenten* (+**plas** latinised, **an**); pl. Trentinney (+**tre**). See also the phrase or compound ***pen-fenten**.

***feryl** 'magician', equivalent to Welsh *fferyll*. The Welsh word is considered to be a late, and learned, borrowing of the Latin name *Vergilius*, and it is first attested in 1632 (GPC); but cf. *Pheryll*, ÉC 14 (1974–75), 454, line 10 (MS c. 1590), and much earlier as a pers. (GBGG, s.v.). Thus it may rather have been a native word or name that was equated with *Vergilius*. At any rate, the place-name seems to show it.

C2. pl. *Fenton Ferilliow* (+**fenten**)

fyn (f.) 'end, boundary'; only in *de(y)th fyn* 'last day, doomsday', PC 724, RD 416; Welsh *ffin*, Breton *fin*. The sense 'boundary', found in the place-names, occurs also in Welsh. With the Cornish Naphene compare Welsh Ffinnant, Mtg., and *finnant iuðhail* (Ann.Camb., s.a. 848); also *fin tref petir*, *fin tref peren* (LLD, pp. 229 and 236); *guoun teirfin*, 'three-bounds moor' (LLD, p. 155). See also ***fynfos** and **fynweth**.

B2. ?Vincgonner, fld. (+**conna**?)

C1. *Fimbol* (+**pol**: Sawyer, no. 450)

C2. Naphene (+**nans**); *Vounder Feene* (+**bounder**); pl. Gunvenna (+**goon**); aj. *Trefinnick* (+**tre**)

***fynfos** 'boundary-dyke', a compound of the preceding and **fos** 'bank, ditch'. The element occurs only in two Anglo-Saxon boundaries.

A. *finfos* (Sawyer, no. 684)

C2. *penfynfos* (+**pen**: Sawyer, no. 810)

fynweth 'end, limit', MC 212d; Breton *finvez*.

C2. Trevenwith (+**tre**)

fok 'blowing-house, furnace', Pryce, s.v. Also *foge*, Pryce, p.Ff1v, *tha an foge* 'to the blowing-house'. Welsh *ffog*.

A. Vogue

B2. Vogue Beloth (+ ?)

C2. Chyvogue (+**chy**); *Whel an Voag*, mine (+**wheyl**, **an**)

fodic (OCo., *o* = /œ/), 'happy, fortunate', Voc. 304 glossing *felix*. Middle Cornish *anfusyk* 'unhappy', RD 1520; *anfugyk*, PC 1424 (Loth, RC 23, 240; 26, 241). Cf. Welsh *ffod(i)og*. Possibly used here as a noun, 'fortunate man', or as a personal name; but 'happy valley' is equally likely.

C2. Nanphysick (+ **nans**)

*fold 'fold, pen' (borrowed from English).

C2. Gwealfolds (+ **guel**); Park-an-fold

forn 'kiln, oven', Voc. 917 glossing *fornax vel clibanus*; *en foarn* 'in the oven', Pryce, p. Ftiv. Welsh *ffwrn*, Breton *forn*. At an earlier stage the word for 'kiln' was **oden**. Park Vorn is the equivalent of the common English Kiln Park, etc. (Field, EFN, p. 117). Cf. Welsh *Kilforn*' 1334, Surv. Denb., p. 43 (= Kilford, Units, p. 43).

C1. A compound *forn-jy (+ **chy**) may occur (as C2) in *Croft en Ferngy* (+ ***croft, an**)

C2. ?*Killivorne* (+ **kelli**); Park Vorn, *Park-an-Vorn*, flds.; *Vithen Vorne*, fld. (+ **budin**)

forth (f.) 'road, way'; *ford*, Voc. 711 glossing *via*; Modern Cornish *for, vor* (Pryce, s.v. *ford*). Welsh *ffordd*. Borrowed from English into Welsh and Cornish, which both changed the sense to 'road', though the original English sense appears in LLD, p. 202, where *ford* and *rit* interchange: *ir ford ar trodi...bet i'r rit ar trodi ubi incepit*, 'the ford on the Trothy... as far as the ford on the Trothy where it began'.

B1. Gwynver (+ **guyn**); see also ***hen-forth**.

B2. Forgue (+ ***kew**); *Foregles*, Park Frigles, *Park Voregles*, Frigleys, Wriggles, Park Wriggles, all flds. (all + **eglos**); ?*Frogabbin*, fld. (+ **cam**?); ?Park Fringey, fld. (+ **hensy**?); Park Vor Trevedra, *Parke Forbelancan*, flds. (both + **park**, p.n.)

fos (f.) 'dyke'; Welsh *ffos* 'ditch, dyke', Breton *foz* 'ditch, trench'. The word always means an upstanding dyke in the plays, never 'ditch', and may be made with cement (OM 2281f., 2317-20 and 2450; *fôs* 'a wall', Lh. 3a) – perhaps as opposed to a **kee**, or drystone hedge. Note *on þa dic to fos no cedu*, 'along the dyke to *Fos no cedu*', Sawyer, no. 755.

A. ?Voss; Harvose (+ ***ar**)

B1. Ruthvoes (+ **ruth**); see also ***fynfos**.

B2. Voskelly (+ **kelli**); Fursnewth (+ **nowyth**); Furswain (+ **guyn**); Vosporth (+ **porth**); *Fosebrase* (+ **bras**); *fos no cedu* (+ ?: Sawyer, no. 755); *fosgall* (+ ?: Sawyer, no. 1019)

C2. Kellivose, Killivose (both + **kelli**); Penvose (+ **pen**; 4 exx.); Halvose (+ **hal**; 2 exx.); Trevose (+ **tre**); *Carfos* (+ ***ker**); Creakavose (+ ?); Marazanvose (+ **marghas, an**); Parn Voose

**fow*

Cove (+**porth, an**); pl. *fos(s)ow*, Trevozah (+**tre**), Colvase (+*****kyl**); dimin. **fosyn*, Hendravossan (+*****hendre**)

fow* 'den, cave', pl. *fowys*, PC 336; Welsh *ffau*. A possible dialect word **vow*, which would come from **fow*, is indicated by the usage as an English word in two names, both of which first appear in the 16th century: Pendeen Vau, Westway. See also **googoo and **vooga** for other words denoting caves.

    A. Vow Cottage, Vow Cave (both +Eng.)

    B2. Vowan Guham, fld. (+**an**, ?)

    C2. Stratton Vow (+*****stret, an**)

Frank (m.) 'freeman; ?Frenchman', only in the Godolphin family motto *Frank ha leal etto ge* 'free and loyal art thou' (Pryce, p. Ee4v; OC 1 (1925–30), i, 19). Pryce also has *Vrink* 'France; a Frenchman' (p. Aa3v), which is of dubious value; if real, it probably represents a plural, **Freynk*, of *Frank* (Welsh pl. *Ffrainc*). Welsh *ffranc*[1] 'mercenary, enemy, Frenchman', *ffranc*[2] 'free'; Breton *frank* 'free (man)' (cf. GMB, p. 245; Piette, p. 123). On the Welsh word, see I. Williams, Beginnings, pp. 95–6 and references there cited; ELl, pp. 33f.; Lloyd-Jones, ElISG, p. 70 (Nant Ffrancon, 2 exx.); B. Rees, BBCS 18 (1958–60), 58f. The Cornish word occurs as a personal epithet: John *Frank*, 1327 SR.

    C2. Trefranck, Trerank (+**tre**); *Ponsfrancke* (+**pons, an**); *Adgyan Frank*, fld. (+*****aswy, an**); pl. **francon*, Bosfranken (+*****bod**), Tranken (+**tre**); pl. **freynk*, Trink (+**tre**)

**fry* 'nose', equivalent to Breton *fri* (cf. Tanguy, p. 74). There is a Cornish diminutive *friic* 'nostril' (Voc. 30 glossing *naris*), Modern Cornish *fridg* (CW 1854), *freyge* (CW 1933). The topographical sense is presumably 'hill, hill-spur' of some sort.

    C2. Treffry (3 exx.), Try (all +**tre**). An aj. **fryek*, Middle Breton *frieuc* (GMB, p. 246), 'nosed', may appear in Trefreock (2 exx.); but it is more likely to be a personal name, like Old Welsh *Friauc*, *Frioc* (LLD, pp. 247, 148, 152), Middle Breton *Frieuc* (Chrest., p. 204).

frot (OCo.) 'stream', Voc. 734 glossing *alveus*; *an frôz* 'the tide', Lh. 42a. Welsh *ffrwd*, Breton *froud* (cf. GMB, p. 248; Chrest., pp. 131 and 204f.; FG, p. 42; ACS, I, 1500f.). JEPNS, 1, 47 cites Winford, Som., Wynford, Dor., and *Wenferð*, Wor., as compounds of **winn* 'white' with this element; another compound is *camfrut* (LLD, p. 228). The Cornish word survived as dialect *froze*, *froase*, 'tide-race, turbulent water' (Nance, Sea-Words, pp. 79f.). In

Sawyer, no. 684, *cofer fros* should not contain this element, even though it refers to a stream, because it would show the sound-change *-d* > *-z* two centuries too early (LHEB, p. 398n.); yet it is tempting to see it as present. Perhaps the sound was already weakly affricate (and was here represented with *-s* by the English), or it changed earlier in that district? On the derivation of **frot**, see Whatmough, Celtica 3 (1956), 249–55; Pokorny, ibid., 308f.; and HPB, pp. 127 n. 1 and 510 n. 5.

A. *difrod* (+**dy-**: Sawyer, no. 951); deriv. ?Fraddon (+ **-an²* suffix)

B1. *sehfrod* (+**segh**: Sawyer, no. 1005)

C2. pl. Trevuzza (+**tre**); aj. Fentafriddle (+**fenten**, **-yel*): cf. Welsh *ffrydiol* 'gushing'.

G

gahen (OCo.) 'henbane', Voc. 633 glossing *simphoniaca* (OE *hennebelle*); derivation and cognates unknown, though cf. Old Irish *gafann* 'simponiaca [etc.]': Stokes, RC 9 (1888), 229.

A. ?Gayan, ?Gahan, flds.

C2. ?Ara Gayan, fld. (+**erw**)

gajah (MnCo.) 'daisy' (Borlase, Vocabulary): derivation unknown.

C2. Dorcatcher, fld. (+**dor**)

ganow 'mouth', OM 1913, etc.; Old Cornish *genau*, Voc. 43 glossing *os*, is odd in its *-au*. Welsh *genau*; Breton *genou*, Old Breton *genou-* (FD, p. 175). Old Welsh *dy genou nant byguan* 'to the mouth of Nant Bywan' (LLD, p. 142); *di genou ir pant* 'to the mouth of the hollow' (LLD, p. 213); Welsh Coedgenau, Bre.

C2. ?Pellagenna (+**pellen**?)

***gar** 'leg', pl. *garrow*, OM 1346 and RD 2501; Welsh, Breton *gar*. The plural *garrow* (and possibly also the singular ***gar**) is confusable with **garow** 'rough'.

A. pl. ?Garah, ?Garras in Kenwyn

garan 'crane', Voc. 500 glossing *grus*. Welsh, Breton *garan* (cf. Tanguy, p. 126). Penhallurick, Birds, I, 97; II, 417. The word is distinguished from **cherhit** 'heron' in the Vocabulary, and likewise in the Welsh Laws (e.g. Iorwerth, §9, line 20); in addition, both Carew in 1602 (p. 35a) and Beeverell in 1707 (Co.Studies 4/5, 29) distinguish cranes and herons as both occurring in Cornwall. Despite all this, it is likely that the two birds were not strictly

garow

distinguished in practice, so that some instances of **garan** may well denote 'heron'. Cf. Welsh Porth-y-garan, Agl.; Pictish *Lin Garan* (Jackson, Antiquity 29 (1955), 78).

C2. Cusgarne (+**cos**); Rosegarden (+**rid**); Hallgarden (+**hal**); Tregarden in St Mabyn (+**tre**); Trenarren (+ ***dyn**); ?Bodgara (+ ***bod**); ?Bosigran (+ ***bod, chy**).

garow 'rough'; Welsh *garw*, Breton *garv*. In the compound *garros, the final *-ow* (< *-w*) has been lost, as in Breton Garros (Tanguy, p. 74: see HPB, pp. 466–70 for other instances of such loss); it is possible that the same loss occurred in other compounds, making **garow** as C1 confusable with ***gar** 'leg'. In addition **garow** is distinguishable from the pl. *garrow* 'legs' only by the presence of *-r-* or *-rr-*, not a very reliable guide: cf. Garo (Tanguy, p. 74) for use as a simplex.

C1. Garras in Gulval and St Mawgan in Meneage, Garrow Tor (+Eng.), *Garros* (all + ***ros**)

C2. Pellengarrow (+ ***pen-lyn**); *Coffin Garrow*, mine (+ ***coffen**); Street-an-Garrow, street (+ ***stret, an**); Park an Garrow, fld. In addition Nancarrow (2 exx.) and Lancarrow (all +**nans**) may contain **garow** with unvoicing of the *g*, instead of **carow**.

***garth** 'enclosure' or 'ridge, promontory'; Welsh *garth* (Loth, RC 36 (1915–16), 174; PKM, p. 293); Breton *garzh* (cf. GMB, p. 254, and Ernault, RC 27 (1906), 55: 'fossé planté'). The sense 'enclosure' appears in ***buorth** and **lowarth**; and the sense 'promontory' or 'ridge' in ***pen-arth** and ***hyr-yarth**. See ***arð** for the confusion between that element and this. The following instances seem to contain the sense 'ridge'.

A. *Garth*

B1. Liggars (+**los**): cf. Welsh Llwydiarth (3 exx., Richards, Units, p. 145), and Llwydarth < *Litgarth* (ibid.; LLD, pp. 124 and 255); also Lydeard, Som. (DEPN, p. 308) and Lydiard (PNWlt., p. 35); see also ***buorth, lowarth** and ***hyr-yarth**; ***pen-arth** might be either a phrase (with ***garth** as C2) or a compound.

gaver (f.) 'goat', OM 126, *gauer*, BM 1625; *gauar*, Voc. 589 glossing *capra vel capella*. Welsh *gafr*; Breton *gavr*, Old Breton *gabr* (FD, p. 173). Old Welsh *di ol i gabr* 'to the goat's track' (LLD, p. 42).

C2. Halgavor (+**hal**); Polgaver Beach (+**pol**: cf. *Polgauer*, PNCmb., II, 360; Ekwall, ERN, p. 329); *Croft Gaver*, fld.

*gawl 'fork', equivalent to Welsh *gafl*, Breton *gaol* (cf. Tanguy, p. 90). The topographical sense is probably 'fork formed by two streams'.

C2. ?Treal (+**tre**)

gawna (MnCo.) 'calfless cow', Lhuyd MS, also *biwh gawna*, glossed *myswynog, vacca effoeta* (see Nance, OC 5 (1951–61), 164–9); Breton *gaonac'h* 'stérile'.

A. Crowner Rocks, ?Gavnas Point (both +Eng.)

ghel (OCo., *gh* = /g/) 'leech', Voc. 619 glossing *sanguissuga*; Welsh *gêl*, etc. (cf. ELlSG, p. 17); cf. Breton *gelaouenn*.

C2. Polgeel Wood (+**pol**)

geler 'coffin, bier', RD 2320, BM 4487; Welsh *elor*, Breton *geler*. See Loth, RC 25 (1904), 289f.

C2. Cargurrel (+**cruc**: cf. **gorhel**, which replaced **geler** by folk-etymology in this name)

geluin (OCo., *u* = /v/) 'beak', Voc. 510 glossing *rostrum*; otherwise only in *gylvinak* 'curlew', Lh. 240c. Welsh *gylfin*, Old Welsh *gilbin* glossing *acumine* (Juvenc., VVB, p. 130); Old Breton *golbinoc* glossing *rostratam* (FD, p. 178), *Golbin* p.n. (Chrest., p. 133); cf. Breton *golvan* 'sparrow', p.n. *Golban* (RC 5 (1881–83), 445). See LHEB, pp. 596 and 608; HPB, p. 296.

D. ?Methers Colling

*genn 'wedge', *gedn* Lh. 53a glossing *cuneus*; pl. *genov*, OM 2318. Welsh *gaing*, Breton *genn* (cf. FD, p. 174). The plural is liable to confusion with **ganow** 'mouth'.

C2. pl. ?*Placengennov* (+***plas, an**); a derivative **genna* (meaning?) may appear in Pengenna (+**pen**).

gentyl, see under J.

gesys 'left', MC 233d, etc., ppp. of *gasa* 'leave'; Welsh *gadu*.

A. ?The Gedges (i.e. a rock left dry by the tide?)

glan (f.) 'bank', Voc. 732 glossing *ripa*; RD 522. Welsh *glan*, Breton *glann*.

A. ?The Ladden, fld.

C2. ?Carrack Gladden, coast (+**karrek**)

glas 'green, blue, grey', OM 1122. Welsh, Breton *glas*. The word is twice glossed with *fulvus* in the early period: *hirglas* glossed *longi fulva* (Wrmonoc, RC 5 (1881–83), 446; FD, p. 212), and by Gildas, *Cuneglase* glossed *lanio fulve* (Loth, RC 33 (1912), 429f.; N. Wright, BBCS 30 (1982–83), 306–9). The phrase *Goon las* 'the

*glasen

green down', could also refer to the sea: so Tonkin (JRIC n.s. 7 (1973–77), 206), and in a late rhyme (Nance, OC 1 (1925–30), ii, 31; ix, 2).

B1. ?Hyrlas Rock (+**hyr**?: cf. Old Breton *hirglas*, above)

C2. Creeglase, Cruglaze, Carclaze (2 exx.) (all +**cruc**: Old Welsh *cruc glas*, LLD, p. 258); Menaglaze (+**meneth**); Goonlaze (+**goon**, 2 exx.); Hallaze (+**hal**); Carn Glaze, Canaglaze (both +**carn**); Borlase (+**bor**); Polglaze (several), Polglase (several), Penglaze (several), Pollglese, fld., Pol Glâs, coast (all +**pol**); Trelease in Ruan Major (+**tre**; here pers., *Glas?); Coyseglase (+**cos**); *Lowarglas*, fld. (+**lowarth**); Porthglaze Cove (+**porth**); ?Carn Venton Lês, coast (+**carn**, **fenten**)

D. Ellenglaze; Pentireglaze

*glasen (f.) 'greensward, verdure', a derivative of **glas**; equivalent to Breton *glazenn*; cf. Welsh *glesin*.

A. ?Glasdon

C2. Chylason (+**chy**, (**an**)); *Nanslason* (+**nans**); Parkan Glasson, fld.

*glasneth 'verdure; ?quagmire', derived from **glas**+noun suffix *-neth* (as in *folneth*, RD 961, *gowegneth*, RD 906). The possible sense 'quagmire' is suggested by Leland's information that the English for Glasney was *Wag Mier* (Itinerary, 1, 197). Note also, with a different suffix, the Welsh river-name Glasnai (2 exx., EANC, p. 25).

A. Glasney; *Glasney Green*

glastan (pl.) 'fruiting oaks', Lh. 240c ('*glastan*, and *glastanan*, an oak'), sg. *glastannen*, Voc. 675 (*quercus vel illex* glossed *glastannen vel dar*: the glossator, being ignorant of holm-oaks, thought that *illex* was roughly synonymous with *quercus*). Breton *glasten* 'chesne qui porte glan' (Catholicon, in GIB); cf. EANC, p. 180.

A. sg. *Glastannen*, r.n. (for sg. of plants as r.ns., cf. ***alaw**, **heligen**, **heschen**, etc.); Gulstatman, fld.

C2. pl. Palestine (+**pen**), Treglasta (+**tre**), Park Glaston, fld.; sg. *Gwyth Glastannan*, mine (+**guyth**[2]), *While anglastannon otherwyse namyd Oke tre*, mine (+**wheyl**, **an**), in *Goon-an-Glastannen* (+**goon**, **an**) and *Pons-glastannen* (+**pons**), the qualifier is the simplex r.n. given under A.

*glynn (m.) 'large valley'; Welsh *glyn*; Breton *glenn* is virtually unknown in place-names (see Tanguy, p. 74, and Quentel, Ogam 6 (1954), 205f.). For the sense, contrast ***comm**; also early Welsh

glyngoet kernyw 'the valley-wood of Cornwall' (= Glynn valley?), BBCS 2 (1923–25), 273 (poem 14th cent., MS 15th cent.). For the use of ***glynn** as a simplex name, cf. Glyn in Wales (8 exx., Units, p. 77) and Glendon (+Eng., PNDev., 1, 203).
 A. Glynn
 C1. See ***glynn-wyth**.
 C2. dimin. ?Carglonnon (+***kyl**)
*glynn-wyth 'valley-trees' (?), a compound of the preceding and **gwyth¹** 'trees'.
 C2. ?Treglinwith (+**tre**); ?Cutlinwith (+**cos**)
*glow¹ 'clear, bright'; Welsh *gloyw*; Old Breton *-gloeu* (FD, p. 199). In the plays, *glev* (PC 2088, RD 2582), *glv* (OM 2062), may be either this word or an equivalent of Welsh *glew* 'brave, fierce, sharp'; Middle Breton *glev* and *Gleu-* in personal names, HPB, p. 241. Welsh *gloyw* is found as the first element of close-compounds: see GPC.
 C1. Galowras (+***red** or **rid**: cf. Welsh Rhyd Loyw, 4 exx.: ÉC 10 (1962–63), 214); ?Gloweth (+**goth¹**?)
glow² 'charcoal, coal', OM 477; *huêl glow*, Lh. 145b. Welsh *glo*, Breton *glaou*.
 C2. *Gweythenglowe*, mine (+**guyth²**, **an**)
*go- (leniting) 'slight, sub-', like Welsh *go-*, Breton *gou-*. See FD, p. 194; WG, p. 267; cf. Old Welsh *Gurimi, uidelicet paruam Rymi* (VSBG, p. 62). Evidence that the prefix was functional in Middle Cornish is strangely lacking in the texts; however, there is some evidence: *goberna* 'to hire' (Pryce, +*prena* 'buy'); *ow koddros* 'menacing' (RD 2408; +*tros* 'noise', Loth, RC 23 (1902), 249); *goheles* 'shun' (BM 3071; +***kel-**, with irregular spirant mutation as in Welsh *gochelyd*). Place-names give better evidence for its existence: see the elements ***gobans**, ***godegh** and ***godre**, and note also the following place-names: Godolphin (+ ?); ?Goodern (+ ?); ?Trebarvah Goodnow (+***tnou**?)
*gobans 'little hollow' (***go-**+***pans**); Welsh *gobant*, Breton *goubant* ('côte', Tanguy, p. 75); Old Welsh *ar hit i gupant* 'along the dingle' (LLD, p. 244).
 A. Gobbens
*godegh (?) 'retreat, lair', equivalent to Welsh *godech*; cf. Breton *tec'h* 'fuite'.
 B2. ?*Goudekining* (+**kynin**?); ?Goada Sessing Close, fld. (+ ?); ?*Good-a-Caurhest*, mine (+ ?)

*godre

*godre (f.) 'homestead, small house' (***go-** + **tre**), Welsh *godref*, pl. *godrefi*.
    A. pl. Godrevy (2 exx.): cf. Welsh Godreddi (Agl.)
    C2. pl. Trewardreva (+ **tre**)

gof (m.) 'smith', Voc. 220; Welsh *gof*, Breton *gov*. Old Breton *Rangof* (Chrest., p. 133); Middle Breton *Ker en gov* (ibid., p. 206). The surname *Angove* 'the smith' occurs.
    C2. Rosegothe (+ ***ros**; Breton Roscoff); Castle Goff (+ **castell**); Hellangove (+ **hel, an**); *Park-an-Gove*, fld.; Trengove (2 exx.), Trengoffe, Trengrove, Tregoiffe (all + **tre, an**); Penscove (+ ***pen cos**); *Kellyengof* (+ **kelli, an**)
    D. *Broncoys Goef*

gofail (OCo.) 'smithy', Voc. 221 glossing *ofinitina* [sic]; Welsh *gefail* (cf. PKM, p. 184), *domus conflatorii, id est, geueil* (Latin Texts, p. 123); Breton *govel*, Old Breton *gobail* glossing *officina* (FD, p. 177).
    A. pl. Goviley

*gogleth (?) 'north', Welsh *gogledd*. (But 'north' is *north*, BM 3427.) The place-name shows a corruption if it contains this word, as seems likely: cf. /ð/ > *s* perhaps also in **cherhit**.
    C2. ?*Vounder Gogglas* (+ **bounder**)

goyf (OCo.) 'winter', Voc. 464 glossing *hyemps*; *gwâv*, Lh. 66c. Welsh *gaeaf*, Breton *goañv*; Old Breton *guoiam* (FD, p. 196). Cf. ***gwavos**.
    C2. Trewoofe (+ **tre**)

gol 'feast, fair', BM 998, 3560 and 4302; *dedh goil* 'feast day', Lh. 59b. Welsh *gŵyl*, Breton *gouel* (cf. FD, pp. 187 and 191f.). See also **guillua**, from this element + ***ma**.
    B2. Golant (+ **nans**); Goldsithney (+ St); ?Woolgarden (+ pers. ?)

goles 'lower', in *pen golas* 'lower end', MC 184b, *y ben goles* 'its lower end', OM 2443. Welsh *gwaelod*, Breton *goueled* (both normally nouns, not adjectives); cf. Tanguy, p. 75, and Gourvil, pp. xx and 79f. (*gouelet* is normally used as B2 in Breton place-names); also Hamp, SC 8/9 (1973–74), 270.
    C2. *Gwellegolas* (+ **guel**); Drewollas (+ **tre**); Parc Wollas, *Park Woolas*, fld. (+ **park**)
    D. Very common (over 60 instances in Gover alone) to denote the lower of the subdivisions of a tenement. Usually now replaced by English 'Lower', but note Grogoth Wollas, Predannack

Wollas, Pennare Wallas, etc. The standard mediaeval spelling is *Wolas* or *Woles*, but note the following variants, some showing the original diphthong: *Mythyanwoeles* 1341 (= Mithian), *Nanslowoeles* 1337 (= Nanceloe), *Treulecyon woeles* c. 1280 (= Treloyan), etc.; *Bosuoylagh goiles* 1313; *Penteyrwoloys* 1439 (= Pentire); *Worvas Collis* 1839 (= Lower Vorvas).

golow (m.) 'light'; Old Cornish *golou*, Voc. 450 glossing *lux*. Welsh *golau*; Breton *goulou*, Old Breton *guolou* (FD, p. 197). Note the Breton place-names Brengoulou, etc., 'beacon hill': Gourvil, Ogam 6 (1954), 90 and 136; Quentel, ibid., 233–6; Old Welsh *bannguolou* 'beacon height' (Ann.Camb. 873); Welsh Bryn Golau (many exx., including PNFli., p. 18).

C2. Burngullow (+**bron**); Tolgullow (+**tal**)

gols 'head of hair', Voc. 32 glossing *cesaries*, Old English *fex*; Welsh *gwallt*, Old Breton *guolt* 'chevelure' (FD, p. 197). In place-names it probably means 'clump of vegetation': cf. the secondary Welsh sense 'leafy tops and small twigs, etc.'; and the use of OE *feax* to mean 'rough grass' (Elements, I, 166); **gols** is cognate with Old English *wald* 'woodland', etc. (VKG, I, 34). It occurs only in the compound *hul-wals 'high-clump' (see **ughel**) which shows unrounding of -*o*- to -*a*-, as in Welsh *gwallt*: cf. Cornish **guas**, **war**, and HPB, pp. 432–7.

googoo (f.) 'cave', Norden (s.v. Polruddon in Powder); Welsh *gogof*, *ogof*; Old Welsh *tig-guocobauc* glossed *speluncarum domus* (Asser), *Guocob*, *Guocof* (LLD, pp. 32, 44 and 157). The Cornish place-names invariably show a form, *ogo*, containing fossilised lenition as in Welsh *ogof*; *ogo* is properly pronounced 'ugga' in the Lizard, and was used as an English dialect word in place-names (Brandy Ogo, Chough's Ogo, Pigeon Ogo, etc.). There was probably some confusion between this element and both ***fow** and **vooga**, both of which also mean 'cave': cf. Loth, ACL 3 (1905–7), 259f. For geological reasons the element occurs mainly in the Lizard peninsula. In the three instances of Resugga, the element may mean 'hollow', rather than 'cave'. The East Cornish dialect word *gug* 'cleft, chine' (equivalent of West Cornish *zawn*, v. **sawn**) is doubtless derived from an earlier form of **googoo**.

B2. Ogo-Dour Cove (+**dour**); Ogo Pons (+**pons**); Ogo Mesul (+**mesclen** pl. ?); Hugga-Dridgee (+**trig** deriv. ?): all coastal

C2. Trenuggo (+**tre**, **war**, **an**); Progo, coast (+**porth**);

goon

Resugga (+ *__ros__; 3 exx.); see *_lugh ogo_ 'seal' (?) (s.v. **lugh**) for a possible phrase containing **googoo**.

goon (f.) 'downland, unenclosed pasture', PC 1552; _gwon_, PC 1544; _guen_, Voc. 723 glossing _campus_; pl. _gonyov_, BM 1037. Welsh _gwaun_ 'bog, marsh' (cf. PNDinas Powys, p. 342), Breton _geun_ 'marais, marécage' (cf. GMB, p. 303; Loth, RC 34 (1913), 143; Fleuriot, ÉC 14 (1974), 52; Tanguy, p. 95; Hamp, BBCS 26 (1974–76), 30–1 and 139). The standard translation since about the 16th century has been English 'down(s)', meaning 'unenclosed upland pasture'; Henderson, Constantine, p. 229: 'the Down is good for nothing but rough summer pasture and is open to the beasts of more than one farmer'. Occasionally it is rendered as 'common' instead (e.g. Bosullow Common was formerly _the Downs of Bosolo, Gunneau Bosolo_). Names of the type Goonhilly Downs, where both _Goon-_ and _Downs_ appear, are quite frequent. Before the 16th century the usual translation was Latin _mora_ (occasionally _vastus, landa_ or _pastura_), but 'Moor' in place-names usually refers to **hal**, not **goon**.

Much the commonest spelling in the mediaeval period is _Goen-_, _-woen_; others are _Gon-_, _-wone_ (all periods); _Guaen-_, _Woyn-_, _-uaeyn_ (13th cent.); _Goyn-_ (14th cent.); _Goun-_ (15th cent.). The great majority of _Goon-_ names are in the four western hundreds, with only a few in East Cornwall. In the west, the form _One-_, _Owen-_, _Ony-_ can appear in the Modern Cornish period, mainly in field-names: note _One vian als. Gonenviean_ 1575 (= Gunvean in Stithians, earlier _Anwone vean_ 1536).

A. Woon (2 exx.; cf. _flumen Gueun_, VSBG, p. 6); pl. Gonew

B1. See the compound *_goon-arð_ (?) s.v. *__arð__.

B2. Qualified by words describing the quality of the ground: Goonzion Downs (+ *__seghan__?); Noongallas (+(**an**), **cales**); Goon Gumpas, Woon Gumpus Common (both +**compes**); Noon Billas (+(**an**), *__pylas__); Gunheath (+**hueth**)

Qualified by words for colours: Gonreeve (+**ruth**); Goonlaze (+**glas**; 2 exx.); Gûnwyn, Gunwen, Goonwin (all +**guyn**: cf. _i guoun guenn_, LLD, p. 258)

Qualified by words denoting animals: _Goengasek_ (+**cassec**); Goonamarth (+**an**, **margh**); Gonnamarris, Noonvares, Noon Veres, fld. (all +(**an**), **margh** pl.)

Qualified by words for position, size, etc.: Goonbell (+**pell**); Goonvrea (+*__bre__); Goonvean (2 exx.), Gunvean, Owen Vean, fld., One-vean, fld. (all +**byghan**); Owan Vrose, *Owen Brose*, flds. (both +**bras**)

Qualified by words denoting plants: Goonhoskyn (+**heschen**?; cf. *Gwayn Hesgog* Crd., Monts.Collns. 56 (1959–60), 178); Goongoose (+**cos**); *Goungylly* (+**kelli**)

Qualified by words denoting nearby features: Goonleigh, Gwealeath (both +***lcgh**); *Goyncrukke* (+**cruc**); Gorrangorras (+**an**, ***cores**); Gonorman Downs (+**naw**, **men**); Condolden (+***tol-ven**); Gunvenna (+**fyn** pl.); Goonhingey (+**hensy**); *Goeneglos* (+**eglos**); *Goencruk du* (+**cruc**, **du**)

Qualified by words denoting use: Goonhilly Downs (+***helgh** vb.); *Goenchase* (+***chas**); Gunnamanning (+**amanen**); Goonhavern (+***havar**)

Qualified by owner: Goonearl (+***yarl**); Goon Prince (+**pryns**); Goenrounsen, Gormellick, Gunvillick, Goonhusband, *Goenspaylard*, Goonpiper, Goonraw, *Own Smith*, fld. (all +pers.); note also Ony Pokis, *Woon Bucka*, flds. (both +**bucka**?)

Qualified by a place-name: *Goenbargothowe* (+Burgotha); *Goen Poldyse* (+Poldice); Goonrinsey (+Rinsey); *Goone Agga Idniall* (+***aswy**, ***enyal**; = Nangidnall, fld.); pl. *Gunneau Bosolo*

C2. Treween, Trengune, Trewoon (3 exx.), Trenoon (2 exx.), Troon (2 exx.), Trewoone (all +**tre**, **(an)**); Chywoon (4 exx.), Choone (2 exx.), Chûn, Chyoone (all +**chy**, **(an)**); Halloon, Halwyn, Hellynoon (all +**hel**, **(an)**); Burnoon (+**bron**: cf. Old Breton *Bren Goen*, ÉC 14 (1974–75), 52; ?Old Welsh *Brangwayn*, LLD, p. 320); Barnoon, Barnewoon (both +**bar**, **an**); Balnoon (+**bal**, **an**); Tor Noon (+**tor**, **an**); Lestoon (+**lost**); Resoon (+**rid**/***ros**); pl. *gonyov*, Tregwinyo (+**tre**, **an**)

D. Tregearwoon; Trebarvah Woon; Boscawen-Noon; *Treryse Wone* (in St Dennis)

***goon-dy** (m.) 'moor-house', a compound of **goon** and **ti** (= **chy**); spellings suggest that this element is not ***gwyn-dy** 'white-house, blessed-house', equivalent to Welsh *gwyndy* (spelt *gundy*, LLD, p. 120).

B2. *Goendywragh* (+**gruah**)

C2. Swingey (+**chy**)

gor-** (+len. or spir.) 'over-, very-', possibly 'high-'; Welsh *gor-*, Breton *gour-* (cf. FD, p. 198). The prefix occurs in Vorvas (+bod**); Worvas (+***bod**?); ?Gurlyn (+**lyn**?). Various elements occurring as C2 may be adjectives containing this prefix; but as it is

*gor-dre

also common in personal names (Chrest, pp. 178f. and 210f.), their status as adjectives is doubtful: **gor-losc* (+**losc**) 'very-burnt' in Bosworlas (+***bod**); **gor-thewl* (+**tewl**?) 'very-dark' in Trethowell (+**tre**); see also next entry and ***gor-ge**, ***gortharap**, and s.v. ***neved**.

*gor-dre (f.) 'over-farmstead' (***gor-**+**tre**); or (with **gorth-*, equivalent to Welsh *gwrdd-* 'fine, strong') **gorth-dre*, Welsh *gwrdd-dref*; cf. Middle Breton *Treffuortre*, Chrest., p. 234 (but Loth thought the first element of that name was probably *tnou*: ibid., p. 233, n. 9).

C2. Trevorder (2 exx.), Treworder (4 exx.), Trevarder (all +**tre**)

*gor-ge 'low or broken-down hedge', dialect *gurgoe, gurgey* 'low hedge; rough fence for waste land' (Courtney and Couch, p. 26): a compound of ***gor-**+**kee**, cf. Breton *gour-gleuz* 'fossé imparfait ou ruiné' (GMB, p. 285), with *kleuz* (= Co. **cleath**).

A. *Gurge Hedge*, fld.

C2. ?Treworga (+**tre**)

gorhel 'boat', OM 950, etc. Not present in any known place-name, but Cargurrel (**cruc**+**geler**) was thought to contain it by folk-etymology, and there was a story that probably went with it, of a boat-burial in a barrow, later transferred to a larger barrow nearby.

*gorm 'dusky, dark', Welsh *gwrm*; not elsewhere in Cornish or Breton, but cf. Old Breton personal names in *Uurm-*, etc. (Chrest., p. 181).

C2. Trecorme (+**rid**)

*gortharap 'very pleasant' or 'unpleasant'; **gorth-* equivalent to Welsh *gwrdd-* 'fine, strong' or to Welsh *gwrth-* 'contra-', and **arap*, equivalent to Welsh *arab* 'pleasant' (cf. Breton *arabad* 'nonsense', *arabat* 'forbidden, vain': GMB, p. 35). The sense 'very pleasant' is intrinsically more probable, but 'unpleasant' has the advantage that Welsh *gurtharab* 'uncivil' is attested once (BBC 45.3).

C2. Tretharrup in Gwennap, Luxulyan and St Martin in Meneage, *Trewartharap*; perhaps Tretharrup in St Cleer and Lanreath and Tredarrup (4 exx.) (all +**tre**)

gortos 'stay, tarry', PC 1497, etc.; Modern Cornish *dho gortha*, Lh. 85b glossing *maneo*. Breton *gortoz*.

C2. *Mene Gurta* 'which signifyeth in English – a staie stone' (+**men**)

*gorweth, either equivalent to Welsh *gorwydd* 'edge of wood; wooded slope', or else the same as Middle Cornish *groweth* 'to lie down' (OM 2759), Welsh *gorwedd*, Breton *gourvez*. The former is topographically more likely, but the word is not attested in Cornish or Breton.
    A. pl. ?Carvoda
    C2. sg. Treswarrow (+ ***ros**); pl. ?Porthgwarra (+ **porth**)
*gosa 'bleed, to bloody', ppp. *gosys* MC 219c; Welsh *gwaedu*, Breton *gwadañ*.
    C2. ?Polgoda (+ **pol**); ?Carvossa (+ ***ker**?)
goth[1] (f. ?) 'water-course', PC 2512, MC 132b; *woth*, OM 1093; pl. *gwyʒy*, MC 183d. Old Cornish *guid*, Voc. 86 glossing *vena*. Welsh *gŵyth*; Breton *gwazh*, etc. (cf. HPB, p. 436; Chrest., p. 206; GMB, p. 267; Tanguy, p. 86; FD, p. 178; Hamp, ÉC 14 (1974–75), 201–4). The element is formally indistinguishable from **goth**[2] 'goose', and only the topography, or general probability, can decide which element a name should appear under. Other words for 'stream' include **gover, streyth, frot, auon** and **dour**. It is unknown how they differed, though each is likely to have had a particular sense. See also perhaps **guyth**[2] and ***gothel**.
    B1. ?Grogoth (+ ***gruk**?); ?Gloweth (+ ***glow**[1]?)
    B2. pl. **gothow*, ?Goodagrane (+ ***crann**?)
    C1. Gothers (+ ***bod**). Two compound words containing **goth**[1] may occur: **goth-nans* 'stream-valley' as C2 in ?Trewannett, ?Tregothnan (both + **tre**); and (OCo.) **goth-red* (+ ***red**), equivalent to Welsh *gwythred* 'channel, brook', in ?Trewethart (+ **tre**; or pers. ?).
    C2. Ponsanooth (+ **pons, an**); pl. **gothow*, Burgotha (+ **bar**), Castle Gotha (+ **castell**); aj. ?Bosvathick (+ ***bod**), Trewothack (+ **tre**); dimin. ?Trevothen (+ **tre**)
goth[2] (f.) 'goose', OM 1195; *goyth*, OM 129; Old Cornish *guit*, Voc. 515 glossing *auca*; Modern Cornish *gûdh*, Lh. 43a, pl. *godho*, Lh. 242c. Welsh *gŵydd*, Breton *gwaz* (Tanguy, p. 127). Formally indistinguishable from **goth**[1] 'stream'.
    C2. Polgooth (+ **pol**; cf. Breton Poulhoas, Tanguy, loc. cit.); Gweal Goose (+ **guel**); *Codnagooth* (+ **conna**: cf. Goose Neck, flds.)
*gothel 'watery ground', equivalent to Breton *gwazhell*: cf. Loth, RC 37 (1917–19), 303: 'lieu fertilisé par des ruisseaux, par eau courante; lieu marécageux'; Middle Breton aj. *Goetheloc* (Chrest.,

p. 206). Possibly from **goth**[1] + *-(y)el. The element is perhaps liable to confusion with ***gwythel** (?) 'thicket'.

    A. Perranarworthal (+ St), Northwethel (both + ***ar**)

    C2. Trengothal, Trewothall, Treworthal (all + **tre**, (**an**)); Croft Gothal, mine (+ ***croft**); Park an Gothal, fld.

gour (m.) 'man'; *gur*, Voc. 205 glossing *vir*. Welsh *gŵr*, Breton *gour*. The phrase **gour guyn* could be either a personal name or a description, used as C2 in Chirgwidden (+ **chy**); cf. Breton *Caer Gorguen*, 1224 (Chrest., p. 211).

gover (m.) 'stream', OM 1845; *guuer*, Voc. 736 glossing *rivus*; pl. *goverov*, BM 1971. Welsh *gofer*, Breton *gouer*.

    A. Gover (2 exx.); pl. Goverrow, Govarrow

    B2. Goverseath (+ **segh**); Gaverigan (+ **guyn**)

    C2. Polgover (+ **pol**); Langore (+ **nans**?); Pengover (+ **pen**); *Ponsgovar* (+ **pons**)

gras 'grace'; Welsh, Breton *gras*, all borrowed from English or French (Piette, p. 126). With the place-name compare *fenten ras*, OM 836.

    C2. Ventonraze (+ **fenten**)

***gre** 'flock, herd', attested only in **grelin**. Welsh, Breton *gre* (cf. GMB, p. 292).

    C2. Lezerea (+ ***lys**); Polgray (+ **pol**); Tregray (+ **rid**); ?*Hall Gre* (+ **hal**?)

***gre-dy** (m.; OCo.) 'cattle-shed' (?), a compound of ***gre** and **ti** (= **chy**); one would have expected **gre-jy*, however, in the western half of Cornwall (where the place-name lies), so the status of the element is slightly doubtful: the written records may have preserved the older form in this manorial name.

    A. pl. Gready

grelin 'cattle-pool', Voc. 742 glossing *lacus*, Old English *sēað*; a compound of ***gre** and **lyn**, Welsh *grelyn*.

    A. pl. Garlidna

gronen (OCo., *o* = /œ/) 'a grain', Voc. 687 glossing *granum*. Welsh *gronyn*, Breton *greunenn*. The use of the singular is odd.

    C2. Trewornan (+ **tre**); pl. *Streatt and Grean*, street (+ ***stret, an**); see also the phrase **polgrean** 'gravel pit', which seems to contain this word, but possibly through confusion with the next element, **grow**.

grow 'gravel', OM 2756; *grou vel trait*, Voc. 738 glossing *harena*; the singular (or collective) would have been **growan*, dialect *growan* 'decomposed granite'. Welsh *graean*, Breton *grouan* (both either plural or collective, singular *graeanen*, *grouanenn*). Compare **gronen**.

C2. collective *Wheal Growan*, mine

gruah, see under Gw.

gruk (*u* = /y/) 'heather', Welsh *grug*; Breton *brug*, pl. *brugou*, may have been influenced by French *bruyère* (which is anyway cognate with *grug*, etc.); see also Bachellery, ÉC 12 (1968–71), 676–8. The Cornish word is attested in the compound *grugyer* 'partridge' (literally 'heath-hen'; OM 132 and 1203); and in the dialect *griglan(s)* 'heather' (Courtney and Couch, p. 26; Welsh *gruglwyn*), which occurs, as either dialect or Cornish, in the place-name and field-name Griglands. One form of Trugo, *Trewrugowe* 1316, shows a by-form ***gwruk**, like Pembrokeshire Welsh *gwrig* (Morris-Jones, WG, p. 98). See also Davey, Flora, pp. 281–3 and Suppl., pp. 84–7.

C1. ?Grogoth (+ **goth**[1] ?)

C2. pl. Trugo (+ **tre**); ?Bedrugga (+ ***bod**)

gubman (MnCo.) 'sea-weed', Lh. 9b and 42b; Welsh *gw(y)mon*, Breton *goumon*; Old Breton or Welsh *gueimmonou* (FD, pp. 186 and 194). Williams (Lexicon, p. 190a) cites an earlier form *gumman*, but without giving the source. Sea-weed would have been put on fields as a fertiliser.

C2. *Park-an-Gubman*, fld.; ?Hall Gommon, fld. (+ **hal**)

guillua (f.; OCo., second *u* = /v/) 'look-out place', Voc. 400 glossing *vigilia*; a compound of ***guil** 'watch' (cf. **gol** 'feast' and LHEB, pp. 462f.; the -*ll*- of Voc. is incorrect) and ***ma** 'place'; Welsh *gwylfa*. With the following names compare Welsh Wylfa Head (Agl.); English Prawle Point (PNDev., I, 319f.).

C2. Penolva, Pedn Olva, coast, Pen Olver, coast (all + **pen**); Carn Olva, coast, Carn Galver, hill (both + **carn**)

Gw

gvak 'empty, weak'; Welsh *gwag*, Breton *gwak*. Perhaps used as a nickname in the place-name.

C2. Boswague (+ ***bod**)

***gwailgi** (f.; OCo.) 'ocean'; equivalent to Welsh *gweilgi* 'ocean', literally 'howl-hound' or 'wolf-hound', Irish *fáelchú* 'wolf' (GPC quotes a Breton *gail* 'wolf'). Cf. the Welsh place-name Cefnwilgi,

guaintoin

Mtg., inland near Welshpool (GPC), and the phrase *uch penn y weilgi* 'above the ocean' (PKM 29.5).
    C2. ?Pennywilgie Point (+**pen, an**?)

guaintoin (OCo.) 'springtime', Voc. 461 glossing *ver*; *guainten*, Lh. 171c. Welsh *gwanwyn*.
    C2. Trengwainton (+**tre, dy-**)

*****gwal** 'wall', equivalent to Welsh *gwal* 'wall' (borrowed from English).
    A. pl. ?Wallow
    B1. ?Bonnal (+**ben**? But cf. Old Welsh *a vado Ponugual*, VSBG, p. 88?)
    C2. ?Rosewall (+**rid**); aj. *****gwalek**, ?Treswallock (+ *****ros**?)

*****gwalader** (m.) 'lord, leader', equivalent to Welsh *gwaladr*; perhaps cf. Old Breton *-uualart* in personal names (with metathesis, cf. HPB, p. 816?). Or the word itself could be a personal name.
    C2. Trewalder (+**tre**)

guan 'weak'; Voc. 379 glossing *debilis*; Welsh, Breton *gwan*. The meaning in place-names is unclear; in some cases it may be a nickname.
    C1. A compound *gwan-dre* (+**tre**; meaning?) seems to appear (as C2) in Trewandra (+**tre**).
    C2. Bodwen in Helland (+*****bod**); ?Skillywadden (+ ?); pl. *gwanyon* (cf. Old Breton *guenion* glossing *mitiores*, FD, p. 188), ?Trewannion (+**tre**)

gwaneth 'wheat', CW 1066; Welsh *gwenith*, Breton *gwinizh*. The field-name Baragwaneth Field is more likely to consist simply of the surname *Baragwaneth* (= English *Whitbread*) than to be a separate instance of **bara**+**gwaneth**.
    C1. Gonitor (+**tyr**: Welsh *gwenithdir*)

guary (m.) 'play': Welsh *chwarae, gwarae*; Breton *c'hoari*, Old Breton *guari-, -huari-* (FD, p. 182): see Vendryes, ÉC 3 (1938), 41; HPB, pp. 429f. It is surprising that the compound, Welsh *chwaraefa, gwaraefa*, Old Breton *guarima*, 'theatre' (cf. Loth, RC 12 (1891), 280f.), does not seem to have been used in Cornish. The phrase *plen an guary* refers to the 'playing place', or open-air theatres where the Cornish plays were performed. See Nance, 'The Plen an Gwary or Cornish Playing Place', JRIC 24 (1933–36), 190–211.
    B2. *Gwarry-Teage*, mine (+**teg**: cf. the hurling-ball inscription, *guary wheage yu guary teage* 'sweet play is fair play')

C2. Castle Wary (+**castell**); Plain-an-Gwarry (2 exx.), Plane-an-Gwarry, *Plain an Quarry*, *Plean-an-Wartha*, etc. (all +**plen, an**)

guartha 'summit; upper', OM 1074; Welsh *gwarthaf* 'top, summit' (also as aj.). The Cornish word is generally used as a noun in the texts; the only possible instance of its use as an adjective (as in all the place-names except those under B2) is *yn trone wartha* 'on the top throne', CW 191. The element is used mainly as D, for sub-divided tenements, usually opposed to *Wollas* (from **goles**); in most cases it has been replaced by 'Upper' on the modern map.

B2. Gwarth-an-drea, *Warthendr'*, *Quarthendrea* (all +**an, tre**); ?Worthyvale (+**auallen**?)

C2. Trewartha in St Agnes (+**tre**); Gwealgwarthas (+**guel**); Halwartha (+**hal**); aj. ('summit-like' or 'possessing a summit'?) ?Trethauke (+**tre**: or **tre**+**worth**+**haf** aj.?); Park Gwarra, Park-Warra, flds.

D. Many instances such as Grogoth Wartha, Pellyn-wartha, Kemyel Wartha, *Scorya wartha* (= Scorrier), *Polwhevererwartha* (= Polwheveral), *Meynwinionwartha* (= Menwinnion), Pennare Wartha and Wallas, Trewoof-wartha (pron. 'Tewarra'), etc.

guarthek 'cattle', OM 1065; *gwarrhog*, Lh. 115c glossing *pecus*; Welsh *gwartheg*.

C2. *Goenwarrack*, fld. (+**goon**); *Park-an-Gwarrack*, fld.; *Porgwarrak*, fld. (+ ?); ?*Polwarrack Mills* (+**pol**)

guas (m.) 'fellow, servant'; Welsh *gwas*, Breton *gwaz*, etc. (cf. FD, pp. 199f.; GMB, pp. 265f.).

C2. Boswase (+ ***bod**)

Personal names of the type *gwas*+St are common in Wales (good selection, BBCS 13 (1948-50), 213) and known in Brittany (cf. FD, loc. cit., and RC 29, 69) and in Cumbria (*Gospatrik*, 11th cent., PNCmb., III, xxviii; *Cwæspatrik*, 10th cent., BCS, no. 1254). They also occur in Cornwall: *Reginaldum Waspeder*, 1288 Ass.R.; *Walterus Guaspeder persona ecclesie sancte Wydole*, 1305 ibid.; and *Wastdewy de Karwarri*, 1227–29, Beauchamp Crtlry., no. 373; and also as C2 in a place-name, *Chestewer* (**chy** + ***Gwas Dewy**), with a *Wastdeui* in the parish, 1201 Ass.R., no. 198. The pers. *Was(s)o* (Bodm.) may be derived from **guas**, and appears as C2 in Trewassa (+**tre**) and *Hensywassa* (+**hensy**). By contrast, Welsh *gwas* 'abode, dwelling' (cf. Watson, CPNS, p. 210; Dumville, Celtica 12 (1977), 27; Old Irish *foss* 'stay, rest') is unknown in Cornwall.

*gwavos (f.) 'winter-dwelling', a compound of **goyf** and ***bod**; Welsh *gaeafod* (only from 1794). The spellings of names containing this element show the variation of the phonology: they include *Guaevos* 1369 (probably archaic at that date), *-waivos* 1298, *Goyvos* c. 1300, *Goevos* 1348. See also Nance, 'The names Gwavas, Hendra and Laity', OC 1 (1925–30), iv, 32–4, and compare ***havos** and ***kyniaf-vod**.
    A. Gwavas (3 exx.)
    C2. Trewavas (+**tre**; 2 exx.)

guel (m.) 'open-field'; the word apparently does not exist in Welsh or Breton, and its derivation is unknown. It cannot be a variant of **wheyl** 'work, mine', with interchange of *gw-/hw-* (cf. VKG, I, 433f.; HPB, pp. 429f.): the two are never confused, and early (13th or early 14th century) spellings of **guel** often show the form *Gwael-*. Connection with Welsh *gwael* 'wretched, lowly', is possible formally but unlikely semantically (cf. Hamp, SC 8/9, 270). It is possible that Welsh *gwäell* 'knitting-needle, skewer, splinter' should be compared: formally it suits the earliest spellings. If so, then the element must originally have meant 'stitch, sliver of land', and the meaning 'open-field' is an extension. Instances in the plays show that as cultivated land, **guel** could be contrasted with woodland and with grazing-land: *yn gveel py yn cos* 'in field or in wood', OM 364; *the wonys guel ha ton* 'to work arable and lay-land', OM 1164; *yn guel nag yn pras* 'in field nor in meadow', OM 1137 and 1151. The phrase 'stitches in Gwelcreege alias Treworracke Field', A.D. 1633 (Doble, St Ewe, p. 44), where *Field* presumably means 'common field' (cf. Elements, I, 167, §2), shows that a **guel** could contain 'stitches'; 'a *landa*...called *Gwael en Whoen*', A.D. 1270 (Payne, Roche, p. 94) again shows **guel** as unenclosed land, though not apparently arable in this case. It is clear that **guel** came in time to mean 'enclosed field', passing through a semantic development similar to English 'field' itself. The change may have occurred by 1611: *han devidgyow oll in gweall* 'and all the sheep in the field', CW 1070, shows it as non-arable grazing land, whether enclosed or not. Like other Cornish field-name generics, the element is found only in the four western hundreds, and is much commoner in the westernmost ones of Penwith and Kerrier. Within west Cornwall, *Gweal-* names are found especially on the outskirts of the mediaeval boroughs (particularly Helston), though

guennol

not exclusively so. *Gweal-* often appears as *Goll-* or similar in late spoken forms. Names refer to tenements unless specified.
  B2. Qualified according to the nature or colour of the ground: *Guaelmeynek* (+ ***meynek**); Gweal-an-Mayn (+ **an, men** sg. or pl.); *Gueldrenek* (+ **dreyn** aj.); *Gwaeldreysec*, fld. (+ **dreys** aj.); Gwealdues (+ **du**); Gwealmellin (+ **melyn**^{2}); Gull-gwidden, fld. (+ **guyn**)
  Qualified according to position or nearby features: Gwealgwarthas (+ **guartha**); *Gwellegolas* (+ **goles**); Gollwest, Goldwest, flds. (+ **west**, or **wast**?); Gwealcarn (+ **carn**); *Gwael en Whoen* (+ **an, goon**); Gwealavellan (+ **an, melin**^{1}); Clydia (+ ***aswy**?); *Gwelcreege*, fld. (+ **cruc**); Gwealfolds (+ ***fold**); Gweal Darras, *Gweal Darros*, flds. (both + **daras**)
  Qualified by a personal name: Gwelmartyn, Gweal Paul, Gwealmayowe, *Gwelbelbouche*
  Qualified by a place-name: Gwealhellis (+ *Hellas* = Helston), *Gwele Cararthen* (+ Carharthen); *Gwelbeauleu*
  Qualified by other words: Gulnoweth, fld. (+ **nowyth**); *Gwael Kephamoc* (+ ***kevammok**); Golden Praunter, fld. (+ **an, pronter**); *Gwell fave*, fld. (+ **fav**); Gwool Pease, fld. (+ **pêz**)
  C2. Chingweal, Chyangweale, Chy-an-Gweal (all + **chy, an**)
guely 'bed', OM 2127; *gueli*, Voc. 801 glossing *lectum vel lectulum*. Welsh *gwely* (*Wele Conws*, ELlSG, p. 35; *gueli banadil* 'broom bed' LLD, p. 214), Breton *gwele* (Guelegoarh, Top.Bret., p. 58). The derivative Welsh meaning 'family, land-owning kinship group' does not occur in Cornish. Compare the rock called King Arthur's Bed, from the shape of a hollow.
  B2. *Guely breteny*, rock (+ ?)
guella 'best'; Breton *gwellañ*, cf. Welsh *gwell* 'better'.
  C2. ?*Park Gwella*, fld.
gwels 'grass', MC 16b; *guels*, OM 712; Welsh *gwellt*, Breton *geot*. The adjective has as parallels Welsh *gwelltog*, Breton *geotek*; Old Breton pl. *gueltiocion* glossing *fenosa*, 'grassy places' (FD, p. 188); Breton rock-name Gualtog 'le chevelu', ÉC 10 (1962–63), 286. There is a possible compound *hul-wels 'high-grass', but the name is more likely to contain **gols** 'hair, vegetation': see **ughel**.
  C2. ?Zawn Wells, coast (+ **sawn**); aj. Carwalsick (+ ***ker**), Illiswilgig, rock (+ **enys**)
guennol (f.) 'swallow', Voc. 512 glossing *hirundo*; Welsh *gwennol*, Breton *gwennili* (a plural form used as singular: PB, p. 247), Old

Breton *guennol* glossing *herundo* (FD, p. 188). The Cornish pl. was the same as Breton *gwennili*.

C2. sg. Tregwindles (+**tre**); pl. Brown Willy (+**bron**)

gweras 'ground, earth', CW 2081; *gwyrras*, ibid. 2084; *gueret*, Voc. 13 glossing *humus*. Welsh *gweryd*; cf. Old Breton cpd. *gueretreou* 'districts' (FD, pp. 188f.), Breton dimin. Guéredic Sant Hervé (Loth, Mél.Arb., pp. 225f.; Ernault, RC 27 (1906), 216). As B1, cf. early Welsh *gwrdweryt* 'greensward' (CA, p. 288), and as B2 (not found in Cornish), early Welsh *Guerit Carantauc* (VSBG, p. 148). Alternatively, some of the names might contain an element *gweres, equivalent to Welsh *gwaered* 'slope, valley-bottom'.

B1. Camerance (+**cam**)

C2. Killiwerris (+**kelli**); Penwerris (+**pen**); Treweese, Treweers, Trewarras, Treverras (all +**tre**)

guern (pl.) 'alders; alder-swamp'; attested only as *guernen*, Voc. 678 glossing *alnus*; *guern*, Voc. 281 glossing *malus* 'mast'; *gvern* 'mast', RD 2331, pl. *wernow* 'masts', CW 2292. Welsh *gwern*, Breton *gwern*, both also meaning 'mast' as well as 'alder-grove, swamp'. Cf. GMB, p. 301; Tanguy, pp. 95 and 107; ElISG, pp. 32 and 112; PNDinas Powys, p. 342; Old Welsh *hal un guernen* 'one-alder moor' (LLD, p. 367). It is unclear whether **guern** could refer to a swamp that did not possess alders; but as there were several other words for 'swamp, marsh', and as alder is 'fairly common' in all parts of Cornwall (Davey, Flora, p. 406), it is at least possible that **guern** was reserved for swamps with alders.

A. aj. Gwarnick; flds. Warneck, Warnick

B1. *penn lidanuwern* (+**pen**, **ledan**: Sawyer, no. 1019)

B2. *Gwernfosov* (+ ?)

C2. Penwarne (3 exx.), Penwarden (all +**pen**); Lewarne (+ ***lann**); Luthergwearne (+ ***lether** ?); sg. ?Porthguarnon (+ **porth**)

gvest 'lodging', OM 356, 361 (see Loth, RC 26 (1905), 222); Welsh *gwest* 'lodging', Old Breton *guest-* 'feast', Breton *-vez* in *banvez* 'feast' (cf. GMB, p. 53). All the examples given here are tentative: it is uncertain whether the word actually occurs in place-names.

B1. ?*Westva*, ?Westway (both + ***ma** ?, cf. Welsh *gwestfa* 'inn, dining-hall'); a compound *gwesty (+**ti** = **chy**), Welsh *gwesty*, may occur as B2 in Westnorth (+ ?).

gwîa 'weave, plait, build', Lh. 163b glossing *texo*; Welsh *gweu*, *gwau*, Breton *gweañ*. Cf. *guiat*, Voc. 828 glossing *tela* 'woven cloth', Welsh *gwead*, Breton *gwiad*. In the first name, the word could refer to an industry carried on at the farm; in the second, the past participle 'plaited' could refer to a set hedge.

C2. v.n. ?Trevia (+**tre**); ppp. ?*Gwyllegwyet* (+**guel**?)

guibeden (f.; OCo.) 'gnat, midge', Voc. 539 glossing *scinifes*; Middle Cornish *webesen*, BM 2421 (so Loth, Mél.Arb., p. 219; Parry, BBCS 8 (1935-37), 127). Welsh pl. *gwybed*; Breton pl. *hwibed, fubu* (PB, pp. 130f., and Loth, RC 40 (1923), 447); on *hw-* for *gw-* cf. **guary**, and HPB, pp. 429f. Some Cornish place-name forms show the same variant. For the adjective, cf. Welsh *gwybedog* 'fly-infested', and the stream-names Gwybedog, Gwybedig, Gwybedyn (EANC, p. 190; Ekwall, ERN, p. 77).

C2. pl. Menawicket, fld. (+**meneth**); Lawhibbet, Lawhippet (both +**nans**); aj. Halabezack (+**hal**), ?Landabethick (+**nans**)

*gwyk 'village' or 'forest'; equivalent to Welsh *gwig* 'forest', Old Breton *guic* 'town part of a parish' (Chrest., p. 210, n. 2). The word is derived from Latin *vicus* 'community', and kept its sense in early Breton place-names: how its meaning changed to 'forest' in Welsh is unknown; so is the meaning in Cornish – whether it followed Breton or Welsh. The sense 'town' (or 'trading-post'?) is suggested by **guicgur** 'trader'; the sense 'forest' by Gweek and perhaps by *Kellewik*, and by the lack of instances as B2, as in the Breton names. It is even possible that the word could have had an intermediate sense in Cornish, 'dwelling in a forest', which would suit Gweek and might give a clue as to how the Welsh meaning arose. As a simplex, it is confusable with Old English *wīc* 'dwelling, hamlet', itself from Latin *vicus* and often used as a simplex name (Elements, II, 261). Gweek itself could even be an English name assimilated to Cornish phonology. In view of all this, all of the following names should be taken as doubtful. For use as C2, cf. Old Welsh *Cair guicou*, VSBG, p. 120.

A. Gweek

C2. Treweege (+**tre**); Castlewitch (+**castell**); ?Polquick (+**pol**); *Kellewik* (+**kelli**)

guicgur (m.; OCo.) 'trader, merchant', Voc. 267 glossing *mercator vel negotiator*; *wecor*, MC 40a; pl. *guycoryon*, PC 331, *guykcoryon*, PC 1304. Cf. Welsh *edwicwr* 'pedlar', and dialect *wicwr* 'a sponger' (BBCS 16 (1954-56), 101). The word was used as a personal

guyn

epithet: John *Gwycor* 1327; perhaps Richard *Wycher*, Baldwin *Wycher* 1327. Middle Breton *guygourr*, Jérusalem, no. 166.
  C2. pl. *Vinten Wicorrian alias the Colliers Well* (+**fenten**)
guillua, see under G.
guyn 'white', Voc. 482 glossing *albus*; Welsh *gwyn*, Breton *gwenn*. In some of the place-names the word may have the meaning 'fair, happy, pleasant' (cf. FD, p. 192); and a personal name equivalent to Welsh *Gwên* is also possible occasionally. Unlike Welsh, Cornish **guyn** did not have a feminine form with -*e*-, but spellings with *y* or *e* occur indiscriminately in both genders: note *Marya wyn* (fem.), MC 221a; *Menethguen* (m.) = Mennergwidden. However, spellings with *e* do seem to be slightly commoner for feminine than for masculine instances (note, e.g., Trewen, below). There was also a plural form **gwynyon* (see Menwinnion, below): cf. Padel, SC 14-15 (1979-80), 237-40.
  C1. Gwendra (2 exx.), Gwendreath (all +**trait**; cf. Welsh Gwendraeth (EANC, p. 113), Irish Fionntráigh, Ventry, etc.); Gonvena (+**meneth**; cf. Old Breton *win monid* glossed *montem candidum*, FD, p. 330); Gwynver (+**forth**); ?Winven Cove (+**men**?); Whimple (+**pol**); see also the compound **gwyn-arð* (?) s.v. **arð*.
  C2. Referring to water-features: Ventonwyn (+**fenten**); Polgwidden, Polgwidden Cove, Polgwyn Beach (all +**pol**); Porthgwidden, Porth Gwidden, Portquin (all +**porth**); Gaverigan (+**gover**); ?Trebullom (+**tre, pol**?); Rosewin in St Enoder, Tredwen (both +**rid**)
  Referring to land-features: Menergwidden (+**meneth**); Gunwen, Gûnwyn, Goonwin (all +**goon**); Rosewin in St Minver, Treswen (both +***ros**); Halwin (+**hal**); Polwin, Pollawyn (both +***pen hal**); *Coeswyn* (+**cos**); *Park gwidden*, fld.; Stennagwyn (+**stean** aj.)
  Referring to stones, etc.: *maen wynn* (+**men**); *carrec wynn* (+**karrek**); Crosswyn (+**crous**); Keigwin (+**kee**); Carnaquidden (+***kernyk**); pl. Menwinnion, Menwidden (both +**men** pl.)
  Referring to habitations (here especially it may be a personal name instead): Bodwen in Lanlivery, Boswyn, Boswin (all +***bod**); Trewen (5 exx.), Trewidden (all +**tre**); Carwen (+***ker**; 2 exx.); Chygwyne, Chegwidden in Constantine (+**chy**); Leswidden (+***lys**); Halwyn (+**hel**; 4 exx.). Note also the phrase or personal

ame *gour guyn* (v. **gour**), occurring as C2 in Chirgwidden (+**chy**).

D. Amalwhidden (cf. Amalveor, Amalebra); *Drinnick guin* (Drannack); *Argelwen* (Argall)

Various words that are probably derivatives of **guyn** appear, taking the forms **gwynnek*, **gwynnyek* or **gwennek* and presumably meaning 'whitish' (?); cf. the Old Welsh r.n. *gunnic, oper guinnic* (LLD, p. 252). A personal name, as in St Winnow, etc. (**Gwennek*, hypocoristic form of **Gwynwaluy*: HPB, p. 149), is likely in some cases as C2.

A. Winnick (2 exx.)

C2. Chywednack (+**chy**); Boswednack (+***bod**); Carvinack in Mylor (+***ker**); Carnwinnick (+**carn**); Menwenick (+**men**); ?Halwinnick (+**hal**?)

guyns (m.) 'wind', PC 1215, RD 2292; *guins*, Voc. 445 glossing *ventus*. Welsh *gwynt*, Breton *gwent*. Cf. also ***melyn wyns** 'windmill'.

A. A derivative (+***-ys**^{2}) or ppp. **gwynsys* of a verb, 'windy place', may appear as rock and field-names (The) Gwinges (3 exx.), Gwineus.

C2. Bosence in St Erth, Boswens (both +***bod**); Chyangwens (+**chy, an**); Trewince (9 exx.), Trewint (7 exx.) (all +**tre**); Menagwins (2 exx.), Minawint, fld., ?Menawink (all +**meneth**); aj. Croft Windjack, fld. (+***croft**), Adjawinjack (+***aswy**)

guirt (OCo., *-t* = /θ/) 'green', Voc. 486 glossing *viridis*; Modern Cornish *guèr, gwêr*, Lh. 18c and 61c. Welsh *gwyrdd*, Breton *gwer* (Middle Breton *guezr*).

C2. Tresquare (+***ros** or **rid**?)

guis 'sow, pig', Voc. 594 glossing *scroffa*; Welsh *gwŷs*, Breton *gwiz* (cf. FD, p. 193; GMB, p. 303).

A. ?Quies, rocks

C2. ?Truas (+**tre**); ?Porqueese, fld. (+**porth**?)

gwyth[1] (f. pl.) 'trees', CW 93; *gweth*, MC 16b; *gveyth*, OM 28; sg. *gvethen*, OM 29, etc. Welsh *gwŷdd* (sg. *gwydden*), Breton *gwez* (sg. *gwezenn*). For the adjectives, cf. Welsh *gwyddog* 'woody', Breton *gwezek* (cf. Tanguy, p. 101). As well as **gwythek*, there was apparently a further adjective **gwythyel* (see ***-yel**).

A. pl. Weeth (3 exx.), Theweeth, The Weeths, The Wyth, fld. (some +**an**); sg. Withan, Withen (referred to as *de Arbore*, 1327); aj. (+***-yel**), Withiel, Gweal (Scilly)

guyth²

B1. See *collwyth, *cren (?), *enwydh* (s.v. **onnen**) and *glynn-wyth.

C2. pl. Trengweath (+**tre, an**); Croftengweeth (+***croft, an**); Woon Weeth, fld. (+**goon**); *Chingwith*, Chyngwith (both +**chy, an**); ?double pl. *gwythow (?), Nanquitho (+**nans**); sg. Brownwithan, Burnwithen (both +**bron**), Manywithan, fld., *Menewithan* (both +**meneth**), Nanswhyden (+**nans**), Trewithen in Stithians (+**tre**); aj. *gwythek, Lawithick (+***lann**), Lawhittack (+**nans**), ?Penwithick (+**pen**), Treswithick (+**rid**/***ros**), Bowithick (+ ***bod**; 2 exx.), Trewethack, Trewithick (2 exx.) (all +**tre**; cf. Old Welsh *treb guidauc*, Chad 3: LLD, p. xlv), *penn hal weðoc* (+***pen hal**); aj. (+*-**yel**), Lostwithiel (+**lost**), ?Bedwithiel (+***bod**)

guyth² 'work', BM 785 and 998; *gueid* (*d* = /θ/), Voc. 231 glossing *opus* (also *gueidwur* 'workman', Voc. 225, etc.). Welsh *gwaith* 'work; mine', Old Breton *gueid-* in *gueidret* and *cintgueith* (see FD, pp. 186f. and 107). Instead of an equation with Welsh *gwaith* 'work; mine', it is attractive to note rather Welsh *gŵyth* 'seam of a mineral; stream, channel'; but, despite the similarity of sense, that does not seem possible, for *gŵyth* is equivalent to Cornish **goth¹**. However, **goth¹** did have a plural *gwyȝy* (MC 183d), and it may be that Cornish **guyth²** is not from Old Cornish *gueid* (= Welsh *gwaith*), but actually a variant of **goth¹**, in a restricted sense; or, more likely, is due to crossing between the two. The word evidently means 'tin-working' when it occurs in Cornish place-names; it is restricted in area and date, being found so far only in Blackmore stannary (St Austell area) and only in the 16th–17th centuries: it is evidently the equivalent of **wheyl** in the districts further west at that date.

B2. All the names refer to tin-works. *Gwyth an bara* (+**an, bara**); *Gwyth en futhen* (+**an, budin**); *Gweythenglowe* (+**an, glow²**); *Gwythancorse* (+**an,** ***cors**); *Gwythanprase* (+**an, pras**); *Gwythengrous* (+**an, crous**); *Gwyth Glastannan* (+**glastan** sg.); *Gweythen-vysten* (+**an,** ?)

***gwythel** 'thicket' or 'Irishman' (?); equivalent either to Welsh *gwyddwal* 'thicket' (pl. *gwyddeli*, which in the 18th century gave a new sg. *gwyddel*), or to *Gwyddel* 'Irishman' (cf. Lhuyd's Cornish *Gwidhili* 'Irishmen', 242c, and Robert *Wydel* 1327 SR, Margaret *Engothall* 1538). The former sense is much more likely, but would rather demand *-wythwal*, etc., in early forms, which does not occur: the names have early forms in *-wydel*, etc. Since *Gwyddelfynydd*,

Mer., was spelt *Gwadel-* in 1292–93 (MerSR, p. 27) and Gwyddelwern was *Gwydel-*, *Gothel-* (ibid., pp. 87 and 84), it is possible that the Welsh sg. *gwyddel* is older than the 18th century, and might have a Cornish cognate ***gwythel** 'thicket'. Note also Old Breton *Guern-uidel* or *-uuital* (Chrest., p. 173), and Middle Breton *Tnouguydel* (now Trevidel), RC 40 (1923), 427. There is also a possibility of confusion with ***gothel** 'watery ground'; and, if it existed in Cornish, with an equivalent of Breton *godell* 'pocket', pl. *godilli* (PB, p. 55).

C2. Menacuddle (+**meneth**); ?Persquiddle, fld. (+**park**, **crous**); ?Delawhidden (+**deyl**); pl. ?*Wheale an Gothilly*, mine

gulas (f.) 'land'; *gulat*, Voc. 715 glossing *patria*. Welsh *gwlad*; Breton *glad* (now 'wealth', formerly 'land'). The one Cornish name is no doubt a translation of English Land's End, and was itself then corrupted into *pen (*an*) *wolas* 'bottom end' (glossed *infimum caput*, Leland, Itinerary, I, 316).

C2. *Pedden an wollas* (+**pen, an**)

***gwleghe** 'moisture, wetness', from *gwlegh 'wet' (equivalent to Welsh *gwlych*, Breton *glec'h*) +noun-suffix *-y* (as in *anfugy*, OM 2328, etc.) or *-e* (from **-eth*, as in *kerense*, PC 549, etc.) or *-a* (as in *nootha*, CW 969).

C2. ?Tregleath (+**tal**); ?Treglith (+**tre**)

***gwlesyk** (m.) 'leader', equivalent to Welsh *gwledig*, Middle Breton *gloedic* 'comte' (RC 33 (1912), 352). But a personal name, equivalent to Old Breton *Uuoletec* (Chrest., p. 176), Old Welsh *Guoleiduc* (LLD, p. 74), is more likely in all cases.

C2. Lellissick, Lowlizzick, ?*Lowlysycke* (all +***lann**); Trelissick in Sithney and St Erth (both +**tre**)

gruah (f.) 'hag, old woman, witch', Voc. 211 glossing *anus*; Welsh *gwrach*, Breton *gwrac'h*. The meaning may have changed in Modern Cornish: Tonkin (c. 1720), in an onomastic tale, spoke of 'a famous Wrath or Giant' (JRIC n.s. 7 (1973–77), 203). Melville Richards (Neath & District, p. 331) says that Welsh *gwrach* is commonly used for rivers and brooks, but there is no particular reason to think that this sense occurred in Cornish; however, some of the names (Nancewrath, Polwrath and *Ponswragh*) could accommodate such a sense.

A. The Wra, (Great/Little) Wrea, Wreathe, rocks (cf. La Vieille, off Pointe du Raz, Brittany, and Ar Wrac'h, rocks, Topon.Nautique 1419, nos. 9559, 9658, and 9679)

gwrek

C2. Borah (+ ***bod**); Crousa (*crouswra(c)h*, Sawyer, nos. 755 and 832/1027), Crows-an-Wray (both + **crous, (an)**); Nancewrath (+ **nans**); Polwrath (+ **pol**); *Goendywragh* (+ ***goon-dy**); Sewrah (+ ?); *Crucwragh* (+ **cruc**); *caer uureh* (+ ***ker**: Sawyer, no. 770); Towanwroath Shaft (+ **toll, an**); *Ponswragh* (+ **pons**); Legereath (+ **les** pl. ?); ?*Loban Rath* (+ **lam**?, **an**; see Thomas, Co.Arch. 3 (1964), 78)

gwrek (f.) 'woman', MC 66c; *gvrek*, OM 386, etc; *grueg*, Voc. 206 glossing *mulier*; plural *gvraget* ($t = /\eth/$), OM 976; *gwregath*, CW 2437. Welsh *gwraig*, pl. *gwragedd*; Breton *gwreg*, pl. *gwragez*. With the place-name compare the lost *Raswraget* (PNCmb., 1, 103) from Cumbric **ros+ *gwrageð*.

C2. pl. *Krucgruageh* (+ **cruc**)

H

ha(g) 'and'; Welsh *a(c)*, Breton *ha(g)*. Occurs in Mevagissey (+ Sts) and perhaps in Tuzzy Muzzy Croft, fld. (+ **tus** ?, **mowes** pl. ?, Eng.); for the former name, compare Welsh (16th-century) *p[arochia] mair a chirig*, Crm. (= Eglwys Fair a Churig, Units, p. 64) and *ll[an] gynin ai waison*, Crm. (= Llangynin, Units, p. 130): RWM, 1, 917.

haf 'summer', Voc. 462 glossing *estas*; etc. Welsh *haf*, Breton *hañv*. For the adjective, ***havek**, cf. Welsh *hafog* 'abundant, bountiful'; Breton *hañvek* 'of summer', Old Breton p.n. *Hamuc* (now Hanvec, Chrest., p. 135), Middle Breton *-haffec*, etc., in p.ns. (Chrest., p. 212). The meaning of the adjective in place-names is presumably 'summer-land'. See also ***havos** 'shieling' and ***havar**.

A. aj. ?Havoc, fld.

C2. aj. Trehawke (+ **tre**), ?Trethauke (+ **tre, worth**: or tre+ **guartha** aj.)

hager 'foul, evil'; (frequently used preceding its noun, as in *hager gowes* 'evil storm', OM 1080; *hager oberou*, Chyanhor 26); Welsh *hagr*, Breton *hakr*.

C1. Haggerowal Croft, fld., Haggarowel (both + **awel**: cf. Modern Cornish *hagar awell* 'foul weather', Pryce, p. Ff1v)

***hay** 'enclosure', borrowed from English: see Elements, 1, 215, §2. But in fact all of the following names may have been originally coined as English names and then made to look Cornish, or half-

hal

translated; there is no certain instance of the use of ***hay** as a Cornish word.

A. Anhay (2 exx.), Hay in Ladock, Lanhay (all +**an**)

B2. *Hay thu* (+**du**; = Blackhay)

hal (f.) 'moor, marsh', OM 1780; *haal*, OM 2708; pl. *hellov*, BM 3411. Welsh *hâl* 'moor' (LLD, passim), Old Welsh pl. *halou* glossing *stercora* 'dungs' (VVB, p. 151); Breton *hal* 'salive'. The original sense seems to have been 'dirty water' (Old Irish *sal* 'dirt', *salach* 'filthy'), whence 'standing water', then 'marsh'. The Cornish word developed further to 'moor', i.e. rough, uncultivated ground (not necessarily marshy), and finally to 'upland', as seen in BM 3411 and in Lhuyd's *hâl bîan* glossing *verruca* 'hillock' (172a) and *halow* 'hills' (245a). The earlier sense of 'marsh' is indicated by the phrase *lyys haal* 'marsh mud', OM 2708, and by *cronek dv...yn hans* [sic MS] *yn hal* 'a black toad...down in a marsh', OM 1778–80. The usual translation in place-names is 'moor' (itself ambiguous between 'marsh' and 'upland'), e.g. *moram de Penhall*, 13th cent.; *Halbothekmore* 1448; Halbullock Moor; Halgavor Moor; *Halgrosse Moore* 1608; *Halvegan Moore* 1660; but 'down' can appear rarely, as in *Halgaverdoune* 1315. Note also Pryce, p. Ff1r, *ker tha'n hâl* 'to go the moor', with a footnote: 'That is, go and work to Tin; they call that especially going *to Moor*, when they work on the Stream Tin.' The main difference from **goon** was probably that **goon** connoted primarily the economic use of the land, as unenclosed grazing, whereas **hal** did not necessarily imply that. See also Charles Thomas, 'The meanings of *hal*, *gun* and *ros*', OC 6 (1961–67), 392–7. Most of the places called Penhale are at the heads of marshy streams, or (in some cases) on valley sides, between the stream and the upland. Names such as Halveor, Halviggan, and Halwidden show the gender, but lenition is by no means universal.

A. Hale in St Kew

B2. Qualified according to size: Halveor, Halmeers, fld. (both +**meur**); Halviggan (+**byghan**)

Qualified by words for animals or birds: Halgavor (+**gaver**); Hallgarden (+**garan**); Halvarras (+**margh** pl.); Halabezack (+**guibeden** aj.)

Qualified by words indicating plants: *Halgrosse Moore*(+ ***cors**); ?Halldrine Cove (+**dreyn**?); Halankene, Helencane Cove (both

+ **an**, ***keun** ?); Halgoss, Halancoose (+ (**an**), **cos**); *Halldrunkard* (+ ***tron-gos**)

Qualified by words for colours: Halwin (+ **guyn**: but Halwyn is usually **hel** + **guyn**); *Haldu* (+ **du**); Hallaze (+ **glas**)

Qualified by words for notable features: Halgarrack (+ **karrek**); Halbathick (+ ***both** aj.); Hallenbeagle (if + **an**, **begel**); Halvoze (+ ***fos**; 2 exx.); ?Hall Dinas, coast (+ ***dynas**)

Qualified by words denoting people: Hallworthy (+ pers.); ?Hayle Kimbro Pool (+ ***Kembro**); Hallenbeagle (if + **an**, **bugel**)

Qualified by other words: Halwartha (+ **guartha**); Halvousack, fld. (+ **mosek**); Halamanning (+ **amanen**); Halstenick, fld. (+ **stean** aj.)

C2. Trenhale, Trenale (both + **tre**, **an**); Chynhale (2 exx.), Chenhall (3 exx.), Chenhale, Chyenhâl (all + **chy**, **an**); Penhale in Davidstow (+ **ben**); Penhole in North Hill (+ **pol**); *Enyshall* (+ **enys**); Rose-an-Hale Cove (+ ***ros**, **an**); see also ***pen hal**; pl. Pontshallow (+ **pons**), and see *pen halou, s.v. ***pen hal**. The aj. *halek (= Welsh *halog*; Old Breton *haloc*, FD, p. 206) may occur in Retallack in Constantine and St Hilary (both + **rid**; cf. Welsh Rhyd Halog, etc., ÉC 10 (1962–63), 214; PNFli., p. 147), though **tal** aj. is also possible; also perhaps in Porthallack (+ **porth**)

hans 'yonder', CW 1547; *yn hans*, OM 1780 (sic MS, ACL 1, 164), *in hans* BM 440, CW 1743, *inhans* BM 3919; *hunt tho* 'beyond', Bosons, p. 27. See Nance, JRIC 23 (1929–32), 352. Lhuyd's *in hauz* 'down', 248c, seems to be a mistake for *in hanz*. Welsh *hwnt*, Breton *hont* (cf. GMB, p. 323; RC 37 (1917–19), 114). The vowel is unexpected, and may be due to confusion with *nans* or *yn nans* (OM 165): *hons would be regular, and the field-names (if they contain the word) seem to reflect that, rather than the form of the texts.

A. ?Hunds, ?The Hunds, flds.
C2. ?Park-an-huns, fld.

hanter 'half', OM 957, etc.; Welsh *hanner*, Breton *hanter*. Not known for certain in any place-name, but in Hantergantick an original ***hendre** has been changed into **hanter**, presumably by folk-etymology. Hantertavis may genuinely contain **hanter** as B2 (+ **taves** or **daves**), the name referring to a notable rock: see Henderson, Mabe, pp. 33f.

*havar 'summer-fallow, land left fallow in summer'; equivalent to Welsh *hafar*, from *haf* 'summer' + *âr* 'ploughed land'.
   A. aj. ?*Haferell*, r.n.
   C2. Goonhavern (+ **goon**)

*havos (f.) 'shieling', a compound of **haf** + ***bod**; Welsh *hafod*: see references cited under ***hendre**, and also ELISG, pp. 64 and 113; M. Richards, Monts Collns. 56 (1959-60), 13-20 and 177-87. According to Henderson, Essays, p. 129, 'in Sancreed as late as the seventeenth century the word [*hewas*] was still used to signify rough pasture'. The place-names from east to west across Cornwall show roughly the development of the word; in the east, *-m-* is the Anglo-Saxon representation of the nasal *v* (LHEB, pp. 486-90). Shielings constitute marginal land, in that they are used only in summer, yet there are two manors in Domesday book of which the names contain the element (Hammett in Quethiock; Hamatethy); they must therefore represent an expansion of farming settlement, at some date before the later 11th century, such that what had previously been marginal land became permanent settlements. For a similar case in Wales by 1392, see M. Richards, op. cit., p. 180. Breton *Hanvod*, Bernier, ÉC 20 (1983), 261-3.
   A. Hammett (2 exx.), Hampt, Havet, Hewas (2 exx.), Hewes Common; Howas, Howes, Hughas, flds.; pl. Halvosso
   B2. Hamatethy (+ ?); *Hewas-an-Grouse*, fld. (+ **an**, **crous**); ?House Oath, fld. (+ ***aweth** ?)
   C2. ?Carnewas (+ **carn**); ?Goonevas (+ **goon**)

*havrek 'arable land'; equivalent to Breton *havreg*, *awrec* 'guéret'. Cf. perhaps Caraverick (PNCmb., I, 202). See Loth, RC 40 (1923), 377-86.
   A. Averack, Haverack, flds.; ?The Avarack, coast
   C2. Treharrock (+ **tre**); *Croft haverek*, fld. (+ ***croft**)
   D. *Penpons Haverack*

*heyl 'estuary'. See the discussion and list of names, Padel, SC 14/15 (1979-80), 240-5. The origin of the word, if it was originally a disyllable, is obscure; but if the evidence making it appear disyllabic is faulty, one could compare the Gaulish river-name *Salia* (DAG, pp. 772 and 728; Ekwall, ERN, pp. 192f.; Nicolaisen, p. 189). Compare **ben**.
   A. Hayle, Hayle Bay, Little Hell, *Hægelmuða* (+ Eng.)
   B2. Helvear Down (+ **meur**)
   C2. Vwgha Hayle, coast (+ **vooga**); Cotehele (+ **cos**); Eglos-

hel

hayle (2 exx.), *Egloshayle* (= Phillack) (all +**eglos**); Trenhayle (+**tre, war, an**); Tywarnhayle (+**ti, war, an**); ?Menalhyl (+**melin**[1]?); dimin. **heylyn*, ?Porth Island, coast (+**porth**)
  D. *Trewelesikwarheil* (+**war**), *Trelowith Heill*

hel (f. ?) 'hall', Voc. 932 glossing *aula*; OM 1501, 2110; *hell*, MC 140a. Borrowed from Old English *heall*. Note that Helnoweth is called *Noua Aula* in c. 1265 (Cart.St M., no. 32).
  A. *Hell alias Anhell*; pl. Ella
  B2. Halloon, Halwyn in St Keverne (both +**goon**); Halwyn (+**guyn**; 4 exx.); Halnoweth (2 exx.), Helnoweth (all +**nowyth**); Hillcoose (+**cos**); Hellangove (+**an, gof**); ?Hellarcher (+pers.)
  C2. Penheale (+**pen**: or as B1 if **pen** = 'chief'); Bethanel (+**budin, an**)

*helgh 'a hunt', only in *helh-*, Voc. 237 and 317, and in the verb **helghy* 'to hunt', *helly* CW 321; ppp. *helhys* OM 709, *helheys* MC 2b, *hellys* CW 1032; 1 pl. ipv. *hellyn* CW 300. Welsh *hela*, *hely* 'hunt', Breton *helc'hiñ* 'fatiguer, épuiser', *(hem-)olchin* 'chasser'; Old Breton verb *olguo* (FD, pp. 68f.; Hamp, ÉC 15 (1976-78), 569). The first name contains the noun **helgh*, the others the verbal noun **helghy*.
  C2. Brill (+***bre**); Goonhilly Downs, *Goenhely* (= Huntingdon, an exact translation) (both +**goon**)

heligen (f.) 'willow-tree, sallow', Voc. 704 glossing *salix*; *helagan*, Lh. 143c, pl. *helak*, ibid. 16b. Welsh *helygen*, pl. *helyg* (cf. ELlSG, p. 26); Breton *halegenn*, pl. *haleg* (cf. Tanguy, p. 107). For the singular as a simplex name, cf. Welsh Helygen r.n. (EANC, pp. 118f.) and Llanfihangel Helygen (Rad.) and Breton Haliguen (Tanguy, loc. cit.; Chrest., p. 134n.; Gourvil, no. 693): it may be that it is originally a stream-name in the Cornish instances. The double plural **helygy* occurs also in Vannetais Breton (GIB, s.v.).
  A. sg. Helligan, Heligan (2 exx.), Hallegan, *Hellegan*; double pl. Halligey, ?Halleggo
  C1. pl. ?Halgolluir (+**lowarth**?: cf. Old Welsh *ir helicluin* 'the willow-grove', LLD, p. 268)
  C2. sg. Penaligon Downs (+**pen**), Selligan (+**rid**), Treleggan (+**tre, an**: cf. Welsh Trefhelygen, Units, p. 118); pl. Penhellick (4 exx.), Penellick, Pennellick (all +**pen**), Retallick in Roche and St Columb Major (both +**rid**), ?Presthilleck (+**pras**)

hen 'ancient, disused, former'; Welsh, Breton *hen*. In all three languages *hen* normally precedes its noun, but it can occasionally follow it in Welsh (and therefore perhaps in Cornish, but no examples have been noted): *Fennun Hen*, glossed *fonte antiquo* and *ueteri fonte* (VSBG, pp. 62 and 90); Lluest-hen, Crd. (Monts. Collns. 56 (1959–60), 179).

C1. Hengar (+ *ker); *Hengastel* (+castell); Hembal (+bal); *Hensens*(+ ?); ?Henscath, rock (+ skath?); ?Engollan (+ *coll sg.). See also *hendre, *hen-forth, *hen-lann, *hen-lys and hensy.
*henkyn (?) 'iron peg for a spindle' (rather than 'icicle'), equivalent to Breton *hinkin* (cf. GMB, p. 320, *Parc-an-Hinquin-Bihan*); supposedly from Breton *enk* 'narrow' (but that is, anomalously, yn in Cornish).

C2. ?Nanjenkin (+nans)

*hendre (f.) 'winter homestead, home farm', a compound of hen+tre; Welsh *hendref* (glossed *mansio hyemalis*, Latin Texts, p. 370). See J. E. Lloyd, 'Hendref and hafod', BBCS 4 (1927–29), 224f.; Nance, 'The names Gwavas, Hendra and Laity', OC 1 (1925–30), iv, 32–4; M. Richards, Monts.Collns. 56 (1959–60), 180–2; Elwyn Davies, 'Hendre and hafod in Caernarvonshire', Trans. Caernarvonshire Historical Soc., 40 (1979), 17–46; Pounds, 'Note on transhumance in Cornwall', Geography, 27 (1942), 34. The shift in meaning from 'ancient farmstead' to 'winter farmstead' is odd, but must be early, since it is shared by Welsh and Cornish. Perhaps it came about through the sense 'original homestead'. The practice of transhumance which the name indicates must also be correspondingly early in Wales and Cornwall. Hendra is one of the commonest Cornish farm-names; the largest number is clustered around Bodmin Moor, and presumably these were settlements that used the Moor for their summer shielings; but the name is found all over the county.

A. Hendra (34 or more instances); Abbott's Hendra; Shrubhendra; dimin. Landrivick

B2. Hendragoth (+kio); Hendraburnick (+bronnen aj.); Hendragreen (+dreyn); Hendravossan (+fos dimin.); Henderweather (+euiter or pers.); Hendrawalls (+pers.?); Hendrawna, Hendersick (both + ?); Hantergantick (+ *cant aj. ?)

C2. Coldhender (+ *kyl); Polhendra (+pol)

*hen-forth 'old road, replaced track', a compound of hen+forth; Welsh *henffordd*.

A. Henvor (2 exx.); Henver; Henforth

C2. Parkhenver (+park); *Goonhenver* (+goon)

*hen-lann (f.) 'old cemetery', a compound of **hen**+***lann**; cf. Welsh Henllan (several: Units, p. 90), and note particularly *Hennlann dibric & lannteliau in uno cimiterio* 'The old cemetery of Dyfrig and the cemetery of Teilo in a single cemetery' (LLD, p. 275).
    A. Helland (3 exx.)
    C2. Kehelland (+**kelli**); Portholland (+**porth**)
*hen-lys 'ancient court; ruins', a compound of **hen** and ***lys**; Welsh Henllys (several: Units, p. 90). Whether the name always implies a former administrative centre is uncertain: it depends on whether ***lys** could also itself mean 'ruin', in which case ***hen-lys** could also have referred to ruins, of indeterminate date. But it is likely that Helston (replaced, presumably, by the Domesday manor of Winnington) and Helstone (replaced presumably by Lesnewth 'new court') did refer to former hundred centres; Helles- near St Ives may possibly have been contrasted with a *Nova aula de Porthya* mentioned in 1429 (Doble, St Ives, p. 39n.); but it is more likely to refer, as a ruin, to the Dark-Age building excavated at Hellesvean (PWCFC n.s. 1 (1953–56), 73–5; 2 (1957–61), 151–5).
    A. Helston (+English *tūn*; = *Hellas*, RD 673); Helstone/Helsbury (+Eng.); Helles-veor/-vean (+**meur/byghan**); ?Helset (+suffix?)
hensy (f.?) 'ancient house', BM 1307 (so Stokes, ACL 1 (1900), 122), a compound of **hen** and **ti** (= **chy**); Welsh *hendy* (cf. Yr Hendy and Hendy-Gwyn = Whitland, both Crm.). The meaning in place-names is probably 'ruin, remains'. As C2 **hensy** may be confusable with **hynse** 'neighbour'.
    A. Hingey
    B2. *Hensywassa* (+pers.); Hensafraen (+**bran**); Hensavisten (+?); Hendywills (+?); Hendawle (+***tawel**?); *Hyngy Espayne*, fld. (+?); *Hengyvghall* (+**ughel**); Ingewidden (+**guyn**)
    C2. Goonhingey (+**goon**); ?Crofthandy (+***croft**); ?Goldingey, fld. (+**guel**); ?Park Fringey, fld. (+**forth**?)
hep 'without'; Welsh *heb*, Breton *hep*. Only in *Bounder Hebwotham* (+**bounder, ethom**).
heschen (f.; OCo., *ch* = /k/) 'sedge, coarse grass; wet ground, bog', Voc. 646 glossing *canna vel arundo*; Welsh *hesgen* (cf. ELlSG, p. 124; Williams, ELl, p. 62); Breton *hesk* (cf. Tanguy, p. 114); see also Loth, RC 46 (1929), 147f. For **heschen** in the sense 'bog', compare Old Irish *sescenn* 'marsh', often used as B2 (Hogan,

Onomasticon, p. 597a), a usage that does not occur in Cornish but is found once in early Welsh, *hescenn iudie*, LLD, p. 143; also Old Breton *hiscent* glossing *uligo .i. humor terrae*, where the *-t* is incorrect (Fleuriot, ÉC 9 (1960–61), 185; cf. ÉC 11 (1964–67), 155–8, and FD, p. 212), and *Penn hischin* (Cart.Land., FD, loc. cit.), where *-in* is probably a spelling of *-en* (cf. HPB, p. 96). For the singular ending extended to mean 'place where X grows', cf. **heligen**, etc.

    A. sg. Heskyn; pl. Sconhoe

    C2. Penhesken, Prenestin (both +**pen**: cf. Old Breton *Penn hischin*, above, and Welsh Penhesgyn, Agl.); *poll hæscen* (Sawyer, nos. 832/1027), Poliskin, Poleskan (all +**pol**); ?Goonhoskyn (+**goon**: or pers.); ?Laneskin Wood (+**nans**/***lann**); pl. *Buthen en hesk* (+**budin, an**)

***heth** 'stag' (?), equivalent to Welsh *hydd*; cf. Breton *heizez* 'biche' (cf. Tanguy, p. 120; GMB, p. 316). Alternatively, the names might represent an element *heth 'barley', equivalent to Welsh *haidd*, Breton *heiz*; and there may also have been a third word *heth 'peace', on which see **hueth**.

    C1. ?Headland (+**lyn**: or **hueth**?)

    C2. ?*Crucheyd* (+**cruc**: not the same as Welsh Cricieth, etc., which is *crug caith*, 'serfs' barrow')

hyly 'brine, salt water', RD 2318 (*nag yn dour nag yn hyly* 'neither in fresh nor in salt water'); Welsh *heli*, Breton *hili*. In Welsh place-names: *Campus Heli*, VSBG, p. 260; Pwllheli; Breton *Porz Hili* (Topon.Nautique 1419, no. 8904).

    C2. Porthilly (+**porth**); Manely (+**myn**)

***hin** (OCo.) 'border', equivalent to Old Welsh *-hin* in *o'r cledhin* glossing *limite levo*; cf. Middle Breton *gourrin* (GMB, p. 288). See Loth, RC 41 (1924), 393f., and I. Williams, BBCS 2 (1923–25), 303–9. With the name compare Old Breton *Hincant*, Modern Breton Hingant (Chrest., p. 137 and n. 4); Middle Breton *Kerhingant* (Bégard, p. 203).

    C1. Hingham (+***cant**)

-hins- 'way', in *camhinsic* 'wrong-wayed', Voc. 306 glossing *iniuriosus* and 403 glossing *iniustus*, and in *eunhinsic* 'right-wayed', Voc. 402 glossing *iustus*; also, spelt *en-*, in *enlidan*, Voc. 651 glossing *plantago*, Old English *wegbræde* (Old Breton *hæntletan*, FD, pp. 205f.). The only occurrence of this word in later texts is in the noun *cammensyth* (Trg. 15), *kammynsoth* (Trg. 15a), 'injustice', the

hynse

abstract noun corresponding to *camhinsic*. Welsh *hynt*, Breton *hent* (cf. Chrest., p. 137); see Elements, I, 276; II, 118; JEPNS 1, 48; 2, 74 for a few possible occurrences of the word in English place-names. The word appears also in the compound ***caubalhint** 'ferry, boat-way', but otherwise there are no clear occurrences of the word in Cornish, either in place-names or in the texts: it may have dropped out of use altogether. Instances of place-names seem to be similarly lacking in Wales, but it occurs in Breton coastal names (e.g. Topon.Nautique 1419, p. 513). In all three languages the word occurs predominantly as the second element in compounds: in addition to those cited, note Breton *karrhent* 'cart-track'; British *Gabrosentum* 'goat-track' (PNRB, p. 364); Welsh Mynydd Epynt, Bre., 'horse-track mountain'. Possibly also as B2 in Henscath, rock (+**coth**, if so: or **hen**+**skath**).

hynse 'neighbour, fellows, kin'; *hynse* 'fellows', OM 2136 (so Loth, RC 26 (1905), 229; Williams, BBCS 4 (1927–29), 339–41); *hense* 'fellows', BM 2925 (so Loth, Mél.Arb., pp. 219–20). Welsh *hennydd* 'friend, enemy, other', Breton *hentez* 'les proches'. The element may be confusable with **hensy**; the sense in the place-name is unclear. Early forms of the name show the etymological final *-th*, which is absent in the occurrences of the word in the mediaeval texts.

C2. Lanhinsworth (+ ***lann**)

hyr 'long'; *hir*, Voc. 948 glossing *longus*. Welsh, Breton *hir*. In all three languages this adjective is often used as C1, preceding its noun: Old Welsh *hirmain* 'long-stone' (Chad 4, LLD, p. xlv: cf. Hervan, below), *hirpant* 'long-valley' (LLD, p. 168); Old Breton *hirglas* glossed *longi fulva* (RC 5 (1881–83), 446; FD, p. 212): cf. Hyrlas Rock and pers. *Hirlas*, RBB, index (but cf. B. Roberts, BBCS 25 (1972–74), 278 and n. 3, 'Hirlas does not seem to occur [outside the Bruts] as a personal name'); Welsh Hirfynydd, Gla. (cf. Halvana, Halvenna); Hirnant, Mtg. (cf. Hernis).

C1. Hervan (+**men**); Hernis (+**nans**); Halvana, Halvenna (both +**meneth**); Herland (+**lyn**; 2 exx.); ?Hyrlas Rock (+**glas**?); Harros (+***ros**); see also ***hyr-drum** and ***hyr-yarth**, next entries.

C2. Retire, *Retier* (both +**rid**); Crowshire (+**crous**); *Savenheer*, coast (+**sawn**); *Curkheir* (+**cruc**); Bowgyheere (+**boudzhi**; 2 exx.); *Doerhyr, Dorheere, Doreheer*, flds. (+**dor**); *Park Heer*,

Parkanheer, Parkenhere, flds. (all +**park**); ?Burniere (+ ***bren**: or **er**[1]); *Goen an Groushire* (+**goon, an, crous**); Holseer Cove (+**als**); see also ***men hyr**.
*****hyr-drum** 'long-ridge', a compound of **hyr** and ***drum** (or rather ***trum**).

C2. Penhedra, Polhorden (both +**pen**)
*****hyr-yarth** 'long-ridge', a compound of **hyr** and ***garth**; Old Breton *Hirgard* (Chrest., p. 131; HPB, p. 517). There are also two Welsh instances in Montgomeryshire, now Hiriaeth and Rhiw Hiriaeth (Richards, Monts.Collns. 54 (1955–56), 102); and Herriard, Hmp., may also contain it, though an English origin has also been suggested for that name (DEPN, s.v., and Elements, I, 198 and 244). For the -*y*- representing lenited *g*, see LHEB, p. 439; HPB, pp. 720f. It appears as -*g*- in the modern forms of some names, which is curious: the -*g*- can hardly be an Anglo-Saxon sound-substitution for -*y*- (though it could be for /γ/). Perhaps /γ/ and /j/ alternated in such words, or the sound was an intermediate one: early spellings contain both -*i*- and -*g*- (cf. ***pen-arth**).

A. Harewood; Herodsfoot (+Eng.)

C2. Penhargard, Penharget (both +**pen**); *Bronhiriard* (+**bron**; = Herodshead)

hiuin, see under I/Y.

*****hoby** 'pony' (borrowed from English, like Welsh *hobi* 'small riding horse, nag'); cf. *hobyhors*, BM 1061 (sic MS: Nance, OC 3 (1937–42), 426: like Middle Welsh *hobi-hors*, the Cornish word is attested considerably earlier than English *hobby-horse*).

C2. Stable Hobba (+ ***stable**)

hoch 'pig', Voc. 592 glossing *porcus*; *hoh vedho* 'a drunken sow', Lh. 243c. Welsh *hwch* 'sow', formerly 'pig' generally; Breton *hoc'h*, Old Breton *hoch* glossing *aper* (FD, p. 212); cf. Tanguy, p. 120.

C2. Carno (+**corn**); Nansough (+**nans**); ?*Lostogh* (+**lost** ?); dimin. ?Bosoughan (+ ***bod**); *Carrack Kine Hoh*, rock (+**karrek, keyn**)

*****hogen** 'heap, pile', equivalent to Breton *hogenn* 'amas, tas'; the word seems to have been specialised in sense in Cornish: dialect *hoggan* means 'lump of baked dough, pasty'. There was also a word ***hogen** meaning 'haw, hawthorn-berry': Breton *hogan*

horn

'cenelle, baie d'aubépine', Middle Breton *hoguen* (GMB, p. 322); Cornish dialect *hoggan* 'haw'. The former word is semantically more likely in the place-name.

C2. Vellanhoggan Mills (+ **melin**[1])

horn 'iron', RD 2162; pl. *hern*, PC 2938; Old Cornish *-hoern*, Voc. 769, 783 and 899. Welsh *haearn*; Breton *houarn*, Old Breton *hoiarn-* (FD, p. 213). The element is frequently used in personal names (Chrest., pp. 139 and 213), and such personal names occur in Cornish place-names. It may be that some of the following place-names contain the simplex **horn** or the adjective *hornek as personal names. For the adjective, cf. Breton Le Nouarnec (Tanguy, p. 59) and Middle Breton *Plo-Harnoc* (Beauport: RC 3 (1876–78), 418). Polhearn Farm (PNDev., II, 508), may be **pol** + **horn**.

A. aj. Hornick

C2. Bushornes, Bosorne (both + ***bod**: or, less likely, **sorn**); ?Rejarne (+ ***ros/rid, an**); perhaps Goldenharn (+ **guel, an**?); aj. Castle Horneck (+ **castell**)

horþ (m.; OCo.) 'ram', Voc. 603 glossing *aries*; MnCo. *hor*', pl. *hÿrroz*, Lh. 243c–4a. Welsh *hwrdd*: E. Evans, BBCS 25 (1972–74), 290–2. A verb *herthye* 'to thrust', RD 2286, etc., occurs: Welsh *hyrddu*, *hyrddio* 'to ram, push'.

A. ?Hor Point

C1. A compound of **horþ** + ***dyn**, presumably *horthyn or *hordyn, 'ram-fort' (meaning?) seems to have existed, occurring as A in Hurdon and as C2 in Benhurden (+ **ben**). A similar compound, **horþ** + ***lys**, seems to appear (as C2) in Carharles (+ ***ker**).

C2. Chyanhor (+ **chy, an**); Park en Hoar, *Park an Horr*, *Parkhour*, flds. (+ **park, (an)**); *Wheal an Hor*, mine (+ **wheyl, an**); ?Carn Hoar, rock (+ **carn**)

hos (m.) 'duck', OM 132; Old Cornish *hoet*, Voc. 506 glossing *aneta*; pl. *heidzhe*, Borlase, Vocabulary. Welsh *hwyad*, Breton *houad* (pl. *houidi*, PB, p. 55).

C2. Resores (+ **rid**); ?Park Ose, fld. (+ **park**); pl. Poligy, Polhigey (both + **pol**: cf. Breton Poul-an-Ouidi, Tanguy, p. 127)

houl (m.) 'sun'; *heuul*, Voc. 6 glossing *sol*; Welsh *haul*, Breton *heol* (Old Welsh and Breton *houl*, FD, pp. 141 and 214). Cf. Breton *Ker-en-heull* 1388 (RC 2 (1873–75), 211). For the adjective compare Welsh *heul(i)og*, Breton *heoliek*.

C2. ?Wheal Howl, mine; aj. Bosoljack (+ ***bod**)

*huel-gos (m.) 'high-wood', a compound of **ughel** and **cos**: the metathesis of *ugh-* to *hu-*, already common, was standardised in this compound. Further instances of **ughel** as C1 are listed under that element. Breton Huelgoat (2 exx., Tanguy, p. 100).

A. Hugus

C2. Trevilgus, Trevilges (both + **tre**); Bodulgate (+ ***bod**); Caruggatt (+ ***ker**); *Castel Uchel Coed* (+ **castell**: ÉC 14 (1974–75), 53; but the form looks Welsh)

hueth (*ue* = /œ/) 'easy, peaceful, happy', MC 225b; Welsh *hawdd*. (The adjective *hutyk*, OM 2818, *huthyk*, RD 2304, 'joyful, gleeful', and the verb, ppp. *huthys* 'gladdened', RD 483, pres. subj. *huththaho* 'may gladden', RD 1877, are derived from this: Loth, RC 23 (1902), 280.) This element may have been confused with, or even have fallen together with, another word, ***heth** 'peace', equivalent to Welsh *hedd*: cf. Middle Cornish *hethy* 'cease' (OM 2696 and PC 2096), Breton *hezañ* 's'arrêter, tarder' (Fleuriot, ÉC 11 (1964–67), 451–3). It is also liable to confusion with ***heth** 'stag' or 'barley'; but the Welsh parallels, or spellings in -*u*- (= /œ/), make it clear that *hueth* is the element concerned in the following names.

C1. Hennett, Henon, Huthnance, *Hethenaunt* (all + **nans**): cf. Welsh *Hoddnant*, several (e.g. EANC, pp. 151–4); glossed *vallis prospera*, VSBG, p. 202; also Hodnet (Shr., DEPN s.v.); and Old Breton *Hudnant* (Chrest., p. 154), Modern Breton Hesnant, 2 exx., Loth, RC 48 (1931), 354; but Loth derives these from Breton *hez = Cornish *heth; ?Headland (+ **lyn**: or ***heth**)

C2. Gunheath (+ **goon**)

hwilen, see under Wh.

I/Y (vocalic)

*-**i**, name suffix (mainly in river-names). See Ekwall, ERN, pp. lxxviif.; Thomas, EANC, pp. 127f.; LHEB, pp. 351–3. The suffix may occur in several names, such as R. Inny (+ **onnen** pl.); R. Fowey (+ ***faw**); Luney, r.n. (+ **leven**?); perhaps Elerkey (+ **elerhc**), ?*Bryghy as C2 in Suffree (+ **rid**, ***brygh**: formerly *Resvreghy*), ?Deli (+ **deyl**?)

-**yk**, adjectival or diminutive suffix, as in Old Cornish *fodic* 'fortunate', Voc. 304 (adjective from *fœd 'fortune', Middle Cornish *anfusyk* 'unhappy', RD 1520); Modern Cornish *temmig* 'a small fragment', Lh. 243b (diminutive of **tam** 'bit'); *s[p]ernic*, Voc. 696

glossing *frutex* 'shrub' (diminutive of **spern**); and *stefenic*, Voc. 48 glossing *palatum* 'palate' (diminutive of **sawn**). Welsh *-ig* (WG, pp. 230 and 257), Old Breton *-ic* (FG, p. 343; Fleuriot, ÉC 18 (1981), 105). The pair of names Kuggar and Corgerrick, a quarter of a mile apart, consists of an unknown word *coger or similar, and its diminutive in -yk; similarly the pair Prislow and Boslowick (formerly *Preslowyk* 1538), also a quarter-mile apart, seems to contain a name *Prysklou* (***prys**+ ***loch**) and its diminutive.

idhio (pl.; MnCo.) 'ivy', Lh. 15c and 65a; Welsh *eiddew*, Breton *iliav* (cf. HPB, pp. 664f., and FD, p. 168).

C2. Bosithow (+ ***bod**: cf. Breton Bodilio, Tanguy, p. 107)

yfarn 'hell'; Welsh *uffern* (cf. LHEB, p. 276), Breton *ifern* (cf. Tanguy, p. 76); note Breton coastal names such as An Ifern, Toull Ifern (Topon.Nautique 1419, nos. 8914 and 9655), and English ones such as Hell's Mouth.

C2. Halsferran (+ **als**)

*ygolen (f.) 'whetstone', equivalent to Welsh *agalen*, *calen*, Breton *higolenn* (cf. GMB, p. 320; also 'anvil' (?), GIB); Old Welsh *ocoluin* glossing *cos* (VVB, p. 198). Owing to the variant forms of this word in Welsh, the Cornish form is slightly uncertain, but the one given agrees both with the mediaeval spellings of the name and with Middle Breton. See Tanguy, p. 59, for Breton place-names containing the word, and see ***menawes** and EANC, p. 100, for names of tools used in place-names.

C2. Nancegollan (+ **nans**)

yn 'narrow', OM 961; Modern Cornish *edn*, Lh. 43a glossing *angustus*, and *idden*, N. Boson (Bosons, p. 25). Welsh *ing*, *yng* 'narrow', Breton *enk* 'étroit' (cf. GMB, p. 211); Old Irish *cumung* 'narrow' (Thurneysen, GOI, p. 115): the *-n* for *-ng* in Cornish is unexplained.

C2. *Street Eden alias Narrow Street* (+ ***stret**); ?*Park Ene*, fld. (or **on** pl.); *Codnidne* (+ **conna**)

ynter, yntre 'between'; Old Welsh *ithr* (Mart.Cap., VVB, p. 169); Old Breton *intr-*, *entr-* (FG, p. 382), Breton *etre*. The preposition occurs in *Croft inter Du-Henvor*, fld. (+ ***croft, dew**, ***hen-forth**); compare Irish Idir dhá Loch, Top.Hib., 1, 87.

*is[1], preposition, 'below'; Welsh *is* (GMW, pp. 41 and 202); Old and Middle Breton *is* (FD, p. 230; GMB, p. 340), Modern Breton *a-is* (Hemon, Grammar, p. 114). No clear instance of the preposition is found in the Cornish texts. It is seen in Welsh and

Breton phrases such as *teir eru iss rit deueit* 'three acres below sheep's ford' (LLD, p. 247), and *Tal-ar-iz-ar-sornigou* 'the brow below the springs' (Tanguy, p. 94; Ernault, RC 22 (1901), 376f.). It occurs in the following names: Trevisquite, Lesquite in Pelynt (both +**tre, cos**); Treskelly (+**tre, kelli**).

*-**ys**², suffix meaning 'place of'; cf. Tanguy, pp. 103-10; Breton Belerit, etc., 'place of cress', ibid., p. 111; *Galvezit* 'coudraie', GMB, p. 533. See G. Bernier, 'Les formations végétales en -*it*, -*ouet*', Ann.Bret. 78 (1971), 661-78. It occurs in such Cornish place-names as Skewes (3 exx.), Treskewes (2 exx.) (all +**scawen** pl.); ?Crannis (+***crann**?); Porthkerris (+***cors**?); Idless, ?Eddless (both +**aidlen** pl.?); ?Treburthes (+***perth**?). The first three of these elements also have adjectival forms in -**ek**, with the same meaning. There are other apparent suffixes, also meaning 'place of' a certain plant: *-*as* (see ***coll**); *-*es* (see **dar, dreyn**).

ysel 'low', OM 373, PC 136, BM 163; Welsh *isel*, Breton *izel* (a derivative of ***is**¹).

 A. Nanjizal (+**nans**)
 C2. spv. *ysa, ?Treeza (+**tre**)

***yslonk** 'abyss, cleft', equivalent to Breton *islonk* 'abîme, gouffre': cf. ***lonk** 'gullet, gorge'. A doubtful element.

 B2. ?Izzacampucca, Scilly (+**an, bucka**?)

yst 'east', BM 664; *yest*, CW 1743. Borrowed from English.

 C2. *Guall Est* (+**guel**); Park East, fld.

hiuin (OCo., *h-* inorganic, *u* = /v/), 'yew(s)', Voc. 676 glossing *taxus*. Welsh sg. *ywen*, pl. *yw*; Breton pl. *ivin*, sg. *ivinenn*. The discrepancy between the Breton and Welsh forms is unexplained (cf. HPB, p. 458n.); the Cornish word, with -*in*, is more likely to be plural, like the Breton, than singular. Note the Breton sg. *Caer'n Iuguinen* (Cart.Land., HPB, p. 456), *Kernivinen* (Bégard, p. 203); cf. Tanguy, p. 108. There is no evidence for a plural *yw: names that appear to contain this could equally well contain **ieu** 'yoke'. As in north-west Breton (HPB, pp. 457f.) the -*w*- became -*v*- in this word. Davey, Flora, p. 419, believed that yews in Cornwall were always introduced by man.

 A. double pl. *ivinow, ?Inow
 C2. pl. Nansevyn (+**nans**), ?Trevivian (+**tre**: or pers. ?)

I/Y (consonantal)

yar (f.) 'hen', Voc. 518 glossing *gallina*; OM 129; plural *yêr*, Lh. 243a. Welsh *iâr* (Old Welsh *iar*, Juvenc., VVB, p. 159), pl. *ieir*; Breton *yar*, pl. *yer* (cf. PB, pp. 71 and 87; Tanguy, p. 127).

C2. pl. Treheer, Treire, Treraire (all +**tre**), Trenear in Wendron, Treneere (both +**tre, an**), *Bodyer* (+ *****bod**); sg. or pl. Lanjore, Landare (both + **nans**: Welsh Nant yr Iar, ELlSG, p. 18)

*****yarl** (m.) 'earl'; *yurl* (read *yarl*), Voc. 171 glossing *comes vel consul*. Welsh *iarll*, borrowed from Old English or Norse (EEW, p. 41).

C2. Goonearl (+**goon**)

yeyn 'cold', PC 1209, etc.; *eyn*, MC 207d; *iein*, Voc. 474 glossing *frigus*. Breton *yen*; cf. Welsh *iäen* 'piece of ice', used as a rivername (2 exx., EANC, p. 119).

C2. Carrine (+ *****ker**); Lanyon (+**lyn**); *fonton gén* (Sawyer, no. 755), Venton Ends, Ventonjean (all +**fenten**); ?Lantyan (+**nans**)

*****-yel**, *****-el**, adjectival suffix equivalent to Welsh -(*i*)*ol* (Middle Welsh -(*y*)*awl*), Old Breton -(*i*)*ol*: see WG, p. 256, and FG, pp. 245 and 359. There is a supposed Celtic element in English place-names, *****ial*, said to mean 'fertile upland' (see Elements, I, 279; BBCS 14 (1950–52), 117f.), which should probably be replaced with this suffix: it is based upon a supposed Welsh word *iâl* (itself a ghost-word, derived from the obscure place-name Iâl, Den.: see GPC, s.v.), and a supposed Gaulish suffix -*oialum*. The equation between the Gaulish suffix and the supposed Welsh word was first made by Thurneysen (*Zeitschrift für romanische Philologie* 15 (1891), 268; cf. RC 12 (1891), 391); and has since been followed by nearly all authorities: Dottin, p. 74; Vendryes, RC 39, 369; Ogam 9 (1957), 271f. Since the Welsh word is almost certainly non-existent, the equation is invalid, and in fact the Gaulish suffix is itself only poorly attested, and ought perhaps instead to be identified with this adjectival ending: on it, see Longnon, RC 13 (1892), 361–7 (the best account), with further examples, RC 20 (1899), 443f., and RC 39 (1922), 334–7. It occurs in the names *Argentoialo* (DAG, p. 596, = Argenteuil), *Nantoialus* (ACS, II, 685f., = Nanteuil) and *Maroialus* (ibid., II, 434f., = Mareuil); but the last two occur also in the forms *Nantogilum/*

Nantoilus and *Marogilum* (ibid., and GPN, pp. 236 and 228), and the exact form of the suffix is thus doubtful. All the other names cited as containing it, notably those containing words for types of tree, are starred forms (e.g. ACS, I, 647, 824 and 1400; III, 224 and 470) – a fact not always made clear by their proponents: they could equally appropriately contain an adjectival suffix, cognate with the one discussed here (and they would thus be precursors of the type containing the different suffix *-ek* in Cornish and Breton, attached especially to words for plants). If so, then the Gaulish suffix would need to have been, not *-oiălon* (which would be needed to give Welsh *iâl), but *-oiālon* (giving Middle Welsh *-(y)awl*, etc.); but this would cause problems with the phonology of the French names. Note that Whatmough (DAG, p. 613) allowed a suffix *-oiolum* 'finis'.

Some of the names listed below show the suffix as ***-el** (rather than ***-yel**), as in the Welsh and Breton forms, or else vary between the two. The meaning is presumably '(place) abounding in X', as with the commoner and more productive **-ek**. It is likely that the suffix is an earlier one than **-ek**, if one judges by its (suggested) occurrences in Gaulish, by the obscurity of some of the Cornish names containing it, by the fact that a high proportion of those names appear to be ancient, being names of districts or of Domesday manors, and by its fair number of (suggested) attestations among English place-names.

The suffix seems to occur in the following names and elements: *gwyth-yel (+**gwyth**[1]: Withiel; Lostwithiel, +**lost**); *devr-el (+***devr-**: Deveral, Derval; cf. Deverill, PNWlt., pp. 6f.); *tyn-yel (+ ?: Tinnel; cf. possibly Gaulish *Tannoialon, RC 39, 334-7, and/or Tindale, PNCmb., I, 36); *dyn-yel (+***dyn** ?: Bodiniel, +***bod**); *bran-el (+ ?: Brannel); *kem-yel (+ ?: Kemyel); **ken-(y)el* (+ ?: Kennall); **kev(e)r-el* (+***kevar** ?: Keveral); *myn-yel (+**myn** ?: Treveniel, +**tre**); ?*cader-yel (+**cadar** ?; cf. OWe. *catteiraul*, VVB, p. 66: Tregatherall, +**tre**); *bryth-yel (+***bryth** ?: Trythall, +**tre**); *ryv-yel (+ ?: Bodrivial, +***bod**); *kew-el (+***kew** ?; Welsh *ceuol*: Tregole, +**tre**; Cargoll, +***ker**); ***gothel**?

-yer, plural ending (LlCC, p. 12) equivalent to early Welsh *-awr* (WG, p. 210; GMW, p. 28), Breton *-(y)er* (FG, pp. 215 and 230f.; Hemon, pp. 33 and 38; PB, pp. 65-7 and 168-70). To Lewis' Cornish example of the ending can be added *syehar*, Trg. 6a, 'sacks'

yet

(plural of OCo. *sach*, Voc. 388; cf. Breton pl. *seier, syher*, GMB, p. 590; PB, pp. 66 and 169), and the following examples from Cornish place-names; they were first identified by Quentel, Ogam 7 (1955), 239-42. For a similar formation in a Welsh place-name, note Sirior (Seriör), Den. (BBCS 11 (1941-44), 148f.; ÉC 11 (1964-67), 383).

The suffix occurs in: Sticker (v. **stoc**); Bryher (v. ***bre**); ?Kerrier (v. ***ker**); possibly **kylyer*, occurring in a number of names (v. ***kyl**).

yet 'gate', OM 743 and 764; plural *yettys*, PC 3039, etc. Borrowed from English, like Welsh *giât* (EEW, p. 221).

C2. flds. Park an Yate, *Park-an-yeat*; pl. Park Yates

ieu (f.) 'yoke', Voc. 345 glossing *iugum*; Welsh *iau* (Old Welsh *iou* glossing *iugum*, VVB, p. 165), Breton *yev*. See Jackson, JCS 1 (1949-50), 75f., and HPB, p. 238, n. 4. Confusable with a possible **yw* 'yew-trees', if that exists (see **hiuin**).

C2. ?Treyew (+**tre**); ?Peloe (+**pol**)

yorch 'roe-deer', Voc. 587 glossing *caprea*. Welsh *iwrch*, Breton *yourc'h* (both masculine, 'roe-buck'). The plural was apparently **yergh*, equivalent to Middle Welsh *yrch* (< **iyrch*, GMW, p. 6), and the diminutive, if present, would have been **yorghell*, equivalent to Welsh *iyrchell*.

C2. Carnyorth (+**carn**); pl. Lanjeth (+**nans**); dimin. ?Crawle (+***ker**)

yow 'Thursday', in *Du yow*, Lh. 232a; *Iow*, PC 2668 (sic MS: Nance, OC 3 (1937-42), 426). Welsh (*Dydd*) *Iau*, Breton *Yaou*. The market implied in the name is contrasted with the adjacent Marazion or **marghas byghan* 'little market', and is referred to as *marcatum...die quinte ferie* in c. 1070 (Cart.St M., no. 1); Leland, who has a unique form including *deyth, Marhasdeythyow*, glosses it as *forum Iovis* (Itinerary, 1, 319).

C2. Market Jew (+**marghas**)

*yuf (?) 'lord', equivalent to early Welsh **udd* (< **iuð*, LHEB, p. 345), Old Breton *Iud-* in personal names (Chrest., pp. 142f.), with early confusion of -ð/-v as in Welsh Cardi*ff*/Caerdy*dd*, and elsewhere in Brittonic (cf. s.v. **skath**, and WG, p. 177; HPB, p. 664). But the same element, where present in personal names, survived as **Yuð-*, as shown in Bodm. *Gyðiccael* and various place-names.

C2. Treeve in Sennen, Treave, Trew in Tresmeer, ?Treeve in Phillack, ?Trereife (all +**tre**); Bodieve, Bogee, Bojea (all + ***bod**)

J

jarden 'garden', CW 1801; *dzharn*, Lh. 66a glossing *hortus*. Breton *jard(r)in*, borrowed from French (Piette, p. 135). The form of the Cornish word shows that it was not borrowed from English *garden*, but comes from the French – presumably through Breton.
  C2. Park an Jarne, Park Jearn, *Park an Jarns*, etc., flds.

gentyl 'noble, well-born', OM 1566, etc.; borrowed from English, like Breton *jentil* from French (Piette, p. 125).
  C2. Wheal Dees Gentle (+**wheyl**, (**an**), **tus**)

L

lakka 'stream, leat, well', Lh. 132c glossing *puteus* 'a well or pit'; Pryce *lakka* 'a well, a pit, or rather a rivulet; which we still call a lake, and leak, or leate'. Borrowed from English.
  B2. Lackavear, fld. (+**meur**)

lader (m.) 'thief', Voc. 299 glossing *latro*; etc.; pl. *laddron*, PC 336 and 2255. Welsh *lleidr*, pl. *lladron* (LHEB, p. 593); Breton *laer*, pl. *laeron*. Both the Welsh and the Cornish words seem especially liable to occur in cliff-names: in Wales, Llam Lleidr, Crn., and Llam y Lladron, Mer., both have onomastic tales about thieves (J. Daniel, *Archaeologia Lleynensis* (Bangor, 1892), pp. 168f.; Williams, Lexicon, p. 232a). It is possible that this is due to some confusion between *lleidr*/*lader* and *llethr*/**lether* 'cliff', though the two ought to be distinguishable. Further instances of the plural in place-names are: Old Welsh *penn Luhin latron* 'the end of thieves' grove' (LLD, p. 146); Welsh Nant y Lladron, Mer. (SH 7839: cf. Lanhadron, below); Middle Breton *Poul-lazron*, Chrest., p. 217.
  C2. sg. ?Chair Ladder, coast (+**cheer**: cf. Padel, Bosons, pp. 12–13: or ***lether**); pl. Lanhadron (+**nans**), Stennack Ladern, mine (+**stean** aj.)

*ladres (?) 'sluice', a possible word which would be equivalent to Breton *laerez* 'bonde d'étang', if that could be from an older **ladres* (cf. HPB, pp. 484–7); alternatively **ladres* would be 'female thief', equivalent to Breton *laerez* 'voleuse' (Le Gonidec).
  C2. ?Polladras (+**pol**)

lam (m.) 'a leap'; Welsh *llam*, Breton *lamm* (meaning 'waterfall' in place-names: Tanguy, p. 90). Place-names beginning with Welsh

llam are quite common: e.g. Llam y Trwsgl 'the bungler's leap' (ELlSG, p. 75); Llamyrewig 'the hind's leap', Mtg. (Rhestr, p. 57).
    A. ?Labham Rock
    B2. Lambledra (+ ***lether**); Lamlavery, Lambsowden (both + ?); ?*Loban Rath* (+**an**, **gruah**? See C. Thomas, Co.Arch. 3 (1964), 78)

lanherch (OCo.: first *h* silent) 'clearing', Voc. 710 glossing *saltus*; Welsh *llannerch* (cf. PNFli., p. 95); for Cumbric examples, see Watson, CPNS, p. 356; PNCmb., I, 71 (Lanercost: cf. I. Williams, 'Lanercost', Cumbs. and Westms.Trans. n.s. 52 (1953), 67–69; Jackson, O'Donnells, p. 80), 72 (*Lanrekaythin*), and 115 (Lanerton). See also Elements, II, 15; JEPNS 1, (1968–69), 48.
    A. Landrake, Larrick (2 exx.), Lanner (5 exx.), Muchlarnick (+ OE *micel*); if Lanarth (2 exx.) is the simplex **lanherch**, it has in both cases undergone an unusual stress-shift; pl. Lanhargy
    B2. ?Lambest (+ ?)

lann** (f.; OCo.) 'enclosed cemetery', Welsh *llan*, Breton *lann (Chrest., pp. 144 and 216). See the discussion, Padel, Co.Studies 4/5 (1976–77), 15–27; about fifty (or just under one in four) of Cornish parish churches are known to have had names in ***lann**, the vagueness arising from the occasional problem of knowing whether a *Lan-* name near a parish church was actually an alternative name for the church-site itself, or merely a nearby place. Of these fifty, twenty have as their qualifiers a form of the patron saint's name, e.g. Lantinning (= St Anthony in Meneage, patron originally St Antoninus or *Entenin*), La Feock (= Feock, patron St Feock), etc. Churches which are known to have had religious communities in the pre-Norman period are especially liable to conform to this type (e.g. Constantine, Kea, St Keverne, etc.). About as many again of the fifty parish-church names contain as qualifier a personal name *not* that of the patron saint, e.g. Linkinhorne (lann** + *Kenhoarn, dedicated to St Mylor), Lanwethenek (= Padstow, ***lann** + *Gwethenek, dedicated to St Petrock), a problem here being that of saying whether a qualifier is actually a personal name or not: in some cases the qualifier may have been a pre-existing place-name, e.g. Lanivet, *Lanuthno* (= St Erth), *Lannaled (= St Germans). A few of these parish-church names have qualifiers that are clearly topographical: Lawithick (= Mylor, ***lann** + **gwyth**[1] aj.), *Landraith* (= St Blazey, ***lann**

+trait). When it comes to *lann in names other than those of parish churches, the matter is much less certain: here it is often impossible to distinguish *lann from nans, or to a lesser extent from lyn. The interchange as B2 between *lann and nans is common to all the Brittonic countries, including Cumberland: Loth, RC 33 (1912), 271n.; Richards, SC 3 (1968), 16; Tanguy, p. 78; Padel, BBCS 29 (1981), 524. There are no criteria by which the two can safely be distinguished. On the whole it is true that *Nans-* became *Lan-* more often than a name in *Lan-* gave forms in *Nans-*; but two good examples can be given of a *Lan-* name with forms in *Nans-*: Landewednack (*Landewenek* 1284) was spelt *Nansowenacke* in 1619; and Lanlawren (*Lanlovern* c. 930) was spelt *Nanslowarn* in 1356. However, clear examples of that change are rare, while the opposite change, *Nans- > Lan-*, is demonstrably common: thus the existence for a given name of forms in *Nans-, Lans-* or *Lant-* should be taken as a strong suggestion that nans was the original element. That change could happen as early as the eleventh century, as is shown by the forms *Lancichuc* 1086 (= Nancekuke, earlier *non cuic* 960: Sawyer, no. 684); *Lantien* 1086 (= Lantyan, *Nantyan* 1296; nans+yeyn?); *Lantmatin* 1086 (= Lametton). Occasionally the sense or topography demands that a name should contain nans rather than *lann, even when no available form suggests it: thus Lambrenny is clearly 'valley of (round) hills', and Lawhibbet and Lawhippet are more likely to be 'valley of gnats', though none of these has forms beginning with *N-*. Phonologically, the unvoicing of the second element can also suggest that the *-s* or *-t* of *Nans-* was originally present: e.g. Lantreise is probably nans+dreys, and Lampetho is probably nans+beth pl. or bedewen pl., though both lack forms in *N-*. However, the presence or absence of lenition is not such a reliable guide, even though *lann is feminine and nans masculine: lenition is often absent where it should grammatically have been present, so that a *lann name need not show it (e.g. Linkinhorne, *lann+*Kenhoarn); and a feminine noun can be lenited (by a lost definite article) even after nans, e.g. Nancewrath or Lewrath (nans+(an)+gruah). On the other hand, lenition of an adjective or proper noun (neither of which would have been preceded by the definite article) is a good indication that the original generic was feminine, therefore *lann: e.g. Lanvean (*lann+byghan), and Langunnett (*lann+ *Kenwyd or *Conwyd, cf. Llangynwyd, Gla.). From all this it will

*lann

be clear that linguistically it is often impossible to say whether a particular name contains one element or the other.

The problem is made worse by another factor: the lack of coincidence, except at parish churches, between apparent *lann names and documented chapel sites or surviving enclosed cemeteries. Of non-parochial chapels with mediaeval documentation, the only ones possessing names in *lann are St Mawes (*Lavousa*), St Michael's in Padstow (*Leveals*) and Lellissick; of non-parochial graveyards (lacking mediaeval documentation), the only ones with names in *lann are Helland in Mabe and Lanvean. Thus the absence of a known chapel or graveyard seems to be almost irrelevant for the purposes of deciding whether a name contains *lann; this in turn raises the awkward possibility that *lann could sometimes have had a secular sense of 'enclosure', as it certainly did in some compounds. But there is no need for that to be so.

A few other details can be added to the discussion cited above. The phenomenon of *lann being qualified by a personal name not that of the patron saint may have a partial explanation in an instance noted by Wendy Davies in the Book of Llandaf, where the name following *Llan* is that of the first incumbent (LLD, p. 162), *Lann Guorboe* (now Garway, Hrc.): *Gvoruodu rex... dedit...agrum...deo et sancto Dubricio; ...fundauit locum in honore sancte trinitatis, et ibi guoruoe sacerdotem suum posuit*: cf. W. Davies, Microcosm, pp. 42, 58 and 143; note also LLD, p. 73, where *podum Iunabui*, now Llandinabo, Hre., was given to St Dubricius and one *iunabui presbiter* was one of the witnesses. A further instance of *llan* glossed *monasterium* occurs in Rhigyfarch, §1, *Depositi monasterium* for the *llann adneu* of RP, 1188, line 20 (cf. Jackson, Gnomic Poems, p. 51). On the derivation of *lann, and a few cognates and compounds, see VKG, II, 3. The only indication that can be given of the date-range of *lann is that it does not seem to become commoner as one goes further west in Cornwall, and therefore was out of use by c. 1200 at the latest; it may well be that it was largely out of use by an earlier date than that. The distribution of names in *lann is fairly even across the county, but there are particular concentrations. They tend to occur more towards the coast than inland, and especially around estuaries or along the major river-valleys, notably in the Fal and Tamar basins; also in Westwivelshire south of the River Fowey.

B1. Talland (+**tal**); ?Sellan (+**segh**: **lyn** would make better sense, but spellings from 1262 have -*lan*); Bohelland (+ ?); see also ***bow-lann**, ***cor-lann** and ***hen-lann**.

B2. Qualified by words for plants: Landreyne, Landrine (both +**dreyn**; or **nans**); Lewarne (+**guern**); Lambessow (+**bedewen** pl.); Lawithick (+**gwyth**[1] aj.)

Qualified by other topographical term: *Landraith* (= St Blazey; +**trait**); Lanvean, Laddenvean (both +**byghan**); Lanlawren (+**lowarn**); ?Levardro (+**bar** ?, **trogh** ?).

Qualified by words denoting people: Lawhyre (+**wuir**); Lanhinsworth (+**hynse**); Lellissick, Lowlizzick, ?*Lowlysycke* (all +***gwlesyk** or pers.); *Lamana* (+**manach**); Lezant (+**sans**); Locrenton, Longcoe, La Feock, Leyowne, Launcells, Laneast, Linkinhorne, Landulph, Lanow, *Lankyp*, Lanreath, Lansallos, Lanhydrock, Lanlovey, etc. (all +pers.)

Owing to the uncertainties, it is impossible to estimate the number of names in ***lann-** in Cornwall, but it cannot be less than 60 (including those of parish churches), and it is most unlikely to be as many as 140. The true figure is probably not more than 100.

C2. Trelan (+**tre**); Clann (+**kelli**)

le 'place'; Welsh *lle*, early Welsh *Matle* glossed *bonus locus* (LLD, p. 79), *guliple* 'wet-place' (LLD, p. 242); early Breton *le* (FD, p. 238). In Cornish place-names the element is not common, and occurs mainly as the generic in compounds.

B1. ?Crylla (+***cryn** ?: the same compound may itself be used as C2 in Tregrill, +**tre**); ?*Bothle* (+***both** ?); Methleigh (+***með**; 2 exx.); Durla, Durlah, flds., Durloe (all +**dour**). See also *mag-le s.v. **maga**, and perhaps ***emle**.

lek 'lay(-man), secular', PC 38 and 681; *leic*, Voc. 119 glossing *laicus*. Welsh *lleyg*, Breton *lik*.

C2. ?Ventonleague, ?Vent-an-League, fld. (both +**fenten**)

ledan 'wide', OM 2261; *leden*, MC 237b; Old Cornish -*lidan*, Voc. 651. Welsh *llydan*; Breton *ledan*, Old Breton *letan*, *litan* (FD, pp. 241 and 244). In west Cornwall, where -*nn* becomes -*dn*, this element is confusable with *lidn (< **lyn** 'pool') unless early forms are available.

C1. *penn lidanuwern* (+**pen, guern**: Sawyer, no. 1019)

C2. Carluddon (+**cruc**); Roseladden (+**rid**: cf. Old Welsh *rit litan*, LLD, p. 258; Modern Welsh *Rhyd Lydan*, ÉC 10 (1962–63), 212); *caer lydan* (+***ker**: Sawyer, no. 770); *Menelidan*

*lefant

(+**meneth**); ?Vounderlidden, fld. (+**bounder**); ?Polledan, ?Pol Ledan, ?Porth Ledden, all coastal (all +**pol/porth**)

***lefant** (OCo.) 'toad', Welsh *llyffant* (Greene, Celtica 2 (1952–54), 148–9); this word had been replaced in Cornish, as early as the 12th century (Voc.), by the circumlocutory **cronek**. Compare the Middle Breton personal epithet *Eudo Leffant*, Beauport 1301 (RC 7 (1886), 61).

C2. Polyphant (+**pol**; cf. Old Welsh *pull lifan*, LLD, p. 229)

***legh** (f.) 'flat stone, slab', Welsh *llech*, Middle Breton *lec'h* (GMB, p. 360); see also the compound ***cromlegh** 'curved-slab, dolmen'. Some Welsh instances: *ad petram...o'r lech* 'to the rock...from the rock' (LLD, p. 122); *super saxum unum...id est Lech meneich* (ibid., p. 128). The normal plural in Cornish seems to have been *leghyon, but one name, Treliggo, may suggest a plural **legh(y)ow*, as in Old Welsh *nant lechou* (LLD, p. 226). Note Lhuyd's Cornish diminutive *lehan*, 13c.

A. derivative *legha 'place of flat stones' (+ *-**a** ?), ?Leah, ?Leha

B1. ?Trilley Rock, ?Trilley (both +**try**^{1} ?)

B2. Lewins (+**guyn**); ?Lethlean (+**lyen**: a drying-stone ?); ?Le Scathe Cove (+**skath**)

C2. Penlee Point (+**pen**; 2 exx.); Goonleigh, Gwealeath (both +**goon**); Fenterleigh (+**fenten**); Treslay (+***ros** or **rid**); Releath, Rillaton (+Eng.), Treslea (all +**rid**: cf. Welsh Rhyd y Llech, ÉC 10 (1962–63), 216); Treleigh in Redruth (+**tre**: cf. Old Welsh *Tremlech*, VSBG, p. 120); Boleigh (+***bod**); Portlegh (+**porth**); ?Penderleath (+***pen (an) dre** ?); pl. Carlyon (2 exx.), Carleon in Morval (all +***ker**), Trelion, Treloyan (both +**tre**), ?*Vougo Lion*, fld. (+**vooga**); pl. **legh(y)ow*, ?Treliggo (+**tre**)

***leyn** 'stitch, strip of land', equivalent to Welsh *llain* 'patch, piece, slip of land' (cf. ELl, p. 81; and Llainmoidir, Lochlann 2, 132). The element is liable to confusion with **lyn** 'pool', even if early forms are available: only the length of vowel in the modern forms might help to distinguish them.

A. Lean (2 exx.: cf. Middle Breton *villa Lein*, Chrest., p. 217)

B2. All of the following are fields: Lean Ridden (+**reden**); ?Lenears (+**hyr** ?); *Lean braus* (+**bras**); *Lene Nyendorwhy* (+Eng., 'near the doorway' ?); *Lene ware an Stenack* (+**war, an, stean** aj.); *Lene an Garrack* (+**an, karrek**); *Lene an Speren* (+**an, spern** ?); *Leane an Trapp* (+**an,** ***trap**); *Lene Toll* (+**toll**)

C2. pl. ?Bolenna, ?Bolenowe (both +***bod**); pl. ?Colona

(+ ***ker**). Some names given under ***pen-lyn** may belong here.

les 'plant', Voc. 628 glossing *herba*; most other forms show the stem *los-*, without *i*-affection (cf. LHEB, p. 595): pl. *losow* 'plants', OM 28 and 77 (= Welsh *llysiau*, Breton *louzoù*: cf. Tanguy, p. 114); double pl. *losowys* 'plants', OM 31 and 1742; singulative *losowen* 'herb, remedy', BM 1483 (= Welsh *llysieuyn*, Middle Breton *lousouenn*, GMB, p. 377). The sg. **les** is confusable with *les* 'benefit' (Voc. 323 glossing *commodum*; Welsh *lles*), and with *les* 'width' (OM 958, etc.; Welsh *lled*, Breton *led*), if those occur in place-names; and perhaps with ***lus** 'bilberries' (Welsh *llus*, Breton *lus*), if that occurs at all; also perhaps with ***lys** 'court' in a few cases. Most of the place-names seem to show a plural, with the form *losyow or *lesyow, with *-sy-* > *-dj-*, rather than the attested *losow*.

    A. pl. ?*Lidgiow*, ?*Lidgey*
    B2. pl. ?*Legereath* (+ **gruah**)
    C2. sg. *les*, ?*Perlees* (+ **pen**); pl. *Trelossa* (+ **tre**), ?*Ponslego* (+ **pons**), ?*Carn Lodgia, coast* (+ **carn**), ?*Bithen Lidgeo*, fld. (+ **budin**); aj. *lesek, ?*Landlizzick Wood* (+ **nans**); aj. *losowek (cf. Breton *louzaouek*), *Porth Joke* (+ **porth**)

leskys 'burnt', OM 433 and 1355, ppp. of *lesky* 'burn', OM 430, etc. (cf. **losc**); Welsh *llosgi*, Breton *leskiñ*; Old Breton ppp. *loscheit* (Chrest., p. 146, FG, p. 314).
    C2. *Carn Leskys, coast* (+ **carn**)

lester 'boat, ship', OM 956, etc.; pl. *listri*, Voc. 270, *lestri* 'vessels', Lh. 116b: Welsh *llestr*, pl. *llestri*; Breton *lestr*, pl. *listri*.
    C2. *Savenlester Cove* (+ **sawn**); *Bolster* (+ ***both**); pl. *Pelistry* (+ **porth**)

leth 'milk', PC 3138; *leyth*, OM 1430; Old Cornish *lait* ($t = /\theta/$), Voc. 867 glossing *lac*. Welsh *llaeth*, Breton *laezh*. Compare **leuerid**.
    A. aj. pl. ('milkinesses, milky places'), ?*Nathaga Rocks, coast*
    C1. See ***lety**.
    C2. *Menedlaed* (+ **meneth**)

***lether** (?) 'cliff, steep slope'; equivalent to Welsh *llethr*, Old Irish *leittir*. Note early Welsh *llethir y brin* 'the hill-side', BBC 68.14. Pryce's Cornish *ledr*, *ledra* 'a cliff, a steep hill' were probably invented in order to explain *Lambledra*, as he cites that place-name under the word. As A, cf. *Lether Hall*, Gla. (PNDinas Powys,

p. 225), and Laughter Tor and Leather Tor (PNDev., I, 195 and 245). Welsh *llethr* in place-names usually seems to take the form *lledr* (EANC, p. 16), and the same applies to the form in the Cornish names. (But Lloyd-Jones, BBCS 2 (1923-25), 112, explains *lledr* as due to the influence of Irish *leittir*, which is not likely in Cornwall.)

A. ?*Lettor*, ?Ledrah

B2. ?Luthergwearne (+**guern**); ?*Ladder Winze*, mine (+**guyns**); ?Latter Scraggan, ?Latereena, flds. (both + ?)

C2. Lambledra (+**lam**); Meledor (+**men** sg. or pl.); ?Chair Ladder, coast (+**cheer**: or **lader**); ?*Pedenleda*, fld. (+**pen**)

***lety** (m.) 'milk-house, dairy', a compound of **leth**+**ti** (= **chy**, with preservation of *t*- by the -*th* of **leth**). Old Welsh *Laithti Teliau* (LLD, pp. 124 and 255), Breton *laezhdi* (cf. Quentel, Ogam 7 (1955), 241f.); Old Breton *Laedti* (Chrest., p. 167); as qualifier, note Welsh Blaenllaethdy, Crm. See also Nance, 'The names Gwavas, Hendra and Laity', OC 1, iv (1926), 32-4. There is apparently no reflex of Welsh *llety* 'inn' in Cornish.

A. Laity (7 exx.)

C2. Polita (+**park, an**)

leven 'smooth, even', Lh. 63c glossing *glaber*; Welsh *llyfn*, Old Breton *limn* (FD, p. 242).

A. ?Luney, r.n. (+ *-**i**: cf. Welsh Llyfni, etc., EANC, pp. 159-62; Ekwall, ERN, pp. 251f. and 271); a possible verb *levna 'to level, smoothen' (Welsh *llyfnu*) may appear as ppp. in Levenus Hill.

C2. Porthleven (+**porth**): here also **leven** may be a simplex r.n.

leuerid (OCo., u = /v/, d = /θ/) 'sweet (i.e. not soured) milk', Voc. 868 glossing *lac dulce*; Welsh *llefrith*, Breton *livrizh*. But 'milk' was normally **leth**.

C2. Trelevra (+**tre**)

lêuiader (MnCo.) 'the master, or pilot of a ship', Pryce, s.v. †*leuint*; Welsh *llywiawdr*. Pryce's word is probably cribbed from Lhuyd's Welsh *lhywiawdwr* (137a glossing *rector*) and his Breton *leviader* (97b glossing *nauclerus*: but the Breton word does not appear in standard dictionaries); nevertheless, the word (or the word used as a personal name) occurs in one place-name.

C2. Trelawder (+**tre**)

leuuit (OCo., -*t* = /ð/) 'pilot', Voc. 274 glossing *gubernator vel nauclerus*; Welsh *llywydd* 'leader'. Or the place-names may contain the same word, but as a personal name, like Old Welsh *Liuit* (Chad 2), if that is for *Llywydd (but cf. CMCS 7, 95).

C2. Treloweth (+**tre**; 3 exx.)

lyen 'cloth', PC 3204, RD 1693; *lien*-, -*lien*, Voc. 792, 805 and 871. Welsh *lliain*, Breton *lien* (cf. FD, p. 242). The word is correctly scanned as a disyllable in Middle Cornish. It is confusable with a different word, *lyn* 'linen', PC 836 (*lin*, Voc. 830 glossing *linum*; Welsh *llin* 'flax', Breton *lin*), which may occur in some of the names. It may also be confusable with **leyn** 'strip' and **lyn** 'pool'.

C2. ?Lethlean (+ ***legh** ?); ?Chellean (+**chy**); Park Lean, fld. (adjoins Flax Field)

lyn (m.) 'pool, pond', MC 221c and BM 3504 (both meaning 'liquid, blood'); Old Cornish -*lin* in *pisclin* 'fish-pond', Voc. 740, and **grelin** 'cattle-pond'. Welsh *llyn*, Old Welsh *linn* (LLD, pp. 146 and 214), pl. *marulinniou* 'dead-pools' (LLD, p. 183); Pictish *Lin Garan* 'crane's pond' (Antiquity 29 (1955), 78); Breton *lenn* (cf. Tanguy, pp. 95f.; Chrest., pp. 144 and 217), Old Breton *lin*, pl. *linnou* (FD, p. 243). Gaulish *linda* 'drinks', Vendryes, ÉC 7 (1955-56), 14-16; Romano-British *Lindum* (= Lincoln), etc., PNRB, pp. 391-3. In some of the place-names, the element probably means 'stream' rather than 'pool'; it may also mean 'bay, cove' occasionally. The word was apparently in use as a dialect word in the 17th century: 'the *lydne* neare Meane Ambeare', 1613. It is often difficult to distinguish **lyn** as B from ***lann**/**nans**, and also (apart from vowel length) from ***leyn**.

A. Lidden; ?Gurlyn (+ ***gor-** ?); Harlyn (+ ***ar**)

B1. Herland (+**hyr**; 2 exx.); ?Sellan (+**segh**: or ***lann** ?); Newlyn in Paul (+**lu** ?); Medlyn (+ ***með**); Headland (+ ***heth** or **hueth**)

B2. *le(i)n broinn*, *lenbrun* (Sawyer, nos. 755 and 832/1027), ?Lambourne (2 exx.) (all + **bronnen** pl.); *lyncenin* (+**kenin**: Sawyer, nos. 832/1027); *Lentrisidyn* (+p.n.: Sawyer, no. 450); Lanjew (+**du**); ?Langarth (+**an**, **cath**); Lannear (+**êr²** ?); ?Legonna (+ ?); Landerio (+ ?); ?Lanagan (+ ?); ?Lemarth (+**margh**); Lanyon (+**yeyn**)

C1. ?Lendra (+**tre** ?)

C2. Trelin, Trelyn (both +**tre**); Roselidden (+ ***ros**); *Corslen* (+ ***cors**); Barlendew (+**bar**, **du**); pl. Fentenluna (+**fenten**), Trelinnoe (+**tre**); aj. ?Carlannick (+ ***ker**); see also the phrase ***pen lyn**.

lynas

lynas (coll.) 'nettles', Trg. 9, *linaz*, Lh. 178a; sg. *linhaden*, Voc. 649 glossing *urtica*. Welsh *danadl*, Breton *linad* (cf. Tanguy, p. 114; Nettlau, RC 10 (1889), 327f.; and LEIA, pp. N9–10).
    C2. ?Park Lines, fld.

*lys (m./f. ?) 'court', also perhaps 'ruin'. Welsh *llys*, Breton *lez*, both 'court'. Note early Welsh *llys*, twice glossed *curia: dignitas curie, id est, breynt llyss* (Latin Texts, p. 121); *curiam Lisarcors* (VSBG, p. 190). On Irish *lios* see D. Flanagan, Onoma 17 (1972–73), 165–7. It is usually assumed that the word implies a former administrative centre, of the period before Cornwall was conquered by the English in 838, but that is not necessarily so. For one thing, the element is by no means evenly distributed: there are no instances at all in the hundreds of Pydar, Stratton and Eastwivelshire, whereas there are three close together (and at least three further ones) in the hundred of Kerrier. Moreover, several names in *Lys-* are not known ever to have been important administratively or manorially. Where ***lys** is qualified by words denoting livestock, or plants, or rocks, the implication is very strong that the word had come to denote 'ruins' of indeterminate date, and that the name, when given, was a retrospective one. That is particularly true of Leskernick in Altarnun and of Lesingey (at both of which the remains can still be seen) and of Hellcs- in St Ives; but also of Lisbue and Lezerea. Lestowder may be a special case where the name was given from folklore (the legendary tyrant Teuder of saints' lives). However, where a name in ***lys** is known to have been a manor at some time in the middle ages, it is perhaps reasonable to assume that the place was also an administrative centre in the pre-English period: this applies to Lesneage, Lizard and Helston in the hundred of Kerrier, to Liskeard in Westwivelshire, to Helstone and to Lesnewth (which presumably replaced Helstone) in the hundreds of Trigg and Lesnewth, and to Arrallas, originally in the hundred of Powder. Loth, RC 24 (1903), 292, says that the 9th and 10th-century Breton charters show local chiefs each based at a *lis*. When used as a suffix, the element may refer to a manorial court, and thus be quite late (after the Norman Conquest) in date.

    B1. Arrallas (+**arghans**); ?*Bogullas* (+**bugel**?); ?Scraesdon (+Eng. *dūn*), ?Crowlas (both +**krow**; or ***crew**+**rid**). Note also the compounds ***cad-lys**, ***hen-lys** and ***horþ-lys** (see **horþ**).

    B2. Lesneage (+p.n., see **manach** aj.); Lesnewth (+**nowyth**); Lesingey, Lanescot, Lestowder (all +pers.); Lizard (+***arð**);

150

Leskernick (+**carn** aj.; 2 exx.); ?Lestrainess (+**dreyn** deriv.?); Lezerea (+***gre**); Lisbue (+***byu**); Leswidden (+**guyn**); Liskeard (+**carow** pl.?)

C2. Trelease in Kea and St Keverne (both +**tre**): cf. Treales (PNLnc., p. 152), Welsh Treflys (Crn.), Middle Breton *Trefles* (Chrest., p. 234); ?Stenalees (+**stean** aj.)

D. Ludgvan Lease (cf. *Lusuon-eglos*); ?Chybarles; ?Pensagillies (also *Pensugemur* 'Great P.')

lyw 'colour'; *liu*, Voc. 481 glossing *color*. Welsh *lliw*, Breton *liv*.

A. ?Lucoombe (= r.n. + Eng. *cumb*): cf. Lew, r.n. (PNDev., I, 8: 2 exx.), Welsh Lliw (several, EANC, p. 121; Ekwall, ERN, p. 253).

C2. Menerlue (+**meneth**); Trelew (+**tre**); Chellew (+**chy**); ?Carclew (+**cruc**): in some of these cases there may be a personal name *Lew instead.

An aj. *lywek (cf. Welsh *lliwiog*, Breton *livek*) may occur as C2 in Trelewack and Trelewick (both +**tre**). A derivative, OCo. *liwet, may also occur (cf. Old Breton *liuuet* 'coloré', Chrest., p. 145), as A in Relowas (= r.n.+ ***ar**?), and as C2 in Hallootts (+**hal**).

***lok** 'chapel'; Middle Welsh *lloc*, BBC 22.5, and Myv.Arch., p. 179a, line 8; Welsh *mynachlog* 'monastery'; Breton *Lok-*, frequent in place-names (cf. Chrest., pp. 145 and 217). Note the Welsh-Latin gloss on *Carrum* (= Carhampton, Som.), *i. locus* 'that is, a monastery' (VSBG, p. 144). See also Loth, RC 39 (1922), 310, n. 5; Largillière, Les Saints, pp. 17–27; Gougaud, Revue Mabillon 15 (1925), 305–7; and Quentel, Ogam 7 (1955), 173–82. In Cornwall the sense 'monastery' for *locus* occurs in the phrase *locusque atque regimen sancti Petroci* 'and the monastery and rule of St Petrock', A.D. 994 (Sawyer, no. 880); and it is not restricted to Celtic: Dimier, 'Le mot *locus* dans le sens de monastère', Revue Mabillon 58 (1970–75), 133–54. Compare also Romano-British *Locus Maponi* 'shrine of Maponus' (*pace* PNRB, p. 395). Despite all this, however, Largillière shows (pp. 175f.) that the element in Brittany generally designates mere chapels, not ancient parishes, and it is interesting to note that the sole Cornish instance also formed a chapelry to the adjacent parish of Lanlivery. He also shows (p. 27) that *Lok-* names date from the 11th century or later, and the Cornish name thus presumably arose between c. A.D. 1000 and 1281, when it is first attested. The almost complete lack of

*loch

Lok- names in Cornwall is the same as in Wales, and, as with **plu**, the one Cornish occurrence, towards the south coast, should be regarded as an outlier of the Breton distribution. In the mediaeval forms of Luxulyan, ***lok** is sometimes replaced by ***lann**, by analogy with other names in *Lan-*. Diminutive *logell* 'tomb', MC 233c.

B2. Luxulyan (+ pers.)

***loch** (OCo., m. ?) 'pool'; Welsh *llwch*, Breton *loc'h* (cf. Tanguy, p. 96); Old Welsh *luch cinahi, luch edilbiu* (LLD, pp. 188 and 191), Old Breton *Luh-guiuuan* (Chrest., p. 147). As occasionally in Breton words (HPB, p. 551), the *ch* in this word became *h* very early, and does not normally appear in place-name forms. All the Cornish instances are coastal, or else (by extension?) river-names; so it may well be that the meaning in Cornish was normally 'creek, tidal pool' (cf. Welsh Amlwch, Agl.). Le Berre (Topon. Nautique 1419, p. 515) gives the specific meaning 'étang de barrage' for Breton coastal names, and that could apply to some of the Cornish instances.

A. Loe (2 exx.); Looe

B1. Duloe (+ **dew**); ?Cowloe, rock (+ ?)

B2. Lo Cabm, coast (+ **cam**)

C2. Portloe, Porloe, Porth Loe (all + **porth**); Nansloe, Landlooe (both + **nans**); Chyvarloc (+ **chy, war**); Prislow (+ ***prys**); in some of these cases, ***loch** is a pre-existing river-name, not an element in its own right.

logaz (pl.) 'mice', Lh. 19b; sg. *logoden*, Voc. 580 glossing *mus*; *logosan*, CW 407; *lygodzhan*, Lh. 96a. Welsh *llygod*, Breton *logod* (cf. Tanguy, p. 121). Note the Breton saint's name *Logot* (NSB, p. 81), which could be present as a personal name in some of the following names.

A. aj. Legossick (cf. Breton Logodec, Tanguy, ibid.)

B1. ?Cans Loggers, fld. (+ **cans** ?)

C2. Treloggas (+ **tre**); Bollogas, Bosloggas (both + ***bod**); Carloggas in St Columb Major and Constantine (both + ***ker**); Tresloggett (+ ***ros** ?); Carloggas in St Mawgan in Pydar and St Stephen in Brannel, Creeglogas (all + **cruc**); Colloggett, Colesloggett (both + **cos** or ***kyl** ?); Porth Logas (+ **porth**); sg. *Parken Legagen*, fld. (+ **park, an**); aj. Trelogossick (+ **tre**)

***lom** 'bare, bleak, poor', Welsh *llwm*; clear instances of this word are lacking. Perhaps John *Lom* (1327 SR) contains it; and the fish-name *lobmaz*, Lh. 41a glossing *abramis* 'a lesser sort of bream', or 'a chad fish' (Pryce) may contain a derivative. A few placenames may contain the word; cf. *Havodlum* 1334 Surv.Denb., p. 1.

B1. A compound **kel-lom* 'shelter-bare' may occur: see ***kel**.

C2. ?Top Lobm (+**top**)

D. Carlumb

*lon 'grove, thicket'; Welsh *llwyn*, Old Welsh *Loyn-garth* 'grove-enclosure' (HB, §71), *helicluin* 'willow-grove' (LLD, p. 268), *penn Luhin latron* 'head of thieves' grove' (LLD, p. 146), *di'r luhinn maur* 'to the great grove' (LLD, p. 262); Old Breton *Loin*, etc. (Chrest., p. 146; Tanguy, p. 101). Cf. D. M. Jones, TPhS 1953, 43–6. There are no instances of this word in the remains of the language, since Pryce's *loinou* 'bushes' looks bogus; and definite examples in place-names are also lacking, but there are one or two possible ones.

A. pl. ?Luna (cf. OBr. *Loeniou*, HPB, p. 207)

*lonk 'gorge, gully'; Welsh *llwnc* 'gulp, gullet' (in place-names: ELl, pp. 60f.), Breton *lonk* 'abîme, précipice, gouffre' (Le Gonidec). Compare the verb *lenky* 'swallow', Lh. 160b, ppp. *lenkis*, BM 3949, 3 sg. pres. *lonk* in the dog's name *Lonk ylo* 'swallows his spoon', BM 3226 (Loth, Mél.Arb., p. 220). See also ***yslonk** (?).

A. Lank; Lonkeymoor (+Eng.)

C2. Trelonk (+**tre**); dimin. ?Bolankan (+***bod**; 2 exx.)

los 'grey', BM 1967, *loys*, OM 72; Old Cornish *-luit*, *lot-*, Voc. 652 and 659; Modern Cornish *lûdzh*, Lh. 46a and 231a. Welsh *llwyd*; Breton *louet*, Old Breton *loit*, FD, p. 246.

C1. Lidcott, Lidcutt (2 exx.), Lydcott, Ludcott, Liskis, Liskey Plantation (all +**cos**: cf. Welsh Llwytgoed (2 exx.), Units, p. 146); Liggars (+***garth**)

C2. Lanhoose (+**nans**); ?Carloose (+***ker**?); Caragloose, Carn Gloose, coast, Carrick Lûz, coast, Carrag-Luz, Cataclews Point (all +**karrek**)

losc 'burning, burnt place', Voc. 285 glossing *arsura vel ustulatio*; Welsh *llosg*, Breton *losk*. See also **leskys**, the ppp. of the corresponding verb; and cf. Newton Arlosh, formerly *Arlosk*, etc. (PNCmb., II, 291, and English and French parallels cited there).

C2. Trelaske (2 exx.), Trelask (all +**tre**); ?**gor-losc* (+***gor-**) in Bosworlas (+***bod**); dimin. ?Treloskan (+**tre**)

lost 'tail', OM 1454; pl. *lostov*, BM 1352; aj. *lostek* 'fox', Lh. 179a glossing *vulpes*. Welsh *llost*, Breton *lost* (cf. Gourvil, pp. xxiv and

lovan

189f.); note the Welsh and Breton names *lost ir inis* (LLD, p. 73), Lostanenez (Tanguy, p. 91), 'the island's tail'.

B2. Lesquite in Lanivet (+**cos**); Lostwithiel (+**gwyth¹** aj.); Lestoon (+**goon**); Longsongarth (+**an, cath**); Rosebargus (+**bargos** 'kite'); ?*Lostogh* (+**hoch** ?); *Lost Kewthyry* (+ ?)

C2. aj. ('fox') ?Polostoc Zawn (+**pol**)

lovan 'rope, cord', MC 105d, etc.; *louan* (*u* = /v/), Voc. 349 glossing *funis vel funiculus*. Cf. Welsh *cynllyfan* 'leash'; Breton *louan* 'thong, strap', Middle Breton *louffan*.

C2. *Reenie Lovan*, fld. (+ *****run** sg. or pl.)

lowarn (m.) 'fox', OM 895; *louuern*, Voc. 563 glossing *vulpes*. Welsh *llywarn*, early Welsh pl. *cruc leuyrn* (LLD, p. 142); Breton *louarn* (cf. Tanguy, p. 121). This word, and derivatives of it, were used as personal names in the Brittonic languages; see Padel, Co.Studies 6 (1978), 24, note 10.

A. aj. ?Lawarnick Cove, ?Lewennick Cove: cf. Breton Louargniec, Tanguy, loc. cit.; Welsh Lavernock (PNDinas Powys, pp. 39–42); Old Welsh *guuer licat laguernnuc* 'the stream of the spring of (the) L.' (LLD, p. 207).

C2. Lanlawren (+*****lann**); Melorn, Millewarn (both +**men**); Park-en-lorn, fld.; Pitslewren (+**pyt** pl.); *Pengellylowarn* (+*****pen (an) gelli**); *Wheal an Lowren*, mine; aj. ?Lanwarnick (+**nans** ?)

lowarth 'garden', MC 140a and 233a; Old Cornish *luworch-* (read *luworth-*), Voc. 684; Modern Cornish *lûar*, Lh. 66a glossing *hortus*: a compound of *****luv** 'herb' (cf. FD, p. 247) and *****garth** 'enclosure'; Old Welsh pl. *luird* 'gardens' (Mart.Cap.; VVB, p. 178), Breton *liorzh* (cf. GMB, p. 369; Chrest., p. 217; ÉC 15 (1976–78), 204).

B1. ?Halgolluir (+**heligen** pl. ?)

B2. All of the following are fields: *Looarth an Men* (+**an, men**); *Loranvellan* (+**an, melin¹**); *Lowarglas* (+**glas**); *Luword en funteyn* (+**an, fenten**); *Lowargh Lower* (+Eng. ?); *Lowarth Thomas* (+pers.); *Lowarth Treuthale* (+p.n.); *Lowerth-Lavender-en-parson* (+Eng., 'the parson's lavender garden')

C2. Park-an-Lower, Park Luar, Park Lore, etc., fields

lowen 'glad, happy', PC 3157, etc.; *louen*, Voc. 939 glossing *letus*; noun *lewene* 'joy', OM 154, etc. Welsh *llawen*, noun *llawenydd*; Breton *laouen*, sb. *levenez*. The noun is used as a personal name: John *Louene*, 1327 SR; cf. Middle Breton pers. *Levenez*, etc. (Chrest., p. 217); Old Breton *Rann-louuinid* (Chrest., p. 147).

A. aj. *Lewannack*, mine

C2. Nanceloan, Nancelone (both + **nans**); Bellowal (+ ***bod**); noun Barlewanna, Barlowennath (both + **bron**: cf. Welsh Bryn Ll(y)wenydd (Agl.), EANC, p. 121, and Middle Breton *Brelevenez*, Chrest., p. 192; ÉC 15 (1976-78), 203)

lu 'host, army', MC 163c; if it is present in the place-name, it must be in the sense 'fleet', as in *luu listri* 'fleet of boats', Voc. 270 glossing *classis*. Welsh *llu*, Old Breton *-lu* (FD, p. 260).

C1. ?Newlyn in Paul (+ **lyn**, with dissimilation of *l-l* to *n-l*)

lugh ($u = /œ/$) 'calf', BM 1557; Old Cornish *loch*, Voc. 600 glossing *vitulus*; Modern Cornish *leauh*, Lh. 230b. Welsh *llo*, Old Welsh *lo* (VVB, p. 177); Breton *leue*. The plural is presumed, from field-names, to have been ***lughy** (cf. Welsh *lloi*).

C2. Park Lew, etc.; pl. ?Parkanloy, etc.; all fields

A phrase ***lugh ogo** 'cave calf' (v. **googoo**), meaning 'seal' (*leuirgo* 'the seale-fish', Borlase, Antiquities), may occur as an English word in Logo Rock.

***luryk** ($u = /y/$) 'coat of mail, breastplate' (?), equivalent to Welsh *llurig*, from Latin *lorica*: as well as 'breastplate', this could mean 'breastwork, rampart' (Lewis and Short, s.v.; cf. continental place-names cited at DAG, p. 782), and that is presumably the meaning in the place-name(s).

C2. Calerick (+ ***ker**); ?Penhalurick (+ ***pen hal**)

***lus** ($u = /y/$) 'bilberries', equivalent to Welsh *llus*, Breton *lus* (aj. ***lusek**, Tanguy, p. 114). Some of the names cited under **les** might contain ***lus** instead.

C2. aj. ?Carnlussack (+ **carn**)

lusew 'ashes', OM 477; *lusow*, OM 1355; Modern Cornish *lidzh(i)w*, Lh. 10b and 48a. Welsh *lludw*, Breton *ludu*. The *e/o* in the Middle Cornish forms is an epenthetic vowel.

A. ?Ludgvan (+ ***-an²** 'place of' ?)

M

***ma** 'place', also 'plain, open country'; Welsh, Breton ***ma**; Irish *magh* 'plain'; Romano-British *Magis* 'at the plains' (PNRB, p. 406). In all three Brittonic languages the word occurs primarily as a suffix, *-va*, in compounds, where it means 'place' simply: Voc. *redegua*, *chetua* and **guillua** (nos. 9, 186 and 400); Old Welsh *Coupalua* 'ferry-place' (LLD, p. 151), *Poguisma* 'resting-place'

155

*mabyar

(LLD, p. 260, etc.); cf. Williams, ELl, p. 40; Old Breton *gronnua* 'endroit marécageux' (FD, p. 180), and others (FD, p. 249); compare the numerous Gaulish names in *-magus* (ACS, II, 384f.). However, there are also some names, especially in Wales, containing *ma as B2, when it takes the spirant mutation as in Ma-chynlleth: for examples see Williams, BBCS 11 (1941–44), 148, ELl, p. 39, and Beginnings, p. 149; Richards, BBCS 23 (1968–70), 324f., ibid. 25 (1972–74), 420f., and ÉC 13 (1972–73), 389–404: in these cases, *ma means 'plain, open country', and is usually followed by a river-name or a personal name. Mahalon is a probable Breton instance of the same (Chrest., p. 220; Smith, Top.Bret., p. 78; HPB, p. 519 and n. 3). See also the element **mes**, composed of ***ma** with a stem suffix.

B1. Bonyalva (+**banathel**); ?Dorothegva, fld. (+ ?); see also ***darva** 'oak-place' (?), **guillua** 'lookout-place', ***morva** 'sea-side', **trygva** 'dwelling-place' and (?) *gwestva (s.v. **gvest**).

B2. Menheniot (+pers. ?)

***mabyar** 'pullet, young hen', a compound of *map* 'son, boy' +**yar** 'hen': dialect *mabyer* 'pullet'.

C2. Park Mabier, fld.

maga 'feed, nourish, rear', PC 71; Welsh *magu*, Breton *magañ*.

C2. ?Gold Maggey, fld. (+**guel**); a compound *mag-le 'rearing-place' (+**le**) may appear in *Park Magla*, fld.

***magoer** (f.; OCo.) 'wall' (probably in the sense 'ruins, remains'); Welsh *magwyr*, Breton *moger* (see HPB, pp. 225f.); Old Breton *Macoer*, glossed *valium* (read *vallum*, Chrest., p. 148), Middle Breton pl. *Magoaerou* (Chrest., p. 219; cf. GMB, p. 422). See also Gourvil, Ogam 7 (1955), 339–46.

A. Maker, Magor

mam (f.) 'mother', Voc. 128 glossing *mater*, etc.; Welsh *mam*, Breton *mamm*. For the sense of the name, cf. Bosoar, etc., s.v. **wuir**.

C2. Bamham (+***bod**)

manach (m.; OCo.) 'monk', Voc. 110 glossing *monachus*; Welsh *mynach*, Breton *manac'h*. In the place-names, the singular perhaps seems to stand for the plural (*menegh), since 'monks' enclosure, dwelling, etc.' is more likely than 'monk's enclosure, etc.'. See also ***meneghy** 'sanctuary'.

A. aj. Meneage; ?Manaccan (+ *-**an**[2] 'place of' ?)

marghas

C2. sg. *Lamana* (+ ***lann**); Busvannah (+ ***bod**); pl. ?Tremaine (+ **tre**); aj. Treveneage (+ **tre**), Lesneage (+ ***lys**)
manal 'sheaf', Lh. 33a (glossing *manipulus*) and 241a. Breton *malan* 'gerbe', *manal* 'gerbier'. The element occurs only as ppp. of a verb *manala 'to stack', corresponding to Breton *manaliñ*, *malanañ* 'engerber, mettre les gerbes en tas'. Some names that appear to contain the simple noun are more likely to contain the later form, *bannal*, of **banathel** 'broom': e.g. *Skebervannel*.

C2. ppp. Carnmenellis, Carn-Mên-Ellas, coast (both + **carn**)
margh (m.) 'horse', OM 124, PC 1658; Welsh *march*, Breton *marc'h*. The plural is *mergh* (OM 1065), *an verh* (Pryce, p. Ff2r), but another plural, *marghes, is shown by some place-names, corresponding to Vannetais *marc'hed*, *marhétt* (PB, pp. 73 and 84; GIB, s.v. *marc'h*; for the ending *-ed*, cf. PB, p. 41; Hemon, Grammar, pp. 31f.; FG, p. 230; and WG, p. 206: Cornish *-es*, in *myrgh* 'daughter', pl. *myrghes*). This plural *marghes is theoretically confusable with **marghas** 'market', but in practice the sense, or the absence or presence of a market, can determine which element is present. Some names that appear to contain *mergh* as the plural may in fact contain it as an old vowel-affected genitive singular, **me(i)rch*, as suggested by Quentel, Rev.Intnl.Onom. 8 (1956), 302–3 (cf. **torch**). There was also a Cornish personal name *Mar(c)h* (Bodm.), which could well be present in some of the place-names. See also Gourvil, 'Le nom propre Marc'h...dans l'Anthroponymie Brittonique', Ogam 7 (1955), 59–62. There is no evidence in Cornish for the Welsh *march* as C1 meaning 'great, large' (M. Richards, Pokorny Festschrift, pp. 257–63).

C2. Polmarth, Polmark (both + **pol**); Lemar, Lemarth (both + **lyn**, ***lann**, or **nans**); Goonamarth (+ **goon**, **an**); Tresmarrow (+ ***ros**; 2 exx.); Kilmarth (+ **keyn** or ***kyl**); Carmar, *Carrack Kine marh*, rock (both + **keyn**); Carn Marth (+ **carn**); Tremar, Trevarth (both + **tre**); *Chyenmargh* (+ **chy**, **an**); Carvath (+ ***ker**); pl. *mergh*, Ventonveth (+ **fenten**), Carveth in Mabe (+ ***ker**), Roseveth (+ **rid**: cf. Welsh Rhyd y Meirch, etc., ÉC 10 (1962–63), 221), ?Park an Veer, fld. (+ **park**, **an**), Savath (+ **enys**); pl. *marghes, Gonnamarris, Noonvares, Noon Veres, fld. (all + **goon**, (**an**)), Halvarras (+ **hal**)

D. Borlasevath, *Gluuyanmargh*
marghas 'market', PC 316, 2419; Welsh *marchnad*, Breton *marc'had*; as B2, note Old Breton *Marchat Rannac* (FG, p. 157).

B2. Marazanvose (+**an, fos**); Marazion (+**byghan**); Market Jew (+**yow**)
  D. *Trewarveneth Varghas*
marrek (m.) 'knight, soldier', OM 2004 and 2150; MC 242a: the adjective of **margh**. Welsh *marchog*, Breton *marc'heg*.
  C2. *Tere marracke* (+**tyr**); *Savan Marake*, coast (+**sawn**)
*__með__ (OCo.) 'middle' occurs only as C1, primarily in the compound ***með-ros**, but also in the other compounds given below. It is equivalent to the Gaulish and British *Medio-*, found in the frequent place-name *Mediolanum* (see Guyonvarc'h, Ogam 13 (1961), 142–58; Le Roux, ibid. 169f.; GPN, p. 215 and note 7; PNRB, pp. 415f.), in the Gaulish tribal name *Mediomatrici* (DAG, p. 769), and in British *Mediobogdum* and *Medionemetum* (PNRB, loc. cit.); Welsh **mei(dd)*- 'middle' (ELl, p. 80; Richards, Monts.Collns. 56 (1959–60), 177: but note SC 8/9 (1973–74), 313, n. 4), Old Welsh *meton* (EWGT, p. 10: $t = /ð/$; cf. LHEB, p. 426: Middle Welsh *mywn* 'in'), Welsh *per-fedd* 'middle'; Old Breton *-med, -met* (FD, p. 252) and p.n. *Medon* (Chrest., p. 150). Cf. Old Irish *medón* and *mide, mid-*, 'middle' (LEIA, pp. M28 and M50). Note also the Old Welsh p.n. *Medgarth* (VSBG, p. 120), and modern Meiarth (3 exx., M. Richards, ÉC 13 (1972–73), 399), and other names in Mei- (ibid., 399–401); early Irish *Midglenn* (Hogan, Onomasticon, p. 538). The names containing *Meth-* would then mean either 'middle-X' or 'middle of the X'. See also BBCS 1, 36–8.

There are theoretically other possibilities for names in *Meth-*, but in practice they are not very likely: one is Cornish *meith* 'whey' (Lh. 149c, glossing *serum*); Welsh *maidd*, Old Breton *meid*[1], *-mid* (FD, p. 253; HPB, pp. 644 and 156f.): cf. Gaulish *mesgus* (DAG, p. 576), Old Irish *medg* (LEIA, p. M28): in favour of this word is the fact that it is attested in Cornish. Another possibility is a word equivalent to Old Breton *meid-*[2] 'soft, softness' (FD, p. 253), Breton *meizh*. (Welsh *mwyth* lacks *i*-affection, like Old Irish *mocht*, LEIA, p. M58.) Finally, *meth* 'mead', OM 2435, Welsh *medd*, Breton *mez*, is phonologically possible and would be semantically possible in some of the names (e.g. a 'mead-hill' would be one producing honey?).

The element was replaced at an early date (during or before the Old Cornish period) by its derivative **perveth**, which was current during the period when *Tre-* names were being formed; and that in turn was replaced by **cres** in the Middle Cornish period.

melhyonen

C1. Methleigh (+**le**; 2 exx.); Medlyn (+**lyn**: situated 'between streams'); Metherin (+ ***rynn**?); ?Lametton (+**nans**, ***dyn**?); see also ***með-ros**.

***með-ros** (OCo.), a compound of ***með** and ***ros**: the meaning is probably '(place situated) in the middle of the hill'. Most of the places with this name are situated in small hollows near the tops of hillsides. Compare Welsh *Meiros* (3 exx., M. Richards, ÉC 13 (1972–73), 401); and, for the sense, Meiarth and Meidrim (ibid., 399–400), both presumably meaning 'middle of the ridge'; also early Irish *Midros* (Hogan, Onomasticon, p. 538).

A. Methrose (3 exx.), Meadrose, Maders, Merrose (2 exx.), Mayrose, Marris, Meres

***meyn-dy** (m.) 'stone-house', a compound of *meyn* (pl. of **men**) and **ti** (= **chy**); Welsh *maendy* (e.g. WM 492.35), Old Welsh *mainti* (LLD, pp. 207 and 242), Breton *Mendy* (Gourvil, Ogam 7 (1955), 151). In Cornish the *d* was apparently prevented from giving $z > dj$ by the adjacent *n* (cf. ***mun-dy**). The spellings of the names suggest that in Cornish, unlike Welsh and Breton, the first word of the compound was plural, *meyn*, not singular.

A. Mendy Pill, coast, Mountjoy

***meynek** 'stony', the adjective from **men**; Welsh *meinog*, Old Welsh *pont meiniauc* 'stony bridge' (LLD, p. 244); Breton *meinek* (cf. GMB, p. 403).

A. Meinek, Manack Point, ?Vinnick Rock, all coast

C2. Carvinack (2 exx.), Carwinnick in Goran (all + ***ker**); Dorminack (+**dor**); *Guaelmeynek* (+**guel**); Porth Minnick (+**porth**); ?Boderwennack (+ ***bod, tre**?); ?Zawn Vinoc, coast (+**sawn**)

***melek** 'honeyed' (?): the adjective of *mel* 'honey', RD 144; Welsh *melog*, Breton *melek*. Both the names are doubtful, because of variable spellings and because of other possible words which could be present instead.

A. ?Millook

C2. *Pollemellecke* (+**pol**)

melhyonen (sg.) 'clover-plant', Voc. 664 glossing *vigila* (read *viola*), Old English *clæfre*. Welsh *meillionen*, Old Welsh pl. *mellhionou* glossing *violas* (Mart.Cap., VVB, p. 184); Breton *melchonenn* (from ***meltjonenn*: VKG, 1, 137). Jackson (JCS 1 (1949–50), 76) pointed out that we would theoretically expect ***meltyonen** in Cornish (cf. LHEB, pp. 400f.), and that the Voc.

melin¹

form ought therefore to be Welsh. However, early forms of the place-names all support the form without -*t*- as being correct Cornish, and no form with -*lt*- occurs for this word, so it is necessary to suppose that it was irregular in Cornish, and that -*lti*- here gave -*lhy*- or voiceless -*lj*-; the case is somewhat similar to Old Breton *guiliat* and *guoliat*, in both of which we would theoretically expect -*lti*-: Jackson, JCS 1 (1949-50), 72f.; HPB, p. 415 n. 1.

A. aj. Molenick, Mellionec, Menheniot in St Stephen by Launceston: compare Breton Melchennec, Melchonnec, etc. (Tanguy, p. 114; Gourvil, p. 194; Chrest., p. 220)

C2. pl. ?Rosemullion (+ *ros)

melin¹ (f.) 'mill', Voc. 910 glossing *molendinum*; Modern Cornish *belin*, Lh. 92c. Welsh *melin*; Breton *milin*, Old Breton *Tnou melin* (Chrest., p. 151), Middle Breton *Millinneuez* (GMB, p. 402). See also the compound *melyn-jy and the phrase *melyn wyns. As qualifier, **melin¹** is formally indistinguishable from **melyn²** 'yellow', though the sense often serves to distinguish the two; however, there are some cases (especially *Ros melyn*, 3 exx.) where ambiguity arises.

B2. Qualified according to location: Melancoose, Mellangoose (2 exx.), Mellingoose (all +(**an**), **cos**); Millendreath, Vellandreath, *Melyntrait* (all +**trait**); pl. Ballaminers (+**meneth**), Lennabray (+*bre)

Qualified according to ownership: Vellyn Saundry (+pers.); *Melynmyhall* (+St); Mellanvrane (+**bran**; 2 exx.)

Qualified by another place-name: *Myllynalsey* (+Alsia); *Melyn Vedle* (+Methleigh); *Melyncroulis* (+Crowlas)

Qualified by use or other feature: Vellanoweth (3 exx.), *Melynnewyth*, Mellanoweth (all +**nowyth**); Mellanzeath (+**segh**); *Melynclap* (+**clap**); Vellenewson (+**usion**); Vellanhoggan Mills (+*hogen); Vellyndruchia (2 exx.), Velandrucia, ?Valley Truckle (all +*trokya); Vellynsaga (+ ?)

C2. Nansmellyn, Nancemellin, Lamellion, Lamellyn (2 exx.), Lamellyon, Lawellen, Halamiling (all +**nans**); Rosemellen, Rosemelling, Rosemellyn (all +**rid** or *ros); Polvellan, Polly Vellyn, fld. (both +**pol**); Porth Mellin, Portmellon (both +**porth**); Tremellin (+**tre**); Gwealavellan (+**guel**, **an**); Goon Mine Mellon (+**goon**, **men** pl.); *Parkenmelyn*, fld. Some of these names may contain **melyn²** instead.

melyn² 'yellow', OM 1965, etc.; *milin*, Voc. 485 glossing *fulvus vel flavus*; Welsh *melyn*, Old Welsh *halmelen* 'yellow moor' (LLD, p. 73); Breton *melen*, Old Breton *milin*, etc. (FD, p. 257). Indistinguishable from **melin¹** except on semantic grounds.

C2. See names listed under **melin¹**; also Carn Mellyn, coast (+**carn**; 2 exx.); Gwealmellin (+**guel**); Mean-Mellin, fld. (+**men**)

***melynder** (m.) 'miller'; *belender* 'a miller', Lh. 240c; a derivative of **melin¹**, cf. Welsh *melinydd* (as a personal epithet, *Seysil Melinyt* 1292, BBCS 13 (1948-50), 227) and Breton *miliner*: the -*d*- in Cornish is unexplained. The Cornish word occurs as a personal epithet: Thomas *Melyneder* 1327 SR (St Keverne); William *Melender* 1562 in Penwith, JRIC n.s. 3 (1957-60), 182.

C2. *Park Belender*, fld.; *Crauft an Mellender*, fld. (next to a mill)

***melyn-jy** 'mill-house', a compound of **melin¹** + **chy**.

A. Melinsey, Molingey, Mellingey (4 exx.), Bolingey (2 exx.), ?Valency, river

C2. Barbolingey (+ ***bar**)

***melyn wyns** 'windmill', a phrase of **melin¹** and **guyns**; Welsh *melin wynt*, Felinwynt (Crm., SN 2250). See Douch, Windmills.

A. *Vellanwens*, Vellanvens, Wheal an Wens Croft, all flds.

***melwhes** 'snails, slugs', the plural of Lhuyd's *molhuidzhon* 'a naked snail' (48c, with variant spellings at 10b and 79c; pl. *molhuez*, Lh. 286a); Old Cornish *melyen* (read *melwen*), Voc. 620 glossing *limax*, is from the same root. Welsh *malwod* 'snails, slugs', sg. *malwen*, *malwoden*; Old Welsh *mormeluet* 'sea-snail' (Mart.Cap., VVB, p. 189); Breton *melc'houed* (HPB, p. 585; cf. Tanguy, p. 121, GMB, p. 401).

C2. Park-an-Velvas, fld.

men (m.) 'stone', OM 1844, etc.; pl. *meyn*, OM 2318, etc. (variant plurals *myyn*, OM 2694, *myn*, RD 401; Modern Cornish *mine*, Pryce, p. Ff3r). Welsh *maen*, pl. *main* (Old Welsh sg. *i main brith* 'the speckled stone', LLD, p. 191); Breton *maen*, pl. *mein* (Old Breton sg. *main*, FD). Owing to imprecision in the spellings, it is often difficult to distinguish the singular and plural of the element in place-names, and some of the names here classed under **men**, sg., may contain *meyn*, pl. The modern map spelling *Maen-* in coastal names is a solecism; it stands for /meːn/ or /mein/.

A. Mean, Mayon, Maen Porth (+ English); the disyllabic forms of Mayon are paralleled by similar ones occasionally for ***ker** as a simplex name.

*menawes

B1. Hervan (+**hyr**: Old Welsh *hirmain*, Chad 4); *deumaen coruan* (+**dew**, ?); ?Trevan Point (+***try-**[2]: or **ban**); ?Mulvin, coast (+***moyl**?); ?Winven Cove (+**guyn**?: cf. Welsh Ffynnon Gwynvaen, Agl., BBCS 8, 89); ?Pleming (+**plu**?). See also the compound ***tol-ven**.

B2. In farm-names: Melorn, Millewarn (both +**lowarn**); Menkee, Mankea (both +**kee**); Menwenick (+pers. or **guyn** deriv.); Menmundy (+***mon-dy**?); Manuels (+*hul-wals?, see **ughel**); Mengearne (+ ?); Marcradden (+**crom**); pl. Menwinnion, Menwidden (both +**guyn** pl.)

In boundary-names and other inland rock-names: *maen wynn* (+**guyn**: Sawyer, no. 770); *maenber* (+**ber**: ibid.); *mayn biw* (+***byu**: Sawyer, no. 832); *Maen tol* (Sawyer, no. 450), *Meane (an) toll*, Mên-an-Tol (all +(**an**), **toll**); *Mahen halen* (+?: Sawyer, no. 450); *Meane glase* (+**glas**); *Mene Gurta*... 'a staie stone' (+**gortos**); Mên Screfys, stone (+**scrife** ppp.); Men-Amber (+ ?); *Menvrause*, fld. (+**bras**); *Mankependoim* (+**kee**, ?: Sawyer, no. 450); pl. *Myne an Downze*, mine (+**an**, ***dons**)

In coastal names: Mên Talhac (+**tal** aj.?); Men Hewel (+**ughel**); Mên-y-grib Point (+**an**?, **krib**); Maen-du Point (+**du**); Maen Dower (+**dour**); Maen Derrens (+ ?); Maenease Point, Maen-lay Rock (both + ?); Men an Mor (+**an**, **mor**?); *Maenenescop* (+**an**, **epscop**; = Bishop Rock)

C1. See the compounds ***men-gleth** and ***meyn-dy**.

C2. Tremayne (3 exx.), Tremain, Demain (all +**tre**); Ponsmain, Ponsmean (both +**pons**); Tresmaine (+***ros**); *Dormayne* (+**dor**); *Looarth an Men*, fld. (+**lowarth, an**); Gonorman Downs (+**goon, naw**); Tal-y-Maen, rock (+**tal, an**); Gweal-an-Mayn (+**guel, an**); Penmayne, Pen-a-maen, coast (both +**pen, (an)**); Polridmouth (+**porth, rid**: cf. Redmain, PNCmb., II, 267; Welsh Rhyd Faen, etc., ÉC 10 (1962–63), 216); pl. Bosvine (+***bod**), Pedn-myin, coast (+**pen**), Dawns Men (+***dons**), *Peele Myne* (+**pyl**), Goon Mine Mellon (+**goon, melin**[1]); Porth Main (+**porth**)

***menawes** (m.) 'an awl'; *benewas*, Lh. 23a, *benewez*, Lh. 157b; Welsh *mynawyd* (south-western *binewid*, Lh. 157b), Breton *minaoued*. In place-names it would mean a hill, stone, or piece of land shaped like an awl: compare Welsh Cae'r Mynawyd and Carn Mynawyd, Agl., Gwaun Mynawyd, Mer., Nant Mynawyd, Mer., and *Porthbinawid* (Owen, Pembs., II, 516): cf. also EANC, p. 100.

meneth

Confusion with a pers. equivalent to Welsh *Manauid* (BBC 94.11) is possible, but unlikely.

C2. Kilminorth, Kilmanant, ?Callymynaws, fld. (all + ***kyl**); Rosebenault, Rosemanowas (both + ***ros**); ?Minmanueth, coast (+ **men**?: cf. perhaps the Breton coastal name Mean Nevez, Topon.Nautique 1399, no. 7202)

*meneghy 'sanctuary' (?), equivalent to early Welsh *Menechi* (LLD, pp. 126 and 255), Middle Breton *Minihi*, etc. (RC 7 (1886), 64f.), Breton *minic'hi* 'refuge, asile'. It is curious that the place-names containing this element do not coincide with the four privileged sanctuaries in Cornwall (see J. Charles Cox, *The Sanctuaries and Sanctuary Seekers of Mediaeval England* (London, 1911), especially Chapter 10; Eccl.Ant., pp. 265-7, 375-6 and 415-16): it may be that, like English 'sanctuary', ***meneghy** came to mean simply 'glebe land' (flds. Sentry, Sanctuary, etc.): if so, the names are insignificant for early church history.

A. Manhay; Menehay; Mennaye, fld.

C2. Bodmin (+ ***bod**); Tremenhee (+ **tre**)

meneth (m.) 'hill'; Old Cornish *menit*, Voc. 717 glossing *mons*. Welsh *mynydd*, Breton *menez* (cf. Tanguy, p. 77). Elements, II, 38 and 41; JEPNS 1 (1968-69), 49.

A. Mena, Menna (2 exx.), Mennor

B1. Halvana, Halvenna (both + **hyr**); Gonvena (+ **guyn**: Old Breton *win monid, id est montem candidum*, FD, p. 330); Colvennor (+ **kal**; 2 exx.); *Cogveneth* (+ **cuic**); Tuelmenna (+ **tewl**); ?Molevenney Quarry (+ ***moyl**?)

B2. Qualified by plant-words: Menniridden (+ **reden**); Many-withan, fld., *Menewithan* (both + **gwyth**[1] sg.); Menaburle (+ **breilu**?); Menacuddle (+ ***gwythel**)

Qualified by colours: Menadue (5 exx.), Menerdue, Menadews (all + **du**: early Welsh *Mynid du*, LLD, p. 42); Menaglaze (+ **glas**); Menergwidden (+ **guyn**); Menerlue (+ **lyw**)

Qualified by size or shape: Minnimeer (+ **meur**); *Menelidan* (+ **ledan**); Menadrum, fld. (+ ***drum** ?)

Qualified by other features: Menagwins (2 exx.), Minawint, fld. (all + **guyns**); Menaclidgey (+ ***clus** pl.); Menadodda (+ ?)

C2. Trewarmenna, Trewarveneth, Trevenner, Trevena in Tintagel, Trevenna in St Neot and St Mawgan in Pydar (all + **tre, war**); Polmenna (4 exx.), Penmennor, Penmenor, Penmenna (all + **pen**); Chywarmeneth (+ **chy, war**); Tolmennor (+ **tal**);

***men-gleth**

Ballaminers (+**melin**¹ pl.); pl. *Penmynytheowe* (+**pen**); aj. ?Trenethick (2 exx., +**tre**)

***men-gleth** 'quarry', a compound of **men** and **cleath**; Welsh *maenglawdd* (obsolete), Breton *mengleuz* (cf. GMB, p. 404; Ernault, RC 27 (1906), 57–60). This word is almost indistinguishable from ***mon-gleth**.

    A. pl. *Maengluthion* (read *-ou*), ?*Pengluthio*
    C2. sg. ?Pits Mingle (+**pyt** pl.)

***men hyr** (m.) 'long stone, standing stone', a phrase of **men** and **hyr** (cf. the compound *hyr-ven, s.v. **men** B1); Welsh *maen hir*, Breton *maen-hir*. Though properly a phrase, the element came to be treated as a single word, with a plural *menhyryon: either that, or that place name stands for *meyn hyryon 'long stones', with a plural adjective (cf. Padel, SC 14/15 (1979–80), 237–40). The one coastal instance refers to a natural feature, not an antiquity.

    A. Menear; Manheirs; Menhire; Mên Hyr, coast; pl. Menherion
    C2. Tremenhere (2 exx.), Tremenheere, Tremenheire (all +**tre**); Goonmenheer, *Goen menhere*, flds. (both +**goon**); Balmynheer (+**bal**)

***merther** 'saint's grave' (?); Welsh *merthyr* 'saint's grave' (cf. Richards, SC 3 (1968), 10f.; James, Cymm.Trans. 1961, p. 178; Units, pp. 156f.), Brcton *merzher* (cf. Chrest., p. 220). See C. Thomas, North Britain, p. 89, for the development of the meaning: the word need not imply anything to do with martyrs. Note that, curiously, there were two different places called *Merther-Euny* (Redruth churchtown and a chapel in Wendron parish), yet neither possessed the body of the saint, which was considered to be at Lelant (Wm. Worcester, p. 114), a place which did not have a name in ***merther**. Similarly, Ruan Major church was called *Merther*, but the saint's body was considered to have been at Ruan Lanihorne (not a ***merther**), before its translation to Tavistock (Doble, St Rumon, pp. 24 and 13). But Sithney (*Merthersithny*) is probably the place intended by 'Sanctus Senseus iacet...iuxta Hellyston' (Wm. Worcester, p. 88).

    A. Merther, *Merther*, Merthyr
    B2. Mertheruny,*Mertherheuny*, *Merthersithny*, Matthiana, Barrimaylor, Menedarva (all +Saints' names)
    C2. Eglosmerther (+**eglos**); Ponsmedda (+**pons**, (**an**)): in these names, *Merther* is probably a pre-existing place-name rather than an element in its own right.

*merthyn 'sea-fort', like Welsh **myrddin* from *Moridunum* (PNRB, p. 422). Only one of the four places actually has a fort: in another case the natural rocks resemble ancient masonry.
    A. Merthen (2 exx.), ?Merthen Point
    C2. ?*crucou mereðen* (+**cruc** pl.: Sawyer, no. 755)

mes (m.) 'open field, open country': the word is attested in Middle Cornish only in the adverbial phrases *yn-mes*, **aves**, *the-ves* 'forth, out, away' (e.g. OM 83, 953 and 1097). Later on, *yn-mes* became a simple adverb *mes*, spelt *meese* 'forth' in 1547 (RC 32 (1911), 443); Lh. 250b, '*mêz* signifies properly an open field, but *a vez* is also the common word for *without*'. Welsh *maes* (cf. PNFli., pp. 104–6; Units, pp. 151f.), Breton *maez* (cf. FD, p. 250; GMB, p. 384); Old Welsh *Lannmaes*, Ann.Camb., s.a. 817; *trui i coit bet i mais* 'through the wood as far as the open country' (LLD, p. 160); glossed *campus* in *mais mail lochou* (LLD, p. 79), called *campo malochu* (LLD, p. 165), and in *campo qui dicitur Maysycros* (Litt.Wall., no. 200, A.D. 1281); Breton names in *Mes-*, etc.: Chrest., p. 219; Gourvil, p. xxiv; Mezahibou, Tanguy, p. 93. See also Elements, II, 34; JEPNS 1 (1968–69), 49; 2 (1969–70), 74. From the same root comes ***ma**. The element is not common in Cornish place-names or field-names; moreover, those that do occur are in the middle rather than in the west of the county (especially around the Camel estuary), suggesting that the element is a fairly early one (perhaps not used after, say, the 13th century?). If the word ever meant 'open field' proper, it was replaced in that sense by the commoner **guel**: on Predannack Wartha in the 17th century, different tenements possessed 'A stitch in Eroger Maze', which thus was apparently a subdivided open field.
    A. aj. Messack (meaning?)
    B2. Messengrose (+**an, crous**: cf. Welsh *Maysycros*, above, and Maes y Groes, PNFli., p. 105); Mesmear (+**meur**: cf. Welsh Maesmawr, Maesmor, Units, p. 151); Meskevammok (+***kevammok**); Messenger, fld. (+**an, *ker**); Maze Dippa, fld. (+***dyppa**)
    C2. Venton Vaise (+**fenten**); *Callyvais*, fld. (+**kelli**); *Eroger Maze*, fld. (+ ?)
    D. Gyllyngvase

mesclen (sg.) 'mussel', Voc. 554 glossing *muscula* [sic]; Lh. 241c *bezlen* 'a muscle' (i.e. the shellfish). Welsh *mesglyn* 'shell, husk', Breton pl. *meskl*. The Cornish plural would have been **meskel*, or

methek

perhaps, on the basis of Lhuyd's form with loss of *c*, *mesel (under English influence?).

C2. sg. ?Carn Veslan (+**carn**); pl. ?Ogo Mesul (+**googoo**); both coastal

methek (m.) 'doctor', RD 1648; *medhec*, Voc. 283 glossing *medicus*. Welsh *meddyg*, Breton *mezeg*.

C2. Tremethick (+**tre**); *Parkenmethek* (+**park, an**)

meur (*eu* = /œ/) 'big, great'; Welsh *mawr*, Breton *meur*.

C2. Tremeer (4 exx.), Treveor (2 exx.), Trevear (4 exx.), Tremear (all +**tre**); Carvear (+***ker**); *Ponsmur* (= Grampound), Ponsmere (both +**pons**); Porth Mear, Porthmeor (2 exx.), Polmear in St Austell, Porthbeer Cove, Porthbeor Beach (all +**porth**); Crugmeer, *crucmur* (both +**cruc**); Dunmere (+***dyn**); Mesmear (+**mes**); Nancemeer, Nancemeor, Lemeers (all +**nans**); Tresmeer (+***ros**); Minnimeer (+**meneth**); Cutmere (+**cos**); Gullivere (+**kelli**); Carnmaer in Stithians (+**carn**); *Ennisveor* (+**enys**); Halveor, Halmeers, fld. (both +**hal**); Polmere, flds. (+**pol**); spv. *moygha*, *Park Moyha*, Park Moya, flds. (+**park**), *Geaw Moyha*, fld. (+***kew**).

D. Delamere, Ardevora Veor, Hellesveor, Amalveor, Bell Veor; and innumerable mediaeval and later instances such as *Nansmeor veor*, *Tregadreythmoer*, *Lanowemeur*, *Polgrunemuer*, *Roscarecmur* (= *Roskarec maior*), *Germoghmur* (= *Magna Germogh*); usually contrasted with *byghan*, but sometimes with other suffixes as well or instead.

midzhar (MnCo.) 'reaper', Lh. 13c and 90a; Breton *meder* 'moissoneur'. Old Cornish *midil* (Voc. 339 glossing *messor*) and Welsh *medelwr* 'reaper' come from the same root but with different terminations.

C2. ?Chymder (+**chy**)

mẏdzhovan (MnCo., *ẏ*= /ə/) 'ridge', Lh. 74a glossing *iugum*. Origin or cognates unknown? Cf. possibly Breton *moudenn* 'mound of earth', and GMB, p. 427; or BBCS 1, 37.

A. Mejuggam, fld.

B1. *An Whea Mogovan*, fld. (+**an, whe**)

C2. Pedn Bejuffin, fld. (+**pen**)

mylgy 'greyhound, hunting-dog', PC 2927 (cf. Loth, RC 26 (1905), 251); Welsh *milgi*.

C2. ?Croft Milgey (+***croft**)

myn (m. ?) 'edge, border', PC 2727; Welsh *min* (cf. Chrest., p. 151 and n. 10).

B2. Mingoose (+**cos**); Manely (+**hyly**); Bydalder (+cpd. **dall**+**dour**?); ?Mendennick (+***dyn** aj. ?). The spelling *veen* (OM 2444) is taken as *myn* in the sense 'tip' by all translators, and the phrase in which it occurs, *y veen mon* 'its slender tip', seems to occur as a field-name: Mean-Moon, Mean Monne, Minny Moon, The Mee Mun, etc. (all +**mon¹**)

C2. ?Lamin (+**nans**); aj. (+***-yel**) ?Treveniel (+**tre**)

***mynster** 'endowed church', borrowed from English *minster*; compare Middle Breton *mostoer* 'monastery', borrowed from French (Piette, p. 146).

A. ?*Menstre* (= Manaccan: or English)

C2. Porthminster (+**porth**)

moelh (f.; OCo.) 'thrush, blackbird', Voc. 503 glossing *merula*; Modern Cornish *mola*, Lh. 89c, 168b and 241b. Welsh *mwyalch*, Breton *moualc'h*.

C2. ?Carvolth (+***ker**: cf. Breton *villa Moalc* 1282, modern Keroualch (Chrest., p. 221); or ***bolgh**)

mogh (pl.) 'pigs, swine', OM 1065; Welsh *moch*, Breton *moc'h*.

C2. Carmouth, Carnemough (both +**carn**); Tremough (+**tre**: cf. Welsh Mochdre, Breton Motrev: Top.Bret., p. 89); *Parkanmo*, *Park-an-Moah*, flds. (both +**park, an**); *Polmogh* (+**pol**); Lamouth Creek (+**nans**, Eng.)

***moyl** 'bald, bare', equivalent to Welsh *moel*, Breton *moal*. In hill-names it means 'smooth-topped, round-topped' (cf. Loth, RC 44 (1927), 293–9). When pretonic, ***moyl-** would have given ***mol-** from an early date (cf. HPB, p. 195), but it is strange that as an individual word, ***moyl** seems not to have become ***mōl** in Middle Cornish as one would have expected (cf. *cuit* > *cos*), but to have kept its diphthong, as ***moyl**: this is shown by the surname Moyle (Nicholas *le Moyl*, John *Moyl*, 1327 SR), unless the modern form represents a spelling-pronunciation of an archaic spelling.

C1. Mulberry, Mulfra, Mulvra (all +***bre**: cf. Welsh Moelfre, LHEB, p. 328; Mellor, PNDrb., 1, 144, and PNLnc., p. 73); Molinnis (+**enys**); ?Molevenney Quarry (+**meneth**?); ?Mulvin, coast (+**men**?); also perhaps a compound (?) ***moyl-arð** (see ***arð**), used (as C2) in Penmillard (+**pen**)

C2. Carn Moyle, ?Carn Boel (both coastal, +**carn**); ?Restormel (+***ros, tor**?)

moyr- (coll.; OCo.) 'blackberries', in *moyrbren* 'bramble-bush', Voc. 702 glossing *morus*; Modern Cornish *moran* 'a bramble berry', Lh. 240c. Welsh *mwyar* (cf. BBCS 16 (1954-56), 28f.), Breton *mouar* (cf. Tanguy, p. 115), Old Breton aj. *Moiaroc* (FG, p. 342). A dubious element: there is too much possibility of confusion with other words.

C2. aj. ?Crigmurrick (+**cruc**: or **mor** aj.)

mols (m.) 'wether', OM 1384; Voc. 604 glossing *verves*; Welsh *mollt*, Breton *maout*.

A. The Mouls, rock: cf. Breton Le Moult (ÉC 11 (1964-67), 154), and other animal-words used as rock-names: **ebol**, etc.

mon¹ 'slender', OM 2444; Old Cornish *muin* glossing *gracilis*, Voc. 947; Breton *moan*. In OM the word occurs in the phrase *y veen mon* 'its slender tip' (see **myn**), which is the same as the phrase in the field-names. Except in that phrase, **mon¹** is confusable with ***mon²** 'ore, mineral', and theoretically with *mon 'dung, manure', if that occurs at all in place-names.

C2. Mean-Moon, *Mean Monne, Minny Moon*, The Mee Mun, etc., flds. (all +**myn**)

***mon²** 'ore, mineral', equivalent to Welsh *mŵn, mwyn*. Clear instances of this are restricted to the compounds ***mon-dy** 'mineral-house' and ***mon-gleth** 'mine-working'.

B2. aj. *Monek de Nanscorlyes*, mine (+p.n.), *Monhek-cam*, mine (+**cam**)

C2. ?Tremoan (+**tre**); ?*Ballamoone* (+**bal, an**?)

***mon-dy** (m.) 'mineral-house', a compound of ***mon²** and **ti** (= **chy**, the *-n* apparently preserving the *d-* from changing to /dʒ/: cf. ***meyn-dy**).

C2. Rosemundy (+***ros**); ?Menmundy (+**men**); Wheal Mundie, fld. (+**wheyl**)

***mon-gleth** 'mine-working', a compound of ***mon²** and **cleath**; Welsh *mwynglawdd*. This element is almost indistinguishable from ***men-gleth** 'quarry'.

A. Mongleath

C2. Trungle (+**tre**; 2 exx.)

mor (m.) 'sea', Voc. 14 glossing *mare*; etc.; Welsh, Breton *mor*.

C1. ?Morvah (+**beth**?); see also ***merthyn** 'sea-fort', ***morrep** 'sea-shore', **morhoch** 'porpoise', and ***morva**.

C2. Mên an Mor, coast (+**men, an**); aj. ?Crigmurrick (+**cruc**; cf. Breton *morek* 'maritime': or **moyr-** aj.)

morhoch 'porpoise, dolphin', Voc. 542 glossing *delphinus*; literally 'sea-hog', a compound of **mor**+**hoch**; Welsh *morhwch*, Breton *morhoc'h* (cf. GMB, p. 426).
    A. ?The Morah, rock (cf. Breton pl. Ar-Morhoc'hed, Tanguy, p. 121)

***moryon** (pl.) 'ants', sg. *mwrrianan* 'an ant or emmet', Lh. 240c (cf. *meuwionen*, Voc. 622 glossing *formica*?); dialect *murrians*, *muryans* 'ants' (also *meryons* and *miryons*); Middle Welsh *myrion* 'ants' (Spurrel-Anwyl), Breton *merien* 'fourmis', *merieneg* 'fourmilière' (cf. Tanguy, pp. 129f.); Old Breton *moriuon* (FD, p. 260); *Morionoc* (Chrest., p. 153). Forms with and without internal *i*-affection may have existed side by side in Cornish, as shown by the place-names and by the dialect forms.
    A. aj. Mornick
    C2. Crigmurrian (+**cruc**); aj. *fonton morgeonec* (+**fenten**: Sawyer, nos. 832/1027)

***morrep** 'sea-shore', equivalent to Welsh *moreb*, of uncertain derivation (Loth, RC 37 (1917–19), 305f.; Richards, BBCS 14 (1950–52), 39f.). Western Cornish parishes were divided, in the 17th and 18th centuries, into the **morrep*, or seaward portion, and the **goonran*, or upland portion: see Borlase in OC 6 (1961–67), 229, and Tonkin in Lake, II, 115.
    A. Morrab; *Moreps*, Murraphs, flds.; Vorrap, coast, etc.

morva** 'sea-marsh', equivalent to Welsh *morfa* 'sea-marsh' (Middle Welsh *morva* 'sea-shore', BBC 100.9), a compound of **mor**+ma**; compare Welsh Morfa (several, Units, p. 160). The word may have been extended to mean 'marsh' simply, as the third name is some way inland.
    A. Morvah in Landrake
    C2. Polmorla in St Breock (+**pol**); Ennis Morvah (+**enys**)

mosek 'stinking', BM 2131; *mowzack*, Carew, p. 56. Cf. Welsh *mws* (obsolete) 'stinking', Breton *mouz* 'dung' (GMB, pp. 432f.; Old Breton verb *admosoi* 'souillerait', FD, p. 54); Old Irish *mosach* 'stinking'. Note Breton Porz mouzek, ÉC 10 (1962–63), 288.
    C2. Halvousack, fld. (+**hal**)

mowes (f.) 'maid, girl', OM 2071 and PC 1876, Modern Cornish *moz*, Lh. 174c; pl. *mowysy*, PC 944, Modern Cornish *mwzi*, Lh. 174c and 242c, *muzi*, Pryce, p. Ff4v. Breton *maouez* 'woman'.
    A. pl. ?Tuzzy Muzzy Croft, fld. (+**tus**?, **ha**?, Eng.)

munys

    C2. pl. Harry Mussy, fld. (+**erw**)
munys 'little', BM 96, etc., Breton *munut*.
    C2. ?Carn Minnis (+**carn**); Park Minnus, *Park Mennes*, flds.

N

nans (m.) 'valley', Voc. 719 glossing *vallis*; Welsh *nant* 'stream' (earlier 'valley'), Breton *ant* 'furrow' (in place-names 'valley': Chrest., p. 154; Tanguy, p. 78). Numerous glosses are available to show that 'valley' was the original meaning of the word: Gaulish *nanto* glossed *valle* (Endlicher's Vienna Glossary: DAG, p. 577); early Welsh *Nant Caruguan* glossed *Uallis Ceruorum* (VSBG, p. 54) and *carbani uallis* (LLD, p. 149); early Welsh *Hodnant* glossed *uallis prospera* (VSBG, p. 202); early Cornish *Nant Funttun* glossed *Vallis Fontis* (An.Boll. 74 (1956), 154). See the remarks given under ***lann** for the confusion between **nans** and ***lann** as B2: a few of the names cited below under B2 may contain ***lann** instead. One of the commonest elements as B2, yet rather rare as C2. The spelling *nand* (and also *land*, by confusion with English ?) is quite common in the mediaeval period; it occurs also in Welsh, e.g. *Canenand* (for *Cauenand*) 1292 (BBCS 9, 68) = Cafnan, Agl.

    A. Nance (4 exx.).

    B1. Hennett, Henon, Huthnance, *Hethenaunt* (all +**hueth**: cf. Welsh and Breton names cited under **hueth** C1); Pensignance (+**pen**), *Signans* (both +**segh**); Hernis (+**hyr**); see also ***kewnans**, ***downans**, and *goth-nans s.v. **goth**[1]. In some of these cases in East Cornwall (Henon, Dannon-) the final *t* of *-nt* has been lost (cf. Welsh *nant pedecou*, LLD, p. 172, but *nan pedecon*, ibid., p. 74; and GMW, p. 120, LHEB, pp. 496f., and HPB, p. 792), but it is clear that at a later date the *-nans* tended instead to lose the *-n-*, giving *-nas*: cf. ***car-bons** > *carbows, *kemmyns > *kemmys* (LlCC, p. 21), and HPB, pp. 794f.

    B2. Qualified by words denoting colour, appearance or atmosphere: Lanhoose (+**los**); Lanyew (+**du**); Lantuel (+**tewl**); Namprathick (+***bryth** deriv.); Nanplough (+**blogh**); Nanceloan, Nancelone (both +**lowen**); Nanteague (+**teg**); Nanphysick (+**fodic**)

    Qualified according to size or shape: Nanpean (4 exx.), Nanspian (all +**byghan**); Nancemeer, Nancemeor, Lemeers (all +**meur**); Nanjizal (+**ysel**)

Qualified according to vegetation: Nansavallan, Landevale Wood (both +**auallen**); Lantreise (+**dreys**); Nantrisack, Nancetrisack (both +**dreys** aj.); Laneskin Wood (+**heschen**?); Landreyne, Landrine (both +**dreyn**; or *****lann**); Nansevyn (+**hiuin**); Nanscow (+**scawen** pl.); Nanscawen, Liscawn (both +**scawen**); Nanswhyden (+**gwyth¹** sg.); Lawhittack (+**gwyth¹** aj.); Nantillio (+**deyl** pl.); Lankelly in Lanreath, Nankilly (2 exx.), Nankelly (all +**kelli**)

Qualified by words for animals: Nancassick (+**cassec**); Nancarrow (2 exx.), Lancarrow (all +**carow**; or **garow**); Nanjewick (+*****ewyk**); Lawhibbet, Lawhippet (both +**guibeden** pl.); Landabethick (+**guibeden** aj.?); Nansough (+**hoch**); Lamouth Creek (+**mogh**)

Qualified by a structure or other feature: Nancorras, Nancollas (both + *****cores**); Lanteglos (2 exx.), Nanzeglos (all +**eglos**); Lancorla (+*****cor-lann**); *nantbuorðtel* (+*****buorth**+**teil**); Nansmellyn, Nancemellin, Lamellyn (2 exx.), Lamellion, Lamellyon, Lawellen, Halamiling (all +**melin¹**); Naphene (+**fyn**); Lamin (+**myn**?); *Nansfonteyn* (+**fenten**); Nancegollan (+*****ygolen**)

Qualified by words denoting persons: Nansawsan (+**Zowzon**); Lanhadron (+**lader** pl.); Nancewrath (+**gruah**); Nancemabyn, Nancledra, Lemail, Lamellen (all +pers.)

Qualified by a river-name: Lanseaton (+Seaton); Lantewey, *Nansdeuy* (both +*****Dewy**); Landlooe, Nansloe (+*****loch**)

Qualified by a place-name: Nancealverne (+Alverton); Nantrelew (+Trelew); *Nanstrelaec, Nanstibragga*

Plus about 70 others, in most of which the qualifier is obscure.

C2. Trenant (6 exx.), Trenance (8 exx.) (all +**tre**); Chynance (+**chy**; 2 exx.); Golant (+**gol**)

*****nath** 'a hewing, a chipping' (?), equivalent to Welsh *nadd* in the phrase *carreg nadd* 'hewn stone, slate pencil' (cf. WVBD, p. 390). The North Cornish and Devon dialect word *nath* 'puffin' (Penhallurick, Birds, I, 188) is rather unlikely to be borrowed from Cornish, because it lacks equivalents in Welsh and Breton, unless one adduces Welsh *nadd* and assumes an odd semantic shift of **nath* from 'something hewn' to 'puffin' (because of its beak?), as suggested by Lockwood, TPhS 1974, 75-6. From its distribution, too, dialect *nath* 'puffin' is not very likely to have been borrowed

from Cornish (or into it, either). Though 'puffin' is theoretically possible and would suit the name well semantically, the similarity of the name to the Welsh phrase 'hewn rock' is too great to ignore.
 C2. Carricknath Point (+**karrek**)
naw 'nine'; Welsh *naw*, Breton *nav* (cf. FD, p. 264); occurs in Gonorman Downs (+**goon, men**). For the name, compare English instances of Nine Stones, etc.
nerth 'strength, force', RD 570, etc.; Welsh *nerth*, Breton *nerzh*. Perhaps used as a nickname, like Welsh *Nerth* WM 461.25.
 C2. Trenerth (+**tre**)
nessa 'nearest', RD 1867, etc. (also *nesse*, BM 239, etc.); the superlative grade of *nes* 'nearer'. Welsh *nesaf*, Breton *nesañ*.
 C2. Park Nessa, fld. (+**park**)
 D. *Kemyelnessa* (contrasted with *Kemyelpella*, v. **pell**: presumably the modern Kemyel Drea 'home K.')
***neth**, river-name of unknown meaning: see Ekwall, ERN, pp. 310f., and (*pace* Ekwall) cf. the River Nedd (English Neath) in South Wales.
 C2. ?*Strætneat* (+ ***stras**; = Stratton)
***neved** (OCo.) 'pagan sacred place, sacred grove', equivalent to Welsh *nyfed*, Middle Breton *Nemet* (a wood: Chrest., p. 222; cf. Tanguy, p. 102); Old Irish *nemed* glossing *sacellum* (LEIA, s.v.); Gaulish *nemēton* and *nimidas* (DAG, pp. 110 and 166f.), and place-names in *Nemet*- and *Nemeto*- (ACS, II, 708–13); in England, Nymet, Dev., and Nympsfield, Glo. (Elements, II, 50; JEPNS I, 50; PNDev., II, 348 and refs.; PNGlo., II, 243). See also PNRB, pp. 254f. and refs. (including *Medionemetum*, p. 416), and D. M. Ellis, BBCS 21 (1964–66), 37–40.
 C2. Lanivet (+***lann**); Carnevas (+**cruc**); aj. Trenovissick (+**tre**); with ***gor-** 'very, over', Trewarnevas (+**tre**: cf. Gaulish *uernemetis* 'fanum ingens', DAG, p. 477; British *Vernemetum*, PNRB, p. 495; Welsh Gwernyfed, D. M. Ellis, loc. cit.; but there is also a pers. *Gurniuet*, LLD, p. 211, *Guornemet*, VSBG, p. 126)
nyth 'nest', BM 3302; Old Cornish *neid*, Voc. 522 glossing *nidus*; Welsh *nyth*, Breton *neizh*.
 B2. Neithvrane (+(**carn**), **bran**: cf. Middle Breton *Neizbran*, Chrest., p. 223)
noth 'naked'; Welsh *noeth*, Breton *noazh*.
 C2. ?Goon Noath, fld. (+**goon**: or **nowyth**)
nowyth 'new' (also *newyth*, MC 233b); Welsh *newydd*, Breton

nevez. In Sawyer, no. 770, the element is spelt variously (*Tref-*) *-neweð*, *-næwð* and *-newið*.

C2. Trenoweth (11 exx.), Trenewth, Trenowah, Trenower, Trenowth (3 exx.), Trenouth (3 exx.), Trenuth, Trenute (all +**tre**); Chynoweth (+**chy**; 7 exx.); Lesnewth (+***lys**); Halnoweth (2 exx.), Helnoweth (all +**hel**); Vellanoweth (3 exx.), Mellanoweth, *Melynnewyth* (all +**melin**¹); Penoweth (+**pons**); Parknoweth, etc. (+**park**); Croftnoweth (+***croft**); *Strete Nowith*, street (+***stret**)

D. Delinuth: *Roskearnoweth*; *Merrose Noweth* (contrasted with *Merrose Coath*)

O

***oden** (f. ?) 'kiln', equivalent to Welsh *odyn*; the word must have been replaced early by **forn**, since it appears in no Cornish placenames, only in one Devon boundary-point.

    B2. *odencolc* (+**kalx**, Sawyer, no. 298: Welsh *odyn galch*, e.g. in 1810, BBCS 9 (1937–39), 307). See Jackson, BBCS 23 (1968–70), 116f.; Padel, Co.Studies 1 (1973), 58f.

odion (m.; OCo.) 'ox', Voc. 598 glossing *bos*; Modern Cornish *wdzheon*, Lh. 45a; Welsh *eidion*, Breton *ejen* (LHEB, p. 596; HPB, p. 302). The word serves as the singular of **oghan**, as in Breton.

    C2. Rosudgeon (+***ros**)

ogas (aj.) 'near'; Old Cornish *car ogos* 'a near kinsman', Voc. 156 glossing *affinis vel consanguineus*; Welsh *agos*, Breton *hogos*.

    A. ?August Rock
    C2. ?*Gwellogas*, fld. (+**guel**)

oghan (coll.) 'oxen', Trg. 27a, *ohan*, Pryce, p. Ff2r; the word serves as the plural of (OCo.) **odion** 'ox'. Welsh *ychen* (plural of *ych*), Old Welsh *hych* glossed *bos* (VSBG, p. 176n.); Welsh Rhyd Ychen 'Oxen's ford' (ÉC 10 (1962–63), 222); Breton *oc'hen* (cf. Tanguy, p. 119), serving as the plural of *ejen* (Hemon, Grammar, p. 38), Old Breton *Penn ohen* glossed *caput boum* (FD, p. 275: cf. Old Welsh *Pennichen*, VSBG, pp. 24 and 76).

    C2. Pollaughan (+**pol**); *Park Anauhan*, fld.; Crohans (+***ker**); ?Cassacawn (+ ?)

oye (m.) 'egg', CW 484, etc.; *uy*, Voc. 521 glossing *ovum*; *wy*, BM 3953 (so Stokes, ACL 1 (1898–1900), 142); Welsh *ŵy*, Breton *vi*.

on

C2. ?*Porthoy* (= Millook Haven) (+ **porth**: meaning?); ?Lanoy (+ ***lann/nans**?)

on (m.) 'lamb', OM 894, PC 707, etc.; also *oan*, PC 697, *oyen*, Trg. 52a; plural *eyn* (Trg. 44), *eyen* (Trg. 43a), *ean, ennes* (Pryce, s.v. *oin*). Welsh *oen*, pl. *ŵyn*; Breton *oan*, pl. *ein* (cf. Tanguy, p. 122). The plural may be confusable with **yn** 'narrow'.

C2. sg. Park Oan; pl. Park Ain, ?Park Nine, *Park-an-Eane*, *Park Ene*: all fields (cf. Lamb Park, etc., Field, EFN, pp. 120f.)

onnen (f.) 'ash-tree', Voc. 674 glossing *fraxus*; *onen*, BM 3289; the plural should have been *onn, but a double plural with vowel-affection as well, plus **gwyth**[1] (cf. the Welsh), is represented by Lhuyd's *enwydh*, 61b glossing *fraximus* (sic). Welsh *onnen*, pls. *onn*, *ynn* and *onwydd*; Breton *onnenn*, pl. *onn* (cf. Tanguy, p. 108).

A. pl. River Inny (+ *-**i**)

C2. sg. **onnen**, Rosenun, Rosenannon (both + ***ros, an**: Breton Rosnonen, Tanguy, loc. cit.); pl. *onn, ?Kilhallen (+ **kelli**?); pl. *enwydh*, ?Lansenwith (+ **nans**?); aj. *onnek (Breton *onneg* 'frênaie'), ?Trenannick (+ **tre, an**)

***op** (?) 'an ope, alley', borrowed from English: the form is uncertain, since the street-name perhaps implies rather a form *opet (= English diminutive?).

B2. *Opetjew*, street (+ **du**)

***orguilus** 'proud, proud man', borrowed from Middle Breton *ourgouilh(o)us* (itself borrowed from French: Piette, p. 150) or from English *orgulous*.

C2. *Croftenorgellous*, fld. (+ ***croft, an**)

ownek 'fearful', MC 77b, *ovnek*, OM 2158: the adjective of *ovn* 'fear'; Welsh *ofnog*, Breton (with a different adjectival ending) *aonik*.

C2. Lansownick (+ **nans**); ?Crownick (+ ***ker**?)

P

padel (f.) 'pan, dish', in the phrase *padelhoern* 'frying-pan', Voc. 899 glossing *sartago*; *padal* 'a pan', Lh. 15b. Welsh *padell*, Breton *pezel*. Note also Breton *padell* 'flat rock' (Vannetais, 20th-century): an equivalent of that sense might be present in the name.

C2. *Porthe an Badall* (+ **porth, an**)

***pans** (f.?) 'a hollow, dingle', equivalent to Welsh *pant*: cf. Elements, II, 59; JEPNS 1 (1968–69), 50; RC 34 (1913), 143. The

word may occur in some Breton field-names (but cf. French *pente* 'a slope'): GMB, p. 459.
    A. Banns (2 exx.)
    B1. ?The Tribbens, coast (+ *try-² ?)
    C2. Trebant (+ **tre**; 3 exx.); *Goenbans* (+ **goon**); ?Park an Bant, fld.

park (m.) 'a field, an enclosure', Pryce; borrowed from English. Welsh *parc* (EEW, p. 76), Breton *park* (Piette, p. 152). The normal word for an enclosed field. The earliest instances so far found of its use are *Parkenmelyn* 1332, *Park en Gellyn* 1335, and *Parkbakou* 1345. As a Cornish element **park** is, like other Cornish field-name elements, almost entirely restricted to the four western hundreds, and commonest in Penwith and Kerrier (though note the remarkable *Park Derras*, fld., in 1696 in Minster parish, north Cornwall). Owing to linguistic confusion, **park** as a Cornish word may be qualified by an English word or phrase, as in Park Little, *Park Saffron*, *Parkenconstable*, Park-an-Bush, etc.; and, whether qualified by English or Cornish, may be followed incorrectly by the definite article before an adjective or proper name (cf. **wheyl**, and see s.v. **an**): e.g. *Parkennowith*, Parkanheer, Park-an-starve-us, Park-an-Patrick, etc. Only a small selection of names in *Park*- is given here. The modern spelling *Parc* in some names is incorrect. In the following list, *Park*- names refer to fields unless starred, when they refer to tenements.

    B2. Qualified according to size or shape: Park Braws* (+ **bras**); Park Bean, etc. (+ **byghan**); *Park Heer*, etc. (+ **hyr**); *Park Try-Corner*, etc., Park Little (both + Eng.); *Park Moyha*, etc. (+ **meur** spv.)

    Qualified according to nearby features: Parc-an-Als Cliff (+ **an, als**); Park-an-Tidno* (+ **fenten** pl.); Persquiddle (+ **crous**, ***gwythel**?); Park an Trap, etc. (+ **an**, ***trap**); Parkhenver* (+ ***hen-forth**); Park Vorn (+ **forn**); Park Chaple, etc. (+ **chapel**); Park an Yate, etc. (+ **an, yet**); Park-Garrack (+ **karrek**); Polita* (+ **an**, ***lety**); *Park Eglos** (+ **eglos**)

    Qualified according to position: Park Bell, *Park Pell* (+ **pell**); Park Crees, etc. (+ **cres**); Park Darras, etc. (+ **daras**); Park-Warra, Park Gwarra (both + **guartha**); *Park Eughandrea* (+ **ugh, an, tre**); Parc Wollas*, etc. (+ **goles**); Park-an-huns (if + **hans**)

    Qualified by words denoting crops or wild plants: *Parke an Onyon* (+ **an**, Eng.); Park Lines (+ **lynas**?); *Park en Gellyn**

**pedreda*

(+**an, kelin**); Park Drysack (+**dreys** aj.); Park an ithan, etc. (+**an, eithin**); *Park-an-Drean* (+**an, dreyn**); Park Spernon, etc. (+**spern** sg.); Park Peas, etc. (+**pêz**); *Park Saffron* (+Eng.); *Parken Reden* (+**an, reden**); Park-an-Bush (+Eng.); Park Fave, etc. (+**fav**)

Qualified according to the quality of the ground: Park-an-starve-us (+Eng.); Park Maine, Park Vine, etc. (+**men** sg. or pl.); Park-Mannin, etc. (+**amanen**); such names appear to be considerably less common in Cornish (at least with **park**) than they are in English.

Qualified by words for animals: *Park an Tarro* (+**an, tarow**); Park Oan (+**on**); *Park an Eball* (+**an, ebel**); Park (an) Bew, etc. (+**an, bugh**); Park (an) Bowen, etc. (+(**an**), **bowyn**); Park-an-Fox* (+**an,** Eng.); *Park an Goag Cuckow* (+**an,** *****cok**)

Qualified by words denoting people: *Parkenmethek** (+**an, methek**); *Park an Arlothas** (+**an, arluth** fem.); *Park Pronter**, Park-an-Prowlter (both +(**an**), **pronter**); *Parkenconstable* (+**an,** Eng.); many instances of **park**+personal name, such as Park Steven, Park Martin, Park Matthews, *Park an Rogers*, Park-an-Patrick, etc.

Qualified by a place-name: *Parke Chypons**, Park Chynoweth, etc.

Qualified by other words: Parknoweth*, etc. (+**nowyth**); *Park gwidden* (+**guyn**)

*****pedreda** (OCo.), element of unknown meaning. In two of the three cases it occurs close to a notable hill-fort; cf., however, the two English river-names Parret (Dor./Som. and Glo./Wor.), both formerly *Pedred-, Pedrid-*, etc. (Ekwall, ERN, pp. 320–2; PNWor., p. 13). As Ekwall saw, a prefix **petru*- (as seen in Gaulish, ACS, II, 977) would fit the names formally, and British **Petru-rit*- could mean 'four-fold stream', 'four-fold ford', or '(stream with) four fords' (see **red** and **rid**): all of these are difficult semantically, and if such a compound occurs in all the various names, it probably had a specialised meaning. (Ekwall compares a Welsh stream Pumryd, apparently '(river with) five fords'.)

A. Patrieda; Perdredda; Pathada

*****pel** (f.) 'ball', equivalent to Welsh *pêl*; this is the word seen in late Modern Cornish *an Bele ma* 'this ball' and *an pelle Arrance ma* 'this silver ball' (Padel, Bosons, pp. 12 and 38), and *dho guare peliow* 'to play at bowls' (Pryce, p. Ff4r). It is confusable in written form with **pell** 'far' when used as C2. Its meaning in place-names would be 'hill, mound', like English *ball* (Elements, I, 18f.). Cf. **pellen** 'ball'.

    C2. ?Penpell (+**pen**; 2 exx.: or **pell**); pl. ?Tempellow (+**tyn** or ***dyn**); aj. ?Penpillick (+**pen**: or **pyl** aj. ?)

pell 'far, distant', MC 140b, etc.; Welsh, Breton *pell*. Comparative and superlative *pella*, and also *pelle* BM 2843. The positive grade, **pell**, is confusable with ***pel** 'ball', though for Cornish speakers the different vowel length would have distinguished them.

    A. ?Bell Veor/Vean; ?The Pell, mine

    C2. ?Penpell (+**pen**; 2 exx.: or ***pel**); ?Trebell (+**tre**; or ***pel**); Goonbell (+**goon**); Park Bell, *Park Pell*, flds. (+**park**)

    D. spv. *Kemyelpella* (presumably = Kemyel Wartha), *Bodryanpella* (= Bodrean), *Chyunwone Pella* (= Choone in Paul), *Kenegy pella alias Bella* (= Kenneggy in Breage); ?Burlorne Pillow

pellen 'ball', Voc. 832 glossing *globus*; Welsh *pellen*, Breton *pellenn*; like ***pel**, this word would mean 'hill' in place-names.

    B2. ?Pellagenna (+**ganow** ?)

    C2. *Creage Pellen* (+**cruc**); ?Trebellan (+**tre**)

pen (m.) 'head, top, end; promontory'; as adjective 'chief, end-'. Welsh *pen*, Breton *penn*; Old Breton *Penn ohen* 'caput boum' (FD, p. 284). Elements, II, 61f.; JEPNS I (1968–69), 50. See the discussion of the syntax and meaning of Welsh *pen* in place-names by I. Williams, BBCS 10 (1939–41), 303–5. In theory as a topographical term it should mean the 'top end' of any feature; and it is true that names appearing to start *Pen-* that apply to the lower end of features usually turn out to contain **ben** 'base' (e.g. Pendower, Pendavey, and Penhale in Davidstow); but one cannot say firmly that **pen** never refers to the lower end of any feature (compare *pen golas* 'bottom end', MC 184b). Another difficulty is that of saying whether **pen** ever means 'hill' simply: that meaning is often given, but more often in English text-books than in Welsh or Breton ones; there are no clear instances of such a meaning in Cornwall. (Of those that might be thought to imply it, **pen**+sg. of a plant-name probably contains sg. for plural of that plant (see **-en**), and **pen**+animal-name probably has a metaphorical sense: see below.) Note, too, that there are no examples of **pen** qualified by a personal name, unlike the common elements denoting 'hill, high ground' (e.g. ***ros**, **goon**). The meaning 'headland, promontory, point' is well-attested, however. It is also sometimes difficult to

pen

say whether *Pen-* in a name is used as B2 ('head, top') or as C1 ('chief, end-'). Several phrases are so common that they are listed as elements in their own right.

A. ?Penhill, fld.

B1. Note the complete lack of instances of **pen** as B1: forms such as *Hyrben and *Moylben (which might be expected if **pen** meant 'hill') do not occur; compare C2 below.

B2. Qualified by words for high ground: Penmenna, Penmennor, Penmenor, Polmenna (4 exx.) (all +**meneth**); *Penmynytheowe* (+**meneth** pl.); Pendrim (+***drum**); Penhedra, Polhorden (both +***hyr-drum**); Penisker (+***esker**?); Penhalt (+**als**); Pengreep (+**krib**); Penvearn (+**bron**); Penhargard, Penharget (both +***hyr-yarth**); Penknight (+***cnegh**); Pednvadan, coast (+**tal, ban**); see also the phrase ***pen ros**.

Qualified by other words denoting land: Polveithan (+**budin**); Penbothidnow (+**budin** pl.); Pentreath, Pentreath Beach (both +**trait**); Penpraze (+**pras**; 2 exx.); Pen Enys Point (+**enys**); Penwerris (+**gweras**); Pen Blue, fld. (+**plu**); Penbro (+**bro**). See also the elements ***pen-arth**, ***pen-ryn** and ***pen-tyr**.

Qualified by words for valleys: Pensignance (+cpd., **segh**+**nans**); Penstrase, Penstrace, Penstraze, Bostraze (all +***stras**); Penstrassoe, Penstroda (both +***stras** pl.); Pencobben (+***comm**); see also ***pen nans** and ***pen hal**.

Qualified by words for plants (usually in the sense 'top of the grove or patch of...'): Penwarne (3 exx.), Penwarden (all +**guern**); *penn lidanuwern* (+cpd., **ledan**+**guern**: Sawyer, no. 1019); Penquean (+***keun**); Pencorse (+***cors**); Pengersick (+***cors** aj.; 2 exx.); Penhellick (4 exx.), Pennellick, Penellick (all +**heligen** pl.); Penaligon Downs (+**heligen** sg.); Penscawn (+**scawen**); Penhesken, Prenestin (both +**heschen**); Penfound (+***faw**?); Penbetha in Creed, Penburthen (both +**bedewen**); Pendarves (+**dar** deriv.?); Pencalenick (+**kelin** aj.); Palestine (+**glastan**); Pendrissick, fld. (+**dreys** aj.)

Qualified by words for woodland: Penperth (+***perth**); see also the phrases ***pen cos** and ***pen (an) gelli**.

Qualified by words denoting water: Pengover (+**gover**); Penhallam (+r.n., see ***alun**); Pengwedna (+r.n. *Gwenna?); *Pednanpill Point* (+**an**, ***pyll**); Pendavey in Minster (+r.n., see ***dewy**); see also the phrase ***pen pol**.

Qualified by words for animals: Pencarrow in St Austell and

pen

Egloshayle, Pencarrow Head (all +**carow**); Pinnick (+*****ewyk**); Penkevil (+*****kevyl**); Penvith (+**bugh**); Penbough, Pemboa (both +**bogh**: cf. Welsh Penbwch, Gla., PNDinas Powys, p. 170; Breton Penboc'h, Tanguy, p. 118); Penfrane (+**bran**); Pedn Kei, rock (+**ky**?); Pedn Tenjack, fld. (+**denjack**). These belong with the widespread class of names composed of 'head' qualified by an animal-name, and noted in English and Germanic by Dickins (PNSur., pp. 403–6), and in Brittonic and Irish by R. J. Thomas, Arch.Camb. 89 (1934), 328–31. The significance of these is unclear (and need not be the same in every case), but the evidence adduced by those authorities for a possible ritual significance in some cases is powerful, and was also the explanation given, in the 9th century, by the Breton Uurmonoc (*Penn ohen*, RC 5 (1881–83), 418); in other cases the usage could well be metaphorical: 'headland (or other feature) resembling the head of...' (cf. **lost**).

Qualified by words for man-made features: Penvose (+**fos**; 4 exx.); Pednvounder, Penny Vounder, etc. (+(**an**), **bounder**); Pedn y coanse (+*****cawns**); Pencrowd (+*****krow-jy**); *Penanchase* (+**an**, *****chas**); *Pencair* (+*****ker**); Pencarrow in Advent (+*****ker** pl.); Penolva, Pedn Olva, coast, Pen Olver, coast (all +**guillua**); *Pendinas*, Pendennis Castle (both +*****dynas**); Penkestle (+**castell**; 2 exx.); Pendeen (2 exx.), Pendine (all +*****dyn**); Penheale (+**hel**). See also the phrases *****pen** (**an**) **dre**, *****pen** (**an**) **pons**.

Qualified by words for natural features: Pengarrock (+**karrek**); Penlee Point (+*****legh**; 2 exx.); Penbeagle, *Penbegle* (both +**begel**); Pen-a-maen, coast, Penmayne (+(**an**), **men**); Pedn-myin, coast (+**men** pl.); Pennywilgie Point (+**an**, *****gwailgi**?); Pen a Gader, coast, *Pengadoer* (+(**an**), **cadar**)

Qualified by a place-name: *Peddenporperre* (+Polpeor); Penkynans (+Kynance); *Penhalemanyn* (+Halamanning)

A word *****penna** seems to appear consistently in the early forms of some names, all in East Cornwall: it might be for *****pen an**, or perhaps simply a lengthened form of **pen**: compare the late *Pedny-*, a by-form of **Pedn-** in some names; and also *Pennhalgar* (DB), *Pennahalgar* (Exon. DB), alternative forms for Penhawger in Menheniot. The second element is obscure in all cases: Pennycrocker (+*****crak**[1] pl.?), Pennygillam, Pengenna (+*****genn** deriv.?), Pennycraddock, Penavartha, Pennydevern (+*****devr-** deriv.?), Penquindle. Some coastal names such as Pen-a-maen may also belong in this group.

*pen (an) dre

C2. Note the lack of cases of **pen** as C2, as of B1: forms such as **Treben, *Chywarben* and **Coyspen* are apparently not possible. This strengthens the idea that **pen** was not a normal word for 'hill'.

***pen (an) dre** 'top of the village', a phrase of **pen, an** and **tre**. The lenition shows that the definite article was understood, even though not normally appearing in early spellings.

A. Pendrift, Pendrea, Pedn-an-drea; Pendray, Pedndrea, flds.; pl. Pendriffey

B2. ?Penderleath (+ ***legh**?)

***pen (an) gelli** 'grove's end', a phrase of **pen, (an)** and **kelli**. Welsh Penygelli, Mtg. (Richards, Units, p. 178), Breton Penguilly, etc. (Tanguy, p. 100). As in ***pen (an) dre** the lenition in the Cornish names shows that the definite article was understood, even though it never appears in early forms: compare Old Welsh *penn i celli* and *penn i cgelli* (LLD, pp. 264f.). For the use of the phrase as B2, compare Old Welsh *penncelli guennuc* (+r.n.) and *penncelli gulible* ('wet-place') (LLD, pp. 242 and 268).

A. Pengelly (12 or more exx.); pl. Penkelly, *Penkelyou*

B2. *Pengellylowarn* (+**lowarn**)

***pen (an) pons** 'bridge's end', a phrase of **pen, (an)** and **pons**; Welsh Pen-y-bont (several: Richards, Units, p. 178), Old Breton *Ran Penpont* (Chrest., p. 157).

A. Penpont (4 exx.), *Penpons* (3 exx.), Penponds, Pednpons, *Penenpons*

B2. *Penponsmethle, Penpons Haverack* (both +p.n.)

***pen-arth** 'promontory, headland', equivalent to Welsh **pen(i)arth* or *penardd; Breton place-names Penharz and (Old Breton) *Penard, Penharth* (Smith, Top.Bret., p. 95; Fleuriot, Ann.Bret. 64 (1957), 530f.). These names are a problem in all three languages: are they compounds or phrases, and do they contain ***arð** 'height' or ***garth** 'height'? All three languages show occasional traces of either an *-h-* or an original lenited *-g-* before the second element in some early forms of some of the names: e.g. Welsh Peniarth (LHEB, p. 439), and *Penharth* 1266, etc. (PNDinas Powys, p. 158), Old Breton *Penharth* (Fleuriot, loc. cit.); Cornish *Pengarth* 1337 (in Wendron: = Penmarth?); *Penharn* 1293 and *Pengard* 1345 (= Penare in Goran); *Pengard* 1296, etc. (= Penair alias Penarth in St Clement); *Penhart* 1238 (= Penarthtown in Morval), etc. It is possible, though unlikely, that two different compounds or phrases are involved: one of **pen**+ ***arö** and one of **pen**+ ***garth**. The stress in Cornish is normally on the second syllable, where ascertainable, and that this is old is shown by such forms as *Nare* c. 1689 (= Penare in St Keverne): cf. PNDinas Powys, p. 346. The problem cannot be resolved here: for discussion, see Williams, PKM, pp. 260f. and 293, and ELl, pp. 22f.; PNDinas Powys, pp. 158-60; Fleuriot, loc. cit. Note also Pennard, Som., and Penyard, Hre. (Turner, BBCS 14 (1950-52), 115f.). Of the eleven Cornish examples, four are well inland (two of them doubtful), two are inland but on estuaries, and five refer to prominent headlands, all on the south coast.

A. Penare (2 exx.), Pennare (2 exx.), Penair or Penarth, Penarthtown (+English), Penearth, *Penarth-Point* (= Gribbin Head), ?Pennards, ?Penmarth, Penare Point

***pen cos** 'wood's end', a phrase of **pen**+**cos**. Welsh Pencoed, etc. (Units, p. 172), Breton Penhoat, etc. (cf. HPB, p. 320 and n. 5); note also, in English counties, *Paunsett* (PNWlt., pp. 12f.), Penge (PNSur., pp. 14f.), Pingewood (PNBrk., 1, 206), Pencoyd (PNHre., p. 149), Penketh (PNLnc., p. 106), and Penquit (PNDev., 1, 272). In Cornwall the earlier, Eastern form, *Penquite*, is commoner than the later, Western form, *Pencoose*, because of the greater amount of woodland in East Cornwall.

A. Penquite (14 exx.), Penquit, Pencoose (4 exx.), Pencoys, ?Pencoose (5 exx.)

B2. Penscove (+**gof**)

***pen-fenten** 'head-spring', i.e. 'source of a stream', a compound of **pen**+**fenten**. It could alternatively in some cases be a phrase **pen fenten* 'spring-head': Welsh Penffynnon and Pen y Ffynnon are of about equal frequency.

A. Penventon (5 exx.); Pedn Venton; Polventon in Lansallos and Warleggan; pl. Penventinue, Penventinnie, Pennytinney, etc.; cf. Breton *Penfeunteuniou* (Tanguy, p. 94)

B2. Pentivale (+r.n. *Fal*)

***pen hal** 'moor's head, marsh's head', a phrase of **pen** and **hal**. The name is extremely common in Cornwall, but apart from Welsh Pennal, Mer. (*Penhale* 1292-93, MerSR, p. 28) does not seem to be common in Wales or Brittany. Most of the Cornish places are at the head of a marshy valley. The phrase ***pen halou** 'head of the marshes' may have been little different in meaning from ***pen hal**.

A. Penhale (c. 26 exx.); pl. Penhallow (8 exx.)

*pen-lyn

B2. Polwin, Pollawyn (both +**guyn**); Penelewey (+pers. or r.n. ***Dewy** ?); Penalguy (+**ky** ?); *pennhal meglar* (+St ?); *penn hal weðoc* (+**gwyth**[1] aj.); Penhalurick (+***luryk** ?)

*pen-lyn 'head-pool' (?), perhaps a compound of **pen**+**lyn**; Welsh Penllyn (several, Units, p. 174). The stress (normally on the second syllable) is wrong for a compound, and moreover the vowel is long in most cases, whereas **lyn** 'pool' has a short vowel. For these reasons it is uncertain whether all the names listed here contain ***pen-lyn** 'head-pool': some may contain a phrase **pen lyn* 'pool's head' (or 'above the stream' ?), and some may contain a different second element (perhaps ***leyn** ?).

A. Pelean, Pelyne, Pelyn, Polean, Polyn ('creek's head'), Penlean, Pellyn-wartha (+**guartha**)

C2. Pellengarrow (+**garow**: meaning ?)

*pen nans 'valley's head', a phrase of **pen** and **nans**; Welsh Pennant (frequent), Breton Pénan (2 exx., Tanguy, p. 78).

A. Pennant (7 exx.), Pennance (7 exx.), Pennans, Penance, Penance Wood

C2. Pennatillie (+**deyl** pl.); Penjerrick (+ ?)

*pennek 'big-headed', the adjective from **pen**; Breton *pennek* 'muni d'une tête' (cf. GMB, p. 478). The Cornish surname *Penneck* (*Pennecke*, etc., 1641) is probably this word. In the place-name, used as a noun, it may mean 'tadpole', though the use of the singular is odd.

C2. Polpidnick (+**pol**)

*pennyn 'little head', a diminutive of **pen**. Like the preceding, the word could mean 'tadpoles' in the place-name, as suggested by Nance (Dictionary, s.v. *penyn*).

C2. pl. ?Poolpanenna (+**pol**)

*pen pol 'creek's head, above the creek; stream's head', a phrase of **pen** and **pol**. Most instances are on the coast, sometimes 'above the creek' rather than exactly at its head; in four cases the name is well inland but at the head of a short tributary stream (see **pol**).

A. Penpol(l) (10 exx.), Penpill

B2. *penpoll lannmoren* (+p.n.: Sawyer, no. 770)

*pen-ryn 'promontory, point of land', a phrase or compound of **pen**+***rynn** 'point'; Welsh *penrhyn* 'promontory' (cf. Units, p. 176). The stress, on the second syllable, suggests a phrase rather than a compound: cf. ***pen-arth** and ***pen-tyr**. The second instance is not coastal, but nor is Welsh Penrhyn always; the third is on a point of land between two streams.

A. Penryn, Pridden

C2. Penvories (+pers. ?)

*pen ros 'hill's end', a phrase of pen and *ros; Welsh Pen-rhos (10 exx., Units, p. 176), Old Welsh *cecin pennros* 'hill's end ridge' or *'Pennros* ridge' (LLD, p. 264); Breton Perros (3 exx., Tanguy, p. 78); Middle Breton *Penros* (RC 7, 203); Penrose (PNDev., I, 159). The exact sense depends upon that of *ros (for the sense of which the Penrose names are part of the evidence); it is obviously 'end of the high ground' in some sense. Most of the names are at the end of fairly well-defined spurs or ridges; if it referred to the vegetation ('end of the heathy hill'), it is irrecoverable now.

A. Penrose (11 exx.), Perrose (2 exx.)

*pen-tyr 'headland' (originally a phrase, *pen tyr* 'land's head', and so stressed); Welsh *pentir* 'headland', Breton *penn-tir in place-names (see Loth, Mél.Arb., p. 227). Note also Kintyre, Arg. (CPNS, p. 92; Hogan, Onomasticon, p. 227; Bannerman, Dalriada, pp. 103, 108 and 111). Whether a *pen-tyr differed from a *pen-arth and a *pen-ryn is unknown. Three of the instances are on the north coast, with two on the south coast and one inland.

A. Pentire (5 exx.), Pedn Tiere, coast

*perth (f.) 'brake, thicket', equivalent to Welsh *perth* 'hedge, bush'; early Welsh *carn perth yr onn* 'the cairn of the ashes' brake' (LLD, p. 143); Welsh Perth Gelyn (ELlSG, p. 23; cf. PNFli., p. 131); not in Breton (see Chrest., p. 156, n. 9; FD, s.v.). Elements, II, 63; JEPNS 1 (1968–69), 50. Note Gaulish *Perta* (DAG, p. 771), Scottish Perth (CPNS, pp. 356f., with other Scottish instances; also Nicolaisen, p. 164), and Peart, Som. (BBCS 15 (1952–54), 17f.); it seems to be an old practice (not attested in Cornwall) for the element to be used as a simplex place-name.

C2. Penberth (+**ben**); Penperth (+**pen**); cf. Welsh Penyberth, Crn. (ELlSG, p. 49); with *-ys² suffix (?), ?Treburthes (+**tre**)

perveth 'middle', surviving only in the phrase *aberveth* or *aperveth* 'within', OM 992, RD 2286 (sic MS: ACL 1, 183), etc. The normal Middle Cornish for 'middle' was **cres**, so that the names **tre**+**perveth** must belong to an earlier period. Welsh *perfedd* (Treberfedd, several, Units, p. 205), Old Breton *permed* (FD, p. 284). See also ***með**.

C2. Trebarva, Trebarvah (2 exx.), Trebarwith, Trebarvath, Trebarveth, Trebarret, Trebarfoot (all +**tre**); Bosparva (+ ***bod**); Tresparrett (+ ***ros**); Resparveth, Resparva, Sparett (all + ***ros** or **rid**); Cutparrett (+**cos**); Callabarrett (+**kelli**)

pêz (pl.) 'peas', Lh. 121a glossing *pisum* 'a pease'; cf. 150b. Borrowed from English, like Welsh *pys* (Parry-Williams, EEW, p. 127); compare Breton *piz* (cf. GMB, p. 482).

C2. Park Peas, Park-an-Pease, flds.; Gwool Pease, fld. (+**guel**)

peswar 'four' (masc.; *pedyr*, etc., fem.). Welsh *pedwar*, Breton *pevar*. The first name should properly contain the feminine form, and the last should properly contain **forth** singular, not plural: the grammar was breaking down in late Modern Cornish.

C1. ?Padgagarrack Cove (+**karrek**?); ?Padgy Vessa, fld. (+ ?); Padge Dinner, fld. (+**dyner**); Padzhuera, fld. (+**erw**); fem. ?Peter Varrow, fld. (+**forth** pl. ?)

***peulvan** 'pillar, standing stone', equivalent to Breton *peulvan* (cf. GMB, p. 483).

C2. ?Bospolvans (+ ***bod**)

***peur** 'pasture', equivalent to Welsh *pawr*, Breton *peur* 'pâturage' (cf. Stokes, Bezz.Beitr. 18 (1892), 109; GMB, p. 484); the existence of this word depends upon the place-names, none of which must definitely contain it.

B2. ?*Porpoder* Field (+**podar** ?)

C2. ?Trembear (+**tre, an**); ?*Parkanboer*, fld.

***pever** 'fair, bright', equivalent to Welsh *pefr* 'radiant' (see Williams, PKM, p. 286, and CA, p. 355). The word occurs in several Scottish river-names, including Peffery, Ros.: see CPNS, p. 452, and Nicolaisen, p. 164.

C2. ?Polpever (+**pol**?)

***pybell** (f.) 'pipe', equivalent to Welsh *pibell*; cf. *pib* 'pipe', Voc. 254 glossing *musa*.

C2. Praze-an-Beeble (+**pras, an**); *Parc-an-bibble*, fld.; *Wheale Bebell*, mine; Gonabibles, fld. (+**goon, an**)

***pybor** (m.) 'piper', pl. *pyboryon*, BM 4563; from *pib* (see previous entry). This word replaced the earlier *pybyth seen in Voc. 253, *tibicen* glossed *piwhit* (read *piphit*), and in Osbert *le Pibith*, Richard *le Pybythe* (1302, Ass.R.), equivalent to Welsh *pibydd* (used also as a bird-name).

A. ?Pibyah Rock, ?Peber, rock: these might alternatively contain the plural of *pib* 'pipe'.

C2. *Wheal an Peber*, mine

*pyk 'point', equivalent to Welsh *pig* 'beak, spout', Breton *pig* 'pick, pickaxe'.
    A. ?Pyg, rock
    C2. ?Porth Pyg, coast (+**porth**)

pyl 'pile, heap', BM 1621 (so Nance, Dictionary, s.v.); Lh. 64b, *pil gwdhar* 'mole-hill' and 154b *pîl tyil* 'dunghill'. Borrowed from English. The word is probably confusable with a possible ***peul** 'pillar' (equivalent to Welsh *pawl*, Breton *peul*: see GMB, p. 483, and WVBD, p. 416, and cf. ***peulvan**); and an adjectival form ***pylek** would be confusable with an adjective from ***pel** 'ball'. Note that Breton *pil* 'rock' is common in coastal rock-names (Topon.Nautique no. 1419, p. 518) and that usage may be present in some of the Cornish examples, though it is not common.
    A. The Peal, ?Peel Cove, both coastal
    B2. *Peele Myne* (+**men** pl.)
    C2. aj. ?Penpillick (+**pen**: or ***pel** aj. ?)

*pylas 'naked oats', equivalent to Breton *pilad* 'avoine à gruau', RC 3 (1876–78), 68n.; both perhaps borrowed from English? The relationship, if any, of ***pylas** 'naked oats' to **pylles** 'bare' is unclear. Cornish dialect *pillas* means 'naked oats, *Avena nuda*', which used to be grown on poor ground: Nance, 'Pillas, an extinct grain', OC 1, xii (1930), 43f.; Pounds, 'Pillas, an extinct grain', DCNQ 22 (1942–46), 199f. Note field-names which contain the dialect word, such as Pillas Croft. Compare Welsh *pilcorn* 'oats' (BBCS 4 (1927–29), 300), and Old English **pil-āte* (DEPN, s.v. Pillaton Hall, Stf.), and in field-names (PNChe., v, i, 305).
    C2. Noon Billas (+(**an**), **goon**); *Croft-Pellas, Croft-and-Peloss*, flds. (+***croft**, (**an**)); Park an Pillars, fld.

*pyll 'creek', cf. English *pill* 'creek'; the English word (Elements II, 75; NCPN, p. 310) is said to be borrowed from a Welsh **pil* or **pyl* (not in Geir.Mawr): so Ekwall, DEPN, s.v.; R. J. Thomas, EANC, p. 58; and G. O. Pierce, BBCS 18 (1958–60), 264. However, the word is so common as an English element in Cornwall, as in Devon (PNDev., II, 669), and so rare as a Cornish one (only one instance), that the supposed Welsh word probably had no direct cognate in Cornish (or in Breton?), and the single place-name example probably represents a loan from English. With the one name, compare Welsh Pen y Pil (3 exx.); the Welsh word, like the

English one, seems to belong predominantly to the estuary of the R. Severn.

C2. *Pednanpill Point* (+**pen, an**)

pylles 'peeled, bare', CW 2318 (*pedn pylles* = Middle English *pilled-pated*, NED s.v. *pilled* 5); *pilez* or *pedn pilez*, Lh. 45c glossing *calvus*; probably a direct borrowing (with substitution of Cornish *-es* for English *-ed*) of English *pilled* 'peeled, stripped', rather than a ppp. of a verb ***pylya** borrowed from English 'peel'.

C2. ?Carrag-a-pilez, coast (+**karrek**); some of the names listed under ***pylas** may contain **pylles** instead, if indeed it is a different word.

pin- 'pine-trees', only in *pinbren* 'pine-tree', Voc. 680 glossing *pinus*; Welsh *pîn*, Breton *pin* (cf. Tanguy, p. 108).

C2. aj. ?Prospidnick (+***prys**)

***pystyll** 'waterfall', equivalent to Welsh *pistyll* (cf. PNDinas Powys, pp. 96f.); the existence of this word depends on the one place-name, which might instead contain English *pistol*. However, the interpretation as 'waterfall' would suit the site.

A. ?Pistil Ogo (+English dialect)

pyt 'pit', RD 2010, etc.; plural ***pyts**. Borrowed from English, like Welsh *pit* (WVBD, p. 431).

B2. *Pit-pry, Pitt Pry*, flds. (+**pry**); pl. *Pitsonbriney*, (*two pitts called*) *Pitts en bryne* (both +**an**, Eng.?); pl. *Pich Kissa*, mine (+?), Pitslewren (+**lowarn**), *Pitshanpassen*, mine (+**an**, ?), Pits Mingle (+***men-gleth**?)

C2. ?Chypit (+**chy**); *Park-an-Pitt*, fld.

plas 'place, spot', BM 4298, etc.; 'mansion, residence', BM 2286, etc. Borrowed from English, like Welsh *plas* 'mansion; open space' (cf. WVBD, p. 433; EEW, p. 84) and (from French) Breton *plas* 'place, space' (cf. Piette, p. 157). In the instances used as A (which may be English *place* 'residence' instead: Elements II, 66) the sense is 'residence', but as B2 it no doubt means 'square, open space surrounded by buildings'.

A. ?Place House (2 exx.), ?Place

B2. *Placengarrett* (+**an, karrek**?); *Placengennov* (+**an**, ***genn** pl.?); *Place an Streat* (+**an**, ***stret**); cf. the partly-latinized *Placea Enfenten* (+**an, fenten**)

plen 'arena, field', OM 2151, etc.; borrowed from English or French.

B2. Plain-an-Gwarry (2 exx.), Plane-an-Gwarry, *Plain an*

Quarry, Plean-an-Wartha, etc. (all +**an**, **guary**); *Pleyn-Goyl-Sithney* (+p.n., **gol**+St)

plos 'filthy'; derivation or cognates unknown?

C2. Tol Plous, coast (+**toll**)

plu (f.) 'parish, the people of a parish'; Old Cornish *plui*, Voc. 106; *plew*, Trg. 25a. Welsh *plwyf*; Breton *ploue* 'campagne' (but, in place-names, 'parish' or 'populace': Largillière, Les Saints). See the note, 'Cornish *plu*, Parish', Co.Studies 2 (1974), 75–8, with addenda, Co.Studies 3 (1975), 19–23; also Quentel, Ogam 6 (1954), 273–6, and Rev.Intnl.Onom. 11 (1959), 34–5. There is only one true place-name containing **plu** as generic (in the manner so common in Brittany but unknown in Wales), and that one instance, like the single instance of ***lok**, is evidently an outlier of the Breton distribution, lying as it does towards the south coast. The other instances of the word as B2 are all properly surnames, and the element probably there means 'populace', so that, e.g., Henry *Pluysie* is 'Henry of the populace of St Issey parish': such names are restricted to the 16th century on present data.

A. Bluestone (2 exx.), *Blew Stone* (2 exx.), Bleu Bridge (all +English). In these instances, **plu** has virtually become an English dialect word used as a qualifier.

B2. Pelynt (+St); *Plu-alyn, Plewe-Golen, Plewgolom, Pluysie, Pluyust, Pluvogan* (all +St)

C1. ?Pleming (+**men**?)

C2. Croft Bloue (+***croft**); Pen Blue, fld. (+**pen**)

podar 'rotten, decayed', Lh. 133a glossing *putridus*; Welsh *pwdr*; cf. possibly Breton *pore* 'maladie' (GMB, pp. 504f.). Note also Cornish derivatives such as *a bodrethes* 'of rottenness', OM 2714; *podrek* 'putrid', BM 3048; etc.

C2. ?*Porpoder Field* (+***peur**?)

pol (m.) 'pit; pool, stream; cove, creek', Voc. 741 glossing *puteus* (OE *pytt*); Welsh *pwll* 'pit, pool', Breton *poull* 'pit, pool'. The meaning 'mud' is also well attested in Cornish: *a bol hag a lyys* 'from mud and slime', OM 1070; pl. *pollov* 'puddles', BM 4190; and *Pol-* glossed *lutum sive puteus* 'mud or pit' in the Glasney Prophecy (Jenner, 96th RCPS (1929), 238–41). It is often difficult to decide which of the various meanings of **pol** is present in a particular place-name; the difficulty is compounded by the fact that in coastal names *Pol-* often comes from an original *Porth-*, as in e.g. Polruan (formerly *Porthruan* 1284), Polperro (formerly

pol

Porthpyra, etc.), and a good many others. Since both **porth** and **pol** could mean 'cove', there was virtually no difference between the two elements in this context. The reverse process, of modern *Porth-* from original *Pol-*, is less common. In addition, there has been a two-way interchange between *Pol-* and *Pen-*, even though the meanings are quite distinct: e.g. Penglaze in Lansallos (formerly *Poleglaze*), Penhole (formerly *Polhal*); and the commoner reverse process, Polmenna (several, all formerly *Penmene*, etc.), Polventon (several, formerly *Penfenten*, etc.), and a few others. In some cases of *pen pol, pol seems to mean 'stream'. Note that **pol** is extremely rare as qualifier.

B1. Blable (+**bleit**); Whimple (+**guyn**); Treamble (+**taran**); *Fimbol* (+**fyn**: Sawyer, no. 450)

B2. Qualified by words for animals: Polmark, Polmarth (both +**margh**); Polgazick, Polgassick Cove (both +**cassec**); Poltarrow (+**tarow**); Polgaver Beach (+**gaver**); *Poole-an-abelly* (+**ebel** pl.); Polgooth (+**goth**²); Polwhele (+**hwilen** pl.); Polbrough, Polbrock (both +**brogh**); Polgeel Wood (+**ghel**); Polyphant (+***lefant**); Poligy, Polhigey (both +**hos** pl.); Polkanuggo, Polkernogo (both +**cronek** pl.); Poolpanenna (+***pennyn** pl.?); Polgray (+***gre**); *Polmogh* (+**mogh**); *Polgath* (+**cath**); *Polcarowe* (+**carow**); Pollaughan (+**oghan**)

Qualified by colour-words: Poldew (2 exx.), Poldue (2 exx.), Polsue in St Erme, St Ewe and Goran, Poldhu Cove (all +**du**); Polgwyn Beach, Polgwidden, Polgwidden Cove (all +**guyn**); ?Polpever (+***pever**?); Polcan (+**can**); Pol Lawrance, coast (+**arghans**); Polglaze (several), Polglase (several), Penglaze (several), Pollglese, fld., Pol Glâs, coast, (all +**glas**)

Qualified by words for people: Polwrath (+**gruah**); Polmaugan, Polkinghorne (2 exx.), Polmartin (all +pers.); Polgorran (+St)

Qualified by words for plants: Poliskin, Poleskan, *poll hæscen* (Sawyer, nos. 832/1027) (all +**heschen**); Polbelorrack (+**beler** aj.); Poltesco, Poltisco (both +***tusk** deriv.?); Polrudden (+**reden**?)

Qualified by a word for 'water' or a description of the water: Polstreath, coast (+**streyth**?); Polshea, Polzeath, Polsue in Lanteglos by Fowey (all +**segh**); Poldower (+**dour** sg. or pl.); Polgover (+**gover**); Polcatt, Polcocks (both +***cagh**); Pooldown, Pulldown (both +**down**); Polstrong (+***stronk**); Polmorla in St Breock (+***morva**); Penhole in North Hill (+**hal**); Polpeor in Uny Lelant, *Poulpere*, Polpeer, fld. (all +**pur**); Polbreen (+ ***breyn**); Polwheveral (+ ***whevrer**)

Qualified by other attributes or by nearby objects: Polmere, flds., Polmear in Tywardreath (all +**meur**); Polledan, Pol Ledan, both coastal (both +**ledan**?); Polvellan, Polly Vellyn, fld. (both +**melin**[1]); Polcreek (+**cruc**); Polladras (+***ladres**?); Polcrebo (+**krib** pl.: meaning?); Polscoe (+**skath**); Polgear (+***ker**); Polhendra (+***hendre**); Polandanarrow Lode (+**an, dyner** pl.?); Porth Kidney Sands (+**cummyas**); Porthcollumb (+***kel**?, ***lom**?). See also the phrases **polgrean**, ***pol ros**, **pwl prî** and **pul stean**.

Qualified by a place-name: Poltreworgy in St Kew; Polskeys (+**cos, schus**); Polpenwith (+**pen, ruth**); Polnare Cove (+ *Penarth); Polcoverack (+r.n.)

Plus about 50 other names where the qualifier is obscure, including Bolatherick, Parbola, Polmesk, etc.

C2. Trepoll (+**tre, an**); Street-an-Pol, street (+***stret, an**); ?Trebullom (+**tre, guyn**?)

D. Delabole (also Delamere, with **meur**, and Delinuth, with **nowyth**)

polgrean 'a gravel pit', Pryce, s.v. *pol*; a phrase of **pol** and **gronen** pl. (rather than *growan = **grow** sg.; cf., however, Welsh *graeanbwll* 'gravel-pit'). There are other possible candidates for the second word in this phrase, such as ***creun** 'dam' (Welsh *crawn*, giving compound *cronbwll* 'reservoir'; but there would be no reason for lenition in the Cornish phrase), and *kren* 'round' (but again, if so, why the lenition?), but Pryce's sense is much the most likely for all the names.

A. Polgrain (2 exx.), Polgrean (2 exx.), Polgreen (3 exx.), Polangrain; Polgrean, Pollgreen, etc., flds.

***pol ros** 'water-wheel pit', a phrase of **pol** and *rôs* 'wheel' (Lh. 32a, etc.); note Breton *poull-rod* (*pul rôt* glossing *molendinarium*, Lh. 154a). The phrase early became a dialect word in English: 'the *polros* of another stamping mill', 1547 (RC 32 (1911), 443f.); '*polrose* (pronounced *pulrose*)' (Symons, p. 144). Some of the field-names may represent this dialect usage, instead of a Cornish usage. The first name, on the other hand, represents a Domesday manor, and since 'water-wheel pit' seems a slightly unlikely sense for such a name, it may be from something else instead (**pol**+ ***rud** 'filthy'?); however, the element would fit phonologically.

pons

    A. ?Polroad; Polrose, flds., Polrose Mine

pons (m.) 'bridge', Voc. 725 glossing *pons*; Welsh *pont* (f.), Breton *pont* (m.). See Henderson, Bridges (esp. pp. 34f.).

    A. Pont, Ponts Mill; pl. Ponjou, Poniou (cf. Breton Ponthou, Gourvil, no. 1731)

    B1. See **car-bons*.

    B2. Qualified by the material or nature of the bridge: *Ponspren* (+**pren**); Ponsmain, *Ponsmean* (both +**men**); Ponsbrital (+**brytyll**); *Ponsmur* (= Grampound), Ponsmere (both +**meur**); Penoweth (+**nowyth**)

Qualified by a nearby feature: Ponson Joppa (+**an**, **shoppa*); *Ponsbeggal Bridge* (+**begel**); Ponson Tuel (+**an**, **tewel*); Ponslego (+**les** pl. ?); *Ponskeyre* (+**ker*?)

Qualified by words for 'stream': Ponsanooth (+**an**, **goth**[1]: or **goth**[2]?); *Ponsgovar* (+**gover**); Pontshallow (+**hal** pl.)

Qualified by words for people: Pantersbridge (+*Iesu*: cf. Breton Pont-Christ, Top.Bret, p. 33); *Ponsfrancke* (+**an, Frank**); *Ponsprontiryon* (+**pronter** pl.); Ponsandane (+**an, den**: or pers.); Polmassick, Polstangey Bridge, Ponsharden, Pons Bennett (all +pers.); *Ponswragh* (+**gruah**)

Qualified by a place-name: *Ponstresilyan*; *Ponsreleubes* (+Relubbus); Ponsmedda (+(**an**), **merther*)

Qualified by an animal: Ponsongath (+**an, cath**)

Qualified by an obscure word: Ponjeravah, *Ponsangavern*, etc.

    C2. Chypons (+**chy**; 6 exx.); Parkenponds, fld. (+**park, an**); Tolponds (+**tal**; cf. Welsh Tal-y-bont: several, Rhestr, p. 109); Ogo Pons, coast (+**googoo**); see also the element **pen* (**an**) **pons**.

porhel 'piglet', BM 1557; *porchel*, Voc. 597 glossing *porcellus*; *porrell* 'porker', Trg. 27a. Welsh *porchell*, Breton *porc'hell* (cf. Tanguy, p. 122). In the second name **porhel** is probably used as a personal name: note Michael *Porghel*, 1286 Ass.Roll.

    C2. *Wheal an Porrall*, mine

    D. Perles (formerly *Caerouporghel*)

porth (m.) 'cove, harbour', also 'gateway'; in the language the word occurs only in the sense 'gate' (Voc. 765 glossing *ianua vel valva*; etc.); Lhuyd, 29a, has Cornish *porh* and *por*, 'a haven', but they may be derived from place-names. Welsh *porth* (m.) 'door'; *porth* (f.) 'harbour' occurs 'only in place-names', WVBD, p. 439 (on the genders, cf. JEPNS 11 (1978–79), 47); Breton *porzh* (m.) 'door, harbour'. (There is no evidence in Cornwall for the Breton sense

porth

of 'courtyard', sometimes even 'manor': Gourvil, p. xxvi.) In the five instances of inland *Porth-* names (Portlooe, Porthkea, Porthkillick, Porkellis and Polmanter), the first two apparently mean 'gateway of, road leading to' (qualified by a place-name); in the other three the meaning is uncertain. Original *Porth-* has been replaced in several names by modern *Pol-* (e.g. Polruan, Polridmouth, Polperro, Polmear, Polmanter), through its early reduction to *Per-*, etc., and without early forms it is hard to say whether a name in *Pol-* contained *Pol-* or *Porth-* originally. The reverse process is rarer, however, and only the inland Porthcollumb (formerly *Polkellom*) and Porth Kidney are attested so far. The change *Porth-* > *Pol-* occurs as early as 1292 in the case of Polruan. It is improbable that a name in *Porth-* necessarily implies that a cove was regularly used for boats: rather it came to mean any cove or bay. Note that in some modern forms the element has become 'Port-', by correct analogy with English *port*: e.g. Port Isaac, Portholland, Portlooe, Portquin, etc. Inland names are starred.

A. Par, Porth (several) (cf. Welsh (Y) Borth, Crd.)

B1. Duporth (+**dew**)

B2. Qualified according to size or shape: Porthmeor (2 exx.), Polmear in St Austell, Porth Mear, Porthbeer Cove, Porthbeor Beach (all +**meur**); Porthbean Beach, Porth Pean, Perbean Beach, Perprean Cove, *Porthpyghan* (= West Looe) (all +**byghan**); Porthcuel (+**cul**); Porth Ledden (+**ledan** ?).

Qualified by words for 'boat': Portscatho (+**skath** pl.); Pelistry (+**lester** pl.)

Qualified by colour: Porthgwidden, Porth Gwidden, Portquin (all+**guyn**); Porthglaze Cove (+**glas**)

Qualified by words denoting people: Polruan (+pers.); Porth Saxon (+**Zowzon**); Porth-cadjack Cove (+pers. ?)

Qualified by a river-name or other word indicating water: Porthilly (+**hyly**); Portloe, Porloe, Porth Loe (all +***loch**); Porthallack (+**hal** aj. ?); Porthluney Cove, *Porthglyvyon*, Porthallow in St Keverne (+***alaw**), Porthleven (all +r.ns.)

Qualified by a place-name: Portlooe*, Portholland (+***henlann**), Polridmouth (+**rid**, **men**), *Porthcovrec, Porthcaswith*. Note also Port Gaverne and Portwrinkle, in both of which the second element appears also in a nearby *Tre-* name.

Qualified by the patron saint of the parish: Porthkea* (+Kea); Porthzennor Cove (+Zennor); Priest's Cove (+Just); *Porth-*

post

mawgan (+Mawgan); *Portsenen* (+Sennen); *Porthia* (= St Ives;) *Porth Perane* (+Perran)

Qualified by plants or animals: Portkillock* (+**kullyek**); Porth Joke (+**les** aj.); Porthguarnon (+**guern** sg.?); Porthkerris (+***cors** deriv.?); *Porthe Brenegan* (+**brenigan**); ?Perbargus Point (+**bargos**?); *Porth Logas* (+**logaz**)

Qualified by a man-made feature: Porth Chapel, *Porthchaple* (both +**chapel**); *Porthencrous* (+**an**, **crous**); Porth Mellin, Portmellon (both +**melin**[1])

Qualified by a natural feature: Porthtowan (+**towan**; 2 exx.); Portreath (+**trait**); *Portlegh* (+***legh**); Progo (+**googoo**); *Porthennis* (+**enys**); Porth-en-Alls (+**an**, **als**); Porth Curno (+**corn** pl.); Porth Cornick (+***kernyk**); Porthgwarra (+***gorweth** pl.?); Porth Pyg (+***pyk**?); *Porth Main* (+**men**); Porth Minnick (+***meynek**)

Qualified by other words: Porkellis* (+**kellys**); Porthminster (+***mynster**); Port Isaac (+**usion** aj.?); *Porthe an Badall* (+**an**, **padel**); *Porthoy* (+**oye**?); Polmanter*, Polperro (both + ?)

C2. *Chyporth* (+**chy**); Vosporth (+**fos**); *Errow-porth*, fld. (+**erw**); *Adjaporth*, fld. (+***aswy**)

Note that **porth** became an English dialect word and occurs thus in various place-names; some of them were formerly Cornish names and have been 'translated' by transposition of the elements: Newporth (2 exx.); Perranporth; Maen Porth; St Columb Porth; Mawgan Porth (formerly *Porthmawgan*); Chapel Porth (formerly *Porthchaple*); Trevellas Porth; *Trevaunance Porth*; Lesceave Por; Tremearne Par.

post 'post, pillar', PC 2058, and Voc. 766 glossing *columpna*; Welsh and Breton *post*, all borrowed from English or French (cf. Piette, p. 159); the early Welsh river-name seen in *aper ipost du* 'the mouth of the black *Post*' (LLD, p. 155) is presumably unconnected.

B2. *Post Mean* (+**men**)

C2. Crosspost (+**cos**); Trebost (+**tre**)

***poth** 'burnt, scorched', equivalent to Welsh *poeth*, Breton *poazh* (cf. HPB, p. 186); note Welsh Bryn Poeth, etc. (PKM, p. 206) and Bryniau Poethion, Mer.; Old Breton *Caer Poeth* (Chrest., p. 157); Middle Breton *Garzpenboez* 1461, now Caspenboih (Chrest., p. 226). The root appears in Cornish only in the word *potvan* 'heat, a scorching', RD 2343 (sic MS: $t = /\theta/$? Miswritten *pocvan*, RD 170 and 2341) and *powan* (< **po'van*?), CW 460 (for the noun suffix cf. *loscvan* 'a burning', RD 1249 (sic MS), and *vrewvan* 'soreness', PC 478: Padel, Co.Studies 7 (1979), 41).

    C2. Brimboyte (+**bron**); ?*Doerpoys*, fld. (+**dor**)

pow 'land, country'; Old Cornish *pou*, Voc. 716 glossing *provintia*. Welsh *pau*, Old Breton *pou* glossed *pagus* (FD, p. 289; Loth, RC 17 (1896), 427f.). The word probably occurs in one hundred-name, as suggested by Picken (DCNQ 30 (1965–67), 38f.) and Quentel (ZCP 39 (1982), 195–200).

    B2. Powder (+**ardar** pl. ?)

pras (m.) 'meadow, pasture', OM 1137 and 1151, both times in the phrase *yn guel nag yn pras* '(neither) in open-field nor in meadow' (meaning 'nowhere'); it is thus complementary to **guel**, the two seemingly making up the whole of the home-fields (but not including the upland grazing, or **goon**). An exact translation would be 'small plots of grazing held in common': note later dialect *prase* 'a small common' (Polwhele, Provincial Glossary), and Henderson's remark: *prase* 'is still used [c. 1930] to denote plots of land by the wayside in St Keverne parish' (101st RCPS (1934), 24). Instances of the simplex might therefore be cases of the dialect word, rather than the Cornish one. Breton *prad* 'pré, prairie', not directly from Latin *prātum* (which would give Breton **preud*, Cornish **preus*), but if it is a late loan the Cornish word has been assimilated to give -*d* > -*z*.

    A. Praze (3 exx.), Prazey; flds. Praze, The Praze

    B2. Praze-an-Beeble (+**an**, **pybell*); Prazeruth (+**ruth**); Prazegooth (+**goth**^{2} or **goth**^{1}); Presthilleck (+**heligen** pl. ?); ?Brass Teague, fld. (+**teg** ?)

    C2. Chypraze (+**chy**; 2 exx.); Penpraze (+**pen**; 2 exx.: cf. Breton *Pem-prat*, etc., FD, p. 204); *Gwythanprase*, mine (+**guyth**^{2}, **an**)

    D. *Trelull pras* (= Trelill in St Breock); Tresahor Praze; Choone Praze

pren (m.) 'timber, tree'; Voc. 706 glossing *lignum*. Welsh *pren*, Breton *prenn*. Cf. Watson, CPNS, p. 351.

    B1. See the compound **crouspren**.

    B2. ?*Prynullin* (+**elaw** sg. ?)

    C2. *Ponspren* (+**pons**: cf. Welsh *Pontbren*, Crm., 1830: BBCS 28 (1978–80), 630; Rhyd y Bontbren, ÉC 10 (1962–63), 218)

pry (m.) 'mud, clay'; Welsh *pridd* 'soil, earth', Breton *pri* 'earth, clay' (Tanguy, p. 60). Note the Breton adjective *priek* 'argileux'.

prif

The Cornish word occurs primarily as C2 in the phrase **pwl prî**, but also in a few other names. It is suggested as C1 in Priddy, Som. (Elements, II, 73; JEPNS 1 (1968–69), 50).
    C1. ?Prideaux (+ ?)
    C2. *Pit-pry*, etc., flds. (+**pyt**); aj. Rospreages (+**rid**); see also **pwl prî**.

prif (m.) 'worm, reptile, vermin', Voc. 581 glossing *vermis*; *hager bref* 'evil reptile', MC 122c; pl. *prevyon* 'reptiles', OM 1160; diminutive *prevan* glossing *tinea* 'worm', Lh. 164a. Welsh *pryf*, Breton *preñv* (cf. Tanguy, p. 130).
    C2. pl. or dimin. Innis Pruen (= Mullion Island; +**enys**)

pryns (m.) 'prince' (borrowed from English); Welsh *prins*, *preins* (EEW, p. 141), Breton *priñs* (Piette, p. 161).
    C2. Goon Prince (+**goon**); Croft Prince (+***croft**); ?Carn Prince, coast (+**carn**)

***prys**, ***prysk** 'copse, thicket', equivalent to Welsh *prys*, *prysg* (cf. PKM, pp. 140f., ELl, p. 66, and ELlSG, p. 114). Note early Welsh *penn i prisc* 'the head of the thicket' (LLD, p. 255), and *di'r prisc* 'to the thicket' (VSBG, p. 132); *prysc katleu* (+pers.: PT, p. 8). Middle Welsh *prysc* (WM 422.27) is equated with *byrgoet* 'underwood, scrub' (WM 420.29). See also Hamp, BBCS 29 (1980–82), 85. The word occurs as a simplex in place-names in England: see Elements, II, 73.
    A. Priske, Preeze; dimin. ?Priscan
    B1. ?Cambrose (+**cam**)
    B2. Presingol (+**an**, ***coll**); Prospidnick (+**pin-** aj. ?); Prislow (+***loch**)
    C2. dimin. ?Trebisken (+**tre**)

pronter (m.) 'priest', BM 785, etc.; *prounder*, Voc. 107 glossing *sacerdos*; pl. *bronteryon*, MC 89a. This must be a different word from Old Welsh *premter*, *primter* 'priest' (Cormac's Glossary), Middle Welsh *prifder* (BT 23.12), which was from British-Latin **premiter* (from *presbiter*; LHEB, p. 128; cf. Old Irish *cruimther*, ogham QRIMITIR, CIIC, no. 145; O'Rahilly, Celtica 1 (1946–50), 347f.). It must instead be derived from Latin *praebendarius*, via Old French *provandier* (so VKG, 1, 198f.; Campanile, p. 90; see also Loth, ACL 3, 251).
    C2. *Park Pronter*, Park-an-Prowlter, fld. (both +**park**, (**an**)); Golden Praunter, fld. (+**guel**, **an**); ?Carn Praunter, rock (+**carn**); pl. *Ponsprontiryon* (+**pons**: Sawyer, no. 450)

pwl prî 'a clay-pit', Lh. 43c glossing *argilletum*; a phrase of **pol** and **pry**; cf. the Old Welsh compound *i pridpull* 'the clay-pit' (LLD, pp. 252 and 265); Breton Poulpri, etc. (Tanguy, p. 60), Middle Breton *poul-pry* 'lieu où l'on fait le mortier' (GMB, p. 507).

A. Polpry, Pulpry, Polpry Cove (2 exx.), Poll Fry; Perpry, *Pollpry*, flds.; *Polpry* (2 exx.), *Polpri*

pul stean 'a tin pit', Pryce, s.v. *pul*: a phrase of **pol** and **stean**.

A. Polstain, Polstein, *Pulstean*

pur 'clean, pure', equivalent to Welsh and Breton *pur*; the word is common in the language, but only in the sense 'very', and the normal phrase for 'clean water' was *dour glan*, PC 629; but the Welsh and Breton words show the meaning in place-names.

C2. Polpeor in Uny Lelant, *Poulpere*, Polpeer, fld. (all +**pol**)

*****puth** ($u = $ /y/), 'a dug well' (as opposed to **fenten** 'a natural spring'); equivalent to Breton *puñs*; Middle Breton *puncc*, etc. (Piette, p. 163), Old Breton *Puz* (FG, p. 106; cf. HPB, pp. 769n. and 774 and Bernier, Ann.Bret. 76 (1969), 654), borrowed from French *puits*. Dialect *peeth* 'a sunk well' (e.g. Jenner, TPhS 1876, 538) occurs frequently in field-names: Peeth Meadow, Peeth Field, etc.

C2. Park Peeth, flds.

R

*****radgel** (MnCo.) 'rocky ground, scree, clitter', dialect *radgell*, *radgil*; this could be a reflex of Breton *radell* 'raft', referring to the flat rocks that emanate from tors in granite country. (Cf. Welsh *gradell*, *radell* 'griddle, bakestone'?) The simplex instances could involve the dialect word instead.

A. ?Radjel, ?*The Radell*

C2. ?Park Rattle, fld.

rag 'before, fore-'. In the language it occurs mainly as a preposition, though note, as a prefix, *rag leueris* 'aforementioned' (MC 224a). Welsh *rhag-* (as leniting prefix, WG, p. 268), Breton *rak* (cf. FD, p. 292, and GMB, p. 559). See also J. E. C. Williams, ÉC 6 (1952–54), 18–20. The element is ambiguous: *Rag-X* can be either 'fore-X, off-X' like Middle Welsh *Rac Ynys* 'adjacent isle, offshore island' (cf. TYP, p. 231), or 'place opposite X', as in Cornish Raginnis. Breton Raguénès is either 'île adjacente' or 'lieu devant l'île' (Gourvil, no. 1889; ÉC 10 (1962–63), 288). The element is used in the following names: Raginnis (+**enys**); **rag-gos* 'opposite

*red

the wood' (+**cos**) as A in ?Raggot Hill, ?Wrecket, and as C2 in Tolraggatt (+**tal**); and possibly in Raftra < *Raghtre* (+**tre**), 'adjacent farm' (?) (though *g* > *gh* = /x/ would be very odd).

*red (OCo.) 'watercourse', equivalent to Welsh *rhed 'course, race, run' (given as a word in Spurrell-Anwyl; cf. *rhedeg* 'to run', *gwyth-red* 'channel, brook'), Breton *red* 'cours, courant' (cf. FD, p. 295; Tanguy, p. 87). Note Voc. 735, *chahenrit* glossing *torrens* (the first part being unexplained, but cf. Welsh *cenlli* 'flood, torrent', < *cefn*+*llif*): here, as in place-names, there is confusion between *red 'watercourse' and **rid** 'ford': cf. GPN, pp. 249–51 (where a list of place-names containing the Gaulish ancestor of *red is also given, pp. 250f.). Some of the names given here under B1 might thus belong under **rid** instead.

A. Race Farm; deriv. Rissick (cf. early Welsh r.n. *ritec*, LLD, pp. 124, etc., EANC, pp. 84f.).

B1. Galowras (+ ***glow**¹); Burras (+ **ber** ?); see also *goth-red (s.v. **goth**¹) and ***kew-rys**.

C2. Penrice (+**pen**: cf. Breton Perret, formerly *Penret* CR, Chrest., p. 156)

reden (pl.) 'bracken', Voc. 666 glossing *filex*; sg. *redanan* 'a brake or fern', Lh. 240c (note that Lhuyd's -*dan*- for earlier -*den* is echoed in some of the place-names). Welsh *rhedyn*, Breton *raden* (cf. Tanguy, p. 115). Note the early Welsh place-names *Redynure* 'bracken-hill', WM 490.23 (with *rhedyn* as C1); *Tref redinauc, .i. villa filicis*, VSBG, p. 72, and *pul retinoc*, Chad 6, both with *rhedyn* aj. as C2; early Breton *Radenec, Rattenuc*, etc., with *raden* aj. as A (Chrest., pp. 159 and 227).

A. aj. Redinnick, Redannack, fld., Redannick

C2. Splattenridden (+ ***splat, an**); Lean Ridden, fld. (+ ***leyn**); Parken Reden, fld.; Menniridden (+**meneth**); Erwereden, fld. (+**erw**); ?Polrudden (+**pol**); aj. Tredinick in St Mabyn, Tredinnick in St Keverne, St Issey and Newlyn East (all +**tre**: contrast **eithin** aj. and ***dyn** aj.)

*rew 'slope', equivalent to Welsh *rhiw* (cf. ELlSG, pp. 72f.; WVBD, p. 464); Loth, RC 43 (1926), 146, 'steep slope'; Elements, II, 82 and JEPNS 1 (1968–69), 50. Note the early Welsh gloss *ad collem* (*uel ad procliuum*) *Morcanti*...*Riu Morgant* (VSBG, p. 88); there are hints of a meaning 'way': *Bochriucarn* glossed *maxilla lapidee uie* (VSBG, p. 26: cf. *kae di y riw* 'you keep closed the *riw*', CLlH, p. 35?). Unlike Welsh, Cornish has no examples of

rid

***rew** as B2 (ELISG, loc. cit.; *Riubrein* 'crows' slope', LLD, p. 257; etc.); nor any as A, like *ecclesia Riu* (LLD, p. 230); but the Cornish names echo the Welsh Trefriw (2 exx., Units, p. 209).

C2. Trerew, Trefrew, Trefrouse (all +**tre**)

rid (OCo., m./f.) 'ford', Voc. 726 glossing *vadum*; the word does not occur elsewhere in the language. Welsh *rhyd* (cf. Richards, ÉC 10 (1962–63), 210–37), Old Breton *rit* and *ret* glossing *uadum* (FD, p. 297: but the word dropped out of use, and the normal Breton words for 'ford' are *roudouz*, etc., in the Middle Breton period (see ***rodwyth**) and *truk* in the Modern period, Tanguy, p. 92). See also Elements, II, 82f., JEPNS 1 (1968–69), 50, and 2 (1969–70), 74. See PNRB, p. 251, for a list of British place-names containing the ancestor of **rid** as B1, and GPN, p. 250, for Gaulish names which contain either this element or the ancestor of ***red** 'course'. Note the translated Cornish instances *vado de Ridchar*, *vadum de Ridmerky* (12th–13th cent.), both in the Launceston Cartulary. The element was confused with ***red** 'watercourse'; some names listed there might belong here, and vice versa. In the Anglo-Saxon charters, where fords feature prominently in the boundary-clauses, **rid** occurs in the following spellings: *hryt* (Sawyer, no. 755 twice; no. 1019 twice), *hryd* (Sawyer, no. 684), *hræt* (Sawyer, no. 770), and *ryt* (Sawyer, nos. 832/1027): all except the last show the voiceless initial fortis *r*- that Cornish must once have possessed (cf. LHEB, p. 478 and HPB, p. 812). As B2, especially in West Cornwall, **rid** (> *res) is confusable with ***ros**. When **rid** was followed by an element beginning with *h*-, the *h* unvoiced the -*d*, giving *Ret*-, and thereby preserved it from the West Cornish -*d* > -*z*: thus we find Retire and *Retier* < **rid**+**hyr** (cf. Welsh Rhyd Hir, etc., ÉC 10, 212, spelt *Redhyr* and *Red Tir*, 1284; Tretire, Hre., DEPN s.v.): in both cases the form *Retyr* was changed, presumably by folk-etymology, to *Restyr*, making it appear to contain **rid**+**tyr**. Retallack (2 exx.) is similar; it could theoretically contain one of three things: **rid**+**tal** aj., 'steep-browed ford', **rid**+*Talek pers., or **rid**+**hal** aj. 'muddy ford'. In view of the Welsh instances of Rhyd Halog, etc. (ÉC 10, 214), the last is much the most probable (despite a lack of spellings such as *Rethalek, *Reshalek or the like), and instances of *Restalek*, etc., are cases of the same folk-etymology as in *Restyr*. Another example of the same process is seen in Retallick in Roche (**rid**+**heligen** pl., spelt *Retelek* 1284, etc., *Restelek* 1370), and in St Columb Major

rid

(*Rettelehc* c. 1250, *Reshelec* c. 1270). Unlike Welsh *rhyd* (f.), Cornish **rid** generally appears as masculine in place-names; however, there are a few which show lenition, indicating it as feminine: Rosewin, Tredwen, Rosewall, Suffree, Roseveth.

A. Rees, Rice; pl. Ridgeo

B1. See names listed under ***red** B1; also ***kew-rys**; ?Scraesdon (+Eng. *dūn*), ?Crowlas (both +***crew**; or **krow**+***lys**); Tretherras, Tretherres (both +**tre, dew**); ?Terras (+**try**[1] fem. ?)

B2. Qualified according to shape or position: Roseladden (+**ledan**: cf. Welsh Rhyd Lydan, ÉC 10, 212); Retanna (+**tanow**); Roskymmer (+***kemer**); ?Resparva, ?Resparveth (both +**perveth**: or ***ros**); Retire, *Retier* (both +**hyr**)

Qualified according to colour: Rosewin in St Enoder, Tredwen (both +**guyn**: cf. Welsh Rhyd Wen, ÉC 10, 213); Trecorme (+***gorm**); Trecan (+**can**); Redruth (+**ruth**: cf. in Wales *rufum uadum*, LLD, p. 172, and *uado rufo*, ibid., p. 155)

Qualified according to the nature of the crossing: Tretawn, Rosedown (both +**down**); Rospreages (+**pry** aj.); Roseath (+**segh**); Retallack in Constantine and St Hilary (both +**hal** aj. or **tal** aj.); Polridmouth (+**porth, men**: cf. Redmain, PNCmb., II, 267; Welsh Rhyd Faen, etc., ÉC 10, 216); Releath, Rillaton (+Eng.), Treslea (all +***legh**); Rosecliston, Relistien (+**cellester** dimin. or sg. ?)

Qualified with plant-names: Redevallen (+**auallen**: cf. Welsh Rhydafallen, etc., ÉC 10, 222); Selligan (+**heligen**); Retallick in Roche and St Columb Major (both +**heligen** pl.); ?Striddicks (+**dreys** aj.: or ***ros**); ?Treswithick (+**gwyth**[1] aj.: or ***ros**)

Qualified with animal-names: Reterth (+**torch**; or **derch**); Rosegarden (+**garan**); Tregray (+***gre**: cf. Welsh Rhyd y Re, etc., ÉC 10, 221); Tretoil (+**hwilen** pl.); Respryn (+**bran** pl. ?); Resores (+**hos**); Roseveth (+**margh** pl.)

Qualified by nearby features: Rescorla in St Ewe (+***corlann**: or ***ros**); Lescrow (+***crew**); Roskear in Camborne, Rezare (both +***ker**); Retyn (+***dyn**; or **tyn**); Rosewall (+***gwal**?); ?Rose-an-Grouse (+**an, crous**: or ***ros**)

Qualified by a personal name: Rosecraddock (+*Caradoc); Resurrance (+*Gerent*); perhaps *Ridmerky*, Rosesuggan, Rôskennals; *hryt eselt*, *hryt catwallon* (both Sawyer, no. 755)

Qualified by a river- or stream-name: Redallen (+***alun**?); Suffree (+***brygh**?, *-**i**? Formerly *Resvreghy*); some of those where the qualifier is obscure may belong in this category, such as Retorrick, Retew, Retanning, Relubbus, Trefrawl, Tretallow, Restineas, Trezare, *hryt ar þugan* (+ ***ar**: for the construction *hryt ar* cf. R. J. Thomas, BBCS 7 (1933-35), 127); some of these are on rivers with other names, however.

C2. Trerice (5 exx.), Trease (2 exx.), Trerise (2 exx.), ?Trefrize (all +**tre**); Chyraᴣc (ɪ **chy**); Boreise (+**bron**); pl. Trefrida (+**tre**)

***rynn** 'point of land', equivalent to Welsh *rhyn* 'point, peak' (apparently occurring, like the Cornish word, only in place-names: ELl, p. 27; ELlSG, p. 53; Williams, CA, p. 92f.); Old Breton *-rinn* 'point' (FD, p. 297). See also Loth, RC 41 (1924), 216-18; Old Irish *rinn* 'point, peak', LEIA, pp. R31f.; Watson, CPNS, pp. 495f. The element is to be distinguished by its short vowel from ***run** 'hill', though that may not be an infallible guide.

 A. pl. ?Ryniau
 B1. ?Metherin (+ ***með**)
 C1. Rinsey (+**ti**)
 C2. ?Jangye-ryn, coast (+ ?); see also the phrase ***pen-ryn**.

ryp, preposition, 'by, beside', MC 208a, RD 266, etc.; not in Welsh or Breton. The word is assumed to be a loan from Latin *ripa* 'bank' (e.g. VKG, index). It occurs in a few field-names, such as *Dore Ribba Vownder* (+**dor, an, bounder**) and *Rebanhale* (+**an, hal**).

***rodwyth** 'ford', equivalent to Welsh *rhodwydd* (Williams, CLlH, pp. 159 and 217; Loth, RC 15 (1894), 97f.; TYP, p. 66); Old Welsh pl. *nant hi rotguidou* 'the valley/stream of the fords', LLD, p. 126; *rodwit*, CLlH, pp. 28-9; note especially *Rodwydd Forlas*, ibid., p. 22 (= *Ryt Uorlas*, p. 4). Breton *roudouz* (RC 8 (1887), 69; GMB, p. 584; FD, pp. 97 and 298); both occurring mainly in place-names and in the early periods. The meaning may be rather 'fortified ford, bank for defending a ford'. One could reasonably expect to find the element in Cornwall, but there are only one or two, doubtful, instances.

 A. *Redwith Mill* (+Eng.)

***ros** (m./f. ?) 'promontory; hill-spur, moor', equivalent to Welsh *rhos* 'moor, high ground', Breton *ros 'hill' ('n'est pas inusité comme le croyait Troude', GMB, p. 583). The various meanings of this element are difficult to establish, especially as the word seems largely to have dropped out of use in Cornish and Breton. The

*ros

original meaning seems to have been 'a promontory, an eminence', and Strachan (Bezz.Beitr. 17 (1891), 301) first compared Sanskrit *prasthas* 'plateau' (cf. WG, p. 139; LP, p. 21). In Irish *ross* means 'promontory, wood' (the second meaning presumably being derivative): for examples see Hogan, Onomasticon, pp. 583–8; Top.Hib., I, 77. The meaning 'promontory' occurs in some Cornish place-names, e.g. Roseland (+ English), *Ros*, Rosemullion, and Restronguet, as well as in some Welsh ones (e.g. Rhosili in Gower; also the cantrefs called Rhos in Den. and Pem., Units, pp. 261 and 306), and, in Scotland, Melrose, Rox., a promontory in a river. However, these usages are rare, and probably belong to the Old Cornish period or earlier, and in the great majority of cases ***ros** means some kind of 'upland, high ground'. In Welsh the meaning is 'rough upland grazing': note Davies, in 1632, 'planities irrigua'; Lhuyd, 32a, 'a mountain-meadow or moss'; WVBD, p. 466, 'a dry, level tract of land, more or less elevated'. This is very similar to Cornish **goon**, and it may well be that ***ros** had this sense in Cornish but was replaced by **goon**; if so, then the names in *Ros-* would date mainly from the Old Cornish period. For the meaning in Breton, note Grégoire de Rostrenen, in 1732: 'petit tertre couvert de fougère ou de bruyère' (cited, GMB, p. 583; cf. Chrest., p. 229, and ELl, p. 27); to which Le Gonidec, 1850, adds, 'terrain en pente, particulièrement lorsqu'il regarde la mer'; Top.Bret., p. 115, 'le versant boisé d'une colline lequel se courbe'. In contrast with Welsh, where the word came to mean 'moorland' in a general way, in Breton the word seems to have developed a specialised sense of 'hill grown with heather and/or bracken'; the other two meanings, 'curved, wooded hill-slope' and 'sloping land, especially facing the sea' are less often cited and may be less important; the latter can easily have developed from the sense of 'promontory'.

The question for Cornish is whether the word denotes primarily the shape of the high ground, or the vegetation on it as in Breton. The problem was studied by J. E. B. Gover, 'The element *ros* in Cornish place-names', *London Mediaeval Studies* 1 (1938), 249–64, who concluded that the normal sense in inland names was 'hill-spur', an obvious development of the coastal sense 'promontory'. The only problem with this conclusion is that it is fairly difficult to find any farm in Cornwall that is *not* near something which could, with a little goodwill, be regarded as some sort of a hill-spur; in addition, it is likely that most farm-names in *Ros-* are rather early ones (because of the lack of occurrence of the element in the remains of the language, and because the element does not become significantly more frequent from east to west), and any vegetational distinction that might have existed when the names were given is more likely than not to have disappeared in the meantime, especially with the drastic loss of moorland in the last two centuries. The instances of Penrose, which are farms situated at the 'end of a *****ros**', are useful: as Gover noted (pp. 262-3), these names (including two not cited by him) are all situated at the ends of hill-spurs or ridges, thus supporting the idea that *****ros** meant 'spur, ridge'. All in all, then, Gover's conclusions can be accepted, though with the reservation that there may have been some idea of the vegetation, perhaps along the lines of the Breton meaning, 'hillside grown with heather and/or bracken', in addition to the sense of the shape alone. The other Breton meaning, 'sloping land facing out to sea', may also be present, especially in some names along the south coast (e.g. Roskorwell, Rosenithon, Rosteague and Boscawen Rôs).

Confusion easily arises between *****ros** and **rid** as B2, and in a good number of names it is impossible to decide between the elements, since the forms can vary between *Ros-* and *Res-*. In the east of the county the presence of spellings in either *Ros-* or *Res-* slightly favours *****ros**, since **rid** there tends not to show -*d* > -*z*; but that is by no means an infallible rule, since **rid** can sometimes appear as *Res-, Rose-*, etc., even in the east (e.g. Rezare, Rosecraddock).

Welsh *rhos* is feminine, and Cornish names quite often show lenition after *****ros**, e.g. Rosewastis, Rosevidney, Rosewin; but more often no lenition appears, either because the word was normally masculine, or because the -*s* prevented lenition of the following consonant.

A few further examples of *****ros** in other Brittonic areas follow. In Cumbric: *Raswraget* 'the moor of the women', PNCmb., I, 103; *ad Rossam*, PNWml., II, 181; **Pren-ros* 'timber-moor' (?), CPNS, p. 351; Melrose 'bald-promontory', CPNS, p. 496. In Welsh: *ros ir eithin*, LLD, p. 221 (cf. Rosenithon, below); *rosulgen*, ibid., p. 239 (= Rhosili in Gower; cf. Roselyon, below): but names in *Ros-* are not common in that source. In Breton: *Roscaroc* (Chrest., p. 163); *Rosiou an Fauh* (Bégard, p. 204) shows *****ros* pl. as B2;

*ros

several names in *Ros-*, Top.Bret., p. 115; personal names *du Roz* or *Rozou* (pl.), GMB, p. 583 (cf. Gourvil, nos. 1944 and 1973): the use of the simplex *Ros as an inland place-name is widespread over all Brittonic areas and occurs in other parts of England: *Ros*, Som. (BBCS 14 (1950-52), 118); Rossmore (+English), PNDor., II, 2; Roose, PNLnc., p. 202; Ross, Ntb., DEPN; Ross-on-Wye, Hre.; and many instances in Wales.

A. Rose (2 exx.), Roose (3 exx.), Rowse; Roseland (+Eng.); *Ros*; Trerose in Minster (+false *tre*); pl. Ruzza

B1. Harros (+**hyr**); Garras in Gulval and St Mawgan in Meneage (+**garow**); Keyrse, Case-(hill) (both + ***kew**); see also ***meŏ-ros**; Tregonce (< *Tregentros*) and Trevance and Trevanters (both < *Trevantros*) all seem to contain (as C2) unknown compounds in *-ros*.

B2. Qualified by a personal name: Rosemaddock (cf. *Moroc*, Chrest., p. 154); Roselyon in St Blazey (cf. *Sulgen*, LLD); Rosegothe (+**gof**: cf. Breton Roscoff, Top.Bret., p. 115); Roseworthy in Kenwyn (+*Wurci*, Bodm.); Roscarrack in Budock (+*Cadoc*); Tresmeak (cf. *Maioc*, LLD and Chrest.); Rosemodress (+*Modred*, Bodm.); Rosemerryn (cf. *Merin*, EWGT); Rosecadghill (cf. *Catgual*, LLD). Considering the meaning of ***ros**, it is striking how frequent this type of name is; compare, amongst others, Cumbric *Raswraget*, above.

Qualified according to vegetation: Rosenannon, Rosenun (both +**an, onnen**); Trescoll (+***coll**); Treskilling (+**kelin**); Rosevallen, Rosevallon in Cuby, Rose in the valley (all +(**an**), **auallen**: or **rid**); Rôspannel (+**banathel**); Rosenithon (+**an**, **eithin**); Rosecassa (+***caswyth**); Roskilly (+**kelli**); Restronguet (+***trongos**); Rosemullion (+**melhyonen** pl. ?). It is striking that in most of the names of this type (or, at least, of those where the type is recognisable), the vegetation is some kind of trees or woodland.

Qualified according to other natural features: Restowrack (+**dour** aj.); Resugga (+**googoo**; 3 exx.); Tresmaine (+**men**); Roscarrock in St Endellion (+**karrek**); Roselidden (+**lyn**); Roscarnon (+**carn** deriv.).

Qualified by colour: Treswen, Rosewin in St Minver (both +**guyn**); Rosemelling (if +**melyn**[2] and not **melin**[1])

Qualified by man-made features: Roseglos (+**eglos**); Rôskestal (+**castell** sg. or pl.); Roskear in St Breock, Rosecare (both

+ *ker); Roskruge (+cruc); Roskrow (+krow); Roscrowgey (+ *krow-jy); Rosemundy (+ *mon-dy).

Qualified by words for animals: Tresmarrow (+margh; 2 exx.); ?Scarsick (+cassec); Rosuic (+ *ewyk); Rosudgeon (+odion).

Qualified according to size, shape, or position: Tresmeer (+meur); Tresparrett (+perveth); Rosebenault, Rosemanowas (both + *menawes); Roscroggan (+crogen).

There are also many names in *Ros-* of which the qualifier is obscure, and many that contain either *ros or rid.

C2. Eglosrose, Eglarooze (both +eglos); Trerose (+tre; 2 exx.); Chyrose (+chy); ?Culdrose, Coldrose (+cul/*kyl; or as B1); ?Balrose (+bal); see also *pen ros, and cf. Old Breton *Lis-ros*, Chrest., p. 163.

D. Boscawen Rôs (also Boscawen-Noon, +an, goon)

*rosan, a diminutive of *ros, probably meaning 'little promontory', or the like. Welsh *rhosan (cf. EANC, p. 84; Owen, Pembrokeshire, II, 407; Williams, ELl, p. 27), and/or *rhosyn (PNChe., I, 151); Breton *rosan in *Quaerosan* (RC 8 (1887), 69), and pers. *Rosan* (Gourvil, nos. 1945 and 1968). Note also Rossendale (PNLnc., p. 92); Rossington (PNYoW., I, 49); Rossen Clough (PNChe., loc. cit.). Both the likely Cornish instances are coastal features.

A. ?Roozen Cove; ?Rosen Cliff

*rud (OCo.) 'foul, filthy', equivalent to Welsh *rhwd*. This word is more likely, as C2, in Polroad (+pol), a Domesday manor, than an older form of *pol ros 'water-wheel pit'; otherwise, however, there are no instances known.

ruen (*ue* = /œ/) 'seal', only in *'n garrek ruen* OM 2464 (= *Caregroyne*, below); Welsh *-rhon* in *moelrhon* 'seal'; Breton diminutive *reunig* 'phoque'; Old Irish *rón* 'seal', borrowed from Old English *hrān* (VKG, I, 21), perhaps via Brittonic (LEIA, p. R42); but cf. Williams, CA, p. 204, and Lloyd-Jones, BBCS 15 (1952–54), 202. See Nance, 'Cornish names of the seal', OC 3 (1937–42), 85–8. Note the Welsh names for The Skerries, Agl., Ynysoedd y Moelrhoniaid or *Rhonynys*, and *Insulae Phocarum* (Units, p. 225). In theory, the second element in both names below could instead be *ruen* 'coarse hair' (BM 1968 and 4443), Welsh *rhawn*, Breton *reun*, which is formally indistinguishable. The second name is glossed by Leland (Itinerary, I, 322) 'insula, vel rupes potius, vitulorum marinorum, alias Seeles'; cf. Bede's gloss on Selsey, Ssx., 'insula vituli marini' (Hist.Eccl., IV, 13).

C2. Cargreen (= *carrecron*, Sawyer, no. 951), *Caregroyne*, coast (= Black Rock) (both +**karrek**)

*run (*u* = /y/) 'hill', only in the plural *runyow*, PC 2654 (*ha why a pys an runyow* 'and you will beseech the hills'), and in the diminutive *runen*, Voc. 718 (*collis* glossed *cruc vel runen*); Breton *run* 'hill' (cf. GMB, p. 587; Tanguy, p. 80); Middle Breton placenames *Runyou*, *Run-*, *Run an*..., (Bégard, p. 204). Even where early forms in -*u*- are lacking, the element should normally be distinguishable from ***rynn** 'point of land' by its long vowel. The meaning in place-names is probably 'slope' more often than 'hill' (so Nance, Dictionary, s.v. *ryn*). Note the use as a dialect word in 'A little Reen under the field' (1649: Henderson, Constantine, p. 240).

A. Reen; flds. *Reens*, The Reens

B2. sg. or pl. *Reenie Lovan*, fld. (+**lovan**)

C2. Chyreen (2 exx.), Chirwyn (all +**chy**, (**war**, **an**)); flds. *Parke Reene, Park Reen*

ruth 'red', PC 2128, etc.; *rud*, Voc. 484 glossing *ruber*; Welsh *rhudd*, Breton *ruz*. In all three languages the word is liable to occur in compound names: see the examples below, and note, in early Welsh, *rudglann* 'red-bank' (Ann.Camb. s.a. 796); *rudpull* 'red-pool' (LLD, p. 156); further examples, Units, p. 190. In early Breton, *Rudfoss* (glossed *rubeam fossatam*, FD, p. 300).

C1. Ruthdower (+**dour**); Ruthvoes (+**fos**); Rooke (+***cok**)

C2. Dowreth (+**dour**); Gonreeve (+**goon**); Prazeruth (+**pras**); Redruth (+**rid**); Wheal Reeth, Wheal Reath, mines (both +**wheyl**); Zawn Reeth, coast (+**sawn**); Polpenwith (+**pol**, **pen**); ?Tregunwith (+**tre**, **goon**); ?Tregatreath (+**tre**, **cos**?)

S/Z

sans 'holy'; also 'saint', pl. *syns*, RD 190 and 2114. Welsh and Breton *sant*.

C1. St Jidgey (+St)

C2. Trezance (+**tyr**); Penzance (+**pen**); Lezant (+***lann**). There are also several names apparently containing a word ******sens*, which could be the same as *syns* 'saints': Nansent (+**nans**;

= St Breock); Colesent (+ *kyl); Bosent (+ *bod); Castlezens (+ castell): cf. Breton *Lan Sent*, CL, no. 19.

D. pl. *Lannowseynt* (= St Kew)

sawn (f.) 'cleft, gully, geo', Lhuyd MS (translated by him with Welsh *porth*: Nance, Sea-Words, p. 176); *sawan* 'a hole in the cliff through which the sea passeth', Pryce; cf. Elements, II, 98; JEPNS I (1968–69), 51. A diminutive, with *st-* instead of *s-*, appears as *stefenic*, Voc. 48 glossing *palatum* 'palate' (Welsh *sefnig*). Welsh *safn* 'mouth'; Breton *saon* 'valley' and *san* 'channel' (Tanguy, p. 80; GMB, pp. 596f.; PB, p. 258; Quentel, Ogam 7 (1955), 181f.; Hamp, BBCS 30 (1982–83), 44). The element occurs only in West Cornwall, where the coastal names are predominantly Cornish; in North Cornwall the corresponding English term is the dialect *gug* (also *gog*, *gut*), as in Waddy Gug, St Illickswell Gug, etc., itself originally borrowed from Cornish **googoo**. The element **sawn** also occurs as an English dialect word in numerous coastal names in West Cornwall, e.g. Chough Zawn, Great Zawn, Cockle Zawn, etc., and one (Barrett's Zawn) in North Cornwall. All of the following are coastal.

A. The Zawn, Zone Point, Sowans Hole

B2. Zawn Reeth (+ **ruth**); Zawn Duel (+ **tewl**); *Savenheer* (+ **hyr**); Zawn Vinoc (+ *meynek?); Zawn Harry, Zawn Susan (both + pers.); *Savan Marake* (+ **marrek**); Zawn Kellys (+ **kellys**); Zawn a Bal (+ **an, bal**); Zawn Brinny (+ **bran** pl.); Zawn Gamper (+ *kemer); Zawn Wells (+ **gwels**?); Zawn Bros (+ **bras**); *Savyn Dolle* (+ **toll**); Zawn Alley, Zawn Organ (both + English?); *Savyn an Skanow* (+ **an**, ?); *Savenlester Cove* (+ **lester**)

C2. *Cribbensaune* (+ **krib, an**)

skath 'large boat', Lh. 33a and 53b; *schath*, RD 2233, etc. See Nance, Sea-Words, p. 139. Welsh *ysgraff* (e.g. in Porthyrysgraff, Agl., JEPNS 11 (1978–79), 49), Breton *skaf* (cf. GMB, p. 601; Piette, p. 173). For *-f-* > *-th* in Cornish, cf. Old Cornish *hanaf* 'cup' (Voc. 875), later *haneth* (Trg. 22a), *hanath* (Lh. 33c and 45c).

A. Scathe, rock

B1. ?Henscath, rock (+ **hen**?: or **-hins-** + **coth**)

C2. Polscoe (+ **pol**); Carn Scathe, rock (+ **carn**); Goon and Scath Craft, fld. (+ **goon, an**, Eng.); Le Scathe Cove (+ *legh?); pl. Portscatho (+ **porth**), Polscatha (+ **pol/porth**)

scawen (f.) 'an elder tree', Bosons, p. 10; *skawan* 'an elder',

skyber

Lh. 240c; pl. **scaw*. Welsh pl. *ysgaw*, Breton *skav* (cf. GMB, p. 603; Tanguy, p. 109; Gourvil, no. 2047); Old Breton *scau* (FD, pp. 302f.). See also Quentel, Rev.Intnl.Onom. 11 (1959), 29f.; Bernier, Ann.Bret. 78 (1971), 667. The adjective in Cornish was **skewyek*: Breton *skaveg*, Squiviec, Squiffiec, etc., Tanguy, loc. cit. (but probably not Welsh Ysgeifiog, PNFli., p. 181: see I. Williams, BBCS 11 (1941–44), 83); the derivative in ***-ys²** was **skewys*: Breton Scaouet, etc. The word survives in Cornwall as dialect *scow*, *scaw*, etc., used in a few place-names (e.g. Scowbuds). Elder is 'generally distributed and very common' throughout Cornwall (Davey, Flora, p. 223).

    A. sg. Scawn, *Scawen*, Scawn Hill, fld.; aj. Skewjack, *Skewyek*; deriv. + ***-ys²**, Skewes (3 exx.)

    C2. sg. Boscawen (+ ***bod**); Nanscawen (+ **nans**); Liscawn (+ ***lann/nans**); Penscawn (+ **pen**); Enniscaven (+ **enys**); pl. Nanscow (+ **nans**), Trescowe (2 exx.), Tresco (all + **tre**), *Inisschawe* 'that ys to sey, the Isle of Elder, by cawse yt bereth stynkkyng elders' (Leland, Itinerary, 1, 318) (+ **enys**); aj. Trescowthick (+ **tre**); deriv. + ***-ys²**, Treskewes (+ **tre**; 2 exx.), Kus Skewes (+ **cos**), Park-Skewis, fld.

skyber (f.) 'barn, hut', PC 638 and 679 (equated with *chy* 'cot', PC 674); Welsh *ysgubor* 'barn', Old Welsh *scipaur* glossing *horrea* 'granary', Juvenc. (VVB, p. 214); Breton *skiber* 'shelter, lean-to', Middle Breton *Squiber Nevez* (Bégard, p. 204).

    A. pl. Skyburriowe, *Skyburrier*, *Skeburia*

    B2. *Skebervannel* (+ **banathel**); Skibber Whidden, fld. (+ **guyn**)

    C2. All of the following are field-names: Goon Skipper (+ **goon**); *Gweale Skeeber*, *Gweal Skeber* (both + **guel**); Park Skiber, Park Skipper, *Park-an-Skeeber*

scorren 'branch', Voc. 685 glossing *remus*; *scoren*, OM 802; pl. *scorennow*, OM 2444. Welsh sg. *ysgwr*, Breton sg. *skourr* (cf. GMB, p. 614). A Cornish **scor* might have been singular, as in Welsh and Breton, with *scorren* a derivative, or else the *-en* could have been taken as a singulative, and **scor* have served as the plural. Or alternatively cf. Welsh *ysgor* 'fort, enclosure' (cf. ELl, p. 79)?

    A. ?Scorran Lode; ?Scor Rock (both + Eng.)

    D. pl. ?*Trewurgyscor* (= Treworgey in St Cleer)

scoul (m.) 'kite', Voc. 498 glossing *milvus*; Welsh *ysgwfl* and *ysglyf* 'prey, bird of prey'; Breton *skoul* 'kite' (cf. GMB, pp. 611f.), Old Breton *scubl* (FD, p. 304).

segh

C1. A compound **scoul-arth* 'kite-height' (+ **arð*) may possibly appear (as C2) in Treskillard (+ **tre**)

C2. *Fentonscroll* (+ **fenten**)

scovern 'ear', MC 71b; *scouarn*, Voc. 34 glossing *auris*. Welsh *ysgyfarn*, Breton *skouarn*. Used as a personal epithet in Thomas Scoverne, in Ludgvan in 1525 (Stoate, Survey, p. 155), and other instances.

A. Scovarn, rock

scrife 'to write', PC 2791, etc.; ppp. *scrifys* 'inscribed', PC 1157 (cf. Voc. 352, *scriuit* glossing *scriptura*). Welsh *ysgrif* 'writing, essay', Breton *skrivañ* 'to write'. The *-f-* for *-v-* in Cornish is unexplained, but well-attested.

C2. ppp. Mên Screfys, stone, *Manskrethes*, fld. (both + **men**)

schus (*ch* = /k/, *u* = /œ/) 'shadow, fear', BM 3233; Old Cornish *scod* 'shade, shelter' (*o* = /œ/), Voc. 492 glossing *umbra*. Welsh **ysgod* (in *cysgod*, *gwasgod*), Breton *skeud*. There is also a *scos* (*o* = /o/) 'shield', PC 22 (sic MS): Welsh *ysgwyd*, Breton *scoed*. If it occurs at all in place-names, it and **schus** may be confusable; but **schus** is more likely in the names cited, probably in the sense 'shade'.

C2. Keskeys, *Caerskes* (both + **ker*); Polskeys (+ **pol**, **cos**); aj. ?Castle Skudzick, fld. (+ **castell**); pl. ?Talskiddy (+ **tal**)

segh 'dry', OM 757, etc.; Welsh *sych*, Breton *sec'h* (cf. Tanguy, p. 58). As in the other Brittonic languages, **segh** is often used as C1: note Welsh Sychnant, frequent, including *Sechnant* (VSBG, p. 290, called a *conuallis* 'ravine') and *ir sichnant* 'the dry-valley' (LLD, p. 182; cf. *siccam vallem*, ibid., p. 168); *in hi sich pull* 'at the dry-pool' (LLD, p. 173); Cumbric *Sechenent* (ERN, p. 355); Breton Séac'h-Ségal (Tanguy, loc. cit.). A **segh-nans* 'dry-valley' was perhaps one that dried up in summer, like an English 'winterbourne'. The first name consists of the ppp. *syghys* 'dried' (OM 756) of the verb *seghe* 'to dry' (PC 876): Welsh *sychu*, Breton *sec'hañ* (FD, p. 304). See also the derivatives **seghan* and **sichor**.

A. ppp. Seghy, rock

C1. Pensignance (+ **pen**), *Signans* (both + **nans**); Sithnoe (+ **tnou*); ?Suffenton (+ **fenten**); *sehfrod* (+ **frot**: Sawyer, no. 1005); Sellan (+ **lyn** ?)

C2. Polzeath, Polshea, Polsue in Lanteglos by Fowey (all + **pol**); Ventonzeth, Park Venton Sah, fld. (both + **fenten**); Mellanzeath (+ **melin**[1]); Roseath (+ **rid**): cf. Welsh Rhyd Sych, ÉC 10

*seghan

(1962–63), 212); ?Dor-se, fld. (+**dor**); ?Park Sea, fld.; *Stagnar Seyth*, *Stenek Segh*, mines (both +**stean** aj.); Goverseath (+**gover**)

***seghan** 'dry place'; **segh**+suffix *-**an**^{2}: compare early Welsh *Brinsychan* (VSBG, p. 120), and other examples (EANC, pp. 87f.); Old Breton *Bronsican* (Chrest., p. 112); and for further Breton examples see Gourvil, Ogam 6 (1954), 87–9. The Welsh and Breton names contain this element qualifying words for 'hill' (*bre*, *bron* and *bryn*), and so does the first Cornish instance, Bozion. In view of that, it is very possible that the other four Cornish names in *Bos-* originally contained **bron** or ***bren**, even though they all show early forms without *r*. The -*gh*- of ***seghan** was liable to disappear quite early: Bozion was spelt -*syan* in 1429, and Lanzion was -*zian* as early as 1284. In the 17th century a word *sian* was held to mean 'sea-shore' (*zîan*, Lh. 81a glossing *litus*), mainly in order to make Marazion (actually **marghas**+**byghan**) mean 'market on the shore': so N. Boson says (Bosons, p. 11), 'marhasion so call'd from a market held by the Sea shore or nigh the breach of the Sea which it aptly signifies'. This word may have contained a memory of ***seghan** 'dry place', but was probably mere invention (so Nance, Sea-Words, p. 192).

C2. Bozion (+**bron**); Bosahan (2 exx.), Bosehan, Boscean (all +***bod**, **bron**, or ***bren**); Lanzion (+**nans**?); ?Goonzion Downs (+**goon**); Polzion Wood (+**pol**)

***Seys** (m.) 'an Englishman', equivalent to Welsh *Sais* (Middle Welsh *Seis*, LHEB, pp. 582f.). This word served as the singular of MnCo. **Zowzon** in place-names, and occurs in the 1327 Subsidy Roll: William *Seys*, William *le Seys* (in different parishes). Compare the Old Cornish variants *crois* (Voc. no. 755) and **crous** (Sawyer, nos. 755 and 832/1027), both from Latin *crux*, and see LHEB, p. 535, for the two treatments of Latin *x*.

C2. Carzise (+***ker**); Tresayes, Trezise (both +**tre**); *Chyseise* (+**chy**)

serth 'sharp, steep', PC 2140; Welsh *serth* 'steep', Breton *serzh* 'steep, sheer' (cf. GMB, p. 624). In the case of Killiserth the spellings have either -*ser* or -*sergh*, showing confusion (or loss) of -*gh*/-*th* as early as the 14th century, which can be paralleled elsewhere.

A. See ***dy-serth**.

C1. ?Sarvan, rock (+**men**)

C2. ?Killiserth (+**kelli**); ?Penzer Point (+**pen**)
seth (f.) 'arrow', MC 223a, etc.; pl. *sethow*, CW 1493. Welsh *saeth*, Breton *saezh*. The meaning of the place-name is unclear, and it may be that the second element was originally a different word: perhaps *seit* ($t = /\theta/$) 'pot', Voc. 889 glossing *olla*, *zeâth* Lh. 106c glossing *olla*. (Old Breton *seitoc* 'potter', ÉC 11 (1964–67), 434; but Ekwall's Welsh *suith* 'cauldron', ERN, p. 355, is very dubious.) However, the name was certainly understood to contain **seth** pl. in the middle ages: see the place-name story in JRIC 6 (1878–81), 217, and cf. William of Worcester's (incorrect) gloss, *puteus sagittarii*, p. 104. Compare the Welsh river-name Saethon (ELlSG, p. 70).

C2. pl. *Polsethow* (+**pol**)
sevi (coll.) 'strawberries', Lh. 61b, etc. Welsh *syfi*, Breton *sivi* (aj. *Siviec*, Tanguy, p. 115; cf. Welsh Brynsifiog, Rad.). For the mine-name, compare another called *Huel Strawberry*.

A. aj. Saveock, ?Sheviock

C2. *Wheal an Seavy* (+**wheyl, an**); ?Carnjewey Point (+**carn**)
sheft 'shaft, rod', OM 2494; borrowed from English. In place-names no doubt the word usually occurs in the sense 'mine-shaft'.

C2. pl. Park Shaftis, fld.; sg. or pl. Park Shafty, fld.
***shoppa** 'workshop', borrowed from English. As a simplex name the element is perhaps indistinguishable from a possible Biblical name, *Joppa* (Acts, 9–11).

A. ?Joppa (2 exx.)

C2. Ponson Joppa (+**pons, an**); Parc Joppa (+**park, an**)
sichor (OCo.) 'dryness', Voc. 477 glossing *siccitas*; a derivative of **segh** 'dry'; *zehar* 'drowth', Lh. 12b. Breton *sec'hor*.

C2. Nanseers (+**nans**)
sỹgal (MnCo., $\dot{y} = /\Lambda/$) 'rye', Lh. 147a glossing *secale*; *sogall*, Bilbao MS. Breton *segal* 'seigle'. Some of the place-names show -*u*- in the first syllable, similar to the -*o*- of the Bilbao form: compare perhaps *chelioc* 'cock' > **kullyek**. Pryce's *sygalek* 'a field of rye' (s.v. *sygal*) is probably taken from Lhuyd's Breton *segalek* (loc. cit.), but could occur in field names, though it has yet to be found.

C1. Segoulder, Chygolder Park, St Golder, flds. (all +**tyr**)

C2. ?Lansugle (+**nans**?); Croft Sugal, fld. (+ ***croft**); Goen Suggalls noweth, fld. (+**goon, nowyth**)
slynckya 'to slink, slither' CW 913 (borrowed from English). The

*soð

word occurs in the rock-name Slinke Dean (+**tyn**), which probably contains the verb as an (exclamatory) imperative, or perhaps the v.n. used as such: cf. **crak²**.

*soð (OCo., *o* = /œ/) 'depression'; equivalent to Welsh *sawdd* 'depth' (Spurrell-Anwyl), 'subsidence' (WG, p. 78). The existence of this element depends on the one place-name only.

B1. pl. ?*Dansotha* (+**down**)

sorn 'nook, corner', PC 3056; RD 539; cf. perhaps Welsh *swrn* 'fetlock'; also 'small space' (Spurrell-Anwyl): CLlH, p. 137. In theory the word might be confusable with a word equivalent to Breton *sourn 'spring, gushing rock' (Tanguy, p. 94), but that is not attested in Cornish and is unlikely to occur in the place-names.

A. Sourne, Sworne, Thorne

C2. *Cheiensorn*, Cheesewarne (both +**chy, (an)**)

zowl (pl.) 'stubble; thatching-straw', Lh. 11c and 155a; Welsh *sofl* 'stubble', Breton *soul* 'chaume'.

B2. ?*Sowlmeor* (+**meur**)

C2. *Park-soule*, fld.; ?*Venton Sowall*, fld. (+**fenten**); ?Goonzoyle (+**goon, an**)

Zowzon (pl.; MnCo.) 'Englishmen', Lh. 242c; Welsh *Saeson*, Breton *Saozon* (cf. Kersauson, etc., GMB, p. 599; Gourvil, nos. 1178 and 1180); cf. Elements, II, 98, and JEPNS I (1968–69), 51. The only singular attested for this word is MnCo. *Zowz*, Lh. 42c, which is not found in place-names; but see ***Seys** for the Middle Cornish singular.

C2. pl. Tresawsan, Tresawsen, Tresawson (all +**tre**); Porth Saxon (+**porth**); Nansawsan (+**nans**); *Cosesawsyn* (+**cos**); *Wheal Zawson*, mine (+**wheyl**); Carsawsan (+***ker** or **cos**); *Pensousen*, coast (+**pen**; = Carn Naun Point)

D. Coswinsawsin; *Treveythin Sauson* (= Trenithon in Probus)

spern (pl.) 'thorns, thorn-trees'; sg. *spernan*, Lh. 240c; *spernan widn* glossing *oxyacantha* 'a white-thorn or haw-thorn', Lh. 110c; *spernan diw* glossing *prunus sylv.* 'a black-thorn', Lh. 131b. Old Welsh *cruc hisbernn* (LLD, p. 155); Breton *spern* (cf. Tanguy, p. 109, and GMB, p. 641); Old Breton *spern* (FD, p. 307). Cf. Gaulish *Sparnacus* and *Sparnomagos (ACS, II, 1623f.). The diminutive *s[p]ernic* (+**-yk**) occurs at Voc. 696 glossing *frutex* 'shrub'. For another word meaning 'thorn-trees' see **dreyn**.

A. sg. Sparnon (3 exx.), Spernon, *Sparnon*, Sparnon Moor; aj. ('spinney') Sparnick; Sparnock, Spernicks, flds., Spernic Cove

C1. Spargo (+*cor²)
C2. sg. Trespearne in Laneast (+tre); Park Spernon, *Park-an-Sparnon*, flds.; pl. Trespearn in Sheviock (+tre), ?Carn Sperm, coast (+carn)

spethas (pl.) 'brambles, briars', Trg. 9; *spethes*, OM 275 and 687; *speras*, CW 947. Welsh *ysbyddad* 'thorn-bushes' (ELl, p. 68), Old Welsh *di'r ispidatenn* 'to the thorn-bush' (LlD, p. 202); Breton *spezad* 'currants' (Tanguy, p. 109), Middle Breton *Spethot*, etc. (Chrest., p. 230). Like Cornish, Breton has a variant *sperat* (GMB, p. 642): for /ð/ > /r/, see HPB, pp. 698–700.

C2. Wheal Sperris, Wheal Sperries, mines (both +**wheyl**)

***splat** 'plot of land', borrowed from English *splott* (Elements, II, 139). With simplex names, such as Splatt Farm, it is impossible to say whether the word occurs as Cornish or English: cf. PNDev., I, 137.

B2. Splattenridden (+**an, reden**)

***stable** 'stable', borrowed from English, like Welsh *ystabl* (see EEW, p. 86).
  A. pl. Stabilyus
  B2. Stable Hobba (+***hoby**)
  C2. *Park-an-stable*, fld.

***stak**, element of unknown meaning, possibly 'mud' or 'pool'. An adverb *stak* 'fixed, on the spot' occurs (BM 1368), and it may have been used also as an adjective: Breton *stag* 'attaché; lien' (cf. GMB, p. 650); but it is hard to see how that could be present in the place-names. Compare rather Lhuyd's *stagen* 'pool', 33a glossing *stagnum*, and perhaps Breton *stank* 'pool' (borrowed from French: GMB, pp. 650f.; Piette, p. 180), though the loss of *n* would be odd; also West-Cornish dialect *stagged* 'stuck in mud, covered in mud' (Courtney and Couch, p. 55), though that might be related instead to *stak* 'fixed'.

C2. *Polstaggs meadow*, Polstags, Pollstack, all fields (all +**pol**)

stampes 'stamping-mill', Pryce, p. Ff1v; borrowed from English *stamps*, with the plural ending assimilated to Cornish -*ys* (LlCC, p. 12). As a simplex name, the element could occur as either Cornish or English, though disyllabic forms indicate the Cornish usage.
  A. Stampers, Stampas Farm
  B2. Stamps and Jowl Zawn, coast (+**an, dyowl**, dialect)
  C2. Park an Stampes, fld.

stean (m.) 'tin', Lh. 154b glossing *stannum*; Welsh *ystaen*, Breton *staen* (cf. Tanguy, p. 60). There seem to have been two adjectival forms, **stenek* and *stênes* (*an stuff stênes* 'the tin stuff', Pryce, p. Ff1v). See Lockwood, *Zeitschrift für Anglistik und Amerikanistik* 13 (1965), 264–5.

    A. aj. Stennack (2 exx.); Stannack, Stennack Bottom, flds.

    B2. aj. Stennagwyn (+**guyn**); Stenalees (+***lys**?); *Stagnar Seyth*, *Stenek Segh*, mines (both +**segh**); *Stennack Ladern*, mine (+**lader** pl.)

    C2. See the phrase **pul stean**; Trestain (+**tre**); aj. *Park Stenack*, fld., Halstenick, fld. (+**hal**)

steren (f.) 'star', Voc. 5 glossing *stella*; etc. Welsh *seren*, Breton *sterenn* (cf. LHEB, p. 530).

    C2. *Wheal Sterren*, mine (+**wheyl**)

steuel (OCo., *u* = /v/), 'room', Voc. 933 glossing *triclinium* (the Cornish word being originally plural); Welsh *ystafell*, Old Welsh pl. *stebill* glossing *limina*, Juvenc. (VVB, p. 216).

    A. ?Steval, rock

stoc 'stump', Voc. 708 glossing *stirbs*: borrowed from English; pl. *stockys* 'the stocks', BM 3554. The first place-name shows a different plural, in **-yer**; the derivation was first suggested by Quentel, Ogam 7 (1955), 239–41.

    A. pl. Sticker

    C2. *Street an Stock*, street (+***stret**, **an**)

strail (OCo.) 'mat, tapestry', Voc. 841 glossing *tapeta*; also *strail elester* 'mat of rushes', Voc. 842; Old Breton *straal* glossing *calamidis* 'cloak' (FD, p. 308).

    C2. Trestrayle (+**tre**)

*stras (m.) 'flat valley, shallow valley', equivalent to Welsh *ystrad* (cf. Williams, ELl, pp. 35f.; Elements, II, 163); Old Irish *srath* 'sward, meadow, valley'. There is only one example, and that from Old Cornish, of ***stras** used as B2 qualified by a river-name, like the frequent Welsh usage (cf. Williams, loc. cit.): this does not seem to have been a normal Cornish construction, and indeed the word itself may have died out in or by the Middle Cornish period. The instances of the plural do not necessarily denote more than one broad valley: the plural may have been used by convention to denote 'low-lying ground', like English *bottoms* in minor and field-names (e.g. PNChe., v, i, 110; PNBrk., III, 852; etc.): cf. Welsh Stradey (= Ystradau), Crm.

A. ?Strase Cliff
B2. *Strætneat* (+r.n., see ***neth**; = Stratton)
C2. Penstraze, Penstrase, Penstrace, Bostraze (all +**pen**); Bostrase (+**pen** or ***bod**: no spellings are available, but in view of all the other names, **pen** is perhaps more likely); pl. Penstroda, Penstrassoe (both +**pen**)

streyth 'stream', OM 772; *stret* (*t* — /θ/), Voc. 743 glossing *latex* and MC 219a. Not in Welsh or Breton. The distinction of a **streyth** from other types of stream is unknown. LEIA, p. S186.
A. Streigh; The Straythe, coast; ?*Lestreth* (+ French *le*?)
C. ?Polstreath, coast (+**pol**)

*stret 'street'; borrowed from English, like Welsh *ystryd*, *stryd* (EEW, p. 127). Most of the older Cornish towns from Truro westwards (i.e. Truro, Penryn, Helston, St Ives, Penzance, Newlyn and Mousehole) can produce Cornish street-names, though it is only in St Ives that they still survive. Some of the Cornish names look as if they were English names which have been translated into Cornish. Occasionally ***stret** is followed by an incorrect definite article, as are **park** and **wheyl**.

B2. All of the following are names of streets. Street-an-Garrow (+**garow**); Street-an-Pol (+**an, pol**); *Street Eden alias Narrow Street* (+**yn**); *Street an Garnick* (+**an, *kernyk**); *Streatt and Grean* (+**an, gronen** pl.); Stratton Vow (+**an, *fow**); *Strete Nowith* (+**nowyth**; = New Street); *Street an Dudden* (+**an, ton**); *Stret-Myhale*, *Strete Seyntjohn* (both +saint); *Stret-Pyddre* (= Pydar Street), *Street Wyndesore* (both +p.n.); *Street an Stock* (+**an, stoc**); *Stret Cowlyn* (+pers.); *Stret-Kedy* (+ ?); Street-an-Nowan (+**an**, ?); *Stret Melenwelen* (+ ?)
C2. *Gweal Street*, fld. (+**guel**); *Place an Streat* (+**plas, an**)

*stronk 'dung, filth', equivalent to Breton *stronk* 'ordure, saleté'.
C2. Polstrong (+**pol**)

*stum (m./f.) 'bend (of a stream, wood, etc.)', equivalent to Welsh *ystum* 'bend', Breton *stumm* 'méandre' (Tanguy, p. 91). On the Welsh and Breton words see Loth, RC 48 (1931), 354–6; ELlSG, p. 119; Williams, ELl, pp. 37f.; Richards, Arch.Camb. 112 (1963), 190f.; early Welsh *ystum Guy*, LLD, p. 270.

B2. Stencoose (2 exx.), Stencooce, Tamsquite (all +**cos**: cf. Welsh *Ystumcoed*, Crd., Units, p. 228); Stephen Gelly, Stephengelly (both +**kelli**); St Ingunger (+saint, *Cuncar*); *Stymwoythegan* (+pers.); *Stymcodde* (= Codda; + ?)

*suant 'level, even', borrowed from English *suant* (cf. dialect *suant, suent*: Courtney and Couch, pp. 57 and 103).
   C2. ?Catasuent Cove (+**karrek**?)

T

*tagell 'throat, constriction' (?), equivalent to Welsh *tagell* 'jowl' (WVBD, p. 520), Breton *tagell* 'collar, snare'; from the root seen in *tage* 'to choke' (PC 1528, etc.). There is slight phonological difficulty in seeing the element in the one name, but it is not insuperable.
   C2. Tintagel (+ ***dyn**)

tal (m./f.?) 'brow, front, end', Voc. 28 glossing *frons*; etc. Welsh, Breton *tal* 'forehead'. For the meaning 'end' in early Welsh, note *tal ir fos* 'the end of the dyke' (LLD, p. 122); *tal ir brinn* 'the brow of the hill' (ibid.); *tal being* 'the end of the bench' (CA, line 537); *tal y noe* 'the end of the trough' (PKM, p. 38.9); *-tel-* glossed *frons* (VSBG, p. 176); also Tallentire 'the end of the land' (PNCmb., II, 324). Though masculine in Welsh and Breton, the word often seems to have been feminine in Cornish, unless a lost definite article can account for the lenition which often follows it. The element is theoretically confusable with **toll** 'hole' unless early forms are available, since *Tal-* often becomes *Tol-*. It is qualified mainly by different words for 'hill' or 'rock'. The adjectival form (used as a noun) occurs as Old Cornish *talhoc* 'roach, dace', Voc. 550 glossing *rocea*, and as a personal name or epithet in John *Talek* 1327 SR (Crantock).
   B2. Tolcarne (several), *Talcarn* (= Minster), Trecarne (2 exx.) (all +**carn**); Tolgarrick (+**karrek**; 3 exx.); Tolmennor (+**meneth**); Tolverne (+**bron** or ***bren**); Tolvadden, Tolvaddon, Trevan (all +**ban**); Pednvadan, coast (+**pen**, **ban**: formerly *Pentalvan*); Tolgroggan, Tregragon (both + ***crak**[1] dimin.); Tolgullow (+**golow**); Trecoogo (+ ***cok** pl.); Tolponds (+**pons**); Talskiddy (+**schus** pl.?); Tolbenny, Tolfrew, Tolpetherwin, Tolzethan (all + ?)
   C1. Talland (+ ***lann**); Tagus (+**cos**)
   C2. Mên-te-heul, coast (+**men**); aj. Botallack (+ ***bod**: or *Talek* pers.), ?Mên Talhac, coast (+**men**), ?Retallack in Constantine and St Hilary (+**rid**; or **hal** aj.)

*talar (f.), meaning either 'headland of a ploughed field' or 'auger':

the first would be equivalent to Welsh and Breton *talar*, literally
'end-tillage' (cf. WVBD, p. 521; GMB, pp. 672f.; Ann.Bret. 50
(1943), 109), Old Breton *talar* glossing *ans* 'line, row' (FD, p. 310;
RC 28 (1907), 53f.): in place-names note, as simplex, Breton An-
Dalard, etc. (Tanguy, p. 80); as generic, Old Breton *Talar Rett*
(Chrest., p. 166) and Welsh Y Dalar Hir and Talar Gerwin
(ELlSG, p. 115). However, while that word is acceptable in field-
names, the Cornish form *Dollar* occurs more often in coastal and
rock-names, usually as a simplex name or qualifying an English
generic (and therefore probably also a simplex name originally).
These are very unlikely to contain ***talar** 'headland of a ploughed
field', but might instead contain a word equivalent to Welsh
taradr (Old Welsh *tarater*, VVB, p. 219), Breton *talar* 'auger',
Middle Breton *talazr* (GMB, p. 673): it could refer to a pointed
rock, for example. Compare EANC, p. 100, on Welsh words for
tools used as river-names. If this is correct, the element was
probably borrowed at a late stage from Breton, for the Middle
Cornish reflex of this word was, regularly, *tardar* (*ov thardar* 'my
auger', OM 1002): see HPB, pp. 487 and 813 for the development
of the word in Breton, and compare *camper (s.v. ***kemer**) for a
similarly late-borrowed coastal element.

    A. Dollar, rock; Dollar Rock, Dolor Point, Dollar Cove,
Dollar Ogo (all +Eng.): all coastal

    C2. *Park Dolar*, fld.

tam (m.) 'morsel'; Welsh *tam* (obsolete), *tamaid*, Breton *tamm*;
compare English *morsels*, used in field-names 'of small pieces of
land' (PNDrb., III, 757).

    B2. *Taban Denty*, fld. (+**dentye**)

tanow 'thin, narrow', Lh. 162a glossing *tenuis*; etc.; (in the plays
usually in the sense 'scarce, few'). Welsh *tenau*, Breton *tanav* (cf.
GMB, p. 676).

    C2. *Retanna* (+**rid**); *Park Tannow*, fld.; *Goldstanna*, fld.
(+**cos**); ?Crig-a-Tana Rocks (+**cruc** pl. ?)

taran 'thunder', Voc. 437 glossing *tonitruum*; etc. Welsh, Breton
taran. The compound is identical to Old Welsh *taranpull* 'thunder-
pool', LLD, p. 166.

    C1. *Treamble* (+**pol**)

tarow (m.) 'bull', OM 123, etc. Welsh *tarw*, Breton *tarv*. Compare
English field-names with *bull* (Field, EFN, p. 32). In Penzance,
Taroveor Road was also known as 'Bull's Lane' (Old Cornwall,

1, vi (1927), 23), but that may have been folk-etymology: the name is obscure.

C2. Poltarrow (+**pol**); *Park an Tarro*, fld.

taves (m.) 'tongue', OM 767; Old Cornish *tauot*, Voc. 47 glossing *lingua*. Welsh *tafod*, Breton *teod*. Both instances are doubtful, the first for lack of early forms, and the second because the element may be present by folk-etymology only: the name seems to have referred to a large rock with a hollow in it, and was certainly thought to mean 'half a tongue' in the 17th–18th centuries (Henderson, Mabe, pp. 33f.).

B2. ?Tavis Vor, coast (+**mor**?)

C2. ?Hantertavis (+**hanter**)

*****tawel** 'quiet', corresponding to Welsh *tawel*; early Welsh *nant tauel*, LLD, p. 146 (cf. EANC, pp. 88f.). The element might be confusable with a plural of Old Cornish *tauolen* 'dock, sorrel' (Voc. 630 glossing *dilla*); Welsh *tafol* (cf. WVBD, p. 526), Breton *teal* 'dock'. However, some of the names seem rather to require an unknown element *****dawel**. Confusion with **tewl** 'dark' is also possible.

C2. Tredole, Tredaule, ?Trethawle (all +**tre**); ?Hendawle (+**hensy**)

teg 'beautiful, fair', Voc. 123 glossing *pulcher*; etc. Welsh *teg*. The element is confusable with the obscure *****daek**.

C2. *Gwarry-Teage*, mine (+**guary**); Nanteague (+**nans**); ?Brass Teague, fld. (+**pras**?); deriv. (as r.n.?) ?Polteggan (+**pol**)

teil (m.) 'dung, manure', Lh. 59c golssing *fimus*; cf. 80a and 154b. Welsh *tail*, Breton *teil*. The word occurs in the compound *****buorth-del** (see *****buorth**), and in an adjectival form, meaning 'manure-heap' as a noun, corresponding to Breton *teileg* (f.) 'tas de fumier'. Theoretically the names could contain instead a word corresponding to Breton *delieg* 'leafy place' (Tanguy, p. 101), but in practice *****teylek** is more likely.

C2. aj. Parkandillack (+**park, an**), ?*Wheal an Dellick* (+**wheyl, an**)

tenewen 'side'. Welsh *tenewyn*. But 'side' is also **tu**.

B2. Transingove (+**cos**)

ternoyth 'ill-clad, half-naked', MC 50a (sic MS); composed of an obscure prefix and **noth**. Compare English field-names containing 'bare', Field, EFN, p. 13.

A. ?Ternooth, fld.

*tescow (pl.) 'sheaves, gleanings' (?), corresponding to Breton *teskoù*, plural of *teskaouenn*. The word might occur in one name, but a form of ***tusk** is also possible.

C2. ?Carntiscoe (+**goon** pl.)

tevys 'grown', OM 78; ppp. of *tevy* 'grow' OM 2034, Welsh *tyfu*. Confusable with ***dywys**: some other names listed there may contain **tevys** instead.

C2. ?Cuttivett (+**cos**: or ***dywys**)

*tewel (m.) 'pipe, conduit', borrowed from English *tewel* 'chimney, conduit'. Possibly confusable with **tewl** 'dark'. Breton *tuell(enn)* 'tap, pipe' is borrowed from Old French *tu(i)el*: Fleuriot, ÉC 18 (1981), 105.

C2. Ponson Tuel (+**pons, an**); Park-an-Tule, fld.

tewl 'dark', RD 539; *tevle*, BM 3680; *teul* 'blind', RD 1274. Welsh *tywyll* (Old Welsh *timuil*, FD, p. 278), Breton *teñval*. The word evidently became a monosyllable early in Middle Cornish, but some of the place-names show a disyllable. Confusion with ***tawel** 'quiet' is possible.

C1. Tuelmenna (+**meneth**)

C2. Lantuel (+**nans**); Zawn Duel, coast (+**sawn**); ?Vounder Duel, fld. (+**bounder**); ?***gor-thewl** (+***gor-**) in Trethowell (+**tre**)

ti (OCo.) 'house', see **chy**.

tyn 'rump, bottom', PC 2105; etc. Welsh *tin*; Breton p.n. Tinduff, ÉC 14 (1974–75), 51. Confusable with ***dyn** 'fort', in its by-form ***tyn**. Presumably **tyn** would have referred to rounded hills; names are given here (rather than under ***dyn**) when there seems no reason to suppose that a name ever referred to a hill-fort.

B2. ?Tempellow (+***pel** pl. ?); ?Penpethy (+ ?)

C2. Slinke Dean, rock (+**slynckya**)

tyr (m.) 'land'; *tir*, Voc. 11 glossing *tellus*. Welsh, Breton *tir*. Used as B2, **tyr** is liable to interchange with **tre**, and may even have been similar in meaning ('estate, holding' ?): cf. Davies, Microcosm, p. 41, n. 3, and note *tir hiernin et tir retoc*, translated as 'duos agros, agrum redoc & agrum hiernin', LLD, p. 150.

B1. Brightor, Brighter (both +***brygh**); Gonitor (+**gwaneth**); Windsor in Cubert (+**guyn**); Knightor (+***crygh**); Segoulder, Chygolder Park, St Golder, flds. (all +**sygal**)

B2. Trezance (+**sans**); Tirbean (+**byghan**); ?Tarewaste,

?*Trewaste* (both +**wast**); *Tere marracke* (+**marrek**); *Tereandreane* (+**an, dreyn**); ?*Trecombe* (+**crom**; or **tre, an**)

    C2. See the phrase ***pen-tyr***.

tyreth 'land', OM 1624; *tereth*, BM 632; etc. This word occurs once (as B2) in the phrase *An Tyreth Vhel* (+**ughel**), BM 2212, 'the High Country', referring to North Cornwall.

***tnou** (m.; OCo.) 'valley', corresponding to Welsh *tyno*, Breton *traoñ*; Old Welsh *tnou* (in *'r tnou guinn* 'the fair valley', LLD, pp. 74 and 172, and *tnou mur*, LLD, p. 32, etc.) and *tonou* (LLD, pp. 126 and 204): cf. CA, p. 235, ELl, pp. 41f., and ELlSG, p. 26n.; Old Breton *tnou* and *tonou* (Chrest., pp. 167f. and 233; Tanguy, pp. 82f.). See further Loth, RC 40 (1923), 426f.; Lloyd-Jones, BBCS 15 (1952–54), 202f.; and M. Richards, 'Tonfannau', *Journal of the Merioneth Historical and Record Society* 4 (1963), 274–6. It is notable that the Old Welsh and Old Breton instances show *tnou* predominantly as B2, a usage of which there are only two instances from Cornwall, one of them lost in the Old Cornish period. The element is rare in Cornwall (only four or five instances) and evidently died out early on. It is unclear how a ***tnou** would have differed from, say, a **nans**: perhaps 'side-valley, tributary valley'. The 1086 spelling of Treknow, *Tretdeno*, shows an epenthetic vowel (and the loss of *-w*), just as in Welsh *tyno* (cf. LHEB, pp. 338 and 379).

    B1. Sithnoe (+**segh**)

    B2. *tnow wæt*' or *tnowpeter* (+ ?: Sawyer, nos. 832/1027); Tremearne in Breage (+ ?: cf. OBr. *Tnou Mern*, Cart.Land., no. 16 and Chrest., p. 151)

    C2. Treknow (+**tre**)

    D. ?Trebarvah Goodnow (+***go-** 'sub-' ?)

***to-** (OCo.), honorific prefix in the pet forms of saints' names; it occurs also in Old Welsh and Old Breton, and the 9th-century life of St Paul Aurelian (edited by Cuissard, RC 5 (1881–83), 413–60) shows how it was used (p. 437): 'Quonoco, quem alii sub additamento, more gentis transmarinae, Toquonocum vocant... Toseocus qui cognomine Siteredus dicebatur, et Woednovius qui alio nomine Towoedocus vocabatur'.

    The 'rule' appears, roughly, to be to take the first syllable of the name, prefix it with *To-*, and add *-oc*: thus *Win*waloe gives ***To-winn-oc**. In the large amount of literature on the subject there has been some dispute as to whether the origin of this prefix is as a demonstrative, or as the possessive adjective, 'thy'; in Irish the corresponding prefix is *Mo-*, certainly 'my', which encourages the second idea. 'My' does occur, though very rarely, in Brittonic: see the references to M. Richards, below, and FG, p. 405. (For *do* in Irish, see RIA, s.v. 2. *do* (b), and C. Plummer, *Vitae Sanctorum Hiberniae* (Oxford, 1910), II, 344f.) Note also, and especially, a 5th-century Gaulish usage, applied to God: *mementobeto to diuo* 'memorare dei', from the *Life* of St Symphorian (5th century?): DAG, p. 587. For further discussion see VKG, II, 63; Loth, RC 32 (1911), 492; Thurneysen, ZCP 19 (1931-33), 354-67; Lewis, ZCP 20 (1933-36), 138-43; Cuillandre, RC 50 (1933), 50f.; Vendryes, ÉC 2 (1937), 254-68; Williams, PKM, p. 265; FG, pp. 403-5; M. Richards, Cymm.Trans. 1965, 32, and WHR 5 (1971), 343; and references given by those authorities. For the change of vowel, OCo. *to- > MlCo. -*de*-, see LHEB, pp. 656f., and HPB, pp. 148-9. The prefix occurs in three Cornish names of parish churches: Towednack; Landewednack (***lann** + ***Towinnoc**), cf. Landevennec in Brittany; Landegea (***lann** + ***Togei** = St Ke), cf. Landkey (PNDev., II, 341), *Lantokai*, Som. (= Leigh-on-Street: BBCS 14 (1950-52), 113), and Llandygai, Crn.

toll (m.) 'hole', MC 182a, etc. Welsh *twll*, Breton *toull*; cf. Elements, II, 181 and 200; JEPNS, 1 (1968-69), 51. The word is also used as an adjective, usually prefixed (as C1), e.g. *tollcorn* 'holed-horn', Voc. 262 glossing *linthuus* 'cornet'; cf. C1 below. The exact meaning in place-names is not always clear; sometimes perhaps 'a hollow'.

A. Toll; Toll Hole, coast

B2. Toldhu (+**du**); Tol-Pedn-Penwith (+p.n.); Tol Toft (+?); Tol Plous (+**plos**): all coastal; Towanwroath Shaft (+**an**, **gruah**)

C1. See the elements ***toll-gos** and ***tol-ven**.

C2. Mên-an-Tol, *Maen tol* (Sawyer, no. 450) (both +**men**, (**an**)); Creeg Tol, Cartole (both +**cruc**); Chitol (+**chy**); Parkentol, fld.

*toll-gos 'hole-wood', a compound of **toll** and **cos**; compare Welsh *Tyllgoed* (several, including *tollcoit*, LLD, pp. 188-9: lists, EANC, p. 171 and Linnard, BBCS 27 (1976-78), 559f.); *Tolchet*, Som. (BBCS 15 (1952-54), 19); Tulketh, Lnc. (PNLnc., p. 146); Middle Breton *Toulgoet* (Chrest., p. 234; Gourvil, no. 2099). But

*tol-ven

Breton *Toulhoat, Toularhoat*, etc., is different, being not a compound but a phrase, 'the wood's hole': Gourvil, nos. 2098 and 2100. The precise meaning of the compound is unknown: it could mean 'wood pierced by a stream', like Welsh *Tyllbrys*, a compound of *twll*+*prys* 'thicket' (EANC, p. 170); or 'wood in a hollow', like the Somerset and (probably) the Lancashire instances; the explanation of W. Linnard (BBCS, loc. cit.) 'wood where holes were bored in the trees for extracting honey', seems contrived, despite the fact of this practice being attested.

A. Tolgus

*tol-ven 'holed-stone' or 'dolmen, cromlech, quoit'; Breton **tol-ven* (*taol-vaen*, GIB). This compound ought to mean 'holed stone', that is, a stone with a hole in it; but in fact it may instead mean 'cromlech', that is, a megalithic monument of three stones supporting a fourth. The derivation is awkward: Loth (RC 46 (1929), 162) suggested that Breton *Tolven* was for earlier **tavl-ven* 'table-stone', which would suit the sense 'cromlech', but does not fit well phonologically: despite the fact that Breton **tavl* became *taol* rather early (perhaps before the Middle Breton period, c. 1100: cf. HPB, pp. 252 and 637), there ought to be spellings such as **taolvaen*, but in fact we find *Tol-* in all sources from Old Breton *Tolmaen* (Chrest., p. 168: but dated 13th cent. at RC 46, 162) onwards. Moreover, the Cornish element is presumably of the same derivation, and it too shows *Tol-* in the spellings of the names. Indeed, one of the Cornish examples actually has a holed stone at the farm (Henderson, Constantine, pp. 8–9). Perhaps two independent elements have been confused. See also Lockwood, *Zeitschrift für Anglistik und Amerikanistik* 13 (1965), 272–5.

A. Tolvan; Tolman (Scilly)

C2. Condolden (+**goon**)

ton (m.) 'lea-land', that is to say, land normally under grass for pasture, but occasionally cultivated; Welsh *ton* 'skin, crust; unploughed land' (ELl, p. 63; WVBD, p. 536; M. Richards, *Lochlann* 4 (1969), 179–92); cf. Breton *tonnen* 'rind, crust' (Le Gonidec; GMB, p. 698), and Old Irish *tonn* 'skin, earth, bog', frequent as B2 (Hogan, Onomasticon, p. 642). The phrase *the wonys guel ha ton* 'to work *guel* and *ton*' (OM 1164) shows that it could be cultivated; *aras an kensa an todn* 'plow first the lay' (Pryce, p. Ff2r). Note the phrase *war ton* 'onto the ground' (RD 2281 and BM 3505), showing the same curious lack of lenition

tor

as in the place-names: it may have been protected by a (lost) definite article, as in *war an ton* (BM 3495 and 3519) and in some of the place-names in **chy + ton**. Two of the names (*Goenandon* and *Street an Dudden*) show the word as feminine instead of masculine. In theory this element would be confusable with a word **tōn* 'hillock', equivalent to Welsh *twyn* (M. Richards, Lochlann 4 (1969), 192–214), Old Breton *tuhen* (FD, p. 324), of which Loth (RC 34 (1913), 144f.) saw a derivative in Cornish *tonek* 'crowd' (MC 257c); but in practice it is improbable that that word occurs in place-names.

B2. Todden Coath (+ **coth**); *Tondre* (+ **tre**); *Tonenbethow* (+ **an, beth** pl.); *Todne Rosemoddress* (+ p.n.): the last three are all fields.

C2. *Goenandon* (+ **goon, an**); Park Todden, *Park-an-Toddan*, flds.; Chytane, Chyton, Chytan (2 exx.), Chytodden (4 exx.), Chiverton (3 exx.), Chyverton, Chyvarton (all + **chy, (war, an)**: cf. Welsh Tyn-y-ton (6 exx.), Lochlann 4, 192); Enys Dodnan, rock (+ **enys**); *Street an Dudden*, street (+ **stret*, **an**)

top 'top', BM 599; etc.; borrowed from English, like Welsh *top* (EEW, p. 180).

B2. Top Tieb (+ ?); Top Lobm (+ **lom* ?); *Toplundie Cove* (+ p.n.; = Lundy Cove): all coastal

C2. Gweal-an-Top (+ **guel, an**)

tor 'belly', Voc. 72 glossing *venter*; BM 80; etc. Welsh, Breton *tor* (cf. GMB, p. 700; Tanguy, p. 81). The meaning of the word in place-names is unclear: Tanguy says the Breton word means 'éminence', which seems most sensible; but in Welsh, Gwenogvryn Evans translated *torr ir allt* (LLD, p. 221) with 'the foot of the *allt*' (ibid., p. 377) while *tor y mynydd* is 'flanc de la montagne' (LEIA, p. T33), and *Torbant* is presumably 'hollow-valley' (ELlSG, p. 114). It is considered that English dialect *tor* 'rocky outcrop', frequent in place-names in Cornwall, Devon, Derbyshire and elsewhere (PNDev., II, 671; Elements, II, 184; PNDrb., III, 709f.) is derived from Brittonic **torr*: if so, it presupposes the meaning 'swelling, protuberance' as having been general in Brittonic at an early date. There are no instances of Cornish **tor** clearly meaning 'tor' (which is **carn** in Cornish), and the rarity of the element in Cornish-language place-names is to be noted.

B2. Tor Noon (+ **an, goon**); ?Tor Balk, rock (+ ?); the second element of Restormel may be a place-name **Tor moyl* 'bald hill' (+ **moyl*), used (as C2) to qualify **ros*.

C1. Torfrey (+ **bre**)

torch (m.; OCo.) 'boar', Voc. 596 glossing *magalis* ('barrow, gelded pig', OE *bearh*). Welsh *twrch*, Breton *tourc'h*, 'boar' (cf. GMB, p. 705; Tanguy, p. 123). If present, the word shows a fossilised *i*-affected genitive singular **terch*, as with **margh** and like several instances of the word in place-names elsewhere: see Quentel, Ogam 7 (1955), 79–81, and note Pentridge (PNDor., II, 235); Pentrich (PNDrb., II, 490); Pentyrch (5 exx., Units, p. 177; *Penntirh*, VSBG, p. 62); Bryntyrch (ElISG, p. 19); Breton Penterc'h, Brenterc'h, Roudouderc'h, etc. (Tanguy, loc. cit.); early Irish *Cenn tuirc* (5 exx., Hogan, Onomasticon, p. 228). The alternative is a stream-name **derch* 'bright, clear' (?).

C2. ?Reterth (+ **rid**)

**torn* 'a quarter of the parish of St Keverne', borrowed from Middle English *tourn* 'sheriff's turn, tour or circuit': see Henderson, 98th RCPS (1931), 55.

B2. *Turn-Bean* (+ **byghan**); *Turn Tregarne*, *Turn-Traboe*, *Turn-Trelan* (all +p.ns.)

**torthell* 'loaf, lump', a diminutive of *torth* 'loaf', RD 1314 and 1490; cf. Welsh *torth*, Breton *torzh* (dimin. *torzhell* 'wen, tumour').

B2. Tater-du, rock (+ **du**)

towan (m.; MnCo.) 'sand-dune', Pryce ('a heap of sand'), Tonkin ('heaps of sand', JRIC n.s. 7 (1973–77), 205), dialect *towans*; Welsh *tywyn* 'sand-dune' (cf. ElISG, p. 117), Breton *tevenn* 'côte de la mer exposée au soleil; dune, falaise' (cf. Tanguy, p. 81). Elements, II, 185; JEPNS 1 (1968–69), 51. The first meaning in Welsh and Breton, 'sea-shore', may be present in some of the Cornish names, especially the simplex ones, like Welsh Tywyn (4 exx., Units, p. 217). Early forms show the form **tewyn* in Middle Cornish.

A. Towan (2 exx.), The Towans (2 exx.); Towan Head, Towan Beach, Tewington (all + English)

B2. Towan Blistra (+ ?)

C2. Porthtowan (2 exx.; + **porth**); Pentewan (+ **pen** or **ben**); Carn Towan, coast (+ **carn**)

towargh (coll.) 'turf, peat'; Welsh *tywarch* 'turves', Breton *taouarc'h*.

B2. ?*Towargh Ane*, mine (+ ?)

C2. Tredower (2 exx.), Tredore, Trethowa (all + **tre**); Parkentower, fld.

trait (m.; OCo., -*t* = /θ/) 'sand, beach', Voc. 738 glossing *harena*; Modern Cornish *treath* (Pryce, p. Ff3r), *dreath*, Lh. 7b ('The Sandy Shore cover'd at high water'), etc. Welsh *traeth*, Breton *traezh*; Old Irish *tracht*, LEIA, p. T121. This word is confusable with ***treth** 'ferry', which is formally identical, but the latter is rare and the majority of names can safely be taken to contain **trait**.

B1. Gwendra (2 exx.), Gwendreath (all +**guyn**; cf. Welsh r.n. Gwendraeth, EANC, p. 113; early Irish *Finntracht*, Hogan, Onomasticon, p. 422; Irish Fionntráigh, Ventry)

C2. Tredreath (+**tre**); Millendreath, *Melyntrait* (= Treesmill), Vellandreath (all +**melin**[1]); Pentreath, Pentreath Beach (both +**pen**; Welsh Pentraeth, Agl.; Breton Pentrez, Chrest., p. 234); Portreath (+**porth**); Tywardreath (+**ti, war**); *Landraith* (+ ***lann**: = St Blazey); *Wheal Dreath*, mine (+**wheyl**); Halldreath (+**hal**); flds. Park Treath, Park-an-trath; Gallentreath, coast (+**guel, an**?)

D. *Pirran in Treth* (= Perranzabuloe)

***trap** 'stile', borrowed from English. Cf. Welsh and Breton *trap* 'snare'.

C2. Park an Trap, etc., flds.; *Leane an Trapp*, fld. (+**leyn, an**)

tre, *tref (f.) 'estate, farmstead', in the texts usually 'town, village' or 'home' (e.g. PC 320, 2997; RD 1381, 2404); Welsh *tref, tre*, Breton *trev*: the basic word denoting an agricultural settlement. In origin it must have meant something to do with agriculture: compare Old Cornish *treuedic*, Voc. 337 glossing *colonus* 'cultivator', Middle Cornish *the drevas* 'thy tillage', OM 425; Old Irish *trebaid* 'cultivates, inhabits'. See also LEIA, pp. T126–8; DAG, p. 171; and FD, p. 318. The usage of **tre** in Cornwall as a placename element is, if anything, more standardised and more widespread than in the other Brittonic areas, though parts of south-west Wales (Cardiganshire and Pembrokeshire) are similarly well-covered. However, the use of **tre** as a standard settlement-term must go back to the Common Brittonic period: that is shown by its use in all Brittonic areas, including southern Scotland and north-west England (see Watson, CPNS, pp. 357–65; PNCmb., I, 116; and Nicolaisen, pp. 166–70). However, while **tre** does occur in some compound names, listed below, the vast majority of names containing the element are of the phrase type, making them probably later than the 5th century in date of formation. (They can, of course, have referred to already-existing settlements when they were

tre

formed as names.) The latest time when most names in *Tre-* were formed is probably c. 1100: the date is given by the fact that so many of them are qualified by a Celtic personal name (and such names were, as far as one can tell, virtually out of use after that date, except for a few rare instances), by the fact that *Tre-* names do not become commoner as one moves westwards across the county (and therefore had ceased to be coined by the time when the Cornish language was moribund in the east), and by the fact that *Tre-* places seem to belong to the native system of land-tenure which preceded the manorial system current under the Anglo-Saxons and Normans, and therefore belong to the period of native administration. This last suggestion is based partly on the parallel evidence from Wales, and especially the Welsh laws (see Ll.Bleg., p. 171; Richards, *Ceredigion* 4 (1960–63), 273–5); and partly on the fact that names in *Tre-* play a particularly large part in the nomenclature of the manorial system in Cornwall from its earliest appearance in Anglo-Saxon charters and Domesday Book; in the latter source, about 93 out of about 347 Cornish manors have names in *Tre-* (the approximation being due to various doubtful cases), or over a quarter. Of named places granted in Saxon charters, 10 out of 42 have names in *Tre-* (or, if grants from the 10th–11th centuries are taken alone, 10 out of 34), again about a quarter (or somewhat over). These figures show a higher proportion of *Tre-* names than over the nomenclature of the county as a whole, and indicate that places called *Tre-* occupied a prominent place in the manorial system. Assuming that the latter was itself based partly upon the native system of land-tenure, it follows that *Tre-* farms had occupied a special place in the Celtic period: that would be in keeping with the Welsh evidence. Another indication of date appears in the names Trebarveth, etc., which use **perveth** 'middle', replaced in the Middle Cornish period by **cres**; *Tre-gres does not occur.

A further reflection of the prominence of **tre** in the manorial system is seen in the names of the mediaeval tithings, which were based on the manors (listed and discussed, Pool 'Tithings'): of approximately 334 places which gave their names to tithings, 83 (slightly less than a quarter) had names in *Tre-*; here, however, there is a contrast between east and west Cornwall, for in the four western hundreds the proportion is 53 tithings in *Tre-* out of 170; it is striking, in east Cornwall, that the hundreds of Trigg and

tre

Lesnewth, full of *Tre-* farms, have rather low proportions of tithings with such names. (It is also most striking how few *Bos-* names there are, especially in Penwith, where such names are particularly common and yet there is not one tithing with such a name.)

It is generally assumed (e.g. Thomas, 'Dumnonia', pp. 97f.) that the **tre** or unfortified farmstead replaced the *****ker** or fortified settlement during the late Roman or early post-Roman period; however, as W. Davies has remarked, 'it has not yet been demonstrated by anyone that there was a change in the terminology of settlement, from *caer* to *tref*, in the post-Roman period' (*History*, n.s. 67 (1982), 117); such a replacement is hypothesis rather than established fact, and the two types of settlement may have coexisted: as mentioned above, the Common-Brittonic nature of **tre** as a settlement-term implies that it is older than the 5th century. However, it is true, as mentioned, that most names containing **tre** are of the phrase type, and therefore should be later than the 5th century. **Tre + nowyth** 'new *tre*' (23 exx.) appears as early as A.D. 969 (Sawyer, no. 770) and as far east as the Tamar (Lezant parish), with two manors of that name in Domesday Book, so that even that formation is likely to be early (pre-1100) when it occurs. In distribution it is similar to **tre** in general, with the largest number, seven, in the hundred of Kerrier, where **tre** is densest. Finally, the *Tre-* places in Devon should be noted as showing that **tre** was already current before the 7th century at the latest, and that *Tre-* farms can be assumed to have existed in areas now thoroughly English in their nomenclature: in PNDev., see Trellick (p. 76), Trebick (p. 211) and Treable (p. 429: *Hyples eald land* of 976, Sawyer, no. 830): see the discussion by Finberg, Lucerna, pp. 116–30.

There are a few cases where a late name in *Tre-* seems to have arisen, the element being qualified by a mediaeval surname: two examples are Tresemple in St Clement (*Tresempel* 1278: cf. William *le Symple*, living within a quarter-mile of the place in 1287), and Trehan in St Stephens by Saltash (*Trehanna* 1306: cf. William *Hanna* in the parish in 1327). There are a few other such cases, but it is always possible instead that the *Tre-* name was older, and that the element had been dropped to provide the inhabitant with his epithet: the continued understanding of the meaning of **tre** could work both ways. However, with **tre** being so common,

tre

one would certainly expect to find the occasional late instance, formed by analogy.

There are also a number of cases where *Tre-* is qualified by an Old English personal name, such as Tredundle (*Tredenewold* 1284, **tre** + *Deneweald*), Trekimletts (*Trekynmound* 1355, **tre** + *Cynemund*), Trebursye (*Trebursi* 1199, **tre** + *Burgsige*) and Trehunsey (*Trehunsy* 1306, **tre** + *Hunsige*). There are about twenty such names (some of them doubtful), all in East Cornwall. They need not represent English incomers, since a practice of Cornishmen taking English names is well-attested, particularly in the 10th century (see Co. Studies 6 (1978), 25, n. 22): for that reason, and because of their different geographical distribution, such names may represent a slightly different phenomenon from the instances of Tresawsen, etc., which are much more western in their distribution, and which clearly do represent incomers.

There are about 1,300 or more genuine *Tre-* names in Cornwall, evenly distributed over the whole county except for some heavily-anglicised areas of the far east and the areas of moorland: the hundred of Stratton has only two or three *Tre-* farms altogether, and the area round Callington and St Mellion has very few. Apart from these areas, the highest density of *Tre-* farms is in the hundred of Kerrier, and the lowest (about half as dense) is in the hundred of West. The hundreds of Trigg, Lesnewth, Powder, Penwith and Pydar fall between these two and are close to one another in density. It is interesting that East Cornish hundreds tend each to contain about a hundred *Tre-* names (Trigg, 93; West, 106; Lesnewth, 128), recalling the Welsh *cantref* (etymologically *cant tref* 'a hundred *trefs*': cf. English 'hundred'); whereas the West Cornish ones contain about double that number (Penwith, 176; Pydar, 213; Kerrier, 228; Powder, 240), though the overall density remains about the same: perhaps they were each originally double hundreds? (Note that all these figures are approximate.) Note also the *cét treb* in early Scotland: Bannerman, Dalriada, pp. 56 and 142–8; Charles-Edwards, *Past & Present* 56 (1972), p. 18 and n. 30.

Further indications of meaning are as follows. In early Welsh, the usual translation is *villa*: e.g. *Tref redinauc .i. uilla filicis* (VSBG, p. 72); *Tref ret*, *uillam ret* (LLD, p. 224; and passim). See also Davies, Microcosm, pp. 38–9. Henderson suggested once that the three villeins recorded in Domesday Book as belonging to the manor of *Langorroc* (= Crantock) represented the three later *Tre*-farms belonging to the manor, viz. Treago, Trevella, and Trevowah (in Doble, *St Carantoc* (2nd edn., 1932), p. 32), but it does not seem possible to follow up this interesting suggestion. On the meaning in Brittany, which is somewhat different, note for instance *quandam tribum nomine Tref Iulitt*, also *tribu Iulitt* (Cart.Land., no. 47), and see Largillière, *Les saints*, especially pp. 27–32, 182–4 and 220 and n. 29. Largillière assumed, rather than proved, a semantic and chronological distinction in Brittany between *Tre*- 'hamlet' and *treff* 'part of a parish': the former is commoner in East Brittany but is supposed to have gone out of use early on; however, the Cornish usage, certainly earlier than 1100 in date, seems to incorporate aspects of both these meanings, and may well be an early sense of the word. For what it may be worth, it is notable that the four Cornish places where a name in *Tre*- has become that of a parish are all in a group in north-east Cornwall, near the heavily-anglicised area.

In later usage within the Cornish language, and as a qualifier in place-names, the word is much vaguer in meaning, and means 'habitation, village, home farm': note *trevow* 'towns' (PC 132: Norris, II, 211); *in trefov hag in gonyov* 'in town and country' (BM 1037); *an trevow marras han trevow trygva* (Trg. 25a, translating 'townes and houses'); there are numerous place- and field-names of the type Pendrea 'top of the village', *Dannandre* '(field) below the farm', where **tre** can refer to any settlement.

In form, **tre** shows the following variations. Its regular spelling in Old Welsh and Old Breton was *treb* (where -*b* = /v/): Old Welsh *treb guidauc*, Chad 3 (LLD, p. xlv); see Chrest., p. 168; FD, p. 318; HPB, pp. 589f. That would have been its regular spelling also in Old Cornish, if the native records had survived in original form, but it is not found. However, it is also quite common in late Old Welsh sources to find the incorrectly archaistic form *Trem*- (where -*m* = /v/), e.g. *Trem Gyllicg*, *Trem carn*, *Trem canus* (LLD, pp. 43, 124 and 125) and *Trem y crucou*, *Tremlech* (VSBG, p. 120); there is a probable Cornish instance of that in Domesday Book, *Tremarustel* (f. 125a, now Treroosel), probably **tre** + pers. *Arwystel (see references s.v. ***arwystel**). The likelihood has wider ramifications, since it implies that Domesday Book was in part compiled from a native Old Cornish written source (a possibility of

tre

which there are other hints as well as this name). In the mediaeval period, a final -*v* or -*f*, since lost, quite often appears preceding the second element of a name, e.g. *Treures* 1262 (FF, no. 207), *Drefvyan* 1398 (FF, no. 810), now Trease and Trevean. In simplex names the final -*f* survived, giving Drift in each case.

Because **tre** is feminine, the initial *t* could be lenited by a definite article (not usually expressed): apart from all three simplex instances (Drift), in only one case, Drewollas, has the lenition survived to the present day; but other instances are found in early spellings, such as *Drefellou* 1460, now Trevelloe; *mora de Drebel* 1391, now Trebell; *Drefvyan* 1398, now Trevean in Newlyn East; *Tregooth als. Andregooth* 1677 (in Colan); and *Andrewartha* 1624, now Trewartha (though *Andrewartha* survives as a surname). An alternative modern treatment is for *Tre-* to be pronounced as 'Te-' (through simplification of the consonant group): the forms of Tembraze (now Trembraze in St Keverne) and Tencreek show this, and oral forms of Trewellard, Trewoof-wartha and others also show it: from its occurrence as far east as Tencreek in Menheniot, this is evidently a modern, English, change rather than a Cornish one.

Because of its extreme frequency as B2, *Tre-* has appeared as a false prefix in a large number of names. These are of two main types: those where a different Cornish first element has been altered to give *Tre-*, and those where *Tre-* has been added to a name, Cornish or English, that originally lacked it. The Cornish ones are of the type Tresmaine, formerly *Rosmaen* (***ros** + **men**); Tregray, formerly *Risgre* (**rid** + **gre**); etc. There are about 66 instances of this type on the modern map, with further instances appearing sporadically in the spellings of other names but not in the modern forms; the change can occur at any period from the 13th century (when *Tresmere* appears, for older *Rosmur*) to the 19th (when places in *Chy-* are liable to appear as *Tre-*). The elements ***ros** and **rid** are the most likely to change to *Tre-*, normally appearing as *Tres-*, but other elements which have also changed to *Tre-* include **tal** (e.g. Trecarne < *Talcarn*, Tregleath < *Talglighy*, etc.), ***dyn** (e.g. Trendeal < *Dyndel*), **ti** (e.g. Trethevey < *Tewardeui*), **tyr** (e.g. Trezance < *Tersant*), **cruc** (e.g. Treconner < *Cruconner*), ***ker** (e.g. Tregorland < *Kaercorlan*), ***tnou** (Tremearne < *Tnoumern*: cf. RC 40 (1923), 426f. for the same in Breton), **tu** (Trecoose < *Tukoys*), and others.

228

tre

This type occurs all over Cornwall, though it is commoner in the eastern half of the county.

The second type of false *Tre-* is restricted to the east; it more often involves an English name than a Cornish one, and is commonest with simplex names. It is of the type Trelay, which (in one case, for instance) was *la Leye* 1305, *Treley* 1427 (OE *leah* 'wood, clearing'), or Tredown, from *Doune* 1370, *Tredowne* 1569. In many of the cases the *Tre-* probably derives from a Middle English *atte(r)* 'at the', as seen in, for instance, Treway (Thomas *atte Weye*, 1327), Treheath (Michael *ata Hethe*, 1327). The same phenomenon occurs in west Devon within a few miles of the boundary with Cornwall: see, in PNDev., Tredown (pp. 79, 146 and 181), Trehill (p. 200), Trelana (p. 163), Treleigh (p. 210) and Trevenn (p. 186), and contrast, in east Devon, Traymill and Terley (ibid., pp. 572–3), incorporating the same phenomenon but without the influence of Cornish *Tre-* affecting the forms: see also ibid., pp. xxxv–xxxvi and li. However, Middle English *atte(r)* cannot be responsible for all the instances of false *Tre-* added arbitrarily to names, often as late as the 19th century, and often with both forms, with and without *Tre-*, surviving side by side. In these cases *Tre-* can be an arbitrary addition, at any date, due to analogy. (The process is helped by the fact that some genuine *Tre-* names drop their first syllable, such as Raven, formerly *Treryvyn* 1401 and still called Treraven on some maps; another example is Reed, spelt *Trevred* 1319, and so up to *Trereed* c. 1728.)

A. Drift (3 exx.): cf. Breton Le Dreff, etc. (e.g. RC 46 (1929), 118); see also ***godre**.

B1. Calendra (+**kelin**); ?Lendra (+**lyn** ?); Raftra (+**rag** ?); ?Candra (+**can** ?); ?Cuddra (+ ?); see also ***hendre**, ***gor-dre**, and *gwan-dre s.v. **guan**.

B2. Of slightly over 1300 instances of **tre** as B2, about 305 have an identifiable personal name as qualifier, while about 392 have a descriptive qualifier; the rest (about 610) have either an obscure or an ambiguous qualifier (mostly the former). There is no apparent difference in the distribution of those qualified by personal names and those qualified by a descriptive word or phrase, except that the slight preponderance of descriptive qualifiers is perhaps more marked in the hundred of Kerrier (where *Tre-* is also densest) than elsewhere. However, since personal names probably account for a good proportion of the obscure names, this preponderance of descriptive qualifiers, both in Kerrier and elsewhere, may be more apparent than real.

Qualified according to size or position: Trebeigh, Trebiffin, Trevine, Treveans, Trebighan, Trebyan, Trevean (10 exx.) (all +**byghan**); Tremear, Tremeer (4 exx.), Trevear (4 exx.), Treveor (2 exx.) (all +**meur**); Trebarfoot, Trebarwith, Trebarret, Trebarvah (2 exx.), Trebarveth, Trebarvath, Trebarva (all +**perveth**: note the lack of the later **cres**, which in Middle Cornish replaced earlier **perveth**); Trewartha in St Agnes (+**guartha**); Drewollas (+**goles**); Trebell (+**pell**?); Trefinnick (+**fyn** aj.); Treveniel (+**myn** aj. ?); Trevenwith (+**fynweth**); Tredrea in St Erth (+ *****treu**)

Qualified according to nearby natural features (the largest category): Trewarveneth, Trewarmenna, Trevenna in St Neot and St Mawgan in Pydar, Trevena in Tintagel, Trevenner (all +**war**, **meneth**); Tregoose (8 exx.), Trequite (2 exx.), Trequites (all +**cos**); Trevisquite (+*****is**[1], **cos**); Trengilly (+**an**, **kelli**); Treskelly (+*****is**[1], **kelli**); Tregilliowe (+**kelli** pl.); Trerise (2 exx.), Trease (2 exx.), Trerice (5 exx.) (all +**rid**); Trefrida (+**rid** pl.); Tretherras, Tretherres (both +**dew**, **rid**); Trenant (6 exx.), Trenance (8 cxx.) (all +**nans**); Treknow (+*****tnou**); Trerose (+*****ros**; 2 exx.); Tregue (3 exx.), Tregew (2 exx.), Treguth (all + *****kew**); Trebrown (+**bron**; 2 exx.); Trenault, Trenalls (both +**an**, **als**); Trenale, Trenhale (both +**an**, **hal**); Tregarrick (+**karrek**; 2 exx.); Trefrouse, Trefrew, Trerew (all + *****rew**); Tregarne (2 exx.), Tregarden in Luxulyan (all +**carn**); Treweese, Treweers, Trewarras, Treverras (all +**gweras**); Trelonk (+*****lonk**); Trelion, Treloyan (both + *****legh** pl.); Tremayne (3 exx.), Tremain, Demain (all +**men**); Trevilges, Trevilgus (+cpd., *****huel-gos**); Trevuzza (+**frot** pl.); Tregaswith, Tregadgwith (both + *****caswyth**); Trenhayle (+**war**, **an**, *****heyl**); Tredreath (+**trait**); Trelin, Trelyn (both +**lyn**); Trelinnoe (+**lyn** pl.); Trengweath (+**an**, **gwyth**[1]); Trewethack, Trewithick (2 exx.) (all +**gwyth**[1] aj.).

Qualified according to a man-made feature: Trencreek (5 exx.), Tencreek in Menheniot, Tencreeks, Trecreege (all +(**an**), **cruc**); Trecrogo (+**cruc** pl.); Trengayor, Tregear (6 exx.), Tregeare (2 exx.), Tregair (all +(**an**), *****ker**); Treneglos, *Treneglos*, Treveglos (4 exx.), Treviglas (2 exx.) (all +(**an**), **eglos**); Trevose (+**fos**);

tre

Trevozah (+**fos** pl.); Trelease in Kea and St Keverne (both + ***lys**); Tremenheere, Tremenhere (2 exx.), Tremenheire (all + ***men hyr**); Trevorder (2 exx.), Treworder (4 exx.), Trevarder (all + ***gor-dre**); Trengrouse (+**an**, **crous**); Trewardreva (+***godre** pl.); Treen (+***dyn**; 2 exx.); Tredenham (+***dynan**; 2 exx.); Trebowland (+(**an**), ***bow-lann**); Tredrea in Perranarworthal (+**tre**?)

Qualified according to the nature or use of the farmland: Trenoon (2 exx.), Trewoon (3 exx.), Troon (2 exx.), Treween, Trewoone, Trengune (all +(**an**), **goon**); Tregwinyo (+**an**, **goon** pl.); Trethowa, Tredower (2 exx.), Tredore (all +***towargh**); Trengothal, Trewothall, Treworthal (all +(**an**), ***gothel**); Treharrock (+***havrek**); Tredivett (+***dywys**); Trelaske (2 exx.), Trelask (all +**losc**); Trevarra (+**erw**); Trewavas (+***gwavos**; 2 exx.)

Qualified by words for particular plants: Trenithan (+**an**, **eithin**; 2 exx.), Tretrinnick, Tredinnick in Morval and Probus (all +**eithin** aj.); Trendrean, Trendrine (both +**an**, **dreyn**); Trendrennen (+**an**, **dreyn** sg.); Tredinick in St Mabyn, Tredinnick in St Issey, St Keverne and Newlyn East (all +**reden** aj.); Tredrizzick, Treisaac (both +**dreys** aj.); Trescowe (2 exx.), Tresco (all +**scawen** pl.); Trescowthick (+**scawen** aj.); Treskewes (+**scawen**, *-**ys**^{2}; 2 exx.); Trespearn (+**spern**); Trespearne (+**spern** sg.); Trevedda, Treveddoe (both +**bedewen** pl.); Treglasta (+**glastan**); Tregallon (+**kelin**); Treglennick (+**kelin** aj.); Tregolls (3 exx.), Tregolds in St Merryn, Tregullas (all +***coll** deriv.)

Qualified by words for animals (mostly birds): Trekee (+**an**, **kio**); Tregwindles (+**guennol**); Trekillick, Treculliacks (both +**an**, **kullyek**); Treire, Treheer, Treraire, Treneere, Trenear in Wendron (all +(**an**), **yar** pl.); Trembleath in St Ervan (+**an**, **bleit**); Tremough (+**mogh**); Treloggas (+**logaz**); Trelogossick (+**logaz** aj.); Tremar, Trevarth (both +**margh**); Trethevas (+**daves** pl.); Tregarden in St Mabyn (+**garan**); Trefula (+**ula**); Tregye (+**ky**)

Qualified by colour, pleasantness or other abstract features: Trewen (5 exx.), Trewidden (all +**guyn**); Trethew, Trew in Breage (both +**du**); Trelease in Ruan Major (+**glas**); Tredaule, Tredole (both +***tawel**); Trewint (7 exx.), Trewince (9 exx.) (all +**guyns**); Tretharrup (3 exx.), *Trewartharap* (all +***gortharap**);

trech

Trenuth, Trenower, Trenute, Trenoweth (11 exx.), Trenowth (3 exx.), Trenouth (3 exx.), Trenewth, Trenowah (all +**nowyth**); Trelease in St Hilary (+ *****dy-les**)

Qualified by ownership (other than personal names): Tregaminion (+*****kemyn** pl.; 4 exx.); Tresawsan, Tresawsen, Tresawson (all +**Zowzon**); Tresayes, Trezise (both +*****Seys**); Treveneage (+**manach** aj.); Tranken, Trink (both +**Frank** pl.); Trerank, Trefranck (both +**Frank**); Trengove (2 exx.), etc. (+**an**, **gof**); Trelissick in Sithney and St Erth (both +*****gwlesyk**); Tremethick (+**methek**); Treeve in Sennen, Treave, Trew in Tresmeer (all +*****yuf**); Trembraze (+**an**, **bras**; 2 exx.)

Qualified by other, miscellaneous, words: Tregatillian (+**cuntell** pl.); Trelevra (+**leuerid**); Trengwainton (+**dy-, guaintoin**)

Qualified by a personal name (a few examples only; personal names are from Bodm. unless otherwise cited): Treworgie (2 exx.), Treworgey (2), Trevorgey (all +*Wurci*); Treworgans in Probus, Trevorgans (+*Gurcant*); Trefingey, *Trefrengy* (both +*Brenci*); Strickstenton, Triggstenton (both +*Costentin*); Trevilley (2 exx.), Trevilly (all +*Beli*: or *****byly** ?); Trevillian (+ *Milian*); Tremoddrett (+*Modred*); Tredellans (+*Telent*); Tregarton (+ *Ylcerthon*); Tresulgan (+ *Sulcæn*); Tresillian (2 exx.; cf. OWe. *Sulgen*, LLD); Tregenfer, Tregenver (cf. cvnomori, CIIC no. 487); Tredruston (cf. OWe. *Tristan*, LLD); Treverbyn (3 exx.; cf. OWe. *Erbin*, EWGT); Tredudwell (2 exx.; cf. OBr. *Tutuual*, etc., Chrest.); Trevaskis, Trevascus (cf. MlBr. *Maelscuet*, Chrest.); Trythance (cf. OBr. *Iudcant*, Chrest.); Triffle (cf. MlBr. *Iuthael*, etc., Chrest.); Tregongeeves (cf. OWe. *Cintimit*, etc., LLD); Tregonebris (cf. OWe. *Conhibrit*, LLD); Tregondean (cf. OWe. *Kyndeyrn*, etc., EWGT); Tregenhorne (cf. OBr. *Conhoiarn*, Chrest.)

C2. Gwarth-an-drea, *Warthendr'*, *Quarthendrea* (all +**guartha, an**); *Tondre*, fld. (+**ton**); *Goentre*, fld. (+**goon**); ?Poldrea (+**pol**); ?Tredrea in Perranarworthal (+**tre**: meaning obscure); *Ewandre*, *Park Eughandrea*, flds. (+(**park**), **ugh, an**); *Dannandre*, etc., flds. (+**dan, an**); *Dorendre*, fld. (+**dor, an**); pl. ?Botreva (+ *****bod**); see also *****pen (an) dre**.

D. Penhale-an-Drea; Kemyel Drea (= *Kemyelnessa*, presumably)

trech 'cut wood', Voc. 707 glossing *truncus* (also no. 903 glossing *fructus*, an error for *frustum* 'piece'); cf. **trogh**, from the same root. There are other words in Welsh or Breton which may be present instead in the name, e.g. Welsh *drech*, variant of *drych* 'form, image'; or Breton *trec'h* 'victorious' (cf. GMB, p. 712), Welsh *trech* 'stronger'; but the word given seems most likely.

C2. Botrea (+ ***bod**)

*treth (m.) 'ferry', equivalent to Breton *treizh* (cf. GMB, p. 714; HPB, p. 517); cf. Welsh *treth* 'toll' (Armes Prydein, line 123n.). The element is formally indistinguishable from **trait**, but the two instances are at either end of a known mediaeval ferry. See ***caubalhint** for another word meaning 'ferry'.

A. Treath (2 exx.)

*treu (= /trœ/) 'beyond, yonder' (?): Welsh *draw*, **traw*, Middle Breton *di-dreu* (GMB, p. 164; cf. Middle Breton *treu* 'ferry', GMB, p. 716). The adverbial use of a preposition (as an adverbial adjective) would be a little odd, but both the forms and the geography suggest it. Compare the use of OE *begeondan*, especially common in Devon (Elements, I, 25), though in names like Yonderlake, Yondercott (PNDev., II, 383, 539), the meaning is '(settlement) beyond the stream, beyond the cottages', whereas the Cornish name means 'farmstead beyond (the river)'.

C2. Tredrea in St Erth (+ **tre**)

try[1] 'three', fem. *ter*; Welsh *tri*, *tair*, Breton *tri*, *teir*. With some names it is hard to tell whether they contain the simple numeral, or the prefix ***try-**[2].

C1. *Trydinner*, mine (+ **dyner**); Dry Carn, rock (+ **carn**); ?Trilley Rock, ?Trilley (both + ***legh**?); fem. ?Terras (+ **rid**?)

*try-[2] 'triple' or 'very'; *tre-* in *tredden* 'three men' (MC 237b), *treddeth* 'three days' (BM 3895), and *vorh trivorh*, Lh. 166c glossing *tridens*: cf. LlCC, p. 23. Welsh *try-*, Breton *tri-*: see LHEB, p. 659; HPB, p. 146; FG, p. 252; Williams, Beginnings, pp. 187–8 and n. 28. As made clear by Jackson, the two meanings are of different etymologies; but they had fallen together in form by the Old Cornish period. Compare names such as Welsh *Tryfan* 'very-pointed' (Williams, ELl, pp. 14f.; Beginnings, p. 188); *trylec bechan*, LLD, p. 156 (= Trelleck, Mon.); Breton Tridour (Tanguy, p. 85); Gaulish *trinanto* 'tres ualles' (DAG, p. 577; and further examples of *tri-*, p. 588; cf. also Old Welsh *nant trineint* 'three-stream valley', LLD, p. 196). With Truro (*Tryuereu*, 12th–13th cents.) compare the Gaulish tribal name *Treveri*

trig

(Vendryes, ÉC 1 (1936), 374f.; DAG, pp. 753f.; LEIA, p. T137), though the ending in the Cornish name is obscure; if the comparison is valid, then the *Try-* in Truro is 'through, very', not 'triple', since that would have been *Tri-*, not *Tre-*, in Gaulish. So also in Trevan, if that is equivalent to the Welsh Tryfan; but in Trigg the meaning is 'triple'.

The prefix occurs in: Trigg (+ ***cor***); Truro (+ ?); Trevan Point, coast (+ **ban** or **men**); ?The Tribbens, coast (+ ***pans**?).

trig (MnCo.) 'ebb-tide', Lh. 136c glossing *maris recessus*; *maoz dho trig* 'go to the strand', Pryce, p. Ff2r. The word is common in West Cornish dialect, especially in *trig-meat* 'shellfish collected at low tide' (Courtney and Couch, p. 60; Nance, Sea-words, p. 166). The *-g* is peculiar; one would not expect any consonant from the reflexes in other Celtic languages: Welsh *trai* (cf. CA, p. 118), Breton *tre* (cf. GMB, p. 712), Old Breton *tre* (FD, p. 318); Old Irish *tráig* (LEIA, pp. T123f.); compare Old Cornish **trait**, above. Note that the *-g* (= /g/) has also behaved oddly in becoming /dʒ/ in one of the names: the name means 'low-tide cave', and may contain a verbal noun (cf. Welsh *treio* 'to ebb'?).

    A. ?Trigg Rocks, coast
    C2. ?Hugga-Dridgee, coast (+ **googoo**)

trygva 'dwelling-place', OM 951; CW 295, etc. Welsh *trigfa*, 'abode': a compound of the root seen in *tryge* 'dwell' (OM 317, etc.) and ***ma** 'place'. Though well-attested in the language, its presence in Cornish place-names is uncertain, for lack of conclusive early forms. Most places with names starting *Trega-* contain **tre**+personal name (see examples under **tre** B2), but one or two cannot be so explained. Even they do not have forms which show **trygva**, but lack the *-v-* from their earliest records, e.g. *Tregamor* 1372: if it is present at all, it must have lost its *-v-* early on, perhaps by analogy with other names in *Trega-*, etc.

    A. ?Trigva (or a modern name?)
    B2. ?Tregamere (+ **meur**)

***tryonenn** (?) 'land left fallow', equivalent to Breton *trionenn* 'partie d'un champ laissée en jachère'. The word may occur in various field-names in West Cornwall; see Nance in Doble, Crowan, p. 29.

    A. ?Tringham, ?Stringham, etc., flds.
    B2. ?Tronghan Flow, fld. (+ ?)

***trokya** 'dip, wash; to full (cloth)'; cf. Lh. 77a, *dho hambrokkya* glossing *lavo* 'wash' (a misprint? perhaps for ***hamdrokkya**); 62a *trýkkiar* glossing *fullo* 'a fuller'. The form is dictated by early forms of the place-names; the word may be the same as Welsh *trochi* 'immerse, dip', or possibly derived from English *tuck* (cf. 'tucking-mill'), with non-original *r* inserted. See Henderson, Essays, pp. 204-8, on Cornish tucking-mills. There seems to have been some confusion with another word, *drwsher* 'a thresher' (Lh. 33c), *drwshier* glossing *tritor* 'a grinder, a thresher of corn' (Lh. 167a), derived from English (cf. dialect *drush* 'thresh', Courtney and Couch, pp. 18 and 84; but note Old Breton *drosion* 'threshings', FD, pp. 152f.): the mills named were all fulling-mills, and their early forms show *-druckya*, etc., but the modern forms and pronunciations show *-druchia*, etc., /druːʃə/, as if the second word were *drwshier* instead.

C2. Vellyndruchia (2 exx.), Velandrucia, ?Valley Truckle (all + **melin**[1])

trogh 'broken', OM 298, PC 2687, etc.; Welsh *trwch* (Old Welsh *truch* glossed *truncatus*, in Barmbtruch glossed *truncate barbe*, EWGT, p. 16), Breton *troc'h* (noun) 'coupe'. Cf. **trech**. With *Carntrough* compare Welsh *Taldrogh* 1334 (Surv.Denb., p. 54), 'broken brow'.

B1. A compound *bar-drogh (+ **bar**) 'summit-broken' (showing a similar usage to the early Welsh *Barmbtruch*) seems to be used (as C2) in Levardro (+ ***lann**) and Trevaddra (+ **tre**), though the meaning is obscure.

C2. *Croustroch* (+ **crous**); ?*Carntrough* (+ **carn**)

tron (m.) 'nose' (Pryce), equivalent to Welsh *trwyn*; Old Cornish *trein*, Voc. 29 glossing *nasus*, seems to be a mistake for *troin. Compare the use of Old English *nos* to mean 'promontory' (Elements, II, 52).

C1. See ***tron-gos**.

C2. ?Chytroon (+ **chy**); ?flds. *Park-an-Troan*, *Park Trone*, *Park-an-Trawn*

***tron-gos** 'nose-wood' (i.e. a wood on a spur of land), a compound of **tron** + **cos**. Compare Welsh Coed-y-trwyn, Mer., and Trunch, Nrf., suggested by Ekwall as consisting of the same compound (DEPN, s.v.); also various Welsh compounds containing *trwyn-*, such as *trwyndrist* (BBCS 2, 318) and *trwyndwn* (BBCS 3, 38).

C2. Restronguet (+ ***ros**); *Halldrunkard* (+ **hal**)

tros (m.) 'foot', PC 98, etc.; Old Cornish *truit*, Voc. 95 glossing *pes*. Welsh *troed*, Breton *troad*.

B2. *Troose-mehall* (+Saint; also called *St Michaell's Foot*)
C2. *Chytroose* (+**chy**)

tu (m.) 'side', MC 163b; etc. Welsh, Breton *tu*. Compare **tenewen**.
B2. Tucoyse (2 exx.), Trecoose (all +**cos**)

tûban 'mound, bank', Lh. 42a glossing *agger*; Welsh *tomen* (ELl, pp. 79f.). Note also West Cornish dialect *tubban* 'a turf' (Courtney and Couch, p. 61). Confusable with **tum**.
A. Tubbon; Tobban Cove; *Toban*, fld.

tum 'warm', BM 1778; Old Cornish *toim*, Voc. 856 glossing *calidam*; Modern Cornish *twbm* (Lh. 45c), *tubm* (Pryce). Welsh *twym*, Breton *tomm*. In its Modern Cornish form, *tubm*, the element is confusable with **tûban**. Some forms show a final vowel after the *-bm*, as also in *Pedny-* < *Pedn-*. All the examples are uncertain, for lack of early forms.
C2. Park Tuban, *Park Tobma*, flds.; *Cairn Tobna*, coast (+**carn**); ?Beagletodn (+**begel**)

tus (coll.) 'people, men', OM 1438, etc.; used as the suppletive plural of **den**. Welsh, Breton *tud*.
A. ?Tuzzy Muzzy Croft, fld. (+**ha**?, **mowes** pl.?, Eng.)
C2. *Wheal Dees Gentle* (+**wheyl**, (**an**), **gentyl**); Bal Dees, *Ballandees*, mines (both +**bal**, (**an**))

*****tusk** 'moss', equivalent to the stem seen in Breton *touskan* 'mousse terrestre'. The Breton word is probably (despite HPB, p. 818, n. 2) a variant of *trousk* 'scab, crust', Cornish *trosgan* 'scab' (Lhuyd MS, glossed 'krachen, scabies'): cf. GMB, pp. 727 and 748. Confusion with *****tescow** 'sheaves' is possible. The two names seem originally to have contained a derivative **tuske* or **tuska*, indicated by the early spellings. Note also Cumbric Poltross Burn (*Poltros* 1169, *Pultrosk* c. 1280: PNCmb., I, 23), and the discussion by Quentel, Ogam 7 (1955), 83f.
C2. deriv. ?Poltisco, ?Poltesco (both +**pol**; cf. *****-a**)

U

ugens 'twenty'. Welsh *ugain*(*t*), Breton *ugent*. Pryce (p. Cc4v) mistranslated the place-name Gaverigan (**gover**+**guyn**) as 'twenty goats'.
C1. ?Eggens-warra, fld. (+**erw**?)

ugh 'above', always *a vgh*, with *a-* prefixed, in the texts (e.g. OM 1136; PC 2808; MC 13d and 237a), but without the prefix in place-names. Middle Welsh *uch* (GMW, p. 212), Middle Breton *a uch* (GIB); compare phrases such as early Welsh *penn ir inis ad huchti* 'the end of the island above it (fem.)' (LLD, p. 242); Middle Breton *Auch an prat bihan* 'above the little meadow' (Chrest., p. 236). The preposition occurs in field-names such as *Ewandre, Park Eughandrea* (both + **an, tre**); *Dorughdre* (+ **dor, tre**).

ughel 'high', normally *vghell, vhel* in Middle Cornish; but the metathesised form *huhel*, etc., is nearly as common (Voc. *huwel*, three times; *huhel*, OM 509, 1368; PC 93, 125; RD 522; Modern Cornish *hual* 'above', Bosons, p. 52); and it is standardised in the compound ***huel-gos**: see VKG, I, 413; HPB, pp. 552–4; FG, pp. 152f. Welsh *uchel*, Breton *uhel, huel*. The superlative *u(g)hella* occurs at PC 2189 and CW 39. As C1, compare other instances where *vghel* precedes its noun, such as *vghel gry* 'a high cry' (MC 207d), *vhel arluth* 'a high lord' (BM 2207); Middle Welsh *uchelloc, uchellann* (Ll.Hendr. 44.14f.); in Cumbric place-names, CPNS, pp. 209 and 378.

C1. See the compound ***huel-gos**; a compound *hul-wals (+ **gols** rather than **gwels**), 'high-clump', is used (as C2) in Manuels (+ **men**).

C2. Men Hewel, coast (+ **men**); Carrickowel Point (+ **karrek**); Carnowall (+ **carn**); *Hengyvghall* (+ **hensy**); spv. Adgewella (+ ***aswy**), ?Chyvelah (+ **chy**).

ula 'owl', Lh. 99c and 241b; Old Cornish *hule*, Voc. 524 glossing *noctualis stix* (read *noctua vel strix*); borrowed from old English *ule*.

C2. Trefula (+ **tre**); *Coffenoola*, mine (+ ***coffen**).

usion (pl.) 'chaff', Voc. 922 glossing *palea* (also *eusion* in margin). Welsh *us* (pl.); Breton *uzien* (cf. GMB, p. 734); Old Breton *eusiniou* (FD, p. 169). Cornish dialect *ishan* 'dust from winnowing' (Courtney and Couch, p. 30): Modern Cornish *ision*, Lh. 111b.

C2. Vellenewson (+ **melin**[1]); an aj. *usek might occur (cf. the Welsh form) in Port Isaac (+ **porth**)

***ussa** 'topmost, outermost' (?), possibly the superlative of a preposition (used adjectivally) *us 'above'; compare Middle Breton *a vz* 'above', Nonne, line 889 (RC 8 (1887), 408), Modern Breton *a us*; or a variant superlative of **ughel**: in either case the origin, like that of the Breton preposition, would be uncertain, but probably due to infection of original *uch, *uchav, by its opposite *is, *isav. The Breton form is thus explained by Ernault, RC 22 (1901), 377,

and 31 (1910), 235 n. 6, and Vendryes, LEIA, p. O32; cf. HPB, p. 553, on Vannetais *ihuel*, and FD, p. 326. The meaning 'outermost' would be parallel to the Welsh meanings of *is* and *uwch*, 'hither' and 'yonder': M. Richards, 'The significance of *is* and *uwch* in Welsh commote and cantref names', WHR 2 (1963-64), 9-18; and compare Eusa (Ushant) in Brittany, from *Uxisama* 'outermost' (Top.Bret., p. 49 and refs.)?
   A. ?Essa
   C2. ?Trevessa in St Erth (+**tre**)

V

vooga (f.) 'cave', Pryce (s.v. *voog* 'smoke'); this word would originally have been *_mooga_, equivalent to Breton *mougev* (f.), but the lenited form evidently became standardised. It is distinct from ***fow** and **googoo**, both also meaning 'cave', but there may well have been mutual influence between the forms of the three words: see the references to discussion under those two words. The initial *f-* in the learned term *fogou* 'souterrain' is presumably due to hyper-correct unvoicing of the initial *v-*. Compare Breton *bougeo* 'petite anse, roche creuse' (RC 27 (1906), 138f.), with interchange of *b-/m-*.
   A. Vugga Cove, Fuggoe Lane (both +English); The Fuggoes, The Vugger, etc., flds.
   B2. *Vougo Lion*, fld. (+***legh** pl.?); Vwgha Hayle, coast (+***heyl**)

W

war 'upon'; Welsh *ar*, Breton *war*; Old Welsh and Breton *guar*. See D. S. Evans, BBCS 17 (1956–58), 1–10, for the use of the preposition in Middle Cornish. Sometimes the expected lenition following **war** fails to appear, as also in the language (e.g. *war ton*, RD 2281 and BM 3505): note the form *Trewarvena alias Trewarmena*, St Mawgan in Pydar, 1650. The preposition is often lost in the modern forms of the names, like other prepositions and like the definite article. It occurs in the following names, among others:
   Tywarnhayle (+**ti**, **an**, ***heyl**); Tywardreath (+**ti**, **trait**); Trethevey in St Mabyn (+**ti**, ***dewy**); *Trewelesikwarheil* (+p.n., ***heyl**); Chyvarloe (+**chy**, ***loch**); Chyverton, Chyvarton,

Chiverton (3 exx.), Chytan in St Austell, Chytane, Chytodden in Breage and Camborne (all +**chy**, **ton**); Trewarmenna, Trewarveneth, Trevenna in St Neot and St Mawgan in Pydar, Trevena in Tintagel, Trevenner (all +**tre**, **meneth**); Chyreen (+**chy**, (**an**), ***run**: 2 exx.); Trenhayle (+**tre**, **an**, ***heyl**); Chywarmeneth (+**chy**, **meneth**); Trenuggo (+**tre**, **an**, **googoo**)

wast 'idle, unused', RD 2155; borrowed from English, like Welsh *wast* (EEW, p. 86); cf. Middle Breton *goast*, borrowed from French (Piette, p. 126). The element is confusable with **west**.

C2. Tarewaste, *Trewaste* (both +**tyr**?)

west 'west', PC 2744; *weyst*, BM 784; borrowed from English. The element is confusable with **wast**, and some of the names may contain that instead.

C2. Balwest (+**bal**); Croft West (+***croft**); Gollwest, Goldwest, flds. (+**guel**); Park West, etc., flds.

worth 'at, against'; Welsh *wrth*, Breton *ouzh*. The preposition may occur in one or two names such as Carevick (*Crowarthevick*), Carines (*Crowarthenys*) (= p.n. *Crou*+**worth**+? and **enys**); but most place-names which appear to contain it actually contain words or personal names starting with the prefix ***gor-**; note, however, Trethauke (*Trewarthavek* 1311), which looks like **tre**+**worth**+**haf** aj. (or **tre**+**guartha** aj. ?).

wuir, see under Wh.

Wh

whe 'six', PC 351; Welsh *chwe*, Breton *c'hwec'h*.

C1. *An Whea Mogovan*, fld. (+**an**, **mẏdzhovan**)

wheyl (m.) 'work, working', OM 1226, 2428, etc.; note *meyn wheyl* 'building-stones', OM 2318; *whyl cref* 'hard work', OM 1490; the meaning in place-names is 'tin-working', at first a place of tin-streaming (as opposed to a ***coffen** or dug work on a lode). Note *weyll* glossed *opus* in A.D. 1507: 'tolnetum operis nostri, quod cornubice vocatur *toll weyll*' (Anc.Deeds, IV, A.9025). The word is presumed to be cognate with Welsh *chwilio* 'seek' (*chwil* 'a search'), Breton *c'hwiliañ* 'chercher, fouiller', or (if it is not the same) with Welsh *chwil*, *chwŷl*, *chwêl* 'a turning' (cf. GPC, s.v. *gorchwyl*). Its first appearance in place-names is in the late 15th or early 16th century, in Stannary Court Rolls. Of the hundreds of names taking the form *Wheal X*, by no means all represent a Cornish-language usage, since the element continued in use during the 19th and 20th centuries for forming the names of mines, qualified by an English word or name (Wheal Fortune, Wheal Busy, Wheal Kitty, etc.); this practice spread as far east as Devon. Like other minor-name elements (cf. **park**, ***stret**), **wheyl** is liable to be followed by an incorrect definite article, as in *Whele an Phelp*, *Wheal an Coats* (both + pers.). As a Cornish element, **wheyl** occurs predominantly in the hundreds of Pydar, Penwith and Kerrier; in Blackmore Stannary (hundred of Powder) the corresponding term is often **guyth**[2]. In 17th-century Breage, **wheyl** and ***bounds** were occasionally interchanged with each other. All of the following names are those of mines.

B2. Wheal Reath, Wheal Reeth (both +**ruth**); *Wheal an Clay* (+**an**, **cleath**); *Wheale an Dowthick* (+**an**, **dewthek**); Wheal Noweth (+**nowyth**); Wheal Owles (+**als**); *Wheal Dees Gentle* (+**an**, **tus**, **gentyl**); *Wheal Sterren* (+**steren**); *While an Attol* (+**an**, **atal**); *Huel Goffen* (+ ***coffen**); *Hwelan Vrân* (+**an**, **bran**); *Hwelan Tshei* (+**an**, **chy**); *Wheal Zawson* (+**Zowzon**); *Wheal an Cullieck* (+**an**, **kullyek**)

whelth 'tale, story', MC 109c (either an error for, or metathesised from, ***whethel**: if the latter, cf. Middle Breton *ararz*, *comparz*, etc., for *arazr*, *compazr*, etc., HPB, p. 491n.); *hwitel*, Lh. 6b; dialect *w(h)iddle* 'whim, nonsensical idea', Courtney and Couch, p. 64; pl. *whethlow*, OM 466, PC 369, etc. Welsh *chwedl*, cf. Middle Breton (*qe*-)-*hezl*, etc. (GMB, p. 527; GIB, s.vv. *keal*, *keloù*). With the manorial suffix compare the motto of the Carminow family, *Cala rag whetlow* 'a straw for tales' (Pryce, s.v. *whetlow*). Alternatively the suffix could theoretically be a personal name, equivalent to Welsh *Hwetheleu*, *Wetheleu*, *Hwiteloue* (in A.D. 1292: BBCS 13 (1948–50), 216, 218 and 225); but since Celtic personal names are rare in Cornwall after the Norman Conquest, that is unlikely.

D. pl. ?*Helygy Whethlowe* (= Halligey in St Martin in Meneage)

***wheth** 'a blow, a puff', cf. *hwetha* 'to blow', Lh. 157c and 245a; *whath*, CW 2299. Welsh *chwyth*, Breton *c'hwezh*. The name may be a translation of English 'blow-hole'.

C2. *Tollan Wheath*, coast (+**tol**, **an**)

***whevrer** 'lively' (?), a possible stream-name from a root ***whevr**- (equivalent to Welsh *chwefr*-), plus an unknown suffix -*er*. The Welsh word is seen in the river-names *guefrduur* (LLD, p. 159; cf. EANC, p. 95), *alt guhebric* (Chad 7), and Chwefri (EANC, p. 137); also in Welsh *chwefrig* 'lively' and the personal name *guebric/huefric* (LLD, pp. 257f.). It may be connected with Welsh *gwefr* 'amber'. The Cornish name was thought to contain *hwevral* 'February' (Lh. 31c and 59a: Welsh *Chwefror, Chwefrol*, Breton *C'hwevrer*), for its early forms are exactly right for this word, and even show a 17th-century change from *-r* to *-l*, just as in the history of that word in Cornish and Welsh (though not in Breton; but cf. HPB, pp. 813f.). However, 'February' would be semantically difficult, and a stream-name with folk-etymology seems more likely in the context of the Welsh names.

C2. Polwheveral (+**pol**)

hwilen (f.; OCo.) 'beetle', Voc. 535 glossing *scarabeus*; Welsh *chwil, chwilen* (both sg.), Breton *c'hwil*, sg. (cf. GMB, p. 328; Tanguy, p. 129). See M. Richards, 'Chwil, Chwiler, Chwilog, etc.', BBCS 19 (1960–62), 91–5. The River Wheelock, Che., should be the same as Welsh Chwilog, but that causes problems of phonology, and Welsh *chwylog or *chwelog 'twisting' has been suggested instead (Ekwall, ERN, pp. 455f. and DEPN, s.v.; PNChe., I, 38f.; *contra*, M. Richards, 'The River Wheelock', Trans. Historic Soc. of Lancs. and Cheshire, 111 (1959), 199f.). The Whyle, Hre., from *Huilech* 1086 (PNHre., pp. 205f.), shows a more regular development of the vowel. Among the Cornish names, Polwhele shows spellings in *-e-* as well as *-y-*, though the latter predominates; similarly in the uncertain Trewhela and Trewhella, earlier spellings in *-y-* are replaced by ones in *-e-* in the 14th or 15th centuries. Confusion with a hypothetical *whel or *whyl (equivalent to Welsh *chwêl, chwŷl* 'turn', and thus perhaps originally the same as **wheyl** 'working') is therefore possible, though the plural of **hwilen** is nevertheless more likely in all names. In the case of Polwhele and Tretoil, it could be that the element was originally a simplex stream-name, like the Welsh 'rivulum qui vocatur *Chwyl*' 1279 (Welsh Assize, p. 301).

C2. pl. Polwhele (+**pol**: cf. Welsh Pwll-y-chwil, Crd., BBCS 19, 92); Tretoil (+**rid**); *Carnwhyl* (+**carn**); a derivative (+ ***-a**?) (meaning 'place of beetles'?) seems to occur in Trewhela, Trewhella (both +**tre**).

wuir (f.) 'sister', Voc. 139 glossing *soror*; *hoer*, CW 1332 and 1338. Welsh *chwaer*, Breton *c'hoar* (Old Breton *guoer*, glossed *soror*, FD, p. 195). (The Voc. form was emended to *huir* by Norris, II, 385f., followed by other authorities; but this is unnecessary: for Voc. *w-* = /χw/, cf. nos. 260 *wibonoul* and 798 *wibanor*.) For the meaning of the names, compare Bamham (***bod**+**mam**), and Middle Breton *Kaerhoer* (Bégard, p. 203).

C2. Lawhyre (+ ***lann**); Besore, Bosoar (both + ***bod**)

INDEX OF CORNISH PLACE-NAMES CITED

This index gives: the place-name; the ancient parish where it is located (or its other status); and those element(s) under which the place-name is cited. (Note that it does *not* necessarily give all the elements in a name, therefore.) When a name is cited under more than one element, they are given in the order in which they occur in that name.

Abbott's Hendra (Davidstow) ***hendre**
Adga Bullocke Bounds, mine 1725 (Lelant) ***aswy**
Adgewella (Camborne) ***aswy, ughel** spv.
Adgyan Frank, fld. 1649 (Constantine) ***aswy, Frank**
Adjaporth, fld. 1614 (Madron) ***aswy, porth**
Adjawinjack (Ruan Major) ***aswy, guyns** aj.
Agahave, fld. (Ruan Major) ***aswy**
Aire Point (St Just in Penwith) ***arð**
Aja-Bullocke, fld. (Perranuthnoe) ***aswy**
Aja-Gai, fld. (Sennen) ***aswy**
Ajareeth 1665 (St Hilary) ***aswy**
Allet (Kenwyn) ***aled**
Alsia (St Buryan) **?als, ?*-a**
Als Roshusyon c. 1214 (St Hilary) **als**
Altarnun (= parish) **alter**
Amalebra (Towednack) ***amal, ?*bre**
Amalveor (Towednack) ***amal, meur**
Amalwhidden (Towednack) ***amal, guyn**
Amble (St Kew) ***amal**
Ammaleglos 1346 (St Kew) **eglos**
Ammalgres 1346 (St Kew) **cres**
an Cogh'n bras, mine 1503 (Breage) ***coffen, bras**
Andennis (St Just in Roseland) ***dynas**
Angarrack (Phillack) **karrek**
Angear (Gwennap) ***ker**
Angew (Gwithian) ***kew**
Angrouse (Mullion) **crous**
Anhay (Gunwalloe; St Keverne) ***hay**
Anhell 1634 (Truro) **hel**
An Tyreth Vhel 1504 (= district) **tyreth**
Anvoaze (Grade) ***bod**
An Whea Mogovan, fld. 1670 (Madron) **whe, mỳdzhovan**
Ara Gayan, fld. (St Ives) **erw, ?gahen**
Ardensaweth (St Levan) **arghans, ?*aweth**
Ardevora (Philleigh) ***ar, *devr-** pl.
Ardevora Veor (Philleigh) **meur**

INDEX OF CORNISH PLACE-NAMES

Argal (Budock) *argel
Argeldu 1345 (Budock) du
Argelwen 1284 (Budock) guyn
Arrallas (St Enoder) arghans, *lys
Arra Venton, fld. (Lelant) erw
Arrowan (St Keverne) ?*ar, ?auon
Arrowe Jellard, fld. 1559 (Paul) erw
Arthia 1481 (St Ives) *arð
Arwenack (Budock) *ar
Assa Govranckowe, fld. 1580 (Gwennap) *aswy, keverang pl.
Assawine, fld. 1582 (Illogan?) *aswy
August Rock (Mawnan) ?ogas
Avarack, The, coast (St Just in Penwith) ?*havrek
Averack, fld. (Paul) *havrek
Awen-Tregare 1698 (Truro) auon
Ayr (St Ives) *arð

Bahavella (St Ives) bagh
Bahow (St Keverne) bagh pl.
Bal Dees, mine (Wendron) bal, tus
Bal dew 1593 (St Ewe?) bal
Baldhu (St Agnes; Kea; Ludgvan) bal, du
Ballaminers (Little Petherick) melin[1] pl., meneth
Ballamoone 1695 (Perranarworthal) bal, ?*mon[2]
Ballandees, mine 1708 (St Agnes) bal, tus
Balleswidden (St Just in Penwith) bal
Balmynheer (Wendron) bal, *men hyr
Balnoon (Lelant) bal, goon
Balrose (Camborne) bal, ?*ros
Balwest (Germoe) bal, west
Bamham (Lawhitton) *bod, mam
Bannel, flds. (Paul) banathel
Banns (St Agnes; St Buryan) *pans
Barbican (St Martin by Looe) bar, byghan
Barbolingey (St Austell) bar, *melyn-jy
Bargoes (Luxulyan) cruc, ?bargos
Bargudul 1330 (Kenwyn?) bar
Bargus (Perranarworthal) bar, cos
Barlendew (Blisland) bar, lyn, du
Barlewanna (Cury) bron, lowen sb.
Barlowennath (St Hilary) bron, lowen sb.
Barncoose (Illogan) bron, cos
Barnewoon (Phillack) bar, goon
Barnoon (St Ives) bar, goon
Barrett's Zawn (St Teath) sawn
Barrimaylor (St Martin in Meneage) *merther
Barteliver (Probus) bagh
Bartinney (St Just in Penwith) *bre

244

INDEX OF CORNISH PLACE-NAMES

Bawdoe (St Winnow) *bod pl.
Beagle, rock (Perranzabuloe) bugel
Beagles Point (St Keverne) bugel
Beagletodn (Towednack) begel, ?tum
Beaulugooth 1314 (St Hilary) coth
Bedrugga (St Columb Minor) ?*gruk pl.
Bedwithiel (Blisland) ?gwyth¹ aj.
Begeledniall, fld. 1686 (St Just in Penwith) begel, *enyal
Beggal, The, fld. (St Keverne) begel
Bellowal (Paul) lowen
Bell Vean (Gwennap) ?pell
Bell Veor (Gwennap) ?pell, meur
Benallack (St Enoder; Probus) banathel aj.
Benbole (St Kew) beth, *bolgh
Benhurden (Goran) ?horþ
Bephillick (Duloe) *bod
Berrangoose (Probus) bron, cos
Bervanjack (Manaccan) bar
Besore (Kenwyn) *bod, wuir
Bessy Benath (Veryan) anneth
Bethanel (Crowan) budin, hel
Bethkile 1490 (Paul) beth, ?*kyl
beð cywrc 960 (S. 684) beth
Biggal of Gorregan, rock (Scilly) bugel
Biggal of Mincarlo, rock (Scilly) bugel
Biscovellet (St Austell) *bod
Bissoe (Perranarworthal) bedewen pl.
Bithen Lidgeo, fld. 1688 (St Levan) budin, ?les pl.
Blable (St Issey) bleit, pol
Blankednick (Perranarworthal) ?blyn
Bleu Bridge (Gulval) plu
Blew Stone 1696 (Lanlivery; Madron) plu
Blouth Point (Veryan) ?blogh
Bluestone (St Erme; Illogan) plu
Bo Cowloe, rock (Sennen) ?bogh
Bodbrane (Duloe) bran
Bodellan (St Levan) ?*elen
Bodelva (St Blazey) ?elaw pl.
Boderlogan (Wendron) *bod
Boderwennack (Wendron) *bod, ?*meynek
Bodgara (Liskeard) ?garan
Bodieve (Egloshayle) *bod, *yuf
Bodiggo (Luxulyan) *bod
Bodiniel (Bodmin) *-yel
Bodinnick (St Stephen in Brannel) *bod
Bodinnick (Lanteglos by Fowey) *dyn aj.
Bodinnick (St Tudy) ?*dyn aj.
Bodmin (= parish) *bod, *meneghy

INDEX OF CORNISH PLACE-NAMES

Bodriancres 1372 (St Clement) **cres**
Bodrivial (Crowan) ***-yel**
Bodryanpella 1485 (St Clement) **pell** spv.
Bodulgate (Lanteglos by Camelford) ***bod, *huel-gos**
Bodwen (Helland) **guan**
Bodwen (Lanlivery) **guyn**
Bodyer 1448 (Launceston) **yar** pl.
Bogee (St Ervan) ***bod, *yuf**
Bogullas 1620 (St Just in Roseland) ?**bugel**, ?***lys**
Bohago (Cuby) **bagh**
Bohelland (St Gluvias) ***lann**
Bohetherick (St Dominick) ***bod**
Bohortha (St Anthony in Roseland) ***buorth** pl.
Bojea (St Austell) ***bod, *yuf**
Bojil (Germoe) ***kel**
Bokelly (St Kew) ***bod, kelli**
Bolankan (St Buryan; Crowan) ?***lonk** dimin.
Bolatherick (St Breward) **pol**
Boleigh (St Buryan) ***legh**
Bolenna (Towednack) ?***leyn** pl.
Bolenowe (Camborne) ?***leyn** pl.
Bolingey (St Mawgan in Pydar; Perranzabuloe) ***melyn-jy**
Bolinnow Cairn 1835 (Camborne) **carn**
Bollogas (Paul) **logaz**
Bolster (St Agnes) ***both, lester**
Bonallack (Constantine) **banathel** aj.
Bonnal (St Martin in Meneage) ?**ben**, ?***gwal**
Bonyalva (St Germans) **banathel, *ma**
Borah (St Buryan) ***bod, gruah**
Boreise (Constantine) **rid**
Borlase (St Wenn) **bor, glas**
Borlasevath (St Columb Major) **margh**
Bosahan (St Anthony in Meneage; Constantine) ***seghan**
Bosanath (Mawnan) ?**anneth**
Boscarn (St Buryan) ***bod, carn**
Boscarne (Bodmin) ***bod, carn**
Boscarnon (St Keverne) ***kernan**
Boscathnoe (Madron) **beth**
Boscawen (St Mawgan in Meneage) ***bod, scawen**
Boscawen-Noon (St Buryan) **goon**
Boscawen Rôs (St Buryan) ***ros**
Boscean (St Just in Penwith) ***seghan**
Boscreeg (Gulval) ***bod, cruc**
Boscregan (St Just in Penwith) **crogen** pl./**cruc** dimin.
Boscrege (Germoe) ***bod, cruc**
Boscudden (St Erth) **cudin**
Bosehan (St Buryan) ***seghan**
Bosence (St Erth) **guyns**

INDEX OF CORNISH PLACE-NAMES

Bosent (St Pinnock) ?**sans** pl.
Bosfranken (St Buryan) ***bod, Frank** pl.
Bosigran (Zennor) ***bod,** ?**garan**
Bosithow (St Mewan) **idhio**
Boskear (Bodmin) ***bod,** ***ker**
Boskell (St Austell) ***kel**
Boskenna (St Buryan) ***kenow**
Boskerris (Lelant) ?***kew-rys**
Bosliven (St Buryan) ***bod**
Bosloggas (St Just in Roseland) **logaz**
Boslowick (Budock) **-yk**
Bosneives (Withiel) ***bod,** ***kyniaf-vod**
Bosoar (Sithney) ***bod, wuir**
Bosoljack (Gulval) **houl** aj.
Bosorne (St Just in Penwith) **horn**
Bosoughan (Colan) ?**hoch** dimin.
Bosparva (Gwinear) **perveth**
Bospolvans (St Columb Major) ?***peulvan**
Bosporthennis (Zennor) ***bod, enys**
Bossava (Paul) ***bod**
Bostrase (St Hilary) ***stras**
Bostraze (St Just in Penwith) **pen,** ***stras**
Bosullow (Madron) ***bod**
Bosulval (Gulval) ***bod**
Bosuoylagh goiles 1313 (St Enoder) **goles**
Bosvathick (Constantine) ?**goth**[1] aj.
Bosvean (Kenwyn) **byghan**
Bosvine (Sennen) **men** pl.
Boswague (Veryan) **gvak**
Boswase (Ludgvan) ***bod, guas**
Boswednack (Zennor) **guyn**
Boswellick (St Allen) ***bod**
Boswens (Sancreed) **guyns**
Boswin (Wendron) **guyn**
Boswinger (Goran) ***bod**
Bosworgey (St Columb Major) ***bod**
Bosworlas (St Just in Penwith) ?***gor-,** ?**losc**
Boswyn (Camborne) **guyn**
Botallack (St Just in Penwith) ?**tal** aj.
Bothle 1350 (lost) ?***both,** ?**le**
Botrea (Sancreed) **trech**
Botreva (Ludgvan) ?**tre** pl.
Botternell (Linkinhorne) **dorn** aj.
Boughie, fld. (Gulval) **boudzhi**
Bounder-en-parson 1330 (Veryan) **bounder**
Bounder Hebwotham 1613 (Gwithian) **bounder, hep, ethom**
Bounder Vean (Camborne) **bounder**
Bounds Begall, mine 1726 (Wendron) ***bounds, begel**

INDEX OF CORNISH PLACE-NAMES

Bounds Brose, mine 1690 (Breage) ***bounds**
Bounds Coath, mine 1651 (Breage) ***bounds, coth**
Bowgey, The, fld. (Crowan) **boudzhi**
Bowgie (Paul) **boudzhi**
Bowgyheere (Ludgvan; St Mawgan in Meneage) **boudzhi, hyr**
Bowithick (Altarnun; Lanteglos by Camelford) **gwyth**[1] aj.
Bownds an Coskar, mine 1672 (Towednack) ***bounds, coscor**
Bozion (Egloshayle) **bron, *seghan**
Brane (Sancreed) **bran**
Brannel (St Stephen in Brannel) ***-yel**
Brass Teague, fld. (St Ives) **?pras, ?teg**
Brawn, The, rock (St Germans) **?bran**
Bray (Altarnun; Morval) ***bre**
Bray Hill (St Minver) ***bre**
Brea (Illogan; St Just in Penwith) ***bre**
Brechiek 1390 (Scilly) **?bregh** aj.
Breglos 1386 (St Keverne) ***bre**
Breney (Lanlivery) ***bren** pl.
Brevadnack (St Hilary) ***bre, *bannek**
Brew (Sennen) **brew**
Brewinney (Paul) ***bre**
Brighter (St Kew) ***brygh, tyr**
Brightor (Landrake) ***brygh, tyr**
Brill (Constantine) ***bre, *helgh**
Brimboyte (Liskeard) **bron, *poth**
Broncoys Goef 1520 (Gwennap) **gof**
Bronhiriard 1284 (Lanreath) **bron, *hyr-yarth**
Brownqueen (St Winnow) **?*keun**
Brownsue (St Minver) ***bren**
Brown Willy (St Breward) **bron, guennol** pl.
Brownwithan (St Columb Major) **gwyth**[1] sg.
Brunnion (Lelant) ***bren** pl.
Bryanick 1884 (St Agnes) ***bre, *bannek**
Bryher (Scilly) ***bre, -yer**
Brynn (Withiel) ***bren**
Bucca's Lane (Camborne) **bucka**
Burghgear (Creed) ***bod, *ker** pl.
Burgois (St Issey) **?bar, cos**
Burgotha (St Stephen in Brannel) **bar, goth**[1] pl.
Burlerrow (St Mabyn) **bor**
Burlorne (Egloshayle) **?elaw** sg.
Burlorne Eglos (St Breock) **eglos**
Burlorne Pillow (Egloshayle) **?pell** spv.
Burn (St Winnow) ***bren**
Burncoose (Gwennap; St Mawgan in Meneage) **bron, cos**
Burngullow (St Mewan) **bron, golow**
Burniere (Egloshayle) ***bren, er**[1]/**hyr**
Burnoon (St Mawgan in Meneage) **bron, goon**

INDEX OF CORNISH PLACE-NAMES

Burnow (Cury) **bron** pl.
Burnuick (St Mawgan in Meneage) **bron**, ***ewyk**
Burnwithen (Gwennap) **bron, gwyth**[1] sg.
Burras (Wendron) **?ber**, ***red/rid**
Burthy (St Enoder) ***bryth** deriv.
Bushornes (Camborne) **horn**
Bussow (Towednack) ***bod** pl.
Busvannah (St Gluvias) **manach**
Busvargus (St Just in Penwith) **?bargos**
Buthan Glebe, fld. 1706 (Truro) **budin**
Buthan Green Grass, fld. 1706 (Truro) **budin**
Buthan Ten Acre, fld. 1706 (Truro) **budin**
Buthen en Hesk 1420 (Probus) **budin, heschen** pl.
Butter Villa (St Germans) ***bod**
Bydalder (St Neot) **myn, ?dall, dour**

Caca-Stull Zawn (Sithney) ***cagh**
Cadedno, coast (Scilly) **cudin** pl.
Cadgwith (Grade) ***caswyth**
Caduscott (Liskeard) **dres**
Caer Brân (Sancreed) **bran**
Caerhays (St Michael Caerhays) **?*ker**
caer lydan 969 (11th) (S. 770) ***ker, ledan**
Caerouporghel 1302 (St Minver) **porhel**
Caerskes 1363 (St Ives) ***ker, schus**
caer uureh 969 (11th) (S. 770) ***ker, gruah**
Cair (St Germans) ***ker**
Cairn Tobna, coast 1860 (Paul) **tum**
Cairo (Otterham) ***ker** pl.
Calamansack (Constantine) ***kyl**
Calendra (Veryan) **kelin, tre**
Calenick (Kea) **?*clun** aj.
Calerick (St Clement) ***luryk**
Callabarrett (Cardinham) **kelli, perveth**
Callenowth (Helland) **kelli, *cnow**
Callevan (Constantine) **kelli, ban**
Callymynaws, fld. (Crowan) **?*menawes**
Callyvais, fld. c. 1800 (Ruan Lanihorne) **kelli, mes**
Calvadnack (Wendron) **kal, *bannek**
Camborne (= parish) **cam, bron**
Cambrose (Illogan) **cam, ?*prys**
Camels (Veryan) **cam, als**
Camerance (St Just in Roseland) **cam, gweras**
Camperdenny, coast (Scilly) ***kemer**
Canaglaze (Altarnun) **glas**
Canavas (Mullion) **?*kyniaf-vod**
Candern Water 1649 (Constantine) **cam, dour**
Candor (Probus) **cam, dour**

INDEX OF CORNISH PLACE-NAMES

Candra (St Breward) ?**can**, ?**tre**
Cannalidgey (St Issey) ***canel**
Cannamanning (St Austell) **carn** aj., **amanen**
Cansford (Otterham) ?**can**
Cans Loggers, fld. (St Just in Penwith) ?**cans**, ?**logaz**
Cant (St Minver) ***cant**
Caragloose (Veryan) **karrek, los**
Carbaglet (Blisland) ***kyl, bagyl** aj.
Carbean (St Austell) **carn, byghan**
Carbilly (Blisland) ***ker**
Carbis (St Austell; St Hilary; Lelant; Roche; Stithians) ***car-bons**
Carblake (Cardinham) ***ker, bleit**
Carbouling (Kea) **corn,** ***bow-lann**
Carcarick (St Cleer) **karrek**
Carclaze (St Austell; St Kew) **cruc, glas**
Carclew (Mylor) ?**lyw**
Cardew (Trevalga; Warbstow) ***ker, du**
Cardinham (= parish) ***ker,** ***dynan**
Cardinham (Crowan) ***ker,** ***dynan**
Careck-an-googe 1660 (St Stephen in Brannel) **karrek,** ***cok**
Caregroyne, coast 1540 (Budock) **karrek, ruen**
Carek Veryasek, rock 1504 (Camborne) **karrek**
Carevick (Cubert) **krow,** ?**worth**
Carfos 1517 (Budock) ***ker, fos**
Cargelley (St Breward) ***ker, kelli**
Cargelly (Altarnun) ***ker, kelli**
Cargentle (St Stephen by Launceston) ?***ker, cuntell**/***cant** aj.
Cargey Gate (Ruan Major) ***car-jy**
Carglonnon (Duloe) ***kyl,** ***glynn** dimin.
Cargloth (St Germans) ?**cleath**
Cargodna, fld. (St Just in Penwith) **crak²**
Cargoll (Newlyn East) ?***kew**, ?***-yel**
Cargreen (Landulph) **karrek, ruen**
Cargurrel (Gerrans) **cruc, geler, (gorhel)**
Carharles (Golant) ?**horþ**
Carharrack (Gwennap) ***arð** aj.
Carharthen (Merther) ***arð** dimin.
Carines (Cubert) **krow,** ?**worth, enys**
Carkeel (St Stephen by Saltash) ?***kyl**
Carkeval (St Kew) ?***kevyl**
Carland (St Erme) ***cor-lann**
Carlannick (Philleigh) ?**lyn** aj.
Carleon (Morval) ***ker,** ***legh** pl.
Carlidna (St Mawgan in Meneage) **kelin** pl.
Carloggas (St Columb Major; Constantine) ***ker, logaz**
Carloggas (St Mawgan in Pydar; St Stephen in Brannel) **cruc, logaz**
Carloose (Creed) ?***ker**, ?**los**
Carluddon (St Austell) **cruc, ledan**

250

INDEX OF CORNISH PLACE-NAMES

Carlumb (St Minver) *lom
Carlyon (Kea; St Minver) *ker, *legh pl.
Carmar (St Kew) keyn, margh
Carmeall-Ball 1584 (St Just in Penwith) bal
Carmouth (St Breward) carn, mogh
Carnantel, fld. (St Keverne) ?antell
Carnanton (St Mawgan in Pydar) carn
Carnaquidden (Gulval) *kernyk, guyn
Carnarthen (Illogan) carn
Carn Barges, coast (St Buryan; St Levan; Sennen) carn, bargos
Carn Bargus, coast (Zennor) carn, bargos
Carn Barra, coast (St Levan) carn, bara
Carn Base, coast (Paul) *bas
Carn Bean, coast (Sennen) carn, byghan
Carn Boel, coast (St Levan) ?*moyl
Carn Boscawen, coast (St Buryan) carn
Carn Brâs, coast (Sennen) carn, bras
Carn Brea (Illogan) *bre
Carn Camborne (Camborne) carn
Carn Cheer, coast (Sennen) carn, cheer
Carn Creagle, rock (St Just in Penwith) carn, cruc dimin.
Carncrees (Stithians) carn, cres
Carn Creis, coast (St Just in Penwith; Sennen) carn, cres
Carn Du, coast (St Just in Penwith) carn, du
Carn-du, coast (Paul) carn, du
Carn-du Rocks (St Keverne) carn, du
Carne (several) carn
Carneadon (Launceston) carn, eithin
Carne Biskey 1695 (Altarnun) carn
Carnegga (St Dennis) *kernyk pl.
Carnegie (Minster) *keun aj. pl.
Carneglos (Altarnun) carn, eglos
Carnelloe (St Ives; Zennor) corn deriv./carn deriv.
Carnemough (Camborne) carn, mogh
Carn Enys, coast (Sennen) enys
Carn Epscoppe 1454 (St Ives) carn, epscop
Carnethick (Fowey) carn deriv.
Carnetrembethow 1410 (Lelant) carn
Carnevas (St Merryn) *neved
Carnewas (St Eval) ?*havos
Carn Fran-Kas, coast (Morvah) carn
Carn Galver, hill (Zennor) carn, guillua
Carn Glaze (St Just in Penwith) carn, glas
Carn Gloose, coast (St Just in Penwith) karrek, los
Carn Greeb, coast (Sennen) carn, krib
Carn Gwavas, coast (Paul) carn
Carn Hoar, rock (Sennen) ?horþ
Carnjewey Point (St Austell) ?sevi

251

INDEX OF CORNISH PLACE-NAMES

Carnkie (Illogan; Wendron) **carn, ky**
Carnkief (Perranzabuloe) ***ker, *kyf**
Carn Leskys, coast (St Just in Penwith) **carn, leskys**
Carn Lodgia, coast (Paul) **carn, ?les** pl.
Carnlussack (Camborne) **?*lus** aj.
Carnmaer (Stithians) **meur**
Carn Marth (Gwennap) **margh**
Carn Mellyn, coast (St Buryan; St Just in Penwith) **carn, melyn**[2]
Carn-Mên-Ellas, coast (Sennen) **carn, manal** vb. ppp.
Carnmenellis (Wendron) **carn, manal** vb. ppp.
Carn Minnis (Towednack) **?munys**
Carn Moyle, coast (Zennor) ***moyl**
Carno (St Keverne) **corn, hoch**
Carn Olva, coast (Sennen) **carn, guillua**
Carnon Crease (Feock) **cres**
Carnowall (Crowan) **carn, ughel**
Carn Pednathan, coast (Veryan) **?ethen**
Carn Peran 960 (S. 684) **carn**
Carn Praunter, rock (St Just in Penwith) **?pronter**
Carn Prince, coast (St Just in Penwith) **?pryns**
Carn Scathe, rock (St Levan) **carn, skath**
Carnsew (St Erth) **carn, du**
Carnsew (Mabe) **dev** pl.
Carn Sperm, coast (Sennen) **carn, ?spern**
Carnstabba (St Ives) **carn**
Carntiscoe (Lelant) **?*tescow**
Carn Towan, coast (Sennen) **towan**
Carntrough 1752 (Wendron) **?trogh**
Carn Venton Lês, coast (St Just in Penwith) **carn, fenten, ?glas**
Carn Veslan, coast (Zennor) **?mesclen**
Carnwhyl 1314 (St Keverne?) **carn, hwilen** pl.
Carnwidden (Stithians) **carn**
Carnwinnick (Probus) **guyn** deriv.
Carnyorth (St Just in Penwith) **carn, yorch**
Carplight (Manaccan) **cruc, bleit**
Carpuan (St Neot) **cos**
Carracawn (St Germans) **?*ker**
Carrack an deeber, rock 1613 (Zennor) **karrek, diber**
Carrack Gladden, coast (Lelant) **karrek, ?glan**
Carrack Kine hoh, rock 1696 (Madron) **keyn, hoch**
Carrack Kine marh, rock 1696 (Mullion) **keyn, margh**
Carracks, coast (St Levan) **karrek**
Carracks, The, coast (Zennor) **karrek**
Carrag-a-pilez, coast (Gunwalloe) **?pylles**
Carrag-Luz (Mullion) **karrek, los**
Carrancarrow (St Austell) **?carn, ?carow**
carrec wynn 1059 (S. 832/1027) **karrek, guyn**
Carrick Calys, rock (Feock) **cales**

INDEX OF CORNISH PLACE-NAMES

Carrick Du, coast (St Ives) **karrek, du**
Carrick Lûz, coast (St Keverne) **karrek, los**
Carricknath Point (St Anthony in Roseland) **karrek, *nath**
Carrickowel Point (St Austell) **karrek, ughel**
Carrine (Kea) ***ker, yeyn**
Carsawsan (Mylor) **Zowzon**
Carthew (St Austell; St Issey; Madron; Wendron) ***ker, du**
Cartole (Pelynt) **cruc, toll**
Caruggatt (Tywardreath) ***ker, *huel-gos**
Carvallack (St Martin in Meneage) ***ker**
Carvannel (Gwennap) ***ker, banathel**
Carvath (St Austell) ***ker, margh**
Carvean (Probus) ***ker, byghan**
Carvear (St Blazey) ***ker, meur**
Carvedras (Kenwyn) ***ker**
Carveth (Cuby) ***bich**
Carveth (Mabe) ***ker, margh** pl.
Carvinack (St Just in Roseland; Kenwyn) ***ker, *meynek**
Carvinack (Mylor) **guyn** deriv.
Carvoda (Lezant) ?***gorweth** pl.
Carvolth (Crowan) ***ker, *bolgh/moelh**
Carvossa (Probus) ?***ker,** ?***gosa**
Carwalsick (St Stephen in Brannel) ***ker, gwels** aj.
Carwarthen (St Just in Roseland) ***ker**
Carwen (Blisland; Lanreath) ***ker, guyn**
Carwinnick (Goran) ***ker, *meynek**
Carwither (St Breward) ***ker**
Carworgie (St Columb Major) ***ker**
Carwynnen Carn (Camborne) **carn**
Carzise (Crowan) ***ker, *Seys**
Cascadden (Gwennap) ?**cassec,** ?**can**
Casehill (St Breward) ***kew, *ros**
Cassacawn (Blisland) ?**oghan**
Cassick-well 1685 (St Stephen in Brannel?) ?**cassec**
Cassock Hill (St Minver) ?**cassec**
Castallack (Paul) **castell** aj.
Castella, rocks (Scilly) **castell** pl.
Castel Uchel Coed 12th (Week St Mary) **castell, *huel-gos**
Castle-an-Dinas (St Columb Major; Ludgvan) **castell**
Castle Canyke (Bodmin) **keyn** aj.
Castledewey (Warleggan) **castell, *dewy**
Castle Gayer (St Hilary) ***ker**
Castle Goff (Lanteglos by Camelford) **castell, gof**
Castle Gotha (St Austell) **castell, goth**[1] pl.
Castle Horneck (Madron) **castell, horn** aj.
Castlemawgan (Lanreath) **castell**
Castle Skudzick, fld. (Zennor) ?**schus** aj.
Castle Wary (Wendron) **castell, guary**

INDEX OF CORNISH PLACE-NAMES

Castlewitch (Callington) *****gwyk**
Castlezens (Veryan) **castell**, ?**sans** pl.
Cataclews Point (St Merryn) **karrek, los**
Catasuent Cove (Goran) ?*****suant**
Cathebedron (Gwinear) ?**cadar**
Caunse (Paul) *****cawns**
Causilgey (Kenwyn) **kee**
Cendefrion 967 (11th) (S. 755) *****kendowrow** or *****devr-**
Chacewater (Kenwyn) *****chas**
Chair Ladder, coast (St Levan) **cheer, lader/*lether**
Chamber an Tresousse 1458 (Sithney) **chammbour**
Chamber Byan 1458 (Sithney) **chammbour**
Chapel Ainger (Lelant) **chapel**
Chapel-an-Grouse (Perranuthnoe) **chapel**
Chapel Curnow (St Levan) **chapel, corn** pl.
Chapel Engarder, coast (Perranzabuloe) **cadar**
Chapel Jane (Zennor) **chapel, *enyal**
Chapelkernewyl 1302 (St Columb Minor) **chapel, kodna huilan** pl.
Chapel Maria 1504 (Camborne) **chapel**
Chapel Porth (St Agnes) **porth**
Cheesewarne (Mevagissey) **chy, sorn**
Chegwidden (Constantine) **chy, guyn**
Chegwin Carn, fld. 1649 (Constantine) **carn**
Cheiensorn 1435 (St Martin in Meneage) **chy, sorn**
Chellean (Gwennap) ?**lyen**
Chellew (Ludgvan) **lyw**
Chenhale (St Keverne) **chy, hal**
Chenhall (St Gluvias; St Martin in Meneage; Wendron) **chy, hal**
Chenhalls (St Erth) **chy, als**
Chepye (Gunwalloe) **chy**
Chestewer 1628 (Illogan) **chy, guas**
Chiawel 1335 (lost) **chy, awel**
Chinalls (Gunwalloe) **chy, als**
Chingenter n.d. (Gwennap) **kenter**
Chingweal (St Enoder) **chy, guel**
Chingwith 1838 (Camborne) **gwyth**[1]
Chirgwidden (Sancreed) **chy, gour, guyn**
Chirwyn (Kea) **chy, *run**
Chitol (St Just in Roseland) **chy, toll**
Chiverton (Perranuthnoe; Perranzabuloe; Sancreed) **chy, war, ton**
Choone (St Buryan; Newlyn East) **chy, goon**
Choone Praze (Newlyn East) **pras**
Chough Zawn (St Buryan) **sawn**
Chûn (Morvah) **chy, goon**
Chyandaunce (Gulval) **chy, *dons**
Chyandour (Gulval) **chy, dour**
Chyandower (Redruth) **chy, dour**
Chy-an-Gweal (Lelant) **chy, guel**

INDEX OF CORNISH PLACE-NAMES

Chyangweale (Towednack) **chy, guel**
Chyangwens (St Buryan) **chy, guyns**
Chyanhor (St Levan) **chy, horþ**
Chyanvounder (Gunwalloe) **chy, bounder**
Chybarles (Ruan Major) **?*lys**
Chybarrett (Kea) **chy**
Chybilly (St Mawgan in Meneage) **?*byly**
Chybucca (Kenwyn) **chy, bucka**
Chycarne (Germoe; St Just in Penwith) **chy, carn**
Chycoanse (Sancreed) ***cawns**
Chycoose (St Clement; Constantine; Feock; Gwennap) **chy, cos**
Chycornekye 16th (Madron) **corn, kee**
Chycowling (Kea) **chy**
Chyenhâl (Paul) **chy, hal**
Chyenmargh 1394 (Kenwyn?) **margh**
Chygolder Park, fld. (Gulval) **sygal, tyr**
Chygwyne (Kea) **chy, guyn**
Chykembro (Zennor) **chy, *Kembro**
Chylason (Towednack) **chy, *glasen**
Chymblo n.d. (St Keverne) **chy, blogh** sb.
Chymder (Gunwalloe) **chy, ?midzhar**
Chynance (Breage; St Buryan) **chy, nans**
Chyngwith (Colan) **gwyth**[1]
Chynhale (Gwennap; Perranzabuloe) **chy, hal**
Chynhalls (St Keverne) **chy, als**
Chynoweth (St Allen; St Austell; Breage; Cubert; St Erth; St Hilary; Mabe) **chy, nowyth**
Chyoone (Paul) **chy, goon**
Chypit (Feock) **chy, ?pyt**
Chypons (Crowan; Cury; St Hilary; Sithney; Towednack; Zennor) **chy, pons**
Chyporth 1416 (Grade) **chy, porth**
Chypraze (St Enoder; Morvah) **chy, pras**
Chyrase (St Hilary) **chy, rid**
Chyreen (St Keverne; Sithney) **chy, war, *run**
Chyrose (St Just in Penwith) **chy, *ros**
Chysauster (Gulval) **chy**
Chyseise c. 1720 (St Agnes) ***Seys**
Chytan (St Austell) **chy, war, ton**
Chytan (Luxulyan) **chy, ton**
Chytane (St Enoder) **chy, war, ton**
Chytodden (St Agnes; Towednack) **chy, ton**
Chytodden (Breage; Camborne) **chy, war, ton**
Chyton (St Erme) **chy, ton**
Chytroon (Perranarworthal) **?tron**
Chytroose n.d. (Redruth) **tros**
Chyunwone Pella 1278 (Paul) **pell** spv.
Chyvarloe (Gunwalloe) **chy, war, *loch**

INDEX OF CORNISH PLACE-NAMES

Chyvarton (St Buryan) **chy, war, ton**
Chyvelah (Kenwyn) **?ughel** spv.
Chyverans (St Keverne) **chy, bran**
Chyverton (Lelant) **chy, war, ton**
Chyvogue (Perranarworthal) **chy, fok**
Chywarmeneth 1613 (Budock) **chy, war, meneth**
Chywednack (St Keverne) **guyn**
Chyweeda (Breage) **chy**
Chywoon (Germoe; St Gluvias; Kenwyn; St Keverne) **chy, goon**
Clahar Garden (Mullion) **carn**
Clann (Lanivet) **kelli, *lann**
Clausiow 1607 (Truro) ***clav-jy** pl.
Cleese (Morval) ***cleys**
Clennick (St Germans) **?kelin** aj.
Clicker Tor (Menheniot) ***cleger**
Clidga (Sennen) ***cleys** pl.
Clies (St Mawgan in Meneage) ***cleys**
Cligga Head, coast (Perranzabuloe) ***cleger**
Clijah (Redruth) ***cleys** pl.
Clinnick (Braddock) **?kelin** aj.
Clise Close, fld. 1649 (Constantine) ***cleys**
Clise Craft, fld. 1649 (Constantine) ***cleys**
Clise Gwenna 1660 (Breage) ***cleys**
Clise Nawiddon 17th (St Allen) ***cleys**
Clodgy Moor (Budock; Paul) ***clav-jy**
Clodgy Point, coast (St Ives) ***clav-jy**
Clugea Lane (St Keverne) ***clav-jy**
Clumyer, fld. (Zennor) ***colomyer**
Clumyers, fld. (Ludgvan) ***colomyer**
Clydia (Feock) **guel, ?*aswy**
Clys Croft, fld. (Paul) ***cleys**
Cockle Zawn (St Just in Penwith) **sawn**
Codde Fowey c. 1540 (Altarnun) **cos**
Codna-coos 1803 (St Agnes) **conna**
Codnagooth 1689 (Sancreed) **conna, goth**[2]
Codnawillan, fld. (Paul) **kodna huilan**
Codna Willy, fld. (St Buryan) **kodna huilan**
Codnidne 1808 (Perranzabuloe) **conna, yn**
Coeswyn 1332 (Gwinear) **cos, guyn**
cofer fros 960 (S. 684) **?frot**
Coffenoola, mine n.d. (Gwennap) ***coffen, ula**
Coffin broaz, mine n.d. (Gwennap) ***coffen, bras**
Coffin Garrow, mine 1822 (St Just in Penwith) ***coffen, garow**
Cogegoes (Camborne) **cos**
Cogveneth 1617 (Lelant) **cuic, meneth**
Coiscuntell 1556 (Wendron) **cuntell**
Colbiggan (Roche) **kelli, byghan**
Coldhender (Duloe) ***hendre**

256

INDEX OF CORNISH PLACE-NAMES

Coldrenick (St Germans) *kyl, dreyn aj.
Coldrinnick (Duloe; Helland) *kyl, dreyn aj.
Coldrose (St Clement) *ros
Coldvreath (Roche) kelli, *brygh
Colesent (St Tudy) ?sans pl.
Colesloggett (Cardinham) logaz
Colgrease (Cubert) krow, cres
Collibeacon (St Winnow) kelli, byghan
Colligeen (Lanlivery) kelli, ?ky pl.
Colloggett (Landulph) logaz
Collon (St Veep) ?*coll sg.
Colona (Goran) ?*leyn pl.
Colquite (Callington; Lanteglos by Fowey; Linkinhorne; St Mabyn) *kyl/cul, cos
Colvannick (Cardinham) kal, *bannek
Colvase (Morval) fos pl.
Colvennor (Cury; Wendron) kal, meneth
Colvithick Wood (Fowey) *collwyth aj.
Colwith (Lanlivery) *collwyth
Comprigney (Kenwyn) cloghprennyer
Conce (Lanlivery) *cawns
Condolden (Minster) goon, *tol-ven
Condurra (St Clement) *kendowrow
Condurrow (St Anthony in Meneage; Camborne) *kendowrow
Coombekeale (Egloskerry) ?*comm, *kel/*kyl
Coosebean (Kenwyn) cos, byghan
Coosehecca (Kea) cos
Coosewartha (St Agnes) cos
Coresturnan 1323 (St Just in Roseland) *cores
Corgee (Luxulyan) corn, kee
Corgerrick (Grade) ?*coger, -yk
Corslen 1383 (Crowan?) *cors, lyn
Corva (St Ives) *cor², bagh
Cosawes (St Gluvias) cos
Cosesawsyn 1556 (St Clement?) cos, Zowzon
Coskallow Wood (South Hill) cos
Coskeyle (St Erme) ?*kel
Cossabnack 1798 (Constantine) cos
Coswinsawsin (Gwinear) Zowzon
Cotehele (Calstock) cos, *heyl
Cowlands (Kea) *kew-nans
Cowloe, rock (Sennen) ?*loch
Cowyjack (St Keverne) *kew, -esyk
Coysbesek 1527 (Ladock) cos
Coyseglase 1550 (Creed) cos, glas
Coyse Laydocke 1547 (Ladock) cos
Coysfala 1337 (Creed) cos
Coyskentueles 1337 (Wendron) cos, cuntell deriv.

INDEX OF CORNISH PLACE-NAMES

Coyspenhilek 1337 (St Clement) **cos**
Coys Penryn c. 1400 (St Gluvias) **cos**
Coysynchase 1560 (Kenwyn) **cos, *chas**
Crackagodna, fld. (Wendron) **crak²**
Crack-an-Godna, fld. (Constantine) **crak²**
Crackington (St Gennys) ***crak¹**
Crane (Camborne) ***ker, bran**
Crankan (Gulval) ***ker, anken**
Crannis (Camborne) ?***crann,** ?*****-ys²**
Crannow (St Gennys) ?***crann** aj.
Crasken (Wendron) ***ker, ascorn**
Crauft an Mellender, fld. 1670 (Madron) ***melynder**
Crawle (Breage) ***ker,** ?**yorch** dimin.
Creage Pellen 1613 (Illogan) **pellen**
Creakavose (St Stephen in Brannel) **fos**
Creeb, coast (Scilly) **krib**
Creegbrawse (Kenwyn) **cruc, bras**
Creegdue (Ruan Major) **cruc, du**
Creeglase (Redruth) **cruc, glas**
Creeglogas (Ruan Major) **cruc, logaz**
Creeg Tol (St Buryan) **cruc, toll**
Creens (Ladock) ?***cren**
Creggan-Geggan, fld. 1809 (Phillack) ?***kegen**
Creggo (Mylor) **karrek** pl.
Cregoe (Ruan Lanihorne) **cruc** pl.
Crellow (St Buryan) ?**elaw** pl.
Crelly (Wendron) ***ker**
Crennick, fld. (South Hill) ***cryn** deriv.
Crenver (Crowan) ***ker**
Creskin (St Buryan) ***ker, ascorn**
Cribba Head, coast (St Levan) ?**krib** pl.
Cribbar Rocks, coast (St Columb Minor) ?**krib** pl.
Cribbensaune, coast 1716 (Madron) **krib, sawn**
Crig-a-Tana Rocks (Grade) ?**tanow**
Criggan (Roche) **cruc** dimin.
Crigmurrian (Philleigh) **cruc, *moryon**
Crigmurrick (St Merryn) ?**moyr-** aj./**mor** aj.
Crill (Budock) ***kel**
Crinnicks, fld. (Bodmin) ***cryn** deriv.
Crinnis (St Austell) ?***ker**
Croan (Egloshayle) **krow** dimin.
Croc-an-codna, fld. 1787 (Sithney) **crak²**
Crockagodna, fld. (Mylor) **crak²**
Crockett (Stoke Climsland) ***cnow, cos**
Croft-an-Askernel, fld. 1649 (Constantine) **ascorn** dimin.
Croft-an-Clyes, fld. 1696 (Wendron) ***cleys**
Croft-an-Creeg (Kea) ***croft, cruc**
Croft-an-Crow, fld. 1621 (St Keverne) **krow**

258

Croft-and-Peloss, fld. 1649 (Constantine) ***pylas**
Croft-an-Garratt, fld. (St Keverne) **karrek**
Croft Bloue (Ludgvan) **plu**
Croft Codanna, fld. (Budock) **?cudin** pl.
Croft Coothe, fld. n.d. (Gwennap) ***croft, coth**
Croft en begel, fld. 14th (Kea?) ***croft, begel**
Croft en Ferngy, fld. 1360 (Cury) ***croft, forn** deriv.
Croftengrous, fld. 1265 (Sithney) ***croft, crous**
Croftengweeth (Gwennap) ***croft, gwyth**[1]
Croftenorgellous, fld. 1333 (St Dennis) ***croft, *orguilus**
Croft Gaver, fld. 1696 (Ludgvan) **gaver**
Croft Gothal, mine (St Hilary) ***gothel**
Crofthandy (Gwennap) ***croft, ?hensy**
Croft haverek, fld. 1513 (Perranuthnoe) ***havrek**
Croft Horspole, fld. 14th (Ladock) ***croft**
Crofthow (St Columb Major) ***croft** pl.
Croft inter Du-Henvor, fld. 1691 (Sithney) **ynter**
Croft Michell (Camborne) ***croft**
Croft Milgey (Wendron) **?mylgy**
Croftnoweth (Ruan Major) ***croft, nowyth**
Croftoe (Morvah) ***croft** pl.
Croft Pascoe (Ruan Major) ***croft**
Croft-Pellas, fld. 1649 (Constantine) ***pylas**
Croft Prince (St Agnes) **pryns**
Croft Sugal, fld. (Zennor) **sygal**
Croftvean (Wendron) ***croft, byghan**
Croft West (Kenwyn) ***croft, west**
Croft Windjack, fld. (Constantine) **guyns** aj.
Crohans (Veryan) ***ker, oghan**
Crosspost (Stithians) **cos, post**
Crosswyn (St Ewe) **crous, guyn**
Crou 1327 (Cubert) **krow**
Crousa (St Keverne) **crous, gruah**
Crousbronnou 1360 (Cury) **crous**
Crouse-Harvey (St Keverne) **crous**
Crouse-widen 1660 (St Stephen in Brannel) **crous**
Crouspren 1335 (lost) **crouspren**
Croustroch c. 1220 (lost) **crous, trogh**
Crowdy Marsh (Davidstow) ***krow-jy**
Crowgey (Constantine; Gwennap; Ruan Minor; Wendron) ***krow-jy**
Crowlas (Ludgvan) ***crew, rid;** or **krow, *lys**
Crowner Rocks (St Ives) **gawna**
Crownick (Mylor) **?ownek**
Crows-an-Wray (St Buryan) **crous, gruah**
Crowse predon bounds, mine 1687 (lost) **crouspren**
Crowshire (Poundstock) **crous, hyr**
Crowsmeneggus (Perranarworthal) **?crous, ?benyges**
Crubzu 1300 (Constantine) **krib, du**

259

INDEX OF CORNISH PLACE-NAMES

cruc drænoc 1059 (S. 832/1027) **cruc, dreyn** aj.
Crucheyd 1300 (Roche) ?***heth**
crucmur 1059 (S. 832/1027) **cruc, meur**
crucou mereðen 967 (11th) (S. 755) **cruc** pl., ?***merthyn**
Crucwragh 1284 (Perranzabuloe?) **gruah**
Cruglaze (Creed) **cruc, glas**
Crugmeer (Padstow) **cruc, meur**
Crugoes (St Columb Major) **cruc** pl.
Crugsillick (Veryan) **cruc**
Crukbargos 1424 (Luxulyan) **cruc**, ?**bargos**
Cruplegh 1428 (lost) **cruc, bleit**
Crylla (St Cleer) ?***cryn,** ?**le**
Cuddra (St Austell) ?**tre**
Cudna Reeth (St Just in Penwith) **conna**
Cudno (Sithney) **conna**
Culdrose (Wendron) ?***ros**
Curkheir 1564 (lost) **cruc, hyr**
Cusgarne (Gwennap) **cos, garan**
Cusveorth (Kea) **cos, *buorth**
Cusvey (Gwennap) **cos**, ?***fe**
Cutbrawn (St Winnow) **cos, bran**
Cutcare (Menheniot) **cos, *ker** pl.
Cutcrew (St Germans) **cos, *crew**
Cutkive Wood (St Cleer) ***kyf**
Cutlinwith (Landrake) **cos**, ?***glynn-wyth**
Cutmadoc (Lanhydrock) **cos**
Cutmere (St Germans) **cos, meur**
Cutparrett (Morval) **cos, perveth**
Cuttivett (Landrake) **cos, *dywys/tevys**

Dal Jo, rock (St Ives) **dall**
Dannandre, fld. 1696 (Towednack) **dan, tre**
Dannett (Quethiock) ***downans**
Dannonchapel (St Teath) ***downans**
Dansotha 1650 (Perranzabuloe) **down**, ?***soð** pl.
Davas, rock (St Keverne) **daves**
Davis Farm (St Gluvias) ***dywys**
Dawna (St Winnow) ***downans**
Dawns Men (St Buryan) ***dons, men** pl.
Dean, The, rock (St Keverne) **den**
Degembris (Newlyn East) **ti** (= **chy**)
Degey, The, 18th (Sithney) ***dyjy**
Delabole (St Teath) **pol**
Delamere (St Teath) **meur**
Delawhidden (St Winnow) **deyl** pl., ?***gwythel**
Deli (St Teath) ?**deyl**, ?***-i**
Delinuth (St Teath) **nowyth**
Demain (Ruan Lanihorne) **tre, men**

INDEX OF CORNISH PLACE-NAMES

Demelza (St Wenn) *****dyn**
Denas (Merther) *****dynas**
Denby (Bodmin) *****dyn, *bich**
Dennis (Padstow) *****dynas**
Dennis Cockers 1650 (St Columb Major) *****dynas**
Dennis Head (St Anthony in Meneage) *****dynas**
Dennis Point (Tintagel) *****dynas**
Denys Azawan 1628 (St Columb Major) *****dynas, asow** sg.
Derese, fld. (Landewednack) **dor, eys**
Derval (Sancreed) *****devr-, *-yel**
deumaen coruan 1059 (S. 832/1027) **dew, men**
Deuy 1326 (river) *****dewy**
Deveral (Gwinear) *****devr-, *-yel**
Devichoys Wood (Mylor) *****dywys, cos**
Devis (Gwennap) *****dywys**
Devoran (Feock) *****devr-** pl.
Dewey (river) *****dewy**
difrod 1018 (S. 951) **dy-, frot**
Digey, The (St Ives) *****dyjy**
Dinas Cove (St Keverne) *****dynas**
Dinas Head (St Merryn) *****dynas**
Dinham (St Minver) *****dynan**
Dinness (St Clement) *****dynas**
Dinnever Hill (St Breward) **?*bre**
Dippa Meadow, fld. (Crowan) *****dyppa**
Dizzard (St Gennys) *****dy-serth**
Doar n.d. (Gwennap) **dor**
Doerhyr, fld. 1437 (Madron?) **dor, hyr**
Doerpoys, fld. 1385 (Grade) **dor, ?*poth**
Dolcoath (Camborne; Morvah) **dor, coth**
Dollar, rock (Sennen) *****talar**
Dollar Cove, coast (Gunwalloe) *****talar**
Dollar Ogo, coast (Grade) *****talar**
Dollar Rock, coast (St Columb Minor) *****talar**
Dolor Point (St Keverne) *****talar**
Domellick (St Dennis) *****dyn**
Dorcatcher, fld. (Zennor) **gajah**
Dorcoath, fld. 1696 (Gulval; Mullion; Ruan Major; Towednack) **dor, coth**
Doreheer, fld. 1649 (Constantine) **dor, hyr**
Dorendre, fld. 1333 (St Mawgan in Pydar) **tre**
Dore Ribba Vownder, fld. 1588 (Sancreed) **ryp**
Dorheere, fld. (Gwennap) **dor, hyr**
Dorheere a Veese, fld. 1688 (St Levan) **aves**
Dorkillier, fld. (Paul) **?*kyl** pl.
Dormayne 1652 (St Kew) **dor, men**
Dorminack (St Buryan) **dor, *meynek**
Dornolds, fld. (Paul) **als**
Dorothegva, fld. (Towednack) **?*ma**

261

INDEX OF CORNISH PLACE-NAMES

Dor-se, fld. (Lelant) **dor, ?segh**
Dorughdre, fld. 1333 (St Dennis) **ugh**
Dounans 1337 (Constantine) ***downans**
Dour Conor c. 1540 (= Connor River) **dour**
Dourragenys 1337 (Paul) **dour**
Dour-Tregoose n.d. (Probus) **dour**
Dovrigger, fld. (Paul) **?dour** aj. pl.
Dower Ithy n.d. (Truro) **dour**
Dower Meor 1613 (Kea) **dour**
Dowgas (St Stephen in Brannel) **dew, cos**
Downas Valley (St Keverne) ***downans**
Downathan (St Minver) **down, ?eithin**
Downinney (Warbstow) **?down**
Dowrack, fld. (Paul) **dour** aj.
Dowran (St Just in Penwith) ***douran**
Dowreth (Feock) **dour, ruth**
Draennek 1337 (St Clement?) **dreyn** aj.
Drannack (Gwinear) **dreyn** aj.
Draynes (St Neot) **dreyn** deriv.
Drewollas (Gwinear) **tre, goles**
Drift (Braddock; Constantine; Sancreed) **tre**
Drinnick (South Petherwin; St Stephen in Brannel) **dreyn** aj.
Drinnick guin 1685 (St Stephen in Brannel) **guyn**
Drisack, fld. 1750 (Madron) **dreys** aj.
Drum Head (Pillaton) **?*drum**
Druse, fld. (Paul) **?eys**
Dry Carn, rock (Sancreed) **try[1], carn**
Drym (Crowan) ***drum**
Duloe (= parish) **dew, *loch**
Dunmere (Bodmin) ***dyn, meur**
Dunveth (St Breock) ***dyn, *bich**
Duporth (St Austell) **dew, porth**
Durla, fld. (Lelant) **dour, le**
Durlah, fld. (Paul) **dour, le**
Durloe (Lelant) **dour, le**
Durva, fld. (Breage) **?dour**
Dynas Ia 1513 (St Ives) ***dynas**

Ebal Rocks (Zennor) **ebel**
Eddless (Perranarworthal) **?aidlen, ?*-ys[2]**
Eggens-warra, fld. (Gerrans) **?ugens, ?erw**
Eggloscraweyn c. 1145 (Crowan) **eglos**
Egglostetha c. 1190 (St Teath) **eglos**
Eglarooze (St Germans) **eglos, *ros**
Egloscrow (St Issey) **eglos, cruc**
Eglosderry (Wendron) **dar** pl.
Egloshayle (= parish) **eglos, *heyl**
Egloshayle (Maker) **eglos, *heyl**

INDEX OF CORNISH PLACE-NAMES

Egloshayle c. 1500 (= Phillack) **eglos**, ***heyl**
Egloskerry (= parish) **eglos**
Eglosmadern 1396 (Madron) **eglos**
Eglosmerther (Merther) ***merther**
Eglospenbro 1302 (Breage) **eglos**
Eglosrose (Philleigh) **eglos**, ***ros**
Eglos-Withiel 1692 (Withiel) **eglos**
Elerkey (Veryan) **elerch**, ?***-i**
Ella (Camborne) **hel** pl.
Ellenglaze (Cubert) **glas**
Embla (Towednack) ***emle**
Endourbyhan 13th (Roche) **dour**
Enestreven (Sancreed) **enys**
Engelley (Perranzabuloe) **kelli**
Engollan (St Eval) ?**hen**, ***coll** sg.
Engoyse (Wendron) **cos**
Ennis (St Dennis; St Enoder; St Erme; Stithians) **enys**
Enniscaven (St Dennis) **enys, scawen**
Ennis Morvah (Constantine) ***morva**
Ennisveor 1632 (St Dennis) **enys, meur**
Ennisworgey (St Columb Major) **enys**
Ennys (St Hilary) **enys**
Enys (St Gluvias) **enys**
Enys, The, island (St Hilary; St Just in Penwith) **enys**
Enys Dodnan, rock (Sennen) **enys, ton**
Enyshall 1747 (Probus) **enys, hal**
Enys Head (Ruan Minor) **enys**
Enysmannen (Sancreed) **enys, amanen**
Enys Vean, rock (Landewednack) **enys, byghan**
Erbyer Gwarra, fld. 1649 (Constantine) ***erber**
Erisey (Grade) ?**ti** (= **chy**)
Eroger Maze 1696 (Mullion) **mes**
Erov Porthm' 1446 (St Ives) **erw**
Erra, fld. (Paul) **erw**
Errow-Brane, fld. 1649 (Constantine) **erw, bran**
Errow-Porth, fld. 1649 (Constantine) **erw, porth**
Eru Marut, fld. 1272 (St Ives) **erw**
Erwereden, fld. 1260 (Roche) **erw, reden**
Essa (Lanteglos by Fowey) ?***ussa**
Ewandre, fld. 1696 (Towednack) **ugh, tre**

Fe Kenel c. 1400 (Stithians) ***fe**
Fe Mareschal 1333 (Paul) ***fe**
Fentafriddle (Tintagel) **fenten, frot** aj.
Fentengo (St Kew) **fenten, *cok**
Fentenluna (Padstow) **lyn** pl.
Fenterleigh (Tintagel) ***legh**
Fentervean (St Gennys) **fenten, byghan**

INDEX OF CORNISH PLACE-NAMES

Fentonadle (Michaelstow) **fenten**
Fenton Ferilliow 1613 (Breage) ***feryl** pl.
Fentongollan (St Michael Penkevil) **fenten, *coll** sg.
Fentonladock (Ladock) **fenten**
Fentonscroll 1702 (Probus) **fenten, scoul**
Fentrigan (Warbstow) **fenten, can**
Fentyn Carensek c. 1510 (Kenwyn) **fenten**
Fimbol c. 939 (14th) (S. 450) **fyn, pol**
finfos 960 (S. 684) ***fynfos**
Foage (Zennor) ***bod**
fonton gén 967 (11th) (S. 755) **fenten, yeyn**
fonton morgeonec 1059 (S. 832/1027) **fenten, *moryon** aj.
Foregles, fld. 1649 (Constantine) **forth, eglos**
Forge (Redruth) ***bod**
Forgue (Ladock) **forth, *kew**
Fosebrase 1654 (Madron) **fos, bras**
fosgall 1049 (S. 1019) **fos**
fos no cedu 967 (11th) (S. 755) **fos**
Fowey (river) ***faw, *-i**
Fraddon (St Enoder) ?**frot, *-an²**
Frightons (St Erth) ***brygh**
Frigleys, fld. (St Ives) **forth**
Frogabbin, fld. 1690 (St Keverne) ?**forth**, ?**cam**
Fuggoe Lane (Lelant) **vooga**
Fuggoes, The, fld. (St Ives) **vooga**
Fursnewth (St Cleer) **fos**
Furswain (St Cleer) **fos**

Gabmas, The, fld. (Lelant) ?***camas**
Gadles (St Gluvias) ***cad-lys**
Gahan, fld. (St Just in Penwith) ?**gahen**
Gaider, The, coast (St Keverne) **cadar**
Gallas, fld. (Roche) **cales**
Gallentreath, coast (St Keverne) **trait**
Galowras (Goran) ***glow¹, *red/rid**
Gam (Michaelstow) **cam** sb.
Gamas Point (St Austell) ***camas**
Gamper, coast (St Just in Penwith; St Levan; Sennen) ***kemer**
Ganinick (Scilly) **kenin** aj.
Gannel, The (river) ***canel**
Gannel Rock (St Levan) ***canel**
Garah (Mullion) ?***gar** pl.
Garder-Wartha c. 1720 (St Agnes) **cadar**
Garder-Wollas c. 1720 (St Agnes) **cadar**
Gare (Lamorran) ***ker**
Garland, fld. (Sithney) ***cor-lann**
Garlenick (Creed) ***cor²**
Garlidna (Wendron) **grelin** pl.

INDEX OF CORNISH PLACE-NAMES

Garras (Kenwyn) ?*gar pl.
Garras (Gulval; St Mawgan in Meneage) garow, *ros
Garros 1668 (St Buryan) garow
Garrow Tor (St Breward) garow
Garth 1343 (St Clement) *garth
Gaverigan (St Columb Major) gover, guyn
Gavnas Point, coast (Goran) ?gawna
Gayan, fld. (St Levan) ?gahen
Gazell, coast (St Buryan) *casel
Gazells, coast (St Buryan) *casel
Gazick, coast (St Just in Penwith) cassec
Gazzle, coast (St Columb Minor) *casel
Gear (8 exx., and flds.) *ker
Geaw Moyha, fld. 1649 (Constantine) meur spv.
Gedges, The, rock (Mawnan) ?gesys
Gelly (St Pinnock) kelli
Germoghmur 1404 (Germoe) meur
Gersick-an-Awn, coast (Mullion) *cors aj.
Gew (Crowan; St Erth; Kea) *kew
Gew, The (St Anthony in Meneage; St Hilary) *kew
Gewan Gampe 1613 (Kea) *kew
Gew-Graze, coast (Mullion) *kew, cres
Gilly (Cury; Gwennap; St Mawgan in Meneage) kelli
Gilly Gabben (St Mawgan in Meneage) kelli, cam
Glasdon (St Germans) ?*glasen
Glasney (Budock) *glasneth
Glasney Green 1709 (Ludgvan) *glasneth
Glastannen, river 1356 (St Clement) glastan sg.
Gloweth (Kenwyn) ?*glow^1, ?goth^1
Gluuyanmargh 1483 (St Columb Major) margh
Glynn (Cardinham) *glynn
Goada Sessing Close, fld. (Roche) ?*godegh
Gobbens (Gulval) *gobans
Godna, fld. (St Anthony in Meneage) conna
Godolphin (Breage) *go-
Godrevy (Gwithian; St Keverne) *godre pl.
Goenandon 1302 (Roche) ton
Goen an Groushire, fld. 1536 (Crowan) crous, hyr
Goenbans 1403 (St Agnes) *pans
Goenbargothowe 1507 (St Stephen in Brannel) goon
Goen Bren 12th (= Bodmin Moor) *bren
Goenchase 1579 (Kenwyn) goon, *chas
Goencruk du 1302 (Ruan Major) goon
Goendywragh 1302 (Wendron) *goon-dy, gruah
Goeneglos 1286 (Roche) goon, eglos
Goengasek 1560 (Lanlivery?) goon, cassec
Goengellom 1538 (Budock) ?*kel cpd.
Goenhely 1300 (Roche) *helgh vb.

INDEX OF CORNISH PLACE-NAMES

Goen menhere, fld. 1696 (Paul) ***men hyr**
Goen Poldyse 1514 (Gwennap) **goon**
Goenrounsen (St Enoder) **goon**
Goenspaylard 1339 (Ladock?) **goon**
Goen Suggalls noweth, fld. 1696 (Ludgvan) **sỳgal**
Goentre, fld. 1409 (Breage) **tre**
Goenwarrack, fld. 1696 (Paul) **guarthek**
Golant (= parish) **gol, nans**
Gold Arish, fld. (Budock) **?daras**
Goldenharn (Sancreed) **?horn**
Golden Praunter, fld. (St Levan) **guel, pronter**
Goldingey, fld. (Madron) **?hensy**
Gold Maggey, fld. (Crowan) **?maga**
Goldsithney (Perranuthnoe) **gol**
Goldstanna, fld. (Constantine) **tanow**
Goldwest, fld. (St Ewe) **guel, wast/west**
Gollen Orchard, fld. (St Minver) **?*coll** sg.
Gollwest (St Gluvias) **guel, wast/west**
Gome 1684 (St Gluvias) ***comm**
Gomm (St Austell) ***comm**
Gonabibles, fld. (St Stephen in Brannel) ***pybell**
Gone Goth, fld. 1546 (Sithney) **coth**
Gonew (Lelant) **goon** pl.
Gonitor (Ruan Lanihorne) **gwaneth, tyr**
Gonnamarris (St Stephen in Brannel) **goon, margh** pl.
Gonorman Downs (Stithians) **goon, naw, men**
Gonreeve (St Gluvias) **goon, ruth**
Gonvena (Egloshayle) **guyn, meneth**
Good-a-Caurhest, mine 1685 (St Stephen in Brannel) **?*godegh**
Goodagrane (Mabe) **?goth**[1] pl., **?*crann**
Goodern (Kea) **?*go-**
Goonamarth (St Mewan) **goon, margh**
Goon and Scath Craft, fld. 1649 (Constantine) **skath**
Goon-an-Glastannen 13th (St Clement) **glastan** sg.
Goonbell (St Agnes) **goon, pell**
Goone Agga Idniall, fld. 1670 (Madron) **goon**
Goonearl (St Agnes) **goon, *yarl**
Goonevas (Sancreed) **?*havos**
Goongoose (Sithney) **goon, cos**
Goon Gumpas (Gwennap) **goon, compes**
Goonhavern (Perranzabuloe) **goon, *havar**
Goonhenver 18th (Wendron) ***hen-forth**
Goonhilly Downs (= district) **goon, *helgh** vb.
Goonhingey (St Gluvias) **goon, hensy**
Goonhoskyn (St Enoder) **goon, ?heschen**
Goonhusband (Wendron) **goon**
Gooninis (St Agnes) **enys**
Goonlaze (St Agnes; Stithians) **goon, glas**

266

INDEX OF CORNISH PLACE-NAMES

Goonleigh (Roche) **goon, *legh**
Goonmenheer, fld. (St Buryan) ***men hyr**
Goon Mine Mellon (Towednack) **men** pl., **melin**[1]
Goon Noath, fld. (St Just in Penwith) **noth/nowyth**
Goonpiper (Feock) **goon**
Goon Prince (St Agnes) **goon, pryns**
Goonraw (Redruth) **goon**
Goonrinsey (Breage) **goon**
Goon Skipper, fld. (Paul) **skyber**
Goonvean (Kenwyn; St Stephen in Brannel) **goon, byghan**
Goonvrea (St Agnes) **goon, *bre**
Goonwin (Wendron) **goon, guyn**
Goonzion Downs (St Neot) **goon, ?*seghan**
Goonzoyle (Camborne) **?zowl**
Gormellick (Liskeard) **goon**
Gorrangorras (St Gluvias) **goon, *cores**
Gorres c. 1720 (St Agnes) ***cores**
Gothers (St Dennis) **goth**[1]
Goudekining 1745 (Wendron) **?*godegh, ?kynin**
Goungylly 1461 (Perranzabuloe) **goon**
Govarrow (St Allen) **gover** pl.
Gover (St Agnes; St Mewan) **gover**
Goverrow (Gwennap) **gover** pl.
Goverseath (St Stephen in Brannel) **gover, segh**
Goviley (Cuby) **gofail** pl.
Goyncrukke 1384 (Kenwyn?) **goon, cruc**
Grambla (Wendron) ***cromlegh**
Grambler (St Agnes; Gwennap) ***cromlegh**
Grankim (Breage) ***ker, anken**
Gready (Lanlivery) ***gre-dy** pl.
Great Zawn (Zennor) **sawn**
Grebe Rock (Mawnan) **krib**
Greeb (Sennen) **krib**
Greeb, The, coast (Perranuthnoe) **krib**
Greeb Point (Gerrans; Goran; Morvah) **krib**
Greenwith (Perranarworthal) **?*cren**
Gribbin Head, coast (Tywardreath) **krib** dimin.
Griglands, fld. (Mabe) ***gruk**
Griglands (Roche) ***gruk**
Grogoth (Cornelly) **?*gruk, ?goth**[1]
Grogoth Wartha (Cornelly) **guartha**
Grogoth Wollas (Cornelly) **goles**
Grumbla (Sancreed) ***cromlegh**
Guaelmeynek 1366 (Merther?) **guel, *meynek**
Guall Est 1463 (St Gluvias) **yst**
Gue Hole (St Anthony in Roseland) ***kew**
Gueldrenek 1424 (Wendron?) **guel, dreyn** aj.
Guely breteny, rock c. 1300 (St Just in Penwith) **guely**

267

INDEX OF CORNISH PLACE-NAMES

Gulchamber, fld. (Crowan) **chammbour**
Gullaveis (St Stephen in Brannel) **?aves**
Gull Gwidden, fld. (St Keverne) **guel**
Gullivere 1734 (St Agnes) **kelli, meur**
Gulnoweth, fld. (Constantine) **guel**
Gulstatman, fld. (Constantine) **glastan** sg.
Gummow (Probus) **?*comm** pl.
Gumm Park, fld. 1696 (St Issey) ***comm**
Gunheath (St Austell) **goon, hueth**
Gunnamanning (Ladock) **goon, amanen**
Gunneau Bosolo c. 1700 (Madron) **goon** pl.
Gunvean (Stithians) **goon, byghan**
Gunvenna (St Minver) **goon, fyn** pl.
Gunvillick (Lesnewth) **goon**
Gunwen (Luxulyan) **goon, guyn**
Gûnwyn (Lelant) **goon, guyn**
Gurge Hedge, fld. 18th (Illogan) ***gor-ge**
Gurland (St Just in Penwith) ***cor-lann**
Gurlyn (St Erth) **?*gor-, ?lyn**
Gurnick (Crowan) ***kernyk**
Guthen Brose, coast (St Levan) **bras**
Gwaeldreysec, fld. 1306 (St Keverne) **guel, dreys** aj.
Gwael en Whoen 1270 (Roche) **guel**
Gwael Kephamoc 1306 (St Keverne) **guel, *kevammok**
Gwarder (St Gluvias) **dour**
Gwarnick (St Allen) **guern** aj.
Gwarry-Teage, mine 1672 (Towednack) **guary, teg**
Gwarth-an-drea (St Mawgan in Meneage) **guartha, tre**
Gwavas (Grade; Paul; Sithney) ***gwavos**
Gweal (Scilly) **gwyth**[1] aj.
Gweal-an-Mayn (St Martin in Meneage) **guel, men**
Gweal-an-Top (Redruth) **top**
Gwealavellan (Camborne) **guel, melin**[1]
Gweal-Bargus, fld. 1649 (Constantine) **bargos**
Gweal Cabben, fld. 1649 (Constantine) **cam**
Gwealcarn (Towednack) **guel, carn**
Gweal Darras, fld. (Mabe) **guel, daras**
Gweal Darros, fld. 1649 (Constantine) **guel, daras**
Gweal Drisack, fld. 1649 (Constantine) **dreys** aj.
Gwealdues (Wendron) **guel, du**
Gwealeath (St Mawgan in Meneage) **goon, *legh**
Gweale Gollas a Choy, fld. 1688 (St Levan) **agy**
Gweale Gollas a Vese, fld. 1688 (St Levan) **aves**
Gweale Skeeber, fld. 1665 (Crowan) **skyber**
Gwealfolds (Wendron) **guel, *fold**
Gweal Goose (Grade) **goth**[2]
Gwealgwarthas (Camborne) **guel, guartha**
Gwealhellis (Wendron) **guel**

INDEX OF CORNISH PLACE-NAMES

Gwealmayowe (Wendron) **guel**
Gwealmellin (Constantine) **guel, melyn²**
Gweal Paul (Redruth) **guel**
Gweal Skeber, fld. 1649 (Constantine) **skyber**
Gweal Street, fld. 1786 (Sithney) ***stret**
Gweek (Constantine) ***gwyk**
Gwelbeauleu 1374 (St Hilary) **guel**
Gwelbelbouche 1398 (Kenwyn?) **guel**
Gwelcreege, fld. 1633 (St Ewe) **guel**
Gwele Cararthen 1485 (Merther) **guel**
Gwellegolas 1643 (St Austell) **guel, goles**
Gwell fave, fld. 1696 (Gulval) **guel, fav**
Gwellogas, fld. 1696 (Towednack) **?ogas**
Gwelmartyn (Probus) **guel**
Gwendra (St Austell; Veryan) **guyn, trait**
Gwendreath (Grade) **guyn, trait**
Gwernfosov 1374 (Roche?) **guern**
Gweythenglowe, mine 1520 (Roche) **guyth², glow²**
Gweythen-vysten, mine 1520 (Roche) **guyth²**
Gwineus, fld. (St Merryn) **guyns** deriv.
Gwinges, The, fld. (St Agnes) **guyns** deriv.
Gwinges, coast (Goran; St Keverne) **guyns** deriv.
Gwool Pease, fld. (St Keverne) **guel, pêz**
Gwyllegwyet 1519 (Roche) **?gwîa** ppp.
Gwynver (Sennen) **guyn, forth**
Gwyth an bara, mine 1505 (lost) **guyth², bara**
Gwythancorse, mine c. 1507 (lost) **guyth², *cors**
Gwythanprase, mine c. 1507 (lost) **guyth², pras**
Gwyth en futhen, mine 1505 (lost) **guyth², budin**
Gwythengrous, mine c. 1507 (lost) **guyth², crous**
Gwyth Glastannan, mine c. 1507 (lost) **guyth², glastan** sg.
Gyllyngvase (Budock) **mes**

Hægelmuða 11th (= Padstow) ***heyl**
Haferell, river 1049 (S. 1019) **?*havar** aj.
Haggarowel (Wendron) **hager, awel**
Haggerowel Croft, fld. (Breage) **hager, awel**
Halabezack (Wendron) **hal, guibeden** aj.
Halamanning (St Hilary) **(pen), hal, amanen**
Halamiling (Lesnewth) **nans, melin¹**
Halancoose (Crowan) **hal**
Halankene (Phillack) **hal, ?*keun**
Halbathick (Liskeard) **hal**
Haldu 1340 (Ladock?) **hal, du**
Hale (St Kew) **hal**
Halgarrack (Crowan) **hal, karrek**
Halgavor (Bodmin) **hal, gaver**
Halgolluir (St Just in Penwith) **?heligan** pl., **?lowarth**

INDEX OF CORNISH PLACE-NAMES

Halgoss (Illogan) **hal**
Halgrosse Moore 1608 (St Winnow) **hal, *cors**
Hallaze (St Austell) **hal, glas**
Hall Dinas, coast (St Levan) **hal, *dynas**
Hall-dreath (Madron) **trait**
Halldrine Cove (Zennor) **?hal, ?dreyn**
Halldrunkard 1748 (Davidstow) **hal, *tron-gos**
Hallegan (Crowan) **heligen**
Halleggo (Kea) **?heligen** pl.
Hallenbeagle (Kenwyn) **hal, bugel/begel**
Hallgarden (Otterham) **hal, garan**
Hall Gommon, fld. (Paul) **?gubman**
Hall Gre 1351 (Madron) **?*gre**
Halligey (St Martin in Meneage) **heligen** pl.
Halloon (St Columb Major) **hel, goon**
Hallootts (Minster) **lyw** deriv.
Hallworthy (Davidstow) **hal**
Halmeers, fld. (Tintagel) **hal, meur**
Halnoweth (St Martin in Meneage; Probus) **hel, nowyth**
Halsferran (Gunwalloe) **als, yfarn**
Halstenick, fld. (St Blazey) **hal, stean** aj.
Halvana (Altarnun) **hyr, meneth**
Halvarras (Kea) **hal, margh** pl.
Halvenna (St Enoder) **hyr, meneth**
Halveor (St Columb Major) **hal, meur**
Halviggan (St Mewan) **hal, byghan**
Halvose (Manaccan; Merther) **hal, fos**
Halvosso (Mabe) ***havos** pl.
Halvousack, fld. (Lelant) **hal, mosek**
Halwartha (St Just in Roseland) **hal, guartha**
Halwin (Wendron) **hal, guyn**
Halwinnick (Linkinhorne) **?guyn** deriv.
Halwyn (St Keverne) **hel, goon**
Halwyn (Crantock; St Issey; Kea; Mylor) **hel, guyn**
Hamatethy (St Breward) ***havos**
Hammett (St Neot; Quethiock) ***havos**
Hampt (Stoke Climsland) ***havos**
Hantergantick (St Breward) ***hendre, ?*cant** aj.
Hantertavis (Mabe) **hanter, ?taves**
Harcourt (Feock) **?*ar, ?*crak**[1]
Harewood (Calstock) ***hyr-yarth**
Harlyn (St Merryn) ***ar, lyn**
Harros (Luxulyan) **hyr, *ros**
Harry Mussy, fld. (Lelant) **erw, mowes** pl.
Harvose (St Stephen in Brannel) ***ar, fos**
Haverack, fld. (Paul) ***havrek**
Havet (Liskeard) ***havos**
Havoc, fld. (St Tudy) **?haf** aj.

INDEX OF CORNISH PLACE-NAMES

Hay (Ladock) ***hay**
Hayle (Phillack) ***heyl**
Hayle Bay (St Minver) ***heyl**
Hayle Kimbro Pool (Ruan Major) **?hal, *Kembro**
Hay thu 1284 (Withiel) ***hay, du**
Headland (Pelynt) ***heth/hueth, lyn**
Helencane Cove (Goran) **hal, *keun?**
Heligan (St Ewe; Warleggan) **heligen**
Hell 1634 (Truro) **hel**
Hellamanning, The, fld. 1696 (St Columb Major) **amanen**
Hellan Conen 1671 (St Ives) **?kynin**
Helland (= parish) ***hen-lann**
Helland (Mabe; Probus) ***hen-lann**
Hellangove (Gulval) **hel, gof**
Hellarcher (Landewednack) **?hel**
Hellegan 1720 (Germoe) **heligen**
Hellesvean (St Ives) ***hen-lys, byghan**
Hellesveor (St Ives) ***hen-lys, meur**
Helligan (St Mabyn) **heligen**
Helliscoth 1337 (Helston) **coth**
Hellynoon (Lelant) **goon**
Helnoweth (Gulval) **hel, nowyth**
Helsbury (Michaelstow) ***hen-lys**
Helset (Lesnewth) **?*hen-lys**
Helston (Wendron) ***hen-lys**
Helstone (Lanteglos by Camelford) ***hen-lys**
Helvear Down (Scilly) ***heyl**
Helygy Whethlowe 1457 (St Martin in Meneage) **?whelth pl.**
Hembal (St Mewan) **hen, bal**
Hendawle (Davidstow) **hensy, ?*tawel**
Hendersick (Talland) ***hendre**
Henderweather (Minster) ***hendre, euiter**
Hendra (34 or more) ***hendre**
Hendraburnick (Davidstow) ***hendre, bronnen** aj.
Hendragoth (Perranzabuloe) ***hendre, kio**
Hendragreen (St Stephen by Launceston) ***hendre, dreyn**
Hendravossan (Perranzabuloe) ***hendre, fos** dimin.
Hendrawalls (Davidstow) ***hendre**
Hendrawna (Perranzabuloe) ***hendre**
Hendywills (Blisland) **hensy**
Henforth (St Martin in Meneage) ***hen-forth**
Hengar (St Tudy) **hen, *ker**
Hengastel 1314 (Scilly) **hen, castell**
Hengyvghall 1586 (Breage) **hensy, ughel**
Hennett (St Juliot) **hueth, nans**
Henon (St Breward) **hueth, nans**
Hensafraen (St Stephen in Brannel) **hensy, bran**
Hensavisten (St Stephen in Brannel) **hensy**

INDEX OF CORNISH PLACE-NAMES

Henscath, rock (Mullion)   ?**hen**, ?**skath**; or **-hins-**
Hensens 1315 (St Minver)   **hen**
Hensywassa 1461 (Wendron)   **hensy, guas** deriv.
Henver (St Allen)   ***hen-forth**
Henvor (Gwinear; St Hilary)   ***hen-forth**
Herland (Breage; Gwinear)   **hyr, lyn**
Hernis (Stithians)   **hyr, nans**
Herodsfoot (Duloe)   ***hyr-yarth**
Hervan (Ruan Minor)   **hyr, men**
Heskyn (St Germans)   **heschen**
Hethenaunt 1356 (St Winnow)   **hueth, nans**
Hewas (St Ewe; Ladock)   ***havos**
Hewas-an-Grouse, fld. 1671 (Sancreed)   ***havos, crous**
Hewes Common (Sancreed)   ***havos**
Hillcoose (Ladock)   **hel**
Hingey (Gunwalloe)   **hensy**
Hingham (Egloshayle)   ***hin**, ***cant**
Holseer Cove (Landewednack)   **als, hyr**
Hornick (St Stephen in Brannel)   **horn** aj.
Hor Point (St Ives)   ?**horþ**
House Oath, fld. (Sancreed)   ?***havos**, ?***aweth**
Howas, fld. (Paul)   ***havos**
Howes, fld. (Paul)   ***havos**
hryt ar þugan 1049 (S. 1019)   **rid**, ***ar**
hryt catwallon 967 (11th) (S. 755)   **rid**
hryt eselt 967 (11th) (S. 755)   **rid**
Huel Goffen, mine c. 1790 (St Just in Penwith)   **wheyl**, ***coffen**
Hugga-Dridgee, coast (Grade)   **googoo**, ?**trig** deriv.
Hughas, fld. (Mabe)   ***havos**
Hugus (Kea)   ***huel-gos**
Hunds, fld. (St Keverne)   ?**hans**
Hunds, The, fld. (St Keverne)   ?**hans**
Hurdon (Launceston)   ?**horþ**
Huthnance (Breage)   **hueth, nans**
Hwelan Tshei, mine c. 1700 (Gwennap)   **wheyl, chy**
Hwelan Vrân, mine c. 1700 (Gwennap)   **wheyl**
Hyngy Espayne, fld. 1492 (Redruth)   **hensy**
Hyrlas Rock (St Keverne)   ?**hyr**, ?**glas**

Idless (Kenwyn)   ?**aidlen**, *-**ys**²
Illiswilgig, rock (Scilly)   **enys, gwels** aj.
Ince (St Stephen by Saltash)   **enys**
Ingewidden (Grade)   **hensy**
Inisschawe c. 1540 (Scilly)   **enys, scawen** pl.
Innis (Luxulyan)   **enys**
Innis Pruen (Mullion)   **enys, prif** deriv.
Inny (river)   **onnen** pl., *-**i**
Inow (Constantine)   ?**hiuin** pl.

INDEX OF CORNISH PLACE-NAMES

Inswork (Maker) **enys**
Izzacampucca, coast (Scilly) ?***yslonk**, ?**bucka**

Jangye-ryn, coast (Gunwalloe) ?***rynn**
Joppa (St Just in Penwith; Sancreed) ?***shoppa**

Kaerclewent 1289 (Crowan) ***ker**
Kasekcan 1331 (Madron) ?**cassec**
Kearn Kee (Gerrans) **corn, kee**
Kebellans 1816 (Feock) ***caubalhint**
Kedue, fld. (Crowan) ?**kee**, ?**du**
Kehelland (Camborne) **kelli**, ***hen-lann**
Keigwin (St Just in Penwith) **kee, guyn**
Keiro (St Minver) ***ker** pl.
Kellewik 1302 (lost) **kelli**, ***gwyk**
Kelley (Blisland) **kelli**
Kellivose (Camborne) **fos**
Kellow (Lansallos; St Martin by Looe) **kelli** pl.
Kelly (Calstock; Egloshayle) **kelli**
Kellyengof 1265 (Sithney) **kelli, gof**
Kellyers, flds. (St Enoder) ?***kyl** pl.
Kellygreen (St Tudy) **kelli**, ***cren**
Kelsters (Kea) **cans, cellester**
Kelynack (St Just in Penwith) **kelin** aj.
Kemyel (Paul) ***-yel**
Kemyel Crease (Paul) **cres**
Kemyel Drea (Paul) **tre**
Kemyelnessa 1427 (Paul) **nessa**
Kemyelpella 1427 (Paul) **pell** spv.
Kemyel Wartha (Paul) **guartha**
Kenegie (Gulval) ***keun** aj. pl.
Kenegy pella 1665 (Breage) **pell** spv.
Kenidjack (St Just in Penwith) **kunys** aj.
Kenketh (Cardinham) ***kenkith**
Kennack (Grade) ***keun** aj.
Kennall (Stithians) ***-yel**
Kenneggy (Breage) ***keun** aj. pl.
Kenython (St Just in Penwith) **kee, eithin**
Kergilliack (Budock) **kee, kullyek**
Kernewas (St Keverne) ***kyniaf-vod**
Kernick (8 exx.) ***kernyk**
Kernock (St Stephen by Saltash) ***kernyk**
Kerrier (= hundred) ?***ker**, ?**-yer**
Kerris (Paul) ?***ker** deriv.
Kerrow (5 exx.) ***ker** pl.
Kerthen (Crowan) **kerden**
Keskeys (St Erth) ***ker, schus**
Kestal (St Hilary) **castell**

INDEX OF CORNISH PLACE-NAMES

Kestelcromlegh c. 939 (14th) (S. 450) **castell, *cromlegh**
Kestle (7 exx.) **castell**
Kestles, fld. (Blisland) **castell**
Kevar, fld. 1696 (Paul) **?*kevar**
Keveral (St Martin by Looe) **?*kevar, *-yel**
Kew Croft, fld. (Lelant) ***kew**
Keyrse (Treneglos) ***kew, *ros**
Kiberick Cove (Veryan) **?keber** aj.
Kilcobben Cove (Landewednack) **?*comm**
Kilcoys 1652 (Mylor) **?*kyl**
Kildown Cove (Ruan Minor) **?down**
Kilgeare (Boconnock) ***ker**
Kilgogue (Tywardreath) **kelli, *cok**
Kilgrew (St Keyne) **krow**
Kilhallen (Tywardreath) **?kelli, ?onnen** pl.
Kilkhampton (= parish) ***kylgh**
Killaworgey (St Columb Major) ***kyl**
Killeganogue (St Wenn) **kelli, *cnow** aj.
Killian 1383 (St Erth) **kelli** dimin.
Killianker (St Mawgan in Meneage) **kelli, cruc**
Killicoff (Egloskerry) **kelli, cough**
Killifreth (Kenwyn) **kelli, *brygh**
Killigam 1332 (Egloshayle?) **kelli**
Killiganoon (Feock) **kelli, *cnow** sg.
Killigarth (Talland) ***kyl, cath**
Killiow (Cornelly; Kea) **kelli** pl.
Killiserth (St Erme) **kelli, ?serth**
Killivorne 16th (lost) **kelli, ?forn**
Killivose (St Allen) **kelli, fos**
Killiwerris (Kea) **kelli, gweras**
Kilmanant (Braddock) ***kyl, *menawes**
Kilmansag (St Pinnock) **?*kyl**
Kilmarth (Tywardreath) **keyn/*kyl, margh**
Kilmar Tor (North Hill) **keyn/*kyl**
Kilminorth (Talland) ***kyl, *menawes**
Kilquite (St Germans) ***kyl/cul, cos**
Kingey (Grade) ***cun-jy**
Kirland (Bodmin) **?*cren**
Kirriers (Withiel) **?*ker** deriv.
Knagat (St Winnow) ***cnow, cos**
Knightor (St Austell) ***crygh, tyr**
Koffan, fld. (St Ewe) ***coffen**
Krucgruageh 1325 (Landewednack?) **cruc, gwrek** pl.
Kuggar (Grade) **?*coger**
Kus Skewes (Crowan) **cos, scawen** deriv.
Kylkethewe 1538 (Liskeard) ***kylgh, du**
Kylwethek 1302 (Kenwyn) ***collwyth** aj.

INDEX OF CORNISH PLACE-NAMES

Kynance (Gwithian; Mullion) *kew-nans
Kyvur Ankou c. 1720 (Gwennap) keverang pl.

Labham Rock (Landewednack) ?lam
Lackavear, fld. (St Keverne) lakka
Ladden, The, fld. (Crowan) ?glan
Laddenvean (St Keverne) *lann, byghan
Ladder Winze, mine 1798 (St Agnes?) ?*lether
La Feock (Feock) *lann
Laity (7 exx.) *lety
Lamana 1664 (Cury) *lann, manach
Lambessow (St Clement) *lann, bedewen pl.
Lambest (Menheniot) ?lanherch
Lambledra (Goran) lam, *lether
Lambo (Gwinear) ?*both
Lambourne (Perranzabuloe; Ruan Lanihorne) ?lyn, bronnen pl.
Lambrenny (Davidstow) *bren pl.
Lambsowden (Goran) lam
Lamellen (St Tudy) nans
Lamellion (Liskeard) nans, melin[1]
Lamellyn (St Blazey; Probus) nans, melin[1]
Lamellyon (Lanteglos by Fowey) nans, melin[1]
Lametton (St Keyne) ?*með
Lamin (Gwinear) nans, ?myn
Lamlavery (Davidstow) lam
Lamouth Creek (Feock) nans, mogh
Lampetho (Tywardreath) beth pl./bedewen pl.
Lanagan (St Teath) ?lyn
Lanarth (St Anthony in Meneage; St Keverne) lanherch
Lancallen (Goran) ?*calon
Lancarrow (Wendron) nans, carow/garow
Lancorla (St Wenn) nans, *cor-lann
Lancrow (Lanlivery) krow
Landabethick (Blisland) nans, ?guibeden aj.
Landare (Duloe) yar
Landegea (Kea) *to-
Landerio (Mylor) lyn
Landevale Wood (St Germans) nans, auallen
Landewednack (= parish) *to-
Landlizzick Wood (Landulph) ?les aj.
Landlooe (Liskeard) nans, *loch
Landraith 1284 (= St Blazey) *lann, trait
Landrake (= parish) lanherch
Landrends (Launceston) dreyn
Landreyne (North Hill) *lann/nans, dreyn
Landrine (Ladock) *lann/nans, dreyn
Landrivick (Manaccan) *hendre dimin.

INDEX OF CORNISH PLACE-NAMES

Landue (Lezant) **du**
Landulph (= parish) ***lann**
Laneast (= parish) ***lann**
Lanescot (Tywardreath) ***lys**
Laneskin Wood (Cardinham) **nans, ?heschen**
Langarth (Kenwyn) **?lyn, cath**
Langore (St Stephen by Launceston) **gover**
Langorthou 1311 (= Fowey) **?*cor¹ pl.**
Langreek (Lansallos) **cruc**
Langunnett (St Veep) ***lann**
Lanhadron (St Ewe) **nans, lader pl.**
Lanhargy (Linkinhorne) **lanherch pl.**
Lanhay (Gerrans) ***hay**
Lanhinsworth (St Columb Major) ***lann, hynse**
Lanhoose (Gerrans) **nans, los**
Lanhydrock (= parish) ***lann**
Lanivet (= parish) ***lann, *neved**
Lanjeth (St Stephen in Brannel) **yorch pl.**
Lanjew (Withiel) **lyn, du**
Lanjore (St Germans) **yar**
Lank (St Breward) ***lonk**
Lankeast (St Neot) **?*kest**
Lankelly (Lanreath) **nans, kelli**
Lankyp 1286 (= Duloe) ***lann**
Lanlawren (Lanteglos by Fowey) ***lann, lowarn**
Lanlovey (Cubert) ***lann**
Lannaled* 10th (= St Germans) ***lann, *aled
Lannear (Lansallos) **lyn, ?êr²**
Lanner (St Allen; Constantine; Gwennap; Kea; Sithney) **lanherch**
Lannowseynt 1549 (= St Kew) **sans pl.**
Lanow (St Kew) ***lann**
Lanowemeur 1443 (St Kew) **meur**
Lanoy (North Hill) **?oye**
Lanreath (= parish) ***lann**
Lansallos (= parish) ***lann**
Lanseaton (Liskeard) **nans**
Lansenwith (Stithians) **?onnen pl.**
Lansownick (Lamorran) **ownek**
Lansugle (South Hill) **?sŷgal**
Lanteglos (= 2 parishes) **nans, eglos**
Lantewey (St Neot) **nans, *dewy**
Lanthorne (St Germans) **bron**
Lantinning (St Anthony in Meneage) ***lann**
Lantivers (St Winnow) **?*evor**
Lantreise (St Neot) **nans, dreys**
Lantuel (St Wenn) **nans, tewl**
Lantyan (Golant) **?yeyn**
Lanuthno 1269 (= St Erth) ***lann**

276

INDEX OF CORNISH PLACE-NAMES

Lanvean (St Mawgan in Pydar) *****lann, byghan**
Lanwarnick (Duloe) **?lowarn** aj.
Lanwethenek 1351 (= Padstow) *****lann**
Lanyew (Kea) **nans, du**
Lanyon (Madron) **lyn, yeyn**
Lanzion (Egloskerry) *****seghan**
Larrick (Lezant; South Petherwin) **lanherch**
Latereena, fld. (Constantine) **?*lether**
Latter Scraggan, fld. (St Keverne) **?*lether**
Launcells (= parish) *****lann**
Laveddon (Bodmin) **bedewen**
Lavousa 1502 (= St Mawes) *****lann**
Lawarnick Cove (Mullion) **?lowarn** aj.
Lawellen (Withiel) **nans, melin**[1]
Lawhibbet (Golant) **nans, guibeden** pl.
Lawhippet (Lanteglos by Fowey) **nans, guibeden** pl.
Lawhittack (Colan) **nans, gwyth**[1] aj.
Lawhyre (Fowey) *****lann, wuir**
Lawithick (Mylor) *****lann, gwyth**[1] aj.
Leah (Phillack) **?*legh, ?*-a**
Lean (Liskeard; St Martin in Meneage) *****leyn**
Lean braus, fld. 1696 (Paul) *****leyn**
Leane an Trapp, fld. 1670 (Sancreed) *****leyn, *trap**
Lean Ridden, fld. (Gwithian) *****leyn, reden**
Ledrah (St Austell) **?*lether**
Legereath (Breage) **?les** pl., **gruah**
Legonna (St Columb Minor) **?lyn**
Legossick (St Issey) **logaz** aj.
Leha (St Buryan) **?*legh, ?*-a**
le(i)n broinn 977 (11th) (S. 755/832) **lyn, bronnen** pl.
Lellissick (Padstow) *****lann, ?*gwlesyk**
Lemail (Egloshayle) **nans**
Lemar (Cardinham) **margh**
Lemarth (St Mawgan in Meneage) **?lyn, margh**
Lemeers (St Just in Roseland) **nans, meur**
lenbrun 1059 (S. 1027) **lyn, bronnen** pl.
Lendra (Callington) **?lyn, ?tre**
Lene an Garrack, fld. 1588 (Sancreed) *****leyn**
Lene an Speren, fld. 1588 (Sancreed) *****leyn**
Lenears, fld. (St Ewe) **?*leyn**
Lene Nyendorwhy, fld. 1588 (Sancreed) *****leyn**
Lene Toll, fld. 1588 (Sancreed) *****leyn**
Lene ware an Stenack, fld. 1588 (Sancreed) *****leyn**
Lennabray (St Agnes) **melin**[1] pl., *****bre**
Lentrisidyn c. 939 (14th) (S. 450) **lyn**
Le Scathe Cove (St Buryan) **?*legh, skath**
Lesceave Por (Breage) **porth**
Lescrow (Fowey) **rid, *crew**

INDEX OF CORNISH PLACE-NAMES

Lesingey (Madron) ***lys**
Leskernick (Altarnun; Golant) ***lys, carn** aj.
Leskinnick (Madron) ***keun** aj.
Lesneage (St Keverne) ***lys, manach** aj.
Lesnewth (= parish) ***lys, nowyth**
Lesquite (Lanivet) **lost, cos**
Lesquite (Pelynt) ***is**[1]**, cos**
Lestoon (Luxulyan) **lost, goon**
Lestowder (St Keverne) ***lys**
Lestrainess (Constantine) ?***lys,** ?**dreyn** deriv.
Lestreth 1442 (Madron) ?**streyth**
Leswidden (St Just in Penwith) ***lys, guyn**
Lethlean (Phillack) ?***legh,** ?**lyen**
Lettor 1664 (Altarnun) ?***lether**
Levalsa (St Ewe) ?**aval, ti** (= **chy**)
Levardro (Probus) ?***lann,** ?**bar,** ?**trogh**
Levenus Hill (Lelant) ?**leven** deriv.
Lewannack, mine 1765 (Lelant) **lowen** aj.
Lewarne (St Neot) ***lann, guern**
Lewennick Cove (Crantock) ?**lowarn** aj.
Lewins (Newlyn East) ***legh**
Leyowne (Golant) ***lann**
Lezant (= parish) ***lann, sans**
Lezerea (Wendron) ***lys, *gre**
Lidcott (Laneast) **los, cos**
Lidcutt (Bodmin; Cardinham) **los, cos**
Lidden (St Just in Penwith) **lyn**
Lidgey (St Gluvias) ?**les** pl.
Lidgiow 1659 (St Ives) ?**les** pl.
Liggars (St Pinnock) **los, *garth**
Linkinhorne (= parish) ***lann**
Linnick (South Petherwin) ?**kelin** aj.
Lisbue (Sancreed) ***lys, *byu**
Liscawn (Sheviock) **nans, scawen**
Liskeard (= parish) ***lys,** ?**carow** pl.
Liskey Plantation (St Clement) **los, cos**
Liskis (Kenwyn) **los, cos**
Little Hell (Tywardreath) ***heyl**
Lizard (Landewednack) ***lys, *arð**
Loban Rath 1770 (Germoe) ?**lam,** ?**gruah**
Lo Cabm, coast (Mullion) ***loch, cam**
Locrenton (St Keyne) ***lann**
Loe (Feock; Sithney) ***loch**
Loggans (Phillack) ***cant**
Logo Rock (Budock) ?**lugh**
Longcoe (St Martin by Looe) ***lann**
Longsongarth (St Clement) **lost, cath**
Lonkeymoor (Altarnun) ***lonk**

278

INDEX OF CORNISH PLACE-NAMES

Looarth an Men, fld. 16th (Wendron?) **lowarth, men**
Looe (St Martin by Looe) ***loch**
Loranvellan, fld. 1803 (St Keverne) **lowarth**
Lost Kewthyry c. 1500 (Redruth) **lost**
Lostogh 1323 (Paul) **?lost, ?hoch**
Lostwithiel (Lanlivery) **lost, gwyth¹, *-yel**
Lowargh Lower, fld. 1570 (Paul) **lowarth**
Lowarglas, fld. 1405 (Breage?) **lowarth, glas**
Lowarth Thomas, fld. 1570 (Paul) **lowarth**
Lowarth Treuthale, fld. 1497 (Redruth) **lowarth**
Lowerth-Lavender-en-parson 1330 (Veryan) **lowarth**
Lowlizzick (Probus) ***lann, ?*gwlesyk**
Lowlysycke 1566 (Feock?) **?*lann, ?*gwlesyk**
Lucoombe (Quethiock) **?lyw**
Ludcott (St Ive) **los, cos**
Ludgvan (= parish) **?lusew, ?*-an²**
Ludgvan Lease (Ludgvan) ***lys**
Luna (St Neot) **?*lon pl.**
Luney (river) **?leven, *-i**
Lusuoneglos 1366 (Ludgvan) **eglos**
Luthergwearne (Madron) **?*lether, guern**
Luword en funteyn, fld. 1270 (Roche) **lowarth**
Luxulyan (= parish) ***lok**
Lydcott (Morval) **los, cos**
lyncenin 1059 (S. 832/1027) **lyn, kenin**

Maders (South Hill) ***með-ros**
maenber 969 (11th) (S. 770) **men, ber**
Maen Derrens, coast (St Ives) **men**
Maen Dower, coast (St Just in Penwith) **men, dour**
Maen-du Point (Perranuthnoe) **men, du**
Maenease Point (Goran) **men**
Maenenescop, coast 1302 (Scilly) **men, epscop**
Maengluthion 1320 (Roche) ***men-gleth pl.**
Maen-lay Rock (Goran) **men**
Maen Porth, coast (Budock) **men, porth**
Maen tol c. 939 (14th) (S. 450) **men, toll**
maen wynn 969 (11th) (S. 770) **men, guyn**
Magor (Illogan) ***magoer**
Mahen halen c. 939 (14th) (S. 450) **men**
Maker (= parish) ***magoer**
Manaccan (= parish) **?manach, ?*-an²**
Manack Point (St Levan) ***meynek**
Manely (St Veep) **myn, hyly**
Manhay (Wendron) ***meneghy**
Manheirs (Creed) ***men hyr**
Mankea (St Gluvias) **men, kee**
Mankependoim c. 939 (14th) (S. 450) **men, kee**

279

INDEX OF CORNISH PLACE-NAMES

Manskrethes, fld. 1696 (Zennor) **scrife** ppp.
Manuels (St Columb Minor) **men, ughel**
Manywithan, fld. (St Winnow) **meneth, gwyth**[1] sg.
Marazanvose (St Allen) **marghas, fos**
Marazion (St Hilary) **marghas, byghan**
Marcradden (St Just in Roseland) **men, crom**
Market Jew (St Hilary) **marghas, yow**
Marris (Cury) ***með-ros**
Matthiana (St Martin in Meneage) ***merther**
Mawgan Porth (St Mawgan in Pydar) **porth**
mayn biw 977 (11th) (S. 832) **men, *byu**
Mayon (Sennen) **men**
Mayrose (Lanteglos by Camelford) ***með-ros**
Maze Dippa, fld. (Lelant) **mes, *dyppa**
Meadrose (St Teath) ***með-ros**
Mean (Constantine) **men**
Meane (an) toll, rocks 1613 (Zennor) **men**
Meane Cadwarth, rock 1613 (Gwithian) **cadar**
Meane glase, rock 1613 (Zennor) **men**
Mean-Mellin, fld. (St Keverne) **melyn**[2]
Mean Monne, fld. 1649 (Constantine) **myn, mon**[1]
Mean-Moon, fld. (Perranuthnoe) **myn, mon**[1]
Meaver-crease (Mullion) **cres**
Medlyn (Wendron) ***með, lyn**
Mee Mun, The, fld. (Towednack) **myn, mon**[1]
Meene-Crouse-an-Especk 1613 (Zennor) **crous, epscop**
Meenkeverango 1580 (St Hilary) **keverang** pl.
Meinek, coast (Sennen) ***meynek**
Mejuggam, fld. (St Just in Roseland) **mẏdzhovan**
Melancoose (Colan) **melin**[1]**, cos**
Meledor (St Stephen in Brannel) ***lether**
Melinsey (Veryan) ***melyn-jy**
Mellangoose (Sithney; Wendron) **melin**[1]**, cos**
Mellanoweth (Phillack) **melin**[1]**, nowyth**
Mellanvrane (St Columb Minor; Phillack) **melin**[1]**, bran**
Mellanzeath (Constantine) **melin**[1]**, segh**
Mellenlappa 1593 (Newlyn East) **?clap**
Mellingey (Constantine; Cubert; St Issey; Perranarworthal) ***melyn-jy**
Mellingoose (Cornelly) **melin**[1]**, cos**
Mellionec (Colan) **melhyonen** aj.
Melorn (Minster) **men, lowarn**
Melynclap 1337 (Creed?) **melin**[1]**, clap**
Melyncroulis c. 1380 (Ludgvan) **melin**[1]
Melynmyhall 1445 (St Keverne) **melin**[1]
Melynnewyth 1366 (Kenwyn) **melin**[1]**, nowyth**
Melyntrait 1235 (Tywardreath) **melin**[1]**, trait**
Melyn Vedle 1390 (Breage) **melin**[1]
Mena (Lanivet) **meneth**

INDEX OF CORNISH PLACE-NAMES

Menabilly (Tywardreath) ?*byly/ebel pl.
Menaburle (Boconnock) meneth, ?breilu
Menaclidgey (Sithney) meneth, *clus pl.
Menacrin Downs (Temple) ?*cryn
Menacuddle (St Austell) meneth, *gwythel
Menadews (St Clement) meneth, du
Menadodda (St Germans) meneth
Menadrum, fld. (St Germans) meneth, ?*drum
Menadue (St Breward; St Cleer; Luxulyan; Talland; Tintagel) meneth, du
Menaglaze (St Neot) meneth, glas
Menagwins (St Austell; Goran) meneth, guyns
Menalhyl (river) ?*heyl
Menallack (Mabe) banathel aj.
Men-Amber, rock (Sithney) men
Mên an Mor, coast (St Ives) men, mor
Mên-an-Tol, rock (Madron) men, toll
Menawicket, fld. (Boconnock) guibeden pl.
Menawink (Lanlivery) ?guyns
Mendennick (Maker) ?myn, ?*dyn aj.
Mendy Pill, coast (St Winnow) *meyn-dy
Meneage (= district) manach aj.
Menear (St Austell) *men hyr
Menedarva (Camborne) *merther
Menedlaed c. 1260 (St Keverne) leth
Mene Gurta 1613 (St Breock) men, gortos
Menehay (Budock) *meneghy
Menelidan c. 1150 (= Trevalga) meneth, ledan
Menerdue (Stithians) meneth, du
Menergwidden (Gwennap) meneth, guyn
Menerlue (Stithians) meneth, lyw
Menewithan 1580 (St Germans) meneth, gwyth¹ sg.
Mengearne (Wendron) men
Menheniot (= parish) *ma
Menheniot (St Stephen by Launceston) melhyonen aj.
Menherion (Wendron) *men hyr pl.
Men Hewel, coast (Mullion) men, ughel
Menhire (Gwennap) *men hyr
Mên Hyr, coast (Landewednack) *men hyr
Menkee (St Mabyn) men, kee
Menmundy (St Stephen in Brannel) men, ?*mon-dy
Menna (Ladock; Stithians) meneth
Mennabroom (St Neot) ?crom
Mennaye, fld. (Madron) *meneghy
Menniridden (St Neot) meneth, reden
Mennor (Lelant) meneth
Mên Screfys, stone (Madron) men, scrife ppp.
Menstre 1275 (= Manaccan) ?*mynster
Mên Talhac (St Keverne) men, ?tal aj.

INDEX OF CORNISH PLACE-NAMES

Mên-te-heul, coast (Mullion) **tal**
Menvrause, fld. 1696 (Stithians) **men, bras**
Menwenick (Trewen) **men, ?guyn** deriv.
Menwidden (Ludgvan) **men** pl., **guyn** pl.
Menwinnion (Illogan) **men** pl., **guyn** pl.
Mên-y-grib Point (Mullion) **men, krib**
Meres (Mullion) ***með-ros**
Merrose (Gerrans; Illogan) ***með-ros**
Merrose Coath 1782 (Illogan) **coth**
Merrose Noweth 1539 (Illogan) **nowyth**
Merthen (St Austell; Constantine) ***merthyn**
Merthen Point (St Buryan) ?***merthyn**
Merther (= parish) ***merther**
Merther 1325 (Ruan Major) ***merther**
Mertherheuny 1302 (Redruth) ***merther**
Merthersithny 1304 (Sithney) ***merther**
Mertheruny (Wendron) ***merther**
Merthyr (Morvah) ***merther**
Meskevammok 1405 (St Issey) **mes, *kevammok**
Mesmear (St Minver) **mes, meur**
Messack (St Just in Roseland) **mes** aj.
Messenger, fld. (St Breock) **mes, *ker**
Messengrose (St Issey) **mes, crous**
Metherin (Blisland) ***með, ?*rynn**
Methers Colling (Gerrans) **?geluin**
Methleigh (St Austell; Breage) ***með, le**
Methrose (Goran; Luxulyan; St Mewan) ***með-ros**
Meudon (Mawnan) ***dyn**
Mevagissey (= parish) **ha(g)**
Meynwinionwartha 1318 (Ludgvan) **guartha**
Millendreath (St Martin by Looe) **melin¹, trait**
Millen Keeve 1600 (Camborne) ?***kyf**
Millewarn (Cury) **men, lowarn**
Millook (Poundstock) ?***melek**
Minawint, fld. (Talland) **meneth, guyns**
Mingoose (St Agnes) **myn, cos**
Minmanueth, coast (Scilly) ?***menawes**
Minnimeer (Tremaine) **meneth, meur**
Minny Moon, fld. 1788 (Crowan) **myn, mon¹**
Molenick (St Germans) **melhyonen** aj.
Molevenney Quarry (St Germans) ?***moyl, ?meneth**
Molingey (St Austell) ***melin-jy**
Molinnis (St Austell) ***moyl, enys**
Monek de Nanscorlyes, mine 1400 (Gwennap) ***mon²** aj.
Mongleath (Budock) ***mon-gleth**
Monhek-cam, mine 1400 (Gwennap) ***mon²** aj., **cam**
Morah, The, rock (St Keverne) ?**morhoch**
Moreps, fld. c. 1720 (Gulval) ***morrep**

INDEX OF CORNISH PLACE-NAMES

Mornick (South Hill) *moryon aj.
Morrab (Madron) *morrep
Morvah (Landrake) *morva
Morvah (= parish) ?mor, ?beth
Mouls, The, rock (St Minver) mols
Mountjoy (Colan) *meyn-dy
Muchlarnick (Pelynt) lanherch
Mulberry (Lanivet) *moyl, *bre
Mulfra (Madron) *moyl, *bre
Mulvin, coast (Landewednack) ?*moyl, ?men
Mulvra (St Austell) *moyl, *bre
Murraphs, fld. (Gunwalloe) *morrep
Myllynalsey 1550 (St Buryan) melin¹
Myne an Downze, mine 18th (St Just in Penwith) men pl., *dons
Mythyanwoeles 1341 (St Agnes) goles

Nampara (Perranzabuloe) bara
Namprathick (Merther) nans, *bryth deriv.
Nancarrow (St Allen; St Michael Penkevil) nans, carow/garow
Nancassick (Feock) nans, cassec
Nance (St Clement; Illogan; Lelant; St Martin in Meneage) nans
Nancealverne (Madron) nans
Nancegollan (Crowan) nans, *ygolen
Nancekuke (Illogan) cuic
Nanceloan (St Mawgan in Meneage) nans, lowen
Nancelone (Perranzabuloe) nans, lowen
Nancemabyn (Probus) nans
Nancemeer (Newlyn East) nans, meur
Nancemellin (Gwithian) nans, melin¹
Nancemeor (St Clement) nans, meur
Nancetrisack (Sithney) nans, dreys aj.
Nancewrath (Kenwyn) nans, gruah
Nancledra (Towednack) nans
Nancollas (Constantine) nans, *cores
Nancor (Creed) corf
Nancorras (St Just in Roseland) nans, *cores
Nangidnall, fld. (Madron) (goon), *aswy, *enyal
Nanjarrow (Constantine) *arð pl.
Nanjenkin (Breage) ?*henkyn
Nanjewick (St Allen) nans, *ewyk
Nanjizal (St Levan) nans, ysel
Nanjulian (St Just in Penwith) elin
Nankelly (St Columb Major) nans, kelli
Nankervis (St Enoder) ?carow pl.
Nankilly (Ladock; Probus) nans, kelli
Nanpean (St Enoder; St Just in Penwith; St Stephen in Brannel; Stithians) nans, byghan
Nanphysick (St Mewan) nans, fodic

INDEX OF CORNISH PLACE-NAMES

Nanplough (Cury) **nans, blogh**
Nanquitho (Sancreed) **?gwyth¹ pl.**
Nansavallan (Kea) **nans, auallen**
Nansawsan (Ladock) **nans, Zowzon**
Nanscawen (Luxulyan) **nans, scawen**
Nansclegy 1267 (St Keverne) ***clav-jy**
Nanscow (St Breock) **nans, scawen** pl.
Nansdeuy 1302 (St Mabyn) **nans, *dewy**
Nanseers (Philleigh) **sichor**
Nansent 1310 (= St Breock) **?sans** pl.
Nansevyn (St Mawgan in Meneage) **nans, hiuin**
Nansfonteyn 1281 (= Little Petherick) **nans, fenten**
Nanslason 1702 (Creed) ***glasen**
Nansloe (Wendron) **nans, *loch**
Nanslowoeles 1337 (Wendron) **goles**
Nansmellyn (Perranzabuloe) **nans, melin¹**
Nansmeor veor 1613 (Newlyn East) **meur**
Nansough (Ladock) **nans, hoch**
Nanspian (Gunwalloe) **nans, byghan**
Nanstibragga 1289 (St Ewe) **nans**
Nanstrelaec 1306 (St Keverne) **nans**
Nanswhyden (St Columb Major) **nans, gwyth¹** sg.
nantbuorðtel 1059 (S. 832/1027) **nans, *buorth, teil**
Nanteague (St Allen) **nans, teg**
Nanterrow (Gwithian) **dar** pl.
Nantillio (St Enoder) **nans, deyl** pl.
Nantrelew (Mylor) **nans**
Nantrisack (Constantine) **nans, dreys** aj.
Nanzeglos (Madron) **nans, eglos**
Naphene (Constantine) **nans, fyn**
Nathaga Rocks, coast (Gwithian) **leth** aj. pl.
Neithvrane (St Keverne) **(carn), nyth, bran**
Newlyn (Paul) **?lu, lyn**
Newporth (St Mawgan in Pydar; Budock) **porth**
Ninnes (St Allen; Lelant; Madron) **enys**
Ninnis (Germoe; Gwennap; St Mewan; Wendron) **enys**
Ninniss Farm (Kenwyn) **enys**
Noon Billas (Towednack) **goon, *pylas**
Noongallas (Gulval) **goon, cales**
Noon Gay (Camborne) **?kee**
Noonvares (Crowan) **goon, margh** pl.
Noon Veres, fld. (Lelant) **goon, margh** pl.
Northwethel (Scilly) ***ar, *gothel**
Numphra (St Just in Penwith) ***bre**

Ogo-Dour Cove (Mullion) **googoo, dour**
Ogo Mesul, coast (Mullion) **googoo, ?mesclen** pl.
Ogo Pons, coast (Mullion) **googoo, pons**

INDEX OF CORNISH PLACE-NAMES

One-vean, fld. (Paul) **goon**
Ony Pokis, fld. (Paul) **goon**, ?**bucka**
Opetjew, street n.d. (Truro) ***op, du**
Owan Vrose (St Buryan) **goon, bras**
Owen Brose, fld. 1665 (Crowan) **goon, bras**
Owen Vean, fld. (Crowan) **goon**
Own Smith, fld. c. 1800 (Ludgvan) **goon**

Padgagarrack Cove (St Anthony in Meneage) ?**peswar**, ?**karrek**
Padge Dinner, fld. (Wendron) **peswar, dyner**
Padgy Vessa, fld. (St Ives) ?**peswar**
Padzhuera, fld. (Zennor) **peswar, erw**
Palestine (Mabe) **pen, glastan**
Pantersbridge (St Neot) **pons**
Par (St Blazey) **porth**
Parbola (Gwinear) **pol**
Parc-an-Als Cliff (Sithney) **park, als**
Parc-an-bibble, fld. 1737 (Illogan) ***pybell**
Parc an gerthen, fld. (Ludgvan) **kerden**
Parcan Growes (Madron) **crous**
Parc Joppa (Ruan Major) ***shoppa**
Parc Wollas (Mullion) **park, goles**
Pargodonnel Rocks (Breage) ?**cudin** pl.
Park-a-dour, fld. (Cardinham) **dour**
Park Ain, fld. (Lelant) **on** pl.
Park-an-Arlothas 1633 (Gunwalloe) **park, arluth** fem.
Park-an-Askerne, fld. 1649 (Constantine) **ascorn**
Park Anauhan, fld. 1696 (Madron) **oghan**
Park an Aule Cliff, fld. 1696 (Lelant) **als**
Park an Bant, fld. (Gerrans) ?***pans**
Park an Bays, fld. (Paul) ?***bas**
Park-an-Beau, fld. 1649 (Constantine) **bugh**
Park (an) Bew, flds. (Paul; Constantine) **park, bugh**
Parkanboer, fld. 1547 (Sithney) ?***peur**
Park (an) Bowen, flds. (Paul; Gwennap; etc.) **park, bowyn**
Park-an-Bowgy, flds. 1649 (Constantine) **boudzhi**
Park-an-Bush, fld. (Mabe) **park**
Park-an-Cady (St Buryan) ***car-jy**
Park an Chamber, fld. 1696 (Ludgvan) **chammbour**
Park an daunce, fld. (Gwinear) ***dons**
Park-an-Davers, fld. 1690 (St Keverne) **daves**
Park-and-Crow, fld. (St Keverne) **krow**
Park an Devers, fld. (Ruan Major) **daves** pl.
Parkandillack (St Dennis) **teil** aj.
Parkandower, fld. 1540 (Wendron) **dour**
Park-an-Drean, fld. 1649 (Constantine) **park, dreyn**
Park-an-Drise, fld. 1649 (Constantine) **dreys**
Park-an-Eane, fld. 1649 (Constantine) **on** pl.

INDEX OF CORNISH PLACE-NAMES

Park an Eball, fld. 1696 (Madron) **park, ebel**
Park an Eglos, fld. 1696 (Gwithian) **eglos**
Park-an-Erbyer, fld. 1649 (Constantine) ***erber**
Park-an-Errow, fld. 1649 (Constantine) **erw**
Park-an-fold (St Mawgan in Meneage) ***fold**
Park-an-Fox (St Keverne) **park**
Park-an-Garratt, fld. (St Keverne) **karrek**
Park an Garrow, fld. (St Ives) **garow**
Park an Gegen, fld. 1696 (Breage) ***kegen**
Parkan Glasson, fld. (St Ewe) ***glasen**
Park an Goag Cuckow, fld. 1696 (St Erth) **park, *cok**
Park an Gollen, fld. (Perranarworthal) ***coll** sg.
Park an Gothal, fld. (Redruth) ***gothel**
Park-an-Gove, fld. 1649 (Constantine) **gof**
Park-an-Gubman, fld. 1649 (Constantine) **gubman**
Park-an-Gwarrack, fld. 1649 (Constantine) **guarthek**
Parkanheer, fld. (Gwennap) **hyr**
Park-an-Herbio, fld. 1696 (Phillack) ***erber**
Parkanhere, fld. (Paul) **hyr**
Park an Horr, fld. n.d. (Gwennap) **horþ**
Park-an-huns, fld. (Feock) **park, ?hans**
Park an ithan, fld. (Paul) **park, eithin**
Park an Jarne, fld. (Crowan) **jarden**
Park an Jarns, fld. 1649 (Constantine) **jarden**
Park an Kellier, fld. 1696 (Sancreed) **?*kyl** pl.
Park an Lastrack, fld. 1696 (Mullion) **elester** aj.
Park-an-Lower, fld. (St Keverne) **lowarth**
Parkanloy, fld. (St Ewe) **?lugh** pl.
Parkanmo, fld. 1745 (Wendron) **mogh**
Park-an-Moah, fld. 1649 (Constantine) **mogh**
Park-an-Munkyer, fld. 1649 (Constantine) **?bynkiar**
Park-an-Nithen, fld. (Constantine) **eithin**
Park-an-Patrick, fld. (St Keverne) **park**
Park-an-Pease, fld. (St Keverne) **pêz**
Park an Pillars, fld. (Lelant) ***pylas**
Park-an-Pitt, fld. 1649 (Constantine) **pyt**
Park-an-Prowlter, fld. (St Keverne) **park, pronter**
Park an Rogers, fld. 1696 (Paul) **park, an**
Park-an-Skeeber, fld. 1649 (Constantine) **skyber**
Park-an-Sparnon, fld. 1649 (Constantine) **spern** sg.
Park-an-stable, fld. 1649 (Constantine) ***stable**
Park an Stampes, fld. (Gwennap) **stampes**
Park-an-starve-us, fld. (St Keverne) **park, an**
Park an Tarro, fld. 1696 (Wendron) **park, tarow**
Park-an-Tidno (St Keverne) **park**
Park-an-Toddan, fld. 1649 (Constantine) **ton**
Park an Trap, fld. (Paul) ***trap**
Park-an-trath, fld. (St Anthony in Roseland) **trait**

286

INDEX OF CORNISH PLACE-NAMES

Park-an-Trawn, fld. 1767 (St Keverne) ?**tron**
Park-an-Troan, fld. 1649 (Constantine) ?**tron**
Park-an-Tule, fld. (St Keverne) ***tewel**
Park an Vave, fld. 1696 (Mullion) **fav**
Park an Veer, fld. (Redruth) ?**margh** pl.
Park-an-Velvas, fld. (Budock) ***melwhes**
Park-an-Vethen, fld. 1649 (Constantine) **budin**
Park-an-Vorn, fld. 1649 (Constantine) **forn**
Park an Yate, fld. (St Ives) **park, yet**
Park-an-yeat, fld. 1649 (Constantine) **yet**
Parkbakou 1345 (Ludgvan) **park**
Park Banker, fld. (Paul) ?**bynkiar**
Park Bannel, fld. (Paul) **banathel**
Park Bean, flds. (Paul; St Keverne) **park, byghan**
Park Belender, fld. 1696 (Mullion) ***melynder**
Park Bell, fld. (Lelant) **park, pell**
Park Brase, fld. (Crowan) **bras**
Park Braws (Landewednack) **park, bras**
Park-Cabben, fld. 1649 (Constantine) **cam**
Park Caregy, fld. 1649 (Constantine) **karrek** pl.
Park Chaple, fld. (St Martin in Meneage) **park, chapel**
Park Cheple, fld. 1660 (Crowan) **chapel**
Park Chynoweth, fld. (Lelant) **park**
Park Connin, fld. (Breage) **kynin**
Park Crane, fld. 1696 (Paul) ?***creun**
Park Crazie, fld. (Perranzabuloe) **cres**
Park Crease, flds. 1649 (Constantine) **cres**
Park Crees, flds. (St Keverne) **park, cres**
Park dannandre, fld. 1696 (Gwinear) **dan**
Park Darras, flds. (Paul) **park, daras**
Park Dolar, fld. 1696 (Luxulyan) ***talar**
Park Drysack, fld. (St Keverne) **park, dreys** aj.
Parke an arlothas, fld. 1613 (Perranuthnoe) **arluth** fem.
Parke an Clibmier, fld. 1630 (St Erth) ***colomyer**
Parke-an-Erbbear, fld. 1574 (St Erth) ***erber**
Parke an Onyon, fld. 1630 (St Erth) **park**
Park East, fld. (St Anthony in Meneage) **yst**
Parke Chypons 14th (St Hilary?) **park**
Parke Forbelancan, fld. 1683 (St Buryan) **forth**
Park Eglos, 1808 (St Mawgan in Pydar) **park, eglos**
Parkenbewgh, fld. 1630 (St Ewe) **bugh**
Parken Chapel, fld. (St Just in Penwith) **chapel**
Parken Clyes, fld. 1630 (St Ewe) ***cleys**
Parkenconstable, fld. c. 1636 (Golant) **park**
Park en Course, fld. (Paul) ***cors**
Parken-Crow, fld. (Constantine) **krow**
Parken Davers, fld. 1696 (Crantock) **daves**
Park Ene, fld. 1696 (Stithians) **on** pl./**yn**

INDEX OF CORNISH PLACE-NAMES

Parkengear (Probus) *ker
Park en Gellyn 1335 (lost) **park, kelin**
Park en Hoar, fld. (Paul) **horþ**
Parken Legagen, fld. 1710 (St Keverne) **logaz** sg.
Park-en-lorn, fld. (Paul) **lowarn**
Parkenmelyn, fld. 1332 (St Erth?) **park, melin**[1]
Parkenmethek 1423 (Colan) **park, methek**
Parkenponds, fld. (Gerrans) **pons**
Parken Reden, fld. 1696 (Cubert) **park, reden**
Parkentol, fld. (Gerrans) **toll**
Parkentower, fld. (Gerrans) ***towargh**
Parke Reene, fld. 1634 (Paul) ***run**
Park Eughandrea, fld. 1684 (Ludgvan) **park, ugh, tre**
Park Fave, fld. (Lelant) **park, fav**
Park Favin, fld. (Lelant) **?fav** sg.
Park Frigles, fld. (Paul) **forth**
Park Fringey, fld. (Lelant) **?forth, ?hensy**
Park-Garrack, fld. (Mabe) **park, karrek**
Park Gernick, fld. (Paul) ***kernyk**
Park Glaston, fld. (St Clement) **glastan**
Park Gorland, fld. 1800 (Feock) ***cor-lann**
Park Gue, fld. (Roche) ***kew**
Park Gwarra, fld. (Paul) **park, guartha**
Park Gwella, fld. 1649 (Constantine) **?guella**
Park gwidden, fld. 1696 (Towednack) **park, guyn**
Park Heer, fld. 1649 (Constantine) **park, hyr**
Parkhenver (Redruth) **park, *hen-forth**
Parkhour, fld. 1791 (Ruan Lanihorne) **horþ**
Park Jearn, fld. (Paul) **jarden**
Park Lean, fld. (Paul) **lyen**
Park Lew, fld. (Paul) **lugh**
Park Lines, fld. (Paul) **park, ?lynas**
Park Little, fld. (St Keverne) **park**
Park Lore, fld. (Paul) **lowarth**
Park Luar, fld. (Paul) **lowarth**
Park Mabier, fld. (Phillack) ***mabyar**
Park Magla, fld. 1696 (Gwinear) **?maga** deriv.
Park Maine, fld. (St Anthony in Meneage) **park**
Park-Manin, fld. (Mabe) **amanen**
Park-Mannin, fld. (Mabe) **park, amanen**
Park Martin, fld. (Crowan) **park**
Park Matthews, fld. (St Austell) **park**
Park Mennes, fld. 1649 (Constantine) **munys**
Park Minnus, fld. (Gerrans) **munys**
Park Moya, fld. 1696 (Lelant) **meur** spv.
Park Moyha, flds. 1649 (Constantine) **park, meur** spv.
Park Nessa, fld. (Gwinear) **nessa**
Park-Nevas, fld. (St Keverne) **?daves**

288

INDEX OF CORNISH PLACE-NAMES

Park Nine, fld. (Redruth) ?on pl.
Parknoweth (Newlyn East) **park, nowyth**
Park Oan, fld. (Lelant) **park, on**
Park Ose, fld. (Roche) ?**hos**
Park Peas, fld. (Lelant) **park, pêz**
Park Peeth, flds. (Crowan) *****puth**
Park Pell, fld. 1649 (Constantine) **park, pell**
Park Pronter 1740 (Crowan) **park, pronter**
Park Rattle, fld. (Paul) ?*****radgel**
Park Reen, fld. 1696 (Breage) *****run**
Park Saffron, fld. 1814 (St Gluvias) **park**
Park Sea, fld. (Crowan) ?**segh**
Park Shaftis, fld. (St Ives) **sheft** pl.
Park Shafty, fld. (Paul) **sheft**
Park-Skewis, fld. (St Keverne) **scawen** deriv.
Park Skiber, fld. (Lelant) **skyber**
Park Skipper, fld. (Paul) **skyber**
Park-soule, fld. 1649 (Constantine) **zowl**
Park Spernon, fld. (Lelant) **park, spern** sg.
Park Stenack, fld. 1649 (Constantine) **stean** aj.
Park Steven, fld. (Gwennap) **park**
Park Tannow, fld. 1649 (Constantine) **tanow**
Park Tobma, fld. 1649 (Constantine) **tum**
Park Todden, fld. (St Ewe) **ton**
Park Treath, fld. (Gerrans) **trait**
Park Trone, fld. (St Ives) ?**tron**
Park Try-Corner, fld. 1715 (St Keverne) **park**
Park Tuban, fld. (Paul) **tum**
Park Venton Sah, fld. (Mullion) **fenten, segh**
Park Vine, fld. (Crowan) **park**
Park Voregles, fld. 1649 (Constantine) **forth**
Park Vorn, fld. (Crowan) **park, forn**
Park Vor Trevedra, fld. (Sennen) **forth**
Park-Warra, fld. (St Keverne) **park, guartha**
Park West, fld. (St Anthony in Meneage) **west**
Park Woolas, fld. 1696 (Ludgvan) **goles**
Park Wriggles, fld. (St Keverne) **forth**
Park Yates, fld. (Lelant) **yet** pl.
Parn Voose Cove (Landewednack) **fos**
Partonvrane (Gerrans) **fenten, bran**
Pathada (Menheniot) *****pedreda**
Patrieda (Linkinhorne) *****pedreda**
Peal, The, coast (Sennen) **pyl**
Peber, rock (Sennen) ?*****pybor**
Pedden an wollas c. 1700 (Sennen) **gulas**
Peddenporperre 1696 (Landewednack) **pen**
Pedenleda, fld. 1696 (Zennor) ?*****lether**
Pedn-an-drea (Redruth) *****pen (an) dre**

INDEX OF CORNISH PLACE-NAMES

Pednanpill Point 1597 (Feock) **pen, *pyll**
Pednanvounder (Ruan Major) **bounder**
Pednavounder (St Keverne; Sithney) **bounder**
Pedn Bejuffin, fld. (Paul) **mẏdzhovan**
Pedn Billy, coast (Constantine) ***byly**
Pedn Crifton, coast (Mullion) **?*crygh**
Pedndrea, fld. (Lelant) ***pen (an) dre**
Pedn Kei, rock (Zennor) **pen, ?ky**
Pedn-myin, coast (St Keverne) **pen, men** pl.
Pedn Olva, coast (St Ives) **pen, guillua**
Pednpons (Madron) ***pen (an) pons**
Pedn Tenjack, fld. (Paul) **pen, denjack**
Pedn Tiere, coast (St Keverne) ***pen-tyr**
Pednvadan, coast (Gerrans) **pen, tal, ban**
Pedn Venton (Madron) ***pen-fenten**
Pednvounder (St Levan) **pen, bounder**
Pedn y coanse (Paul) **pen, *cawns**
Peel Cove (Tywardreath) **?pyl**
Peele Myne 1613 (Zennor) **pyl, men** pl.
Pehel-Carnawhenis 1660 (St Stephen in Brannel) **kodna huilan**
Pelean (Tywardreath) ***pen-lyn**
Pelistry (Scilly) **porth, lester** pl.
Pell, The, mine (St Agnes) **?pell**
Pellagenna (St Cleer) **?pellen, ?ganow**
Pellengarrow (St Kew) ***pen-lyn, garow**
Pellyn-wartha (Perranarworthal) ***pen-lyn, guartha**
Peloe (Crowan) **?ieu**
Pelyn (Lanlivery) ***pen-lyn**
Pelyne (Lanreath) ***pen-lyn**
Pelynt (= parish) **plu**
Pemboa (St Mawgan in Meneage) **pen, bogh**
Penadlick (Lanreath) **banathel** aj.
Pen a Gader, coast (St Agnes) **pen, cadar**
Penair (St Clement) ***pen-arth**
Penalguy (Constantine) ***pen hal**
Penaligon Downs (Bodmin) **pen, heligen**
Pen-a-maen, coast (Goran) **pen, men**
Penance (St Issey) ***pen nans**
Penance Wood (St Breock) ***pen nans**
Penanchase 1447 (Kenwyn) **pen, *chas**
Penare (Goran; St Keverne) ***pen-arth**
Penare Point (Mevagissey) ***pen-arth**
Penarth (St Clement) ***pen-arth**
Penarth-Point c. 1540 (Tywardreath) ***pen-arth**
Penarthtown (Morval) ***pen-arth**
Penavartha (St Winnow) **pen**
Penbeagle (St Ives) **pen, begel**
Penbegle 1696 (Merther) **pen, begel**

INDEX OF CORNISH PLACE-NAMES

Penberth (St Buryan) **ben, *perth**
Penbetha (Creed) **pen, bedewen**
Penbetha (Probus) **beth pl./bedewen** pl.
Pen Blue, fld. (St Enoder) **pen, plu**
Penbothidnow (Constantine) **pen, budin** pl.
Penbough (St Stephen in Brannel) **pen, bogh**
Penbro (Breage) **pen, bro**
Penbugle (Bodmin; Duloe) **bugel**
Penburthen (Lanivet) **pen, bedewen**
Pencabe, coast (Gerrans) **capa**
Pencair c. 1540 (Germoe) **pen, *ker**
Pencalenick (St Clement) **pen, kelin** aj.
Pencarrow (Advent) **pen, *ker** pl.
Pencarrow (St Austell; Egloshayle) **pen, carow**
Pencarrow Head (Lanteglos by Fowey) **pen, carow**
Pencobben (Camborne) **pen, *comm**
Pencoose (St Ewe; Gwennap; Kenwyn; Perranarworthal) ***pen cos**
Pencoose (Cuby; St Erme; St Gluvias; Probus; Wendron) **?*pen cos**
Pencoose (Stithians) **bron, cos**
Pencorse (St Enoder) **pen, *cors**
Pencoys (Wendron) ***pen cos**
Pencrennow (Perranzabuloe) **?*crann** pl.
Pencrowd (Menheniot) **pen, *krow-jy**
Pendarves (Camborne) **pen, ?dar** deriv.
Pendavey (Egloshayle) **ben, *dewy**
Pendavey (Minster) **pen, *dewy**
Pendavy (St Breward) **?ben, *dewy**
Pendeen (St Just in Penwith; Sheviock) **pen, *dyn**
Pendennis Castle (Budock) **pen, *dynas**
Penderleath (Towednack) **?*pen (an) dre, ?*legh**
Pendewey (Bodmin) **?ben, *dewy**
Pendinas c. 1540 (St Ives) **pen, *dynas**
Pendine (Roche) **pen, *dyn**
Pen Diu, coast (Tintagel) **du**
Pendoggett (St Kew) **dew, cos**
Pendower (Philleigh) **ben, dour**
Pendray, fld. (St Ewe) ***pen (an) dre**
Pendrea (St Buryan) ***pen (an) dre**
Pendriffey (Pelynt) ***pen (an) dre** pl.
Pendrift (Blisland) ***pen (an) dre**
Pendrim (St Martin by Looe) **pen, *drum**
Pendriscott (Duloe) **dres**
Pendrissick, fld. (Morval) **pen, dreys** aj.
Penearth (Menheniot) ***pen-arth**
Penelewey (Kea) ***pen hal**
Penellick (Tywardreath) **pen, heligen** pl.
Penenpons 1315 (Merther) ***pen (an) pons**
Pen Enys Point (St Ives) **pen, enys**

INDEX OF CORNISH PLACE-NAMES

Penfound (Poundstock) **pen,** ?***faw**
Penfrane (St Pinnock) **pen, bran**
penfynfos c. 963 (14th) (S. 810) ***fynfos**
Pengadoer 1302 (Mevagissey?) **pen, cadar**
Pengarrock (St Keverne) **pen, karrek**
Pengegon (Camborne) ***kegen**
Pengelly (12 or more exx.) ***pen (an) gelli**
Pengellylowarn 1321 (Perranzabuloe) ***pen (an) gelli, lowarn**
Pengenna (St Kew) **pen,** ?***genn** deriv.
Pengersick (Breage; Mullion) **pen, *cors** aj.
Penglaze (Kenwyn; Lansallos; Perranzabuloe) **pol, glas**
Pengluthio 1575 (Ladock) ?***men-gleth** pl.
Pengover (Menheniot) **pen, gover**
Pengreep (Gwennap) **pen, krib**
Pengwedna (Breage) **pen**
Penhaldarva (Kenwyn) ?***darva**
Penhale (Davidstow) **ben, hal**
Penhale (c. 26 exx.) ***pen hal**
Penhale-an-Drea (Breage) **tre**
Penhalemanyn 1651 (St Hilary) **pen**
Penhallam (Jacobstow) **pen, *alun**
Penhallick (St Keverne) **banathel** aj.
Penhallow (8 exx.) ***pen hal** pl.
Penhalt (Poundstock) **pen, als**
Penhalurick (Stithians) ***pen hal,** ?***luryk**
Penhargard (Helland) **pen, *hyr-yarth**
Penharget (St Ive) **pen, *hyr-yarth**
Penhawger (Menheniot) **pen**
Penheale (Egloskerry) **pen, hel**
Penhedra (St Austell) **pen, *hyr-drum**
Penhellick (St Clement; Illogan; St Pinnock; St Wenn) **pen, heligen** pl.
Penhesken (Ruan Lanihorne) **pen, heschen**
Penhill, fld. (St Mabyn) ?**pen**
Penhole (North Hill) **pol, hal**
Penimble (St Germans) ?***emle**
Penisker (St Mewan) **pen,** ?***esker**
Penjerrick (Budock) ***pen nans**
Penkelly (Pelynt) ***pen (an) gelli** pl.
Penkelyou 13th (Roche) ***pen (an) gelli** pl.
Penkestle (Braddock; St Neot) **pen, castell**
Penkevil (St Michael Penkevil) **pen, *kevyl**
Penknight (Lanlivery) **pen, *cnegh**
Penkynans 1613 (Mullion) **pen**
Penlean (Poundstock) ***pen-lyn**
Penlee Point (Paul; Rame) **pen, *legh**
Penmarth (Wendron) ?***pen-arth**
Penmayne (St Minver) **pen, men**
Penmenna (Philleigh) **pen, meneth**

292

INDEX OF CORNISH PLACE-NAMES

Penmennor (St Buryan) **pen, meneth**
Penmenor (Stithians) **pen, meneth**
Penmillard (Rame) ?***moyl**, ?***arð**
Penmynytheowe 1461 (St Clement?) **pen, meneth** pl.
Pennance (7 exx.) ***pen nans**
Pennans (Creed) ***pen nans**
Pennant (7 exx.) ***pen nans**
Pennards (St Breock) ?***pen-arth**
Pennare (St Allen; Veryan) ***pen-arth**
Pennare Wallas (Veryan) **goles**
Pennare Wartha (Veryan) **guartha**
Pennatillie (St Columb Major) ***pen nans, deyl** pl.
Pennellick (Pelynt) **pen, heligen** pl.
pennhal meglar 977 (11th) (S. 832) ***pen hal**
penn hal weðoc 969 (11th) (S. 770) ***pen hal, gwyth**[1] aj.
penn lidanuwern 1049 (S. 1019) **pen, ledan, guern**
Pennycraddock (St Cleer) **pen**
Pennycrocker (St Juliot) **pen,** ?***crak**[1] pl.
Pennydevern (St Clether) **pen,** ?***devr-** deriv.
Pennygillam (Launceston) **pen**
Pennytinney (St Kew) ***pen-fenten** pl.
Penny Vounder (Goran) **pen, bounder**
Pennywilgie Point (St Minver) **pen,** ?***gwailgi**
Penolva (Paul) **pen, guillua**
Pen Olver, coast (Landewednack) **pen, guillua**
Penoweth (Mylor) **pons, nowyth**
Penpell (Cornelly; Lanlivery) ***pel/pell**
Penperth (St Just in Roseland) **pen,** ***perth**
Penpethy (Lanteglos by Camelford) ?**tyn**
Penpill (Stoke Climsland) ***pen pol**
Penpillick (Tywardreath) ***pel** aj./**pyl** aj.
Penpol(l) (10 exx.) ***pen pol**
penpoll lannmoren 969 (11th) (S. 770) ***pen pol**
Penponds (Camborne) ***pen (an) pons**
Penpons 1608 (St Hilary) ***pen (an) pons**
Penpons n.d. (Kenwyn) ***pen (an) pons**
Penpons c. 1470 (Probus) ***pen (an) pons**
Penpons Haverack 1727 (Madron) ***pen (an) pons, *havrek**
Penponsmethle 1394 (Breage) ***pen (an) pons**
Penpont (Altarnun; St Breward; St Kew; St Mawgan in Pydar) ***pen (an) pons**
Penpraze (Illogan; Sithney) **pen, pras**
Penquean (St Breock) **pen,** ***keun**
Penquindle (Egloshayle) **pen**
Penquit (St Pinnock) ***pen cos**
Penquite (14 exx.) ***pen cos**
Penrice (St Austell) ***red**
Penrose (11 exx.) ***pen ros**

INDEX OF CORNISH PLACE-NAMES

Penryn (St Gluvias) ***pen-ryn**
Pensagillies (St Ewe) ?***lys**
Penscawn (St Enoder) **pen, scawen**
Penscove (Mylor) ***pen cos, gof**
Pensignance (Gwennap) **pen, segh, nans**
Pensousen, coast 1582 (Towednack) **Zowzon**
Penstrace (Lanivet) **pen, *stras**
Penstrase (Kenwyn) **pen, *stras**
Penstrassoe (St Ewe) **pen, *stras** pl.
Penstraze (Roche) **pen, *stras**
Penstroda (Blisland) **pen, *stras** pl.
Pentewan (St Austell) **towan**
Penteyrwoloys 1439 (St Eval?) **goles**
Pentire (Crantock; St Eval; St Minver; St Tudy; Wendron) ***pen-tyr**
Pentireglaze (St Minver) **glas**
Pentivale (Roche) ***pen-fenten**
Pentreath (Breage) **pen, trait**
Pentreath Beach (Landewednack) **pen, trait**
Penvearn (Cury) **pen, bron**
Penventinnie (Kenwyn) ***pen-fenten** pl.
Penventinue (Fowey) ***pen-fenten** pl.
Penventon (St Austell; Braddock; Gwennap; St Juliot; Sithney) ***penfenten**
Penvith (St Martin by Looe) **pen, bugh**
Penvories (St Mawgan in Meneage) ***pen-ryn**
Penvose (Cornelly; St Mawgan in Pydar; St Tudy; Veryan) **pen, fos**
Penwarden (South Hill) **pen, guern**
Penwarne (Cuby; Mawnan; Mevagissey) **pen, guern**
Penwerris (Budock) **pen, gweras**
Penwithick (St Austell) ?**gwyth**[1] aj.
Penzance (Madron) **sans**
Penzer Point (Paul) ?**serth**
Perbargus Point (Veryan) ?**porth**, ?**bargos**
Perbean Beach (St Michael Caerhays) **porth, byghan**
Perdredda (St Germans) ***pedreda**
Perlees (St Breock) ?**les**
Perles (St Minver) **porhel**
Perprean Cove (St Keverne) **porth, byghan**
Perpry, fld. (St Ewe) **pwl prî**
Perranarworthal (= parish) ***ar, *gothel**
Perranporth (Perranzabuloe) **porth**
Perrose (St Enoder; Lanlivery) ***pen ros**
Persquiddle, fld. (Goran) **park, crous**, ?***gwythel**
Peter Varrow, fld. (Lelant) ?**peswar** fem.
Pibyah Rock (Gerrans) ?***pybor**
Pich Kissa, mine n.d. (Gwennap) **pyt** pl.
Pinnick (Fowey) **pen, *ewyk**
Pirran in Treth 1425 (= Perranzabuloe) **trait**

INDEX OF CORNISH PLACE-NAMES

Pistil Ogo (Landewednack) ?*pystyll
Pit-pry, fld. 1649 (Constantine) pyt, pry
Pitshanpassen, mine c. 1700 (Gwennap) pyt pl.
Pitslewren (Kenwyn) pyt pl., lowarn
Pits Mingle (Roche) pyt pl., ?*men-gleth
Pitsonbriney n.d. (Gwennap) pyt pl.
Pitt Pry, fld. 1649 (Constantine) pyt
Pitts en bryne 17th (St Allen) pyt pl.
Place (Padstow) ?plas
Placea Enfenten 1502 (St Just in Roseland) plas, fenten
Place an Streat 1632 (lost) plas, *stret
Place House (St Anthony in Roseland; Fowey) ?plas
Placengarrett 1392 (St Ives) plas
Placengennov 1392 (St Ives) plas, ?*genn pl.
Plain-an-Gwarry (St Hilary; Redruth) plen, guary
Plain an Quarry 1816 (St Ives) plen, guary
Plane-an-Gwarry (Sithney) plen, guary
Plean-an-Wartha 1696 (Constantine) plen, guary
Pleming (Gulval) ?plu, ?men
Plewe-Golen 1501 (= Colan) plu
Plewgolom 1543 (= St Columb Major?) plu
Pleyn-Goyl-Sithney 1396 (Perranuthnoe) plen
Plu-alyn 15th (= St Allen) plu
Pluvogan 1523 (= St Mawgan in Meneage?) plu
Pluysie 1548 (= St Issey) plu
Pluyust 16th (= St Just in Penwith?) plu
Polandanarrow Lode (St Just in Penwith) pol, ?dyner pl.
Polangrain (Wendron) polgrean
Polbelorrack (St Keverne) pol, beler aj.
Polborder (Pillaton) *boðour
Pol Bream, coast (Mullion) ?*breyn
Polbream Cove (Landewednack) ?*breyn
Polbream Point (Grade) ?*breyn
Polbreen (St Agnes) pol, *breyn
Polbrock (Egloshayle) pol, brogh
Polbrough (St Just in Roseland) pol, brogh
Polcan (Lanlivery) pol, can
Polcarowe 1430 (Camborne) pol, carow
Polcatt (St Germans) pol, *cagh
Polcocks (Lezant) pol, *cagh
Polcoverack (St Keverne) pol
Polcrebo (Crowan) pol, krib pl.
Polcreek (Veryan) pol, cruc
Polcrybowe Grees c. 1480 (Crowan) cres
Poldew (Lanlivery; Liskeard) pol, du
Poldhu Cove (Mullion) pol, du
Poldower (St Erme) pol, dour
Poldowrian (St Keverne) ben, *douran

295

INDEX OF CORNISH PLACE-NAMES

Poldrea (Tywardreath) ?**tre**
Poldue (Advent; Blisland) **pol, du**
Polean (Pelynt) ***pen-lyn**
Poleskan (Cury) **pol, heschen**
Polgassick Cove (St Ives) **pol, cassec**
Polgath 1502 (Towednack) **pol, cath**
Polgaver Beach (St Austell) **pol, gaver**
Polgazick (Lansallos) **pol, cassec**
Polgear (Wendron) **pol, *ker**
Polgeel Wood (Egloshayle) **pol, ghel**
Pol Glâs, coast (Mullion) **pol, glas**
Polglase (Crowan; Cury; St Erme; Wendron) **pol, glas**
Polglaze (Altarnun; St Austell; Cuby; Fowey; St Mabyn; Mylor; Philleigh; St Veep) **pol, glas**
Polgoda (Luxulyan) ?***gosa**
Polgooth (St Ewe) **pol, goth²**
Polgorran (Goran) **pol**
Polgover (Morval) **pol, gover**
Polgrain (St Michael Caerhays; St Wenn) **polgrean**
Polgray (Altarnun) **pol, *gre**
Polgrean (Cury; Ludgvan) **polgrean**
Polgrean, flds. (Constantine; Crowan; Paul) **polgrean**
Polgreen (St Mawgan in Pydar; Newlyn East; St Veep) **polgrean**
Polgrunemuer 1345 (Madron) **meur**
Polgwidden (Grade) **pol, guyn**
Polgwidden Cove (Constantine) **pol, guyn**
Polgwyn Beach (St Austell) **pol, guyn**
Polhendra (Gerrans) **pol, *hendre**
Polhigey (Ludgvan) **pol, hos** pl.
Polhorden (Lanlivery) **pen, *hyr-drum**
Poligy (Wendron) **pol, hos** pl.
Poliskin (St Erme) **pol, heschen**
Polita (St Columb Major) **park, *lety**
Polkanuggo (Stithians) **pol, cronek** pl.
Polkernogo (St Keverne) **pol, cronek** pl.
Polkerris (Tywardreath) ?**cherhit**
Polkerth (Gerrans) ?**cherhit**
Polkinghorne (Gulval; Gwinear) **pol**
Polkirt (Mevagissey) ?**cherhit**
Polladras (Breage) **pol, ?*ladres**
Pollaughan (Gerrans) **pol, oghan**
Pol Lawrance, coast (St Keverne) **pol, arghans**
Pollawyn (Colan) ***pen hal, guyn**
Polledan, coast (Landewednack) **pol, ?ledan**
Pol Ledan, coast (St Levan) **pol, ?ledan**
Pollemellecke 1565 (Cardinham) ?***melek**
Poll Fry (Menheniot) **pwl prî**
Pollglese, fld. (Blisland) **pol, glas**

INDEX OF CORNISH PLACE-NAMES

Pollgreen, fld. (Lansallos) **polgrean**
poll hæscen 1059 (S. 832/1027) **pol, heschen**
Pollpry, fld. 1649 (Constantine) **pwl prî**
Pollstack, fld. (St Breock) ***stak**
Polly Vellyn, fld. (St Gluvias) **pol, melin**[1]
Polmanter (St Ives) **porth**
Polmark (St Merryn) **pol, margh**
Polmarth (Wendron) **pol, margh**
Polmartin (Lanreath) **pol**
Polmassick (St Ewe) **pons**
Polmaugan (St Winnow) **pol**
Polmear (Tywardreath) **pol**
Polmear (St Austell) **porth, meur**
Polmenna (St Enoder; Liskeard; St Neot; Veryan) **pen, meneth**
Polmere, flds. (St Germans; Lanteglos by Camelford) **pol, meur**
Polmesk (Philleigh) **pol**
Polmogh c. 1500 (Kenwyn) **pol, mogh**
Polmorder 1731 (St Austell) ?***boðour**
Polmorla (St Breock) **pol, *morva**
Polnare Cove (St Keverne) **pol**
Polostoc Zawn (St Levan) ?**lost** aj.
Polpeer, fld. (Tintagel) **pol, pur**
Polpenwith (Constantine) **pol, ruth**
Polpeor (Landewednack) (**pen**)
Polpeor (Lelant) **pol, pur**
Polperro (Lansallos) **porth**
Polpever (Duloe) ?**pol**, ?***pever**
Polpidnick (St Keverne) ***pennek**
Polpri c. 1250 (Warleggan) **pwl prî**
Polpry (Blisland) **pwl prî**
Polpry 1314 (Antony) **pwl prî**
Polpry 1374 (St Hilary) **pwl prî**
Polpry Cove (St Just in Penwith; St Levan) **pwl prî**
Polpuckey (Talland) **bucka**
Polquick (St Clement) ?***gwyk**
Polridmouth (Tywardreath) **porth, rid, men**
Polroad (St Tudy) ?***pol ros** or ***rud**
Polrose, flds. (Lanteglos by Fowey; St Veep) ***pol ros**
Polrose Mine (Breage) ***pol ros**
Polruan (Lanteglos by Fowey) **porth**
Polrudden (St Austell) **pol, ?reden**
Polscatha (Mylor) **skath** pl.
Polscoe (St Winnow) **pol, skath**
Polsethow 1478 (Budock) **seth** pl.
Polshea (St Tudy) **pol, segh**
Polskeys (Roche) **pol, cos, schus**
Polstaggs meadow 1672 (Lanlivery) ***stak**
Polstags, fld. (St Kew) ***stak**

INDEX OF CORNISH PLACE-NAMES

Polstain (St Allen) **pul stean**
Polstangey Bridge (Ruan Minor) **pons**
Polstein (Kenwyn) **pul stean**
Polstreath, coast (Mevagissey) **pol**, **?streyth**
Polstrong (Camborne) **pol**, ***stronk**
Polsue (St Erme; St Ewe; Goran) **pol**, **du**
Polsue (Lanteglos by Fowey) **pol**, **segh**
Polsue (Philleigh) **du**
Poltarrow (St Mewan) **pol**, **tarow**
Polteggan (Madron) **?teg** deriv.
Poltesco (Ruan Minor) **pol**, **?*tusk** deriv.
Poltisco (St Clement) **pol**, **?*tusk** deriv.
Poltreworgy (St Kew) **pol**
Polurrian (Mullion) **?beler** pl.
Polveithan (St Veep) **pen**, **budin**
Polvellan (Fowey) **pol**, **melin**[1]
Polventon (Lansallos; Warleggan) ***pen-fenten**
Polwarrack Mills 1616 (Perranarworthal) **?guarthek**
Polwhele (St Clement) **pol**, **hwilen** pl.
Polwheveral (Constantine) **pol**, ***whevrer**
Polwhevererwartha 1474 (Constantine) **guartha**
Polwin (Cury) ***pen hal**, **guyn**
Polwrath (St Cleer) **pol**, **gruah**
Polyn (Gerrans) ***pen-lyn**
Polyphant (Lewannick) **pol**, ***lefant**
Polzeath (St Minver) **pol**, **segh**
Polzion Wood (Pelynt) ***seghan**
Poniou (Zennor) **pons** pl.
Ponjeravah (Constantine) **pons**
Ponjou (Gulval) **pons** pl.
Ponsandane (Gulval) **pons**, **den**
Ponsangavern n.d. (Madron) **pons**
Ponsanooth (St Gluvias) **pons**, **goth**[1]
Ponsbeggal Bridge c. 1700 (Cury) **pons**, **begel**
Pons Bennett (Madron) **pons**
Ponsbrital (Camborne) **pons**, **brytyll**
Ponsfrancke 1571 (St Keverne) **pons**, **Frank**
Pons-glastannen 13th (St Clement) **glastan** sg.
Ponsgovar c. 1260 (St Enoder) **pons**, **gover**
Ponsharden (Budock) **pons**
Ponskeyre 1430 (Creed?) **pons**
Ponslego (Perranzabuloe) **pons**, **?les** pl.
Ponsmain (Feock) **pons**, **men**
Ponsmean 1617 (Perranzabuloe) **pons**, **men**
Ponsmedda (Ruan Major) **pons**, ***merther**
Ponsmere (Perranzabuloe) **pons**, **meur**
Ponsmur 1302 (Creed) **pons**, **meur**
Ponsongath (St Keverne) **pons**, **cath**

INDEX OF CORNISH PLACE-NAMES

Ponson Joppa (Ruan Major) **pons, *shoppa**
Ponson Tuel (St Mawgan in Meneage) **pons, *tewel**
Ponspren 1549 (Perranzabuloe) **pons, pren**
Ponsprontiryon c. 939 (14th) (S. 450) **pons, pronter** pl.
Ponsreleubes 1298 (St Hilary) **pons**
Ponstresilyan 1386 (Merther) **pons**
Ponswragh 1506 (St Keverne) **pons, gruah**
Pont (Lanteglos by Fowey) **pons**
Pontshallow (Madron) **pons, hal** pl.
Ponts Mill (Lanlivery) **pons**
Pooldown (Sennen) **pol, down**
Poole-an-abelly 1660 (St Stephen in Brannel) **pol, ebel** pl.
Poolpanenna (St Buryan) **pol, ?*pennyn** pl.
Pordenack Point (Sennen) ***dyn** aj.
Porgwarrak, fld. 1649 (Constantine) **guarthek**
Porkellis (Wendron) **porth, kellys**
Porloe (Mylor) **porth, *loch**
Porpoder Field 1649 (Constantine) **?*peur, ?podar**
Porqueese, fld. (St Minver) **?guis**
Port Gaverne (St Endellion) **porth**
Porth (several) **porth**
Porthallack, coast (Mawnan) **?porth, ?hal** aj.
Porthallow (St Keverne) **porth, *alaw**
Porthbean Beach (Gerrans) **porth, byghan**
Porthbeer Cove (St Keverne) **porth, meur**
Porthbeor Beach (St Anthony in Roseland) **porth, meur**
Porth-cadjack Cove (Illogan) **porth**
Porthcaswith 1360 (Grade) **porth**
Porthcew, coast (Breage) **?*kew**
Porth Chapel, coast (St Levan) **porth, chapel**
Porthchaple 1516 (St Agnes) **porth, chapel**
Porthcollumb (St Erth) **pol, ?*kel** cpd.
Porth Cornick, coast (Gerrans) **porth, *kernyk**
Porthcovrec 1262 (St Keverne) **porth**
Porthcuel (Gerrans) **porth, cul**
Porth Curno (St Levan) **porth, corn** pl.
Porthe an Badall, coast 1580 (Illogan?) **porth, padel**
Porthe Brenegan, coast 1580 (Illogan) **porth, brenigan**
Porth-en-Alls (St Hilary) **porth, als**
Porthencrous 1324 (Paul) **porth, crous**
Porthennis c. 1700 (Paul) **porth, enys**
Porthglaze Cove (Zennor) **porth, glas**
Porthglyvyon 1334 (St Mawgan in Pydar) **porth**
Porthguarnon, coast (St Buryan) **porth, ?guern** sg.
Porthgwarra (St Levan) **porth, ?*gorweth** pl.
Porthgwidden (Feock) **porth, guyn**
Porth Gwidden, coast (St Ives) **porth, guyn**
Porthia 1356 (St Ives) **porth**

299

INDEX OF CORNISH PLACE-NAMES

Porthillieglos 1355 (St Minver) **eglos**
Porthilly (St Minver) **porth, hyly**
Porth Island, coast (Cubert) ?***heyl** dimin.
Porth Joke, coast (Cubert) **porth, les** aj.
Porthkea (Kea) **porth**
Porthkerris (St Keverne) **porth, ?*cors, *-ys**[2]
Porth Kidney Sands (Lelant) **pol, cummyas**
Porth Killier, coast (Scilly) ***kyl** pl.
Porth Ledden, coast (St Just in Penwith) **porth, ?ledan**
Porthleven (Sithney) **porth, leven**
Porth Loe, coast (St Levan) **porth, *loch**
Porth Logas, coast 1580 (Zennor) **porth, logaz**
Porthluney Cove (St Michael Caerhays) **porth**
Porth Main, coast 1580 (St Levan) **porth, men**
Porthmawgan 1755 (St Mawgan in Pydar) **porth**
Porth Mear, coast (St Eval) **porth, meur**
Porth Mellin (Mullion) **porth, melin**[1]
Porthmeor (St Ives; Zennor) **porth, meur**
Porth Minnick (Scilly) **porth, *meynek**
Porthminster (St Ives) **porth, *mynster**
Porth Navas (Constantine) **daves**
Portholland (St Michael Caerhays) **porth, *hen-lann**
Porthoy 1481 (Poundstock) **porth, ?oye**
Porth Pean (St Austell) **porth, byghan**
Porth Perane, coast 1580 (Perranuthnoe) **porth**
Porth Pyg, coast (Mullion) **porth, ?*pyk**
Porthpyghan 1399 (Talland) **porth, byghan**
Porth Saxon, coast (Mawnan) **porth, Zowzon**
Porthtowan (St Austell; Illogan) **porth, towan**
Porthylly Gres 1541 (St Minver) **cres**
Porthzennor Cove (Zennor) **porth**
Port Isaac (St Endellion) **porth, ?usion** aj.
Portkillock (St Minver) **porth, kullyek**
Portlegh 1345 (Breage) **porth, *legh**
Portloe (Veryan) **porth, *loch**
Portlooe (Talland) **porth**
Portmellon (Mevagissey) **porth, melin**[1]
Portquin (St Endellion) **porth, guyn**
Portreath (Illogan) **porth, trait**
Portscatho (Gerrans) **porth, skath** pl.
Portsenen 1461 (Sennen) **porth**
Portwrinkle (Sheviock) **porth**
Post Mean n.d. (Gwennap) **post**
Poulpere c. 1540 (Constantine) **pol, pur**
Poundergourth, fld. (Crowan) **coth**
Powder (= hundred) **pow, ?ardar** pl.
Praze (St Erth; Grade; Sithney) **pras**
Praze, fld. (St Breock) **pras**

INDEX OF CORNISH PLACE-NAMES

Praze, The, fld. (St Gluvias) **pras**
Praze-an-Beeble (Crowan) **pras, *pybell**
Prazegooth (Grade) **pras**
Prazeruth (Wendron) **pras, ruth**
Prazey (St Dennis) **pras**
Predannack Wollas (Mullion) **goles**
Preeze (Cardinham) ***prys**
Preish Gelley 1618 (Mullion) **kelli**
Prenestin (Goran) **pen, heschen**
Presingol (St Agnes) ***prys, *coll**
Presthilleck 1567 (Creed) **pras, ?heligen** pl.
Pridden (St Buryan) ***pen-ryn**
Prideaux (Luxulyan) **?pry**
Priest's Cove (St Just in Penwith) **porth**
Priscan (St Keverne) **?*prys** dimin.
Priske (Mullion) ***prys**
Prislow (Budock) ***prys, *loch**
Progo, coast (St Just in Penwith) **porth, googoo**
Prospidnack-an-Chaple n.d. (Sithney) **chapel**
Prospidnick (Sithney) ***prys, ?pin-** aj.
Prynullin 1768 (Gwennap) **?pren, ?elaw** sg.
Pulldown (Breage) **pol, down**
Pulpry (St Just in Roseland) **pwl prî**
Pulstean c. 1720 (St Agnes) **pul stean**
Pyg, rock (Landewednack) **?*pyk**

Quarthendrea 1702 (Padstow) **guartha, tre**
Quies, rocks (St Merryn) **?guis**

Race Farm (Camborne) ***red**
Radell, The, 1613 (Zennor) **?*radgel**
Radjel (Constantine) **?*radgel**
Raftra (St Levan) **?rag, tre**
Raggot Hill (North Tamerton) **?rag**
Raginnis (Paul) **rag, enys**
Rebanhale, fld. 1680 (Sancreed) **ryp**
Redallen (Breage) **rid, ?*alun**
Redannack, fld. (Paul) **reden** aj.
Redannick (Truro) **reden** aj.
Redevallen (Trevalga) **rid, auallen**
Redinnick (Madron) **reden** aj.
Redruth (= parish) **rid, ruth**
Redwith Mill 1650 (Lanlivery) ***rodwyth**
Reen (Perranzabuloe) ***run**
Reenie Lovan, fld. 1634 (Paul) ***run, lovan**
Reens, fld. 1649 (Constantine) ***run**
Reens, The, fld. (Lelant) ***run**
Rees (Perranzabuloe) **rid**

INDEX OF CORNISH PLACE-NAMES

Rejarne (Lelant) ?**horn**
Releath (Crowan) **rid, *legh**
Relistien (Gwinear) **rid,** ?**cellester** deriv.
Relowas (St Mawgan in Meneage) ?***ar,** ?**lyw** aj.
Relubbus (St Hilary) **rid**
Reperry (Lanivet) ?***bery**
Rescorla (St Ewe) **rid/*ros, *cor-lann**
Resoon (Gulval) **goon**
Resores (Gerrans) **rid, hos**
Resparva (St Enoder) **rid/*ros, perveth**
Resparveth (Probus) **rid/*ros, perveth**
Respryn (Lanhydrock) **rid,** ?**bran** pl.
Restineas (St Blazey) **rid**
Restormel (Lanlivery) ?**tor,** ?***moyl**
Restowrack (St Dennis) ***ros, dour** aj.
Restronguet (Mylor) ***ros, *tron-gos**
Resugga (St Austell; St Erme; St Stephen in Brannel) ***ros, googoo**
Resurrance (St Enoder) **rid**
Retallack (Constantine; St Hilary) **rid, hal** aj./**tal** aj.
Retallick (St Columb Major; Roche) **rid, heligen** pl.
Retanna (Wendron) **rid, tanow**
Retanning (St Mewan) **rid**
Reterth (St Columb Major) **rid, *derch/torch**
Retew (St Enoder) **rid**
Retier 1596 (Ladock) **rid, hyr**
Retire (Withiel) **rid, hyr**
Retorrick (St Mawgan in Pydar) **rid**
Retyn (St Enoder) **rid,** ?***dyn**
Rezare (Lezant) **rid, *ker**
Rice (Goran) **rid**
Ridgeo (Gulval) **rid** pl.
Ridmerky 13th (Altarnun?) **rid**
Rillaton (Linkinhorne) **rid, *legh**
Rinsey (Breage) ***rynn, ti** (= **chy**)
Rissick (Perranarworthal) ***red** deriv.
Rooke (St Kew) **ruth, *cok**
Roose (Laneast; Otterham; Treneglos) ***ros**
Roozen Cove (Lansallos) ?***rosan**
Ros c. 839 (14th) (= Rame?) ***ros**
Roscarecmur c. 1290 (St Endellion) **meur**
Roscarnon (St Keverne) ***ros, carn** deriv.
Roscarrack (Budock) ***ros**
Roscarrec bian 1249 (St Endellion) **byghan**
Roscarrock (St Endellion) ***ros, karrek**
Roscollas (Mabe) ***coll** deriv.
Roscroggan (Illogan) ***ros, crogen**
Roscrowgey (St Keverne) ***ros, *krow-jy**
Rosculiangoth 1350 (Little Petherick) **coth**

INDEX OF CORNISH PLACE-NAMES

Rose (St Breward; Davidstow) ***ros**
Rose an Castle (Breage) **castell**
Rose-an-Grouse (St Erth) **rid/*ros, crous**
Rose-an-Hale Cove (Zennor) **hal**
Roseath (Stithians) **rid, segh**
Rosebargus (Goran) **lost, bargos**
Rosebenault (Davidstow) ***ros, *menawes**
Rosecadghill (Madron) ***ros**
Rosecare (St Gennys) ***ros, *ker**
Rosecassa (St Just in Roseland) ***ros, *caswyth**
Rosecliston (Crantock) **rid, ?cellester** deriv.
Rosecraddock (St Cleer) **rid**
Rosedown (Camborne) **rid, down**
Rosegarden (Kenwyn) **rid, garan**
Roseglos (Budock) ***ros, eglos**
Rosegothe (Tywardreath) ***ros, gof**
Rose in the valley (St Kew) **rid/*ros, auallen**
Roseladden (Sithney) **rid, ledan**
Roseland (= district) ***ros**
Roselidden (Wendron) ***ros, lyn**
Roselyon (St Blazey) ***ros**
Rosemaddock (Liskeard) ***ros**
Rosemanowas (Stithians) ***ros, *menawes**
Rosemellen (Warbstow) **melin**[1]
Rosemelling (Luxulyan) ***ros, melin**[1]
Rosemellyn (Roche) **melin**[1]
Rosemerryn (Budock) ***ros**
Rosemodress (St Buryan) ***ros**
Rosemullion (Mawnan) ***ros, ?melhyonen** pl.
Rosemundy (St Agnes) ***ros, *mon-dy**
Rosenannon (St Wenn) ***ros, onnen**
Rosenbeagle, fld. (Paul) **bugel/begel**
Rosen Cliff (Veryan) **?*rosan**
Rosenithon (St Keverne) ***ros, eithin**
Rosenun (Liskeard) ***ros, onnen**
Roserrow (St Minver) **?erw**
Rosesuggan (St Columb Major) **rid**
Rosevallen (St Stephen in Brannel) **rid/*ros, auallen**
Rosevallon (Cuby) **rid/*ros, auallen**
Roseveth (Kenwyn) **rid, margh** pl.
Rosevidney (Ludgvan) ***ros**
Rosevine (Gerrans) **?*breyn**
Rosewall (Towednack) **rid, ?*gwal**
Rosewastis (St Columb Major) ***ros**
Rosewin (St Minver) ***ros, guyn**
Rosewin (St Enoder) **rid, guyn**
Roseworthy (Kenwyn) ***ros**
Roskear (Camborne) **rid, *ker**

INDEX OF CORNISH PLACE-NAMES

Roskear (St Breock) *ros, *ker
Roskearnoweth 1830 (Camborne) nowyth
Rôskennals (Sancreed) rid
Rôskestal (St Levan) *ros, castell
Roskief (St Allen) *kyf
Roskilly (St Keverne) *ros, kelli
Roskorwell (St Keverne) *ros
Roskrow (St Gluvias) *ros, krow
Roskruge (St Anthony in Meneage) *ros, cruc
Roskymmer (St Mawgan in Meneage) rid, *kemer
Rôspannel (St Buryan) *ros, banathel
Rospeath (Ludgvan) *bich
Rospreages (St Austell) rid, pry aj.
Rosteague (Gerrans) *ros, ?*daek
Rosudgeon (St Hilary) *ros, odion
Rosuic (St Keverne) *ros, *ewyk
Rowse (Pillaton) *ros
Ruthdower (Breage) ruth, dour
Ruthvoes (St Columb Major) ruth, fos
Ruzza (Lanlivery) *ros pl.
Ryniau (Mullion) ?*rynn pl.

St Agnes, island (Scilly) enys
St Columb Porth (St Columb Minor) porth
St Dennis (= parish) *dynas
St Golder, fld. (Paul) sýgal, tyr
St Ingunger (Lanivet) *stum
St Jidgey (St Issey) sans
Sarvan, rock (Crantock) ?serth
Savan Marake, coast 1580 (Sennen) sawn, marrek
Savath (Luxulyan) enys, margh pl.
Savenheer, coast 1597 (St Anthony in Roseland) sawn, hyr
Savenlester Cove, coast 1597 (Budock) sawn, lester
Saveock (Kea) sevi aj.
Savyn an Skanow, coast 1580 (Sennen) sawn
Savyn Dolle, coast 1580 (Towednack) sawn
Scarsick (Trengelos) ?*ros, cassec
Scathe, rock (Landewednack) skath
Scawen 1337 (Perranzabuloe) scawen
Scawn (St Pinnock) scawen
Scawn Hill, fld. (Pillaton) scawen
Sconhoe (Mevagissey) heschen pl.
Scorran Lode (St Just in Penwith) ?scorren
Scor Rock (Goran) ?scorren
Scoryacres 1404 (Gwennap) cres
Scorya wartha 1404 (Gwennap) guartha
Scovarn, rock (Mullion) scovern
Scraesdon (Antony) *crew, rid; or krow, *lys

INDEX OF CORNISH PLACE-NAMES

Scrawsdon (St Ive) ?**crogen**
Screek Wood (St Martin by Looe) **cruc**
Screws (Davidstow) **cres**
Seghy, rock (St Levan) **segh** deriv.
Segoulder, fld. (Madron) **sȳgal, tyr**
sehfrod 1044 (S. 1005) **segh, frot**
Sellan (Sancreed) **segh,** *****lann/lyn**
Sellanvean (Sancreed) **byghan**
Selligan (Redruth) **rid, heligen**
Sewrah (Stithians) **gruah**
Sheviock (= parish) ?**sevi** aj.
Shrubhendra (St Endellion) *****hendre**
Signans 1538 (Newlyn East?) **segh, nans**
Silver Carn (Scilly) **carn**
Sithnoe (Breage) **segh, *tnou**
Skebervannel 1725 (Crowan) **skyber, banathel**
Skeburia 1753 (Sithney) **skyber** pl.
Skewes (Crowan; Cury; St Wenn) **scawen, *-ys²**
Skewjack (Sennen) **scawen** aj.
Skewyek 1329 (Goran) **scawen** aj.
Skibber Whidden, fld. (Lelant) **skyber**
Skillywadden (Towednack) ?**guan**
Skyburrier n.d. (Gwennap) **skyber** pl.
Skyburriowe (St Mawgan in Meneage) **skyber** pl.
Slinke Dean, rock (Paul) **slynckya, tyn**
Sourne (Ladock) **sorn**
Sowans Hole, coast (Mawnan) **sawn**
Sowlmeor 1626 (Redruth) ?**zowl**
Sparett (St Cleer) **perveth**
Spargo (Mabe) **spern, *cor²**
Sparnick (Kea) **spern** aj.
Sparnock, fld. (Liskeard) **spern** aj.
Sparnon (Budock; St Buryan; Redruth) **spern** sg.
Sparnon 1730 (St Blazey) **spern** sg.
Sparnon Moor (St Mewan) **spern** sg.
Spernic Cove (St Keverne) **spern** aj.
Spernicks, fld. (St Pinnock) **spern** aj.
Spernon (St Buryan) **spern** sg.
Splattenridden (Lelant) *****splat, reden**
Stabilyus (Phillack) *****stable** pl.
Stable Hobba (Paul) *****stable, *hoby**
Stagnar Seyth, mine 1549 (Gwennap) **stean** aj., **segh**
Stampas Farm (Perranzabuloe) **stampes**
Stampers (St Stephen in Brannel) **stampes**
Stamps and Jowl Zawn, coast (St Just in Penwith) **stampes, dyowl**
Stannack, fld. (Crowan) **stean** aj.
Stenalees (St Austell) **stean** aj., ?*****lys**
Stencooce (Ladock) *****stum, cos**

INDEX OF CORNISH PLACE-NAMES

Stencoose (St Agnes; Kenwyn) *stum, cos
Stenek Segh, mine 1498 (Breage) stean aj., segh
Stennack (Camborne; St Ives) stean aj.
Stennack Bottom, fld. (Paul) stean aj.
Stennack Ladern, mine 1832 (Roche) stean aj., lader pl.
Stennagwyn (St Stephen in Brannel) stean aj., guyn
Stephen Gelly (Lanivet) *stum, kelli
Stephengelly (St Neot) *stum, kelli
Steval, rock (Scilly) ?steuel
Sticker (St Mewan) stoc, -yer
Strætneat c. 880 (11th) (S. 1507) *stras, ?*neth
Strase Cliff (St Mawgan in Pydar) ?*stras
Stratton Vow (St Ives) *stret, *fow
Straythe, The, coast (Veryan) streyth
Streatt and Grean, street 1580 (Penzance) *stret, gronen pl.
Street an Dudden, street 1795 (Penzance) *stret, ton
Street an Garnick, street 1588 (Mousehole) *stret, *kernyk
Street-an-Garrow, street (St Ives) *stret, garow
Street-an-Nowan, street (Paul) *stret
Street-an-Pol, street (St Ives) *stret, pol
Street an Stock, street 1704 (Helston) *stret, stoc
Street Eden, street 1748 (Truro) *stret, yn
Street Wyndesore, street 1423 (Helston) *stret
Streigh (Lanlivery) streyth
Stret Cowlyn, street 1537 (Truro) *stret
Strete Nowith, street 1713 (Penzance) *stret, nowyth
Strete Seyntjohn, street 1429 (Helston) *stret
Stret-Kedy, street 1537 (Truro) *stret
Stret Melenwelen, street 1381 (Penryn) *stret
Stret-Myhale, street 1593 (Helston) *stret
Stret-Pyddre, street 1464 (Truro) *stret
Strickstenton (Lanlivery) tre
Striddicks (St Martin by Looe) ?rid, dreys aj.
Stringham, fld. (Paul) ?*tryonenn
Stymcodde 1523 (Altarnun) *stum
Stymwoythegan 1294 (St Michael Penkevil?) *stum
Suffenton (St Teath) ?segh, fenten
Suffree (Probus) rid, ?*brygh, ?*-i
Swingey (Constantine) chy, *goon-dy
Sworne (St Martin in Meneage) sorn

Taban Denty, fld. 1786 (Sithney) tam, dentye
Tagus (St Breock) tal, cos
Talcarn 1463 (= Minster) tal, carn
Talland (= parish) tal, *lann
Talskiddy (St Columb Major) tal, ?schus pl.
Tal-y-Maen, rock (Sennen) men
Tamsquite (St Tudy) *stum, cos

306

INDEX OF CORNISH PLACE-NAMES

Tarewaste (Redruth) ?**tyr, wast**
Taroveor Road (Penzance) ?**tarow**
Tater-du, rock (St Buryan) ***torthell, du**
Tavis Vor, coast (Paul) ?**taves**
Tempellow (Liskeard) ?**tyn**, ?***pel** pl.
Tencreek (Menheniot) **tre, cruc**
Tencreek (Talland) **keyn, cruc**
Tencreeks (St Veep) **tre, cruc**
Tendera (St Anthony in Meneage) ***dyn**, ?**dar** pl.
Tereandreane c. 1720 (St Agnes) **tyr, dreyn**
Tere marracke 1578 (Gerrans) **tyr, marrek**
Ternooth, fld. (Goran) ?**ternoyth**
Terras (St Stephen in Brannel) ?**try**[1] fem., ?**rid**
Tewington (St Austell) **towan**
Te Wyn Ha Vaes 1349 (Sennen) **aves**
Theweeth (Germoe) **gwyth**[1]
Thorne (Warleggan) **sorn**
Tinnel (Landulph) ***-yel**
Tintagel (= parish) ***dyn**, ***tagell**
Tinten (St Tudy) ***dyn**
Tirbean (Breage) **tyr, byghan**
tnow wæt' 977 (11th) (S. 832) ***tnou**
Toban, fld. 1696 (St Hilary) **tûban**
Tobban Cove (Illogan) **tûban**
Todden Coath (Paul) **ton, coth**
Todne Rosemoddress, fld. 1640 (Madron) **ton**
Tolbenny (St Stephen in Brannel) **tal**
Tolcarne (several) **tal, carn**
Toldavas (St Buryan) ?**daves** pl.
Toldhu, coast (Mullion) **toll, du**
Tolfrew (Creed) **tal**
Tolgarrick (Camborne; Kenwyn; St Stephen in Brannel) **tal, karrek**
Tolgroggan (St Allen) **tal, *crak**[1] dimin.
Tolgullow (Gwennap) **tal, golow**
Tolgus (Redruth) ***toll-gos**
Toll (Gunwalloe) **toll**
Tollan Wheath, coast 1716 (Gwithian) ***wheth**
Toll Hole, coast (Mawnan) **toll**
Tolman (Scilly) ***tol-ven**
Tolmennor (Breage) **tal, meneth**
Tol-Pedn-Penwith, coast (St Levan) **toll**
Tolpetherwin (South Petherwin) **tal**
Tol Plous, coast (St Levan) **toll, plos**
Tolponds (Sithney) **tal, pons**
Tolraggatt (St Endellion) **rag**
Tol Toft, coast (St Buryan) **toll**
Tolvadden (St Hilary) **tal, ban**
Tolvaddon (Illogan) **tal, ban**

307

INDEX OF CORNISH PLACE-NAMES

Tolvan (Constantine) *tol-ven
Tolverne (Philleigh) tal, ?bron
Tolzethan (Gwithian) tal
Tondre, fld. 1567 (Gerrans) ton, tre
Tonenbethow, fld. 1330 (Veryan) ton, beth pl.
Top Lobm, coast (St Ives) top, ?*lom
Toplundie Cove 1584 (St Minver) top
Top Tieb, coast (St Hilary) top
Tor Balk, rock (Landewednack) ?tor
Torfrey (Golant) tor, *bre
Tor Noon (St Just in Penwith) tor, goon
Towan (St Agnes; St Merryn) towan
Towan Beach (St Anthony in Roseland) towan
Towan Blistra (St Columb Minor) towan
Towan Head (St Columb Minor) towan
Towans, The (Gunwalloe; Phillack) towan
Towanwroath Shaft (St Agnes) toll, gruah
Towargh Ane, mine 1502 (Towednack) ?*towargh
Towednack (= parish) *to-
Tranken (Madron) tre, Frank pl.
Transingove (Cury) tenewen, cos
Treago (Crantock) tre
Treal (Ruan Minor) ?*gawl
Treamble (Perranzabuloe) taran, pol
Trease (Cury; St Just in Penwith) tre, rid
Treath (Constantine; Manaccan) *treth
Treave (St Buryan) tre, *yuf
Trebant (Altarnun; Lanreath) *pans
Trebarfoot (Poundstock) tre, perveth
Trebarret (Boconnock) tre, perveth
Trebarva (St Columb Minor) tre, perveth
Trebarvah (Constantine; Perranuthnoe) tre, perveth
Trebarvah Goodnow (Constantine) ?*go-, ?*tnou
Trebarvah Woon (Constantine) goon
Trebarvath (St Keverne) tre, perveth
Trebarveth (Stithians) tre, perveth
Trebarwith (Tintagel) tre, perveth
Trebeigh (St Ive) tre, byghan
Trebell (Lanivet) tre, pell/*pel
Trebellan (Cubert) ?pellen
Trebiffin (Lesnewth) tre, byghan
Trebighan (Landrake) tre, byghan
Trebisken (Cubert) ?*prys dimin.
Treblethick (St Mabyn) bleit aj.
Trebost (Stithians) post
Treboul (St Germans) *bolgh
Trebowland (Gwennap) tre, *bow-lann
Trebrown (St Germans; Quethiock) tre, bron

308

INDEX OF CORNISH PLACE-NAMES

Trebullom (Altarnun) ?**pol**, ?**guyn**
Trebursye (South Petherwin) **tre**
Treburthes (Veryan) ?*****perth**, ?*****-ys**^{2}
Trebyan (Lanhydrock) **tre, byghan**
Trecaine (Creed) *****keun**
Trecan (Lanreath) **rid, can**
Trecarne (Advent; Tintagel) **tal, carn**
Trecombe (Constantine) ?**tyr, crom**
Trecoogo (St Kew) **tal,** *****cok** pl.
Trecoose (St Martin in Meneage) **tu, cos**
Trecorme (Quethiock) **rid,** *****gorm**
Trecorner (Altarnun) **corn** pl.
Trecreege (St Endellion) **tre, cruc**
Trecrogo (South Petherwin) **tre, cruc** pl.
Treculliacks (Constantine) **tre, kullyek**
Tredarrup (Michaelstow; St Neot; Warbstow; St Winnow) ?*****gortharap**
Tredaule (Altarnun) **tre,** *****tawel**
Tredeague (Gwennap) *****daek**
Tredellans (St Just in Roseland) **tre**
Tredenham (Lanivet; Probus) **tre,** *****dynan**
Tredinick (St Breock) *****dyn** aj.
Tredinick (St Mabyn) **tre, reden** aj.
Tredinnick (St Issey; St Keverne; Newlyn East) **tre, reden** aj.
Tredinnick (Morval; Probus) **tre, eithin** aj.
Tredivett (South Petherwin) **tre,** *****dywys**
Tredole (Trevalga) **tre,** *****tawel**
Tredore (St Issey) **tre,** *****towargh**
Tredorn (Minster) ?**dorn**
Tredower (St Martin in Meneage; St Minver) **tre,** *****towargh**
Tredrea (St Erth) **tre,** *****treu**
Tredrea (Perranarworthal) **tre,** ?**tre**
Tredreath (Lelant) **tre, trait**
Tredrizzick (St Minver) **tre, dreys** aj.
Tredruston (St Breock) **tre**
Tredudwell (St Germans; Lanteglos by Fowey) **tre**
Tredundle (Egloskerry) **tre**
Tredwen (Davidstow) **rid, guyn**
Treen (St Levan, Zennor) **tre,** *****dyn**
Treeve (Phillack) ?*****yuf**
Treeve (Sennen) **tre,** *****yuf**
Treeza (Sithney) ?**ysel** spv.
Treffry (St Gluvias; Lanhydrock; Merther) *****fry**
Trefingey (Tywardreath) **tre**
Trefinnick (South Hill) **tre, fyn** aj.
Trefranck (St Clether) **tre, Frank**
Trefrawl (Lanreath) **rid**
Trefrengy 1396 (St Clether) **tre**
Trefreock (St Endellion; St Gennys) ?*****fry** aj.

309

INDEX OF CORNISH PLACE-NAMES

Trefrew (Lanteglos by Camelford) **tre, *rew**
Trefrida (Jacobstow) **tre, rid** pl.
Trefrize (Linkinhorne) ?**rid**
Trefronick (St Allen) **bron** aj.
Trefrouse (Week St Mary) **tre, *rew**
Trefula (Redruth) **tre, ula**
Tregadgwith (St Buryan) **tre, *caswyth**
Tregadreythmoer 1302 (St Mawgan in Meneage) **meur**
Tregair (Newlyn East) **tre, *ker**
Tregallon (St Mabyn) **tre, kelin**
Tregamere (St Columb Major) ?**trygva**
Tregaminion (St Keverne; Landewednack; Morvah; Tywardreath) **tre, *kemyn** pl.
Treganoon (Lanlivery) ***kenow** dimin.
Tregantle (Antony; Lanlivery) ***arghantell**
Tregarden (Luxulyan) **tre, carn**
Tregarden (St Mabyn) **tre, garan**
Tregarland (Morval) **cruc, *alun**
Tregarne (St Keverne; Mawnan) **tre, carn**
Tregarrick (Menheniot; Roche) **tre, karrek**
Tregarton (Goran) **tre**
Tregaswith (St Columb Major) **tre, *caswyth**
Tregatherall (Minster) ?**cadar,** ?***-yel**
Tregatillian (St Columb Major) **tre, cuntell** pl.
Tregatreath (Mylor) ?**ruth**
Tregavarras (Goran) ***kevar** deriv.
Tregays (St Winnow) ?***ker** deriv.
Tregea (Illogan) **kee**
Tregeage (St Keverne) ?**kee** aj.
Tregear (Bodmin; Crowan; St Eval; Gerrans; Ladock; St Mawgan in Meneage) **tre, *ker**
Tregeare (Egloskerry; St Kew) **tre, *ker**
Tregearwoon (Gerrans) **goon**
Tregembo (St Hilary) ?**dy-,** ?***kemer**
Tregenfer (Braddock) **tre**
Tregenhorne (St Erth) **tre**
Tregenna (Blisland; St Ewe) ?***kenow**
Tregenver (Budock) **tre**
Tregerthen (Zennor) ?**kerden**
Tregew (Feock; Mylor) **tre, *kew**
Tregidden (St Keverne) ?**cudin**
Tregidgio (Creed) ***crys** pl.
Tregilliowe (St Hilary) **tre, kelli** pl.
Treglasta (Davidstow) **tre, glastan**
Tregleath (Egloshayle) ?***gwleghe**
Treglennick (St Ervan) **tre, kelin** aj.
Treglinwith (Camborne) ?***glynn-wyth**
Treglith (Treneglos) ?***gwleghe**

310

INDEX OF CORNISH PLACE-NAMES

Treglyn (St Minver) ?*dyn, ?*clun
Tregoiffe (Linkinhorne) gof
Tregolds (St Merryn) tre, *coll deriv.
Tregole (Poundstock) ?*kew, ?*-yel
Tregolls (St Clement; Stithians; St Wenn) tre, *coll deriv.
Tregonce (St Issey) *ros
Tregondean (Goran) tre
Tregonebris (Sancreed) tre
Tregongeeves (St Austell) tre
Tregoniggie (Budock) *keun aj. pl.
Tregoose (St Columb Major; St Erth; Feock; St Mawgan in Meneage; Probus; Sithney; Stithians; Wendron) tre, cos
Tregorland (St Just in Roseland) *ker, *cor-lann
Tregoss (Roche) *cors
Tregothnan (St Michael Penkevil) ?goth[1]
Tregowris (St Keverne) ?*kew-rys
Tregragon (St Teath) tal, *crak[1] dimin.
Tregray (Otterham) rid, *gre
Tregrehan (St Blazey) ?*crygh pl.
Tregrill (Menheniot) ?*cryn, ?le
Tregue (Altarnun; Lansallos; Minster) tre, *kew
Tregullas (Kea) tre, *coll deriv.
Tregunwith (Mylor) ?ruth
Tregurno (Probus) corn pl.
Treguth (Cubert) tre, *kew
Tregwindles (St Breock) tre, guennol
Tregwinyo (St Ervan) tre, goon pl.
Tregye (Feock) tre, ky
Tregyffian Vrose 1732 (St Just in Penwith) bras
Treharrock (St Kew) tre, *havrek
Trehawke (Menheniot) haf aj.
Treheer (Liskeard) tre, yar pl.
Treheveras (Kenwyn) *devr- deriv.
Trehunsey (Quethiock) tre
Treire (Lanreath) tre, yar pl.
Treisaac (St Columb Minor) tre, dreys aj.
Trekee (St Teath) tre, kio
Trekeive (St Cleer) *kyf
Trekelland (Lezant) *kew-nans
Trekennick (Altarnun) cruc, keyn aj.
Trekernell (North Hill) ?corn dimin.
Trekillick (Lanivet) tre, an, kullyek
Trekimletts (Lezant) tre
Treknow (Tintagel) tre, *tnou
Trelan (St Keverne) *lann
Trelanvean (St Keverne) byghan
Trelask (St Cleer) tre, losc
Trelaske (Lewannick; Pelynt) tre, losc

311

INDEX OF CORNISH PLACE-NAMES

Trelawder (St Minver) **lêuiader**
Trelease (St Hilary) **tre, *dy-les**
Trelease (Ruan Major) **tre, glas**
Trelease (Kea; St Keverne) **tre, *lys**
Treleggan (Constantine) **heligen**
Treleigh (Redruth) ***legh**
Trelevra (Budock) **tre, leuerid**
Trelew (St Buryan) **lyw**
Trelewack (St Ewe) **lyw** aj.
Trelewick (St Allen) **lyw** aj.
Treliggo (Breage) ?***legh** pl.
Trelin (Altarnun) **tre, lyn**
Trelinnoe (South Petherwin) **tre, lyn** pl.
Trelion (St Stephen in Brannel) **tre, *legh** pl.
Trelissick (St Erth; Sithney) **tre, *gwlesyk**
Treloggas (Kea) **tre, logaz**
Trelogossick (Veryan) **tre, logaz** aj.
Trelonk (Ruan Lanihorne) **tre, *lonk**
Treloskan (Cury) **losc** dimin.
Trelossa (Philleigh) **les** pl.
Treloweth (St Erth; Illogan; St Mewan) **leuuit**
Trelowith Heill 1466 (St Erth) ***heyl**
Treloyan (St Keverne) **tre, *legh** pl.
Treludrou Gres 1327 (Newlyn East) **cres**
Trelull pras 1554 (St Breock) **pras**
Trelyn (St Keverne) **tre, lyn**
Tremain (Pelynt) **tre, men**
Tremaine (= parish) ?**manach** pl.
Tremar (St Cleer) **tre, margh**
Tremayne (St Columb Major; Crowan; St Martin in Meneage) **tre, men**
Trembath (Madron) **bagh**
Trembear (St Austell) ?***peur**
Trembethow (Lelant) **beth** pl.
Trembleath (St Ervan) **tre, bleit**
Trembleath (St Mawgan in Pydar) **beth**
Trembraze (St Keverne; Liskeard) **tre, bras**
Tremear (St Ive) **tre, meur**
Tremearne (Breage) ***tnou**
Tremearne Par (Breage) **porth**
Tremeer (St Clether; Lanivet; Lanteglos by Fowey; St Tudy) **tre, meur**
Tremellin (St Erth) **melin**[1]
Tremenhee (Mullion) ***meneghy**
Tremenheere (Ludgvan) **tre, *men hyr**
Tremenheire (Wendron) **tre, *men hyr**
Tremenhere (St Keverne; Stithians) **tre, *men hyr**
Tremethick (Madron) **tre, methek**
Tremoan (Pillaton) ?***mon**[2]
Tremoddrett (Roche) **tre**

INDEX OF CORNISH PLACE-NAMES

Tremough (Mabe) **tre, mogh**
Trenadlyn (Tywardreath) ?**banathel** sg.
Trenale (Tintagel) **tre, hal**
Trenalls (St Hilary) **tre, als**
Trenance (St Austell; St Columb Minor; St Issey; St Keverne; St Mawgan in Pydar; Mullion; Newlyn East; Withiel) **tre, nans**
Trenannick (Warbstow) ?**onnen** aj.
Trenant (Duloe; Egloshayle; Fowey; Menheniot; St Minver; St Neot) **tre, nans**
Trenarlett (St Tudy) **arluth**
Trenarren (St Austell) ***dyn, garan**
Trenault (Trewen) **tre, als**
Trencreek (Blisland; St Columb Minor; Creed; St Gennys; Veryan) **tre, cruc**
Trencrom (Lelant) **crom**
Trencruk 1405 (St Issey) **cruc**
Trendeal (Ladock) ***dyn**, ?**deyl**
Trendrean (Newlyn East) **tre, dreyn**
Trendrennen (St Levan) **tre, dreyn** sg.
Trendrine (Zennor) **tre, dreyn**
Trenear (Wendron) **tre, yar** pl.
Treneere (Madron) **tre, yar** pl.
Treneglos 1755 (Ludgvan) **tre, eglos**
Treneglos (= parish) **tre, eglos**
Trenerth (Gwinear) **nerth**
Trenethick (St Germans; Wendron) ?**meneth** aj.
Trenewth (Michaelstow) **tre, nowyth**
Trengayor (St Gennys) **tre, *ker**
Trengilly (Constantine) **tre, an, kelli**
Trengoffe (Warleggan) **gof**
Trengothal (St Levan) **tre, *gothel**
Trengove (Constantine) **tre, an, gof**
Trengove (Illogan) **tre, gof**
Trengrouse (Veryan) **tre, crous**
Trengrove (Menheniot) **gof**
Trengune (Warbstow) **tre, goon**
Trengwainton (Madron) **tre, dy-, guaintoin**
Trengwainton Carn (Madron) **carn**
Trengweath (Redruth) **tre, gwyth**[1]
Trenhale (Newlyn East) **tre, hal**
Trenhayle (St Erth) **tre, war, *heyl**
Trenithan (St Enoder; Probus) **tre, eithin**
Trenoon (St Mawgan in Pydar; Ruan Major) **tre, goon**
Trenouth (St Cleer; St Ervan; Tintagel) **tre, nowyth**
Trenovissick (St Blazey) ***neved** aj.
Trenowah (St Austell) **tre, nowyth**
Trenower (Roche) **tre, nowyth**
Trenoweth (St Agnes; Budock; Crowan; Cury; Gunwalloe; Gwinear; St Keverne; Lelant; Mabe; Mylor; Scilly) **tre, nowyth**

313

INDEX OF CORNISH PLACE-NAMES

Trenowth (St Columb Major; Luxulyan; Probus) **tre, nowyth**
Trentinney (St Kew) **fenten** pl.
Trenuggo (Sancreed) **war, googoo**
Trenute (Lezant) **tre, nowyth**
Trenuth (Davidstow) **tre, nowyth**
Trepoll (Veryan) **pol**
Trequean (Breage) **?*keun**
Trequite (St Germans; St Kew) **tre, cos**
Trequites (St Mabyn) **tre, cos**
Treraire (St Eval) **tre, yar** pl.
Trerank (Roche) **tre, Frank**
Trereife (Madron) **?*yuf**
Trerew (Crantock) **tre, *rew**
Trerice (St Allen; St Breock; St Dennis; Newlyn East; Sancreed) **tre, rid**
Trerise (Crowan; Ruan Major) **tre, rid**
Treroosel (St Teath) **tre, ?*arwystel**
Trerose (St Endellion; Mawnan) **tre, *ros**
Trerose (Minster) ***ros**
Treryse Wone 1545 (St Dennis) **goon**
Tresahor Praze (Constantine) **pras**
Tresawsan (Merther) **tre, Zowzon**
Tresawsen (Perranzabuloe) **tre, Zowzon**
Tresawson (Lanreath) **tre, Zowzon**
Tresayes (Roche) **tre, *Seys**
Trescavyghan 1323 (Breage) **byghan**
Tresco (Scilly) **tre, scawen** pl.
Trescoll (Luxulyan) ***ros, *coll**
Trescowe (Breage; St Mabyn) **tre, scawen** pl.
Trescowthick (Newlyn East) **tre, scawen** aj.
Tresemple (St Clement) **tre**
Tresillian (Merther; Newlyn East) **tre**
Treskelly (St Germans) **tre, *is[1], kelli**
Treskewes (St Keverne; Stithians) **tre, scawen, *-ys[2]**
Treskillard (Illogan) **?scoul, ?*arð**
Treskilling (Luxulyan) ***ros, kelin**
Treskinnick (Poundstock) **keyn** aj.
Treslay (Davidstow) ***legh**
Treslea (Cardinham) **rid, *legh**
Tresloggett (St Mabyn) **logaz**
Tresmaine (Altarnun) ***ros, men**
Tresmarrow (Davidstow; South Petherwin) ***ros, margh**
Tresmeak (Altarnun) ***ros**
Tresmeer (= parish) ***ros, meur**
Tresparrett (St Juliot) ***ros, perveth**
Trespearn (Sheviock) **tre, spern**
Trespearne (Laneast) **tre, spern** sg.
Tresquare (St Tudy) **guirt**
Trestain (Ruan Lanihorne) **stean**

INDEX OF CORNISH PLACE-NAMES

Trestrayle (Probus) **strail**
Tresulgan (St Germans) **tre**
Treswallock (St Breward) ?***gwal** aj.
Treswarrow (St Endellion) ***gorweth**
Treswen (Warbstow) ***ros, guyn**
Treswithick (Cardinham) ?**rid, gwyth**[1] aj.
Tretallow (St Veep) **rid**
Tretawn (St Kew) **rid, down**
Trethake (St Cleer; Lanteglos by Fowey) ***daek**
Tretharrup (Gwennap; Luxulyan; St Martin in Meneage) **tre**, ***gortharap**
Tretharrup (St Cleer; Lanreath) ?***gortharap**
Trethauke (St Minver) **worth, haf** aj.; or **guartha** aj.
Trethawle (Menheniot) ?***tawel**
Tretheague (Stithians) ***daek**
Tretheake (Veryan) ***daek**
Trethella (Ruan Lanihorne) ?***dyllo**
Trethem (St Just in Roseland) ***drum**
Tretherras (St Columb Minor) **tre, dew, rid**
Tretherres (St Allen) **tre, dew, rid**
Trethevas (Landewednack) **tre, daves** pl.
Trethevey (St Mabyn) **ti** (= **chy**), **war**, ***dewy**
Trethew (Menheniot) **tre, du**
Trethowa (Probus) **tre**, ***towargh**
Trethowell (Kea) ***gor-**, ?**tewl**
Trethurgy (St Austell) **doferghi**
Tretoil (Lanivet) **rid, hwilen** pl.
Tretrinnick (St Issey) **tre, eithin** aj.
Treulecyon woeles c. 1280 (St Keverne) **goles**
Trevaddra (Manaccan) ?**bar,** ?**trogh**
Trevadlock (Lewannick) ?**aidlen** aj.
Trevagecbyghan 1300 (Constantine) **byghan**
Trevan (Probus) **tal, ban**
Trevance (St Issey) ***ros**
Trevan Point (St Minver) ***try-**[2], **ban/men**
Trevanters (St Clether) ***ros**
Trevarder (Lanteglos by Fowey) **tre**, ***gor-dre**
Trevarra (St Minver) **tre, crw**
Trevarrick (St Austell) ?***arð** aj.
Trevarth (Gwennap) **tre, margh**
Trevascus (Goran) **tre**
Trevaskis (Gwinear) **tre**
Trevaunance Porth c. 1720 (St Agnes) **porth**
Trevean (Kea; St Keverne (2); St Levan; Madron; St Merryn; Morvah; Newlyn East; Perranuthnoe; Sancreed) **tre, byghan**
Treveans (St Teath) **tre, byghan**
Trevear (St Issey; St Merryn; Sennen; St Stephen in Brannel) **tre, meur**
Trevedda (Lanteglos by Fowey) **tre, bedewen** pl.
Treveddoe (Warleggan) **tre, bedewen** pl.

315

INDEX OF CORNISH PLACE-NAMES

Treveglos (Grade; St Mabyn; St Merryn; Zennor) **tre, eglos**
Trevella (Crantock) **tre, ?elaw**
Trevellas Porth (St Agnes) **porth**
Trevelloe (Paul) **tre, ?elaw**
Trevena (Tintagel) **tre, war, meneth**
Treveneage (St Hilary) **tre, manach** aj.
Trevenen Bal (Wendron) **bal**
Treveniel (North Hill) **tre, ?myn, *-yel**
Trevenna (St Mawgan in Pydar; St Neot) **tre, war, meneth**
Trevenner (St Hilary) **tre, war, meneth**
Trevenwith (St Keverne) **tre, fynweth**
Treveor (Goran; Merther) **tre, meur**
Treverbyn (St Austell; St Neot; Probus) **tre**
Treverras (St Just in Roseland) **tre, gweras**
Trevessa (St Erth) **?*ussa**
Trevethoe (Lelant) **beth** pl.
Treveythin Sauson 1293 (Probus) **Zowzon**
Trevia (Lanteglos by Camelford) **?gwîa**
Treviddo (Menheniot) **?bedewen** pl.
Treviglas (St Columb Minor; Probus) **tre, eglos**
Trevilges (Wendron) **tre, *huel-gos**
Trevilgus (St Issey) **tre, *huel-gos**
Trevilley (Sennen; St Teath) **tre, ?*byly**
Trevillian (Warbstow) **tre**
Trevilly (St Columb Minor) **tre, ?*byly**
Trevine (St Minver) **tre, byghan**
Trevisquite (St Mabyn) **tre, *is[1], cos**
Trevivian (Davidstow) **?hiuin**
Trevolen c. 1587 (Mevagissey) **?*bolgh** dimin.
Trevolland (Probus) **?*bolgh** dimin.
Trevone (Padstow) **?auon**
Trevorder (St Breock; Warleggan) **tre, *gor-dre**
Trevorgans (St Buryan) **tre**
Trevorgey (St Eval) **tre**
Trevose (St Merryn) **tre, fos**
Trevothen (St Keverne) **?goth[1]** dimin.
Trevowah (Crantock) **tre**
Trevozah (South Petherwin) **tre, fos** pl.
Trevuzza (St Enoder) **tre, frot** pl.
Trew (Breage) **tre, du**
Trew (Tresmeer) **tre, *yuf**
Trewalder (Lanteglos by Camelford) ***gwalader**
Trewandra (Landrake) **guan**
Trewannett (St Juliot) **?goth[1]**
Trewannion (Lesnewth) **?guan** pl.
Trewardreva (Constantine) **tre, *godre** pl.
Trewarmenna (Creed) **tre, war, meneth**
Trewarnevas (St Anthony in Meneage) ***neved**

INDEX OF CORNISH PLACE-NAMES

Trewarras (St Mewan) **tre, gweras**
Trewartha (St Agnes) **tre, guartha**
Trewartharap 1315 (St Issey) **tre, *gortharap**
Trewarveneth (Paul) **tre, war, meneth**
Trewarveneth Varghas 1343 (St Hilary) **marghas**
Trewassa (Davidstow) **?guas** deriv.
Trewaste n.d. (St Keverne) **?tyr, wast**
Trewavas (Breage; Wendron) **tre, *gwavos**
Treweege (Stithians) ***gwyk**
Treween (Altarnun) **tre, goon**
Treweers (Lansallos) **tre, gweras**
Treweese (Quethiock) **tre, gweras**
Trewelesikwarheil 1356 (St Erth) **war, *heyl**
Trewellard (St Just in Penwith) **tre, ?*arð**
Trewen (= parish) **tre, guyn**
Trewen (Budock; Lanreath; Liskeard; St Tudy) **tre, guyn**
Trewethack (St Endellion) **tre, gwyth¹** aj.
Trewethart (St Endellion) **?goth¹**
Trewhela (St Enoder) **?hwilen, ?*-a**
Trewhella (St Hilary) **?hwilen, ?*-a**
Trewidden (Madron) **tre, guyn**
Trewince (St Anthony in Meneage; St Columb Minor; Constantine; Gerrans; St Issey; Ladock; St Martin in Meneage; Probus; Stithians) **tre, guyns**
Trewinnard (St Erth) **?*arð**
Trewint (Advent; Altarnun; St Endellion; St Erney; Menheniot; St Minver; Poundstock) **tre, guyns**
Trewithen (Stithians) **gwyth¹** sg.
Trewithick (Breage; St Stephen by Launceston) **tre, gwyth¹** aj.
Trewonnard (Treneglos) **?*arð**
Trewoofe (St Buryan) **goyf**
Trewoof-wartha (St Buryan) **tre, guartha**
Trewoon (Budock; St Mewan; Mullion) **tre, goon**
Trewoone (Phillack) **tre, goon**
Treworder (Blisland; Egloshayle; Kenwyn; Ruan Minor) **tre, *gor-dre**
Treworga (Ruan Lanihorne) **tre, ?*gor-ge**
Treworgans (Probus) **tre**
Treworgey (Duloe; St Cleer) **tre**
Treworgie (St Gennys; Manaccan) **tre**
Trewornan (St Minver) **gronen**
Treworthal (Philleigh) **tre, *gothel**
Trewothack (St Anthony in Meneage) **goth¹** aj.
Trewothall (Crantock) **tre, *gothel**
Trewurgyscor 1311 (St Cleer) **?scorren** pl.
Treyew (Kenwyn) **?ieu**
Trezance (Cardinham) **tyr, sans**
Trezare (Fowey) **rid**
Trezelland (Altarnun) **?*elen**

317

INDEX OF CORNISH PLACE-NAMES

Trezise (St Martin in Meneage) **tre, *Seys**
Tribbens, The, coast (Sennen) **?*try-², ?*pans**
Triffle (St Germans) **tre**
Trigg (= hundred) ***try-², *cor¹**
Trigg Rocks, coast (Breage) **?trig**
Triggstenton (St Endellion) **tre**
Trigva (Sithney) **?trygva**
Trilley (St Neot) **?try¹, ?*legh**
Trilley Rock (Zennor) **?try¹, ?*legh**
Tringham, fld. (St Just in Penwith) **?*rynntone**
Trink (Lelant) **tre, Frank** pl.
Tronghan Flow, fld. (Crowan) **?*tryonenn**
Troon (Breage; Camborne) **tre, goon**
Troose-mehall 1698 (Illogan) **tros**
Truas (Tintagel) **?guis**
Trugo (St Columb Major) ***gruk** pl.
Trungle (Gwinear; Paul) ***mon-gleth**
Truro (= city) ***try-²**
Truro Vean (St Clement) **byghan**
Truscott (St Stephen by Launceston) **dres**
Trusell (Tremaine) **?*arwystel**
Trussall (Wendron) **?*arwystel**
Try (Gulval) ***fry**
Trydinner, mine 1708 (St Agnes) **try¹, dyner**
Trythall (Gulval) **?*bryth, *-yel**
Trythance (St Keverne) **tre**
Tubbon (Stithians) **tûban**
Tucoyse (Constantine; St Ewe) **tu, cos**
Tuelmenna (Liskeard) **tewl, meneth**
Turn-Bean 18th (St Keverne) ***torn**
Turn-Traboe (18th) (St Keverne) ***torn**
Turn Tregarne 1660 (St Keverne) ***torn**
Turn-Trelan 18th (St Keverne) ***torn**
Tuzzy Muzzy Croft, fld. (St Just in Penwith) **?tus, ?ha, ?mowes** pl.
Tybesta (Creed) **ti (= chy)**
Tywardreath (= parish) **ti (= chy), war, trait**
Tywarnhayle (Perranzabuloe) **ti (= chy), war, an, *heyl**

Valency (river) **?*melyn-jy**
Valley Truckle (Lanteglos by Camelford) **?melin¹, ?*trokya**
Vans, The, coast (St Austell) **ban**
Var, The, coast (Mullion) **?bar**
Velandrucia (Stithians) **melin¹, *trokya**
Vellandreath (Sennen) **melin¹, trait**
Vellanhoggan Mills (Gulval) **melin¹, *hogen**
Vellanoweth (St Agnes; Constantine; Ludgvan) **melin¹, nowyth**
Vellanvens, fld. (St Keverne) ***melyn wyns**
Vellanwens, fld. 1690 (St Hilary) ***melyn wyns**

INDEX OF CORNISH PLACE-NAMES

Vellenewson (Crowan) **melin¹, usion**
Vellyndruchia (St Buryan; Madron) **melin¹, *trokya**
Vellynsaga (St Buryan) **melin¹**
Vellyn Saundry (Camborne) **melin¹**
Vent-an-League, fld. (St Keverne) **fenten**, ?**lek**
Ventin Gay, fld. (Paul) ?**kee**
Venton Ariance (Mullion) **fenten, arghans**
Ventonberron (Probus) **fenten**
Ventonear n.d. (Camborne) **fenten,** ?**ēr²**
Venton Ends (St Issey) **fenten, yeyn**
Venton Gannal (St Mawgan in Meneage) **fenten, *canel**
Ventongassick (St Just in Roseland) **fenten**
Ventongimps (Perranzabuloe) **fenten, compes**
Ventonglidder (Probus) **fenten**
Ventongoose (Kea) **fenten, cos**
Ventonjean (Madron) **fenten, yeyn**
Venton kelinack 1696 (St Clement) **fenten, kelin** aj.
Ventonleague (Phillack) **fenten,** ?**lek**
Ventonraze (Illogan) **fenten, gras**
Venton Sowall, fld. 1649 (Constantine) ?**zowl**
Ventontinny (Probus) **fenten**
Ventontrissick (St Allen) **fenten, dreys** aj.
Venton Vadan (Veryan) **fenten, ban**
Venton Vaise (Perranzabuloe) **fenten, mes**
Ventonvedna (Sithney) **fenten,** ?***fenna**
Venton Veor (Liskeard) **fenten**
Ventonveth (Veryan) **fenten, margh** pl.
Ventonwyn (Creed) **fenten, guyn**
Ventonzeth (Kenwyn) **fenten, segh**
Vinegonner, fld. (Lelant) ?**fyn,** ?**conna**
Vinnick Rock (St Minver) ?***meynek**
Vinten Wicorrian 1613 (Kenwyn) **fenten, guicgur** pl.
Vithen Cornett, fld. 1696 (Breage) **budin**
Vithen Vorne, fld. 1696 (Breage) **budin, forn**
Vithin Thomas, fld. n.d. (Gwennap) **budin**
Vogue (Gwennap) **fok**
Vogue Beloth (Illogan) **fok**
Vorrap, coast (St Levan) ***morrep**
Vorvas (Lelant) ***gor-, *bod**
Vorvas Crease (Lelant) **cres**
Vose (St Ewe) ***bod**
Voskelly (St Just in Roseland) **fos, kelli**
Vosporth (Crantock) **fos, porth**
Voss (Duloe) ?**fos**
Vougo Lion, fld. 1696 (Cubert) **vooga,** ?***legh** pl.
Vounder (St Blazey; Mullion) **bounder**
Vounder Bullock, fld. 1643 (Truro) **bounder**
Vounder Duel, fld. (St Levan) ?**tewl**

319

INDEX OF CORNISH PLACE-NAMES

Vounder Feene 1630 (St Erth) **bounder, fyn**
Vounder Gogglas 1613 (St Just in Penwith) **bounder, ?*gogleth**
Vounderlidden, fld. (Paul) **bounder, ?ledan**
Voundervour Lane (Penzance) **bounder**
Vowan Guham, fld. (St Just in Penwith) ***fow**
Vow Cave (Paul) ***fow**
Vow Cottage (St Ives) ***fow**
Vro, The, rock (Mullion) **?brogh**
Vugga Cove (Crantock) **vooga**
Vugger, The, fld. 1696 (Lanivet) **vooga**
Vwgha Hayle, coast (St Agnes) **vooga, *heyl**

Wallow (St Cleer) **?*gwal** pl.
Warneck, fld. (Camborne) **guern** aj.
Warnick, fld. (Crowan) **guern** aj.
Warthendr' 1327 (St Anthony in Roseland) **guartha, tre**
Weeth (Camborne; Gwinear; Sithney) **gwyth**[1]
Weeths, The (Madron) **gwyth**[1]
Westnorth (Duloe) **?gvest** deriv.
Westva 1260 (Roche) **?gvest**
Westway (Lelant) **?gvest**
Wheal an Clay, mine c. 1720 (St Agnes) **wheyl, cleath**
Wheal an Coats, mine 1708 (St Agnes) **wheyl**
Wheal an Cullieck, mine 1712 (St Agnes) **wheyl, kullyek**
Wheal an Dellick 1712 (St Agnes) **?teil** aj.
Wheal an Dinner, mine 1708 (St Agnes) **dyner**
Wheal an Gogue, mine 1712 (St Agnes) ***cok**
Wheal an Harbier, mine 1708 (St Agnes) ***erber**
Wheal an Hor, mine n.d. (Gwennap) **horþ**
Wheal an Lowren, mine 1712 (St Agnes) **lowarn**
Wheal an Peber, mine 1708 (St Agnes) ***pybor**
Wheal an Porrall, mine 1712 (St Agnes) **porhel**
Wheal an Seavy, mine 1712 (St Agnes) **sevi**
Wheal an Wens Croft, fld. (Breage) ***melyn wyns**
Wheal Arrans, mine c. 1720 (St Agnes) **arghans**
Wheal Bal Hill (St Just in Penwith) **bal**
Wheal Busy, mine (Kenwyn) **wheyl**
Wheal Callice, mine n.d. (Gwennap) **cales**
Wheal Dees Gentle, mine 1712 (St Agnes) **wheyl, tus, gentyl**
Wheal Dreath, mine c. 1720 (St Agnes) **trait**
Wheale an Dowthick, mine 1687 (lost) **wheyl, dewthek**
Wheale an Gothilly, mine 1741 (St Just in Penwith) **?*gwythel** pl.
Wheale Bebell, mine 1734 (Wendron?) ***pybell**
Wheal Fortune, mine (several) **wheyl**
Wheal Growan, mine c. 1741 (Breage) **grow**
Wheal Howl, mine (Kea) **?houl**
Wheal Kitty, mine (St Agnes) **wheyl**
Wheal Mundie, fld. (St Hilary) ***mon-dy**

INDEX OF CORNISH PLACE-NAMES

Wheal Noweth, mine (Sennen) **wheyl**
Wheal Owles, mine (St Just in Penwith) **wheyl, als**
Wheal Reath, mine (Lelant) **wheyl, ruth**
Wheal Reeth, mine (Germoe) **wheyl, ruth**
Wheal Sperries, mine (Kea) **spethas**
Wheal Sperris, mine (Towednack) **spethas**
Wheal Sterren, mine 18th (Illogan) **wheyl, steren**
Wheal Varizick, mine 1712 (St Agnes) **barthesek**
Wheal Zawson, mine n.d. (Gwennap) **wheyl, Zowzon**
Whel an Voag, mine 1699 (Gwennap) **fok**
Whele an Phelp, mine n.d. (Gwennap) **wheyl, an**
Whele Arantall, mine 1508 (lost) ***arghantell**
While an Attol, mine 1502 (Sithney) **wheyl, atal**
While anglastannon, mine 1542 (St Gluvias?) **glastan** sg.
Whimple (Calstock) **guyn, pol**
Windsor (Cubert) **tyr**
Winnick (St Austell; St Veep) **guyn** deriv.
Winven Cove (Zennor) **?guyn, ?men**
Withan (St Martin in Meneage) **gwyth¹** sg.
Withen (Lelant) **gwyth¹** sg.
Withiel (= parish) **gwyth¹, *-yel**
Withiel-eglos 1305 (Withiel) **eglos**
Withielgoose (Withiel) **cos**
Woolgarden (St Clether) **?gol**
Woon (Luxulyan; Roche) **goon**
Woon Bucka, fld. c. 1790 (Kenwyn) **goon, bucka**
Woon Gumpus Common (St Just in Penwith) **goon, compes**
Woon Weeth, fld. (Paul) **gwyth¹**
Worthyvale (Minster) **?guartha, ?auallen**
Worvas (Grade) ***gor-, ?*bod**
Worvas Collis (Lelant) **goles**
Wra, The, rock (St Just in Penwith) **gruah**
Wrea, Great/Little, rocks (St Keverne) **gruah**
Wreathe, rock (St Keverne) **gruah**
Wrecket (St Winnow) **?rag**
Wriggles, fld. (St Keverne) **forth, eglos**
Wyth, The, fld. (Lelant) **gwyth¹**

Yellow Carn (Zennor) **carn**

Zawn, The, coast (Veryan) **sawn**
Zawn a Bal, coast (St Just in Penwith) **sawn, bal**
Zawn Alley, coast (Morvah) **sawn**
Zawn Brinny, coast (St Just in Penwith) **sawn, bran** pl.
Zawn Bros, coast (Towednack) **sawn, bras**
Zawn Carve, coast (St Keverne) **?corf**
Zawn Duel, coast (Zennor) **sawn, tewl**
Zawn Gamper, coast (St Buryan) **sawn, *kemer**

INDEX OF CORNISH PLACE-NAMES

Zawn Harry, coast (St Hilary) **sawn**
Zawn Kellys, coast (St Levan) **sawn, kellys**
Zawn Organ, coast (Paul) **sawn**
Zawn Reeth, coast (Sennen) **sawn, ruth**
Zawn Susan, coast (St Hilary) **sawn**
Zawn Vinoc, coast (St Keverne) **sawn, ?*meynek**
Zawn Wells, coast (Sennen) **sawn, ?gwels**
Zone Point, coast (St Anthony in Roseland) **sawn**

INDEX OF NON-CORNISH PLACE-NAMES CITED

Aberalaw, Agl. 3–4
Aber Daugleddyf, Pem. 82
Aber Deunant, Crn. 82
Adwy-wynt, Fli. 12
Airget-, Irl. 11
Alaw, Agl. 3
Aled, Den. 4
Aletanus pagus, Gau. 4
Aleth, Brt. 4
Aletum, Gau. 4
Alt, Lnc. 4
Alt Bough, Hre. 26
alt guhebric, Wal. 240
Altgundliu, Wal. 4
Amalburnan, Sfk. 5
Amlwch, Agl. 152
aper ipost du, LLD 192
Arddau'r Myneich, Crn. 9
Ardd Machae, Irl. 10
ard ir allt, LLD 9
Aremorici, Gau. 8
Arfon, Crn. 8
Argantomagus, Gau. 11
Argentilla, Gau. 11
Argentoialo, Gau. 138
Argentorate, Gau. 11
Argoad, Brt. 8
Argoed, Wal. 8
Ariannell, Wal. 11
Arlosh, Cmb. 8, 153
Auch an prat bihan, Brt. 237

bachlatron, LLD 14
Bach(y)sylw, Crm. 14
-*Banalec*, Brt. 16
Banastère, Brt. 20
Bannec, Brt. 17
bannguolou, Wal. 16, 107
Bannhedos, Brt. 50
Barnbougle, WLo. 34
Barrock Fell, Cmb. 17
Bedd Captain Morgan, Fli. 20
Beddgelert, Crn. 20
Bedd y Ci Du, Wal. 20

Dedowe (lo), Den. 19
Beef Close, Eng. 29
Belerit, Brt. 137
-*Benalec*, Brt. 16
Bénodet, Brt. 20
Berkshire, Brk. 17
Blaenllaethdy, Crm. 148
Blaidd-bwll, Wal. 22
Blencogo, Cmb. 61
Bloc'h (Ar), Brt. 23
Bochriucarn, Wal. 39, 196
Bodgate, Dev. 24
Borth (Y), Crd. 191
Bot Frisunin, Brt. 24
Botha, Irl. 25
Botis, RB 25
Bot Tahauc, Brt. 24
Brangwayn, LLD 109
Branodunum, RB 29
Break Back, Eng. 68
Brechfa, Crm. 32
Brelevenez, Brt. 155
Brenbagle, Den. 15
Bren Goen, Brt. 109
Brengoulou, Brt. 33, 107
Brenterc'h, Brt. 222
brinn bucelid, LLD 34
brinn cornou, LLD 66
brinn i cassec, LLD 31, 42
Brinsychan, Wal. 208
bronn ir alt, LLD 7, 32
Bronsican, Brt. 208
Brwynog, Agl. 33
Bryn Barlwm, Gla. 17
Bryn Ceffyl, Mer. 57
Bryn Golau, Wal. 107
Bryniau Poethion, Mer. 192
Bryn Ll(y)wenydd, Agl. 155
Brynn Kyuergyr, Wal. 56
Bryn Poeth, Wal. 192
Brynsifiog, Rad. 209
Bryntyrch, Crn. 222
Bryn-yr-Eryr, Mer. 94
Buceles, Wal. 34

INDEX OF NON-CORNISH PLACE-NAMES

Bugeildy, Rad. 34
Buguélès, Brt. 34
Buorht, Brt. 35

Caerau, Gla. 52
Caer Banhed, Brt. 50
Caerdydd, Gla. 140
Caer Gorguen, Brt. 112
Cae'r Mynawyd, Agl. 162
Caer'n Iuguinen, Brt. 137
Caer Poeth, Brt. 192
Cafnan, Agl. 170
cairduicil, LLD 50
Cair guicou, Wal. 119
Cair Trigguid, Wal. 50
Calchuynid, Wal. 36
Calfannog, Bre. 17, 36
Camddwr, Wal. 37, 87
cam dubr, Wal. 37
cam lann, Wal. 36
campull, LLD 36
Camros, Pem. 36
Candover, Hmp. 49
Candubr, Wal. 37
Capquern, Brt. 38
Caraverick, Cmb. 127
Cardew, Cmb. 53, 90
Cardiff, Gla. 140
Carhaix, Brt. 53–4
Carley, Dev. 52
Carningli, Wal. 39
Carn Mynawyd, Agl. 162
carn perth yr onn, LLD 183
Carn Tylyuguay, Wal. 39
Carn Ymenyn, Pem. 5
Carpont, Brt. 38
Carrégui, Brt. 41
Carreg y Bwch, Crn. 26
Caspenboih, Brt. 192
Caubal hint, Brt. 44
Cayr i coc, Wal. 45, 50
Ceann an Bhóthair, Irl. 28
cecin i minid, LLD 45
cecin pennros, LLD 183
Cefnwilgi, Mtg. 113
Ceinmeirch, Den. 45
Celynen, Den. 46, 93
ceninuc, LLD 50
Cenn tuirc, Irl. 222
Charles, Dev. 40
Chwefri, Bre. 240
Chwil, Crd. 241
Chwiler, Wal. 241

Chwilog, Crn. 241
Clap Mill, Dev. 59
Clecher, Brt. 60
Clegyr, Mer. 60
Cleker, Brt. 60
Cleuziou, Brt. 60
Clynnog, Crn. 46
Cnoch, Brt. 61
Coat-ar-Guéo, Brt. 57
Coatascorn, Brt. 12
Cocker, Cmb. 62
Codanew, Agl. 67
Codragheyn, Den. 67
Coedgenau, Bre. 101
Coed-y-Trwyn, Mer. 235
Coegnant, Wal. 75
Coetqueff, Brt. 58
Coker, Som. 62
Co(o)mbe, Dev. 63–4
Corn ar Gazel, Brt. 42
Corsley, Wlt. 66
Coum (Le), Brt. 64
Coum bras/bihan, Brt. 64
Co(u)mou, Brt. 64
Coupalua, LLD 44, 155
Crénard, Brt. 10
Cricieth, Wal. 131
Croftbathoc, Cmb. 71
Croftbladen, Cmb. 71
Croftmorris, Cmb. 71
cruc glas, LLD 74, 104
cruc hisbernn, LLD 210
Crucin, Wal. 74
Cruckebleith, Wal. 23
cruc leuyrn, LLD 154
crucou, LLD 74
Crugiau, Pem. 74
Crugyll, Agl. 74
Crynfenyht, Wal. 71
Cuckoo Pen, Eng. 62
Culcheth, Lnc. 75
Culgaith, Cmb. 75
Cunbleid, Wal. 22
Cwmblwch, Crm. 23
Cylchau, Crm. 59

Dalard (An-), Brt. 215
Dalar Hir (Y), Crn. 215
daudour, Brt. 82
deri emreis, LLD 80
Derwgoed, Mer. 80
Désert, Brt. 85
Deverill, Wlt. 82, 139

INDEX OF NON-CORNISH PLACE-NAMES

Dinan, Brt. 85
Dinan, Wal. 85
Dinard, Brt. 10
Dinas Brân, Wal. 29
Dinbrain, Den. 29
Dinbych, Den. 21
Dindryfol, Agl. 84
dinduicil, LLD 50
Din Efwr, Crm. xv, 96
Diserth, Fli. 86
Dollerline, Cmb. 8
Dork, Eng. 82
Dornoch, Sco. 87
Dornock, Sco. 87
dou pull, LLD 82
Dreff (Le), Brt. 229
Drostre (Y), Bre. 89
Drumburgh, Cmb. 26
Dryll yr Ymenyn, Agl. 5
dubnnant du, LLD 88
dún mbrain, Irl. 29
Dwyfech, Wal. 21
Dwyryd, Wal. 82

Eburacum, RB 96
Eburodunum, Gau. xv, 96
Elerch, Crd. 93
Ennavallen, Brt. 13
Eruguenn, Wal. 95
Eusa, Brt. 238
Evercreech, Som. 96

Faill an Bhric, Irl. 7
Fau, Brt. 96
Feidr Fach, Pem. 28
Felinwynt, Crm. 161
Fennun Hen, Wal. 129
Ffinnant, Mtg. 98
Ffrith y Menyn, Den. 5
Ffynnon-cyff, Crd. 58
Ffynnon Gwynvaen, Agl. 162
finnant iuðhail, Wal. 98
finnaun he collenn, LLD 62, 97
Finntracht, Irl. 223
fin tref peren, LLD 98
fin tref petir, LLD 98
Fionntráigh, Irl. 120, 223
Foidir, Pem. 28
Foredoor Field, Eng. 81

Gabrosentum, RB 132
Galvezit, Brt. 137
Gamallt, Rad. 4, 37

Garros, Brt. 102
Garway, Hre. 144
Garzpenboez, Brt. 192
Gauze Brook, Wlt. 66
Gazek (Ar), Brt. 42
Gelliwig, Crn. 47
Gelli-wig, Mon. 47
genou ir pant, LLD 101
genou nant byuguan, LLD 101
Glasnai, Wal. 104
Glendon, Dev. 105
Glyn, Wal. 105
Glyn Bwch, Bre. 26
Godreddi, Agl. 106
Goetheloc, Brt. 111
Golban, Brt. 103
Golbin, Brt. 103
Gorzit, Brt. 66
Gribin, Den. 70
Gualtog, Brt. 117
guefrduur, LLD 240
Guelegoarh, Brt. 117
gueli banadil, LLD 117
Guéredic Sant Hervé, Brt. 118
Guerit Carantauc, Wal. 118
Guern-Condy, Brt. 76
guern i drution, LLD 7
Guern-uidel, Brt. 123
Gueun (flumen), Wal. 108
guinnic, LLD 121
gunnic, LLD 121
Guocob, Guocof, LLD 107
guoun guenn (i), LLD 108
guoun teirfin, LLD 98
Gurimi, Wal. 105
Gwaun Mynawyd, Mer. 162
Gwayn Hesgog, Crd. 109
Gwendraeth, Crm. 120, 223
Gwernyfed, Bre. 172
Gwybedig, Crd. 119
Gwybedog, Bre. 119
Gwybedyn, Wal. 119
Gwyddelfynydd, Mer. 122-3
Gwyddelwern, Mer. 123

Haliguen, Brt. 128
halmelen, LLD 161
hal un guernen, LLD 118
Hamuc, Brt. 124
Havodlum, Den. 153
Heli, Wal. 131
helicluin, LLD 128, 153
Helygen, Crd. 128

325

INDEX OF NON-CORNISH PLACE-NAMES

Hendy (Yr), Crm. 130
Hendy-Gwyn, Crm. 130
Henllan, Wal. 130
Henllys, Wal. 130
Hennlann dibric, LLD 130
Herriard, Hmp. 133
hescenn iudie, LLD 131
Hesnant, Brt. 135
Hingant, Brt. 131
Hirfynydd, Gla. 132
Hirgard, Brt. 133
hirglas, Brt. 103–4, 132
Hiriaeth, Mtg. 133
hirmain, Wal. 132, 162
Hirnant, Mtg. 132
hirpant, LLD 132
Hoddnant, Wal. 135, 170
Hodnet, Shr. 135
Hog Back, Sur. 46
Hogs Back, Sur. 46
Hudnant, Brt. 135
Huelgoat, Brt. 135
Hyples eald land, Dev. 225

Iaen, Gla., Mtg. 138
Iâl, Den. 138
Idir dhá Loch, Irl. 136
Ifern (An), Brt. 136
Ile d'Arz, Brt. 10

Kaerhoer, Brt. 242
Karreg an Awezed, Brt. 14
Karreg ar Bara, Brt. 17
Kastil Cadoci, Wal. 43
Kelennec, Brt. 46
Kerbannalen, Brt. 17
Ker en gov, Brt. 106
Ker-en-heull, Brt. 134
Kerguntuill, Brt. 76
Kerhingant, Brt. 131
Kerivaladre, Brt. 54
Kernivinen, Brt. 137
Keroualch, Brt. 167
Kerregou, Brt. 41
Kersauson, Brt. 210
Kestel, Brt. 43
Kilford, Den. 99
Kiln Park, Eng. 99
Kintyre, Arg. 183
Kinver, Stf. 30
Kleier Arc'hantell, Brt. 11

Laedti, Brt. 148

Laithti Teliau, LLD 148
Lamb Park, Eng. 174
Lamplough, Cmb. 23
Landcross, Dev. 66
Landevennec, Brt. 219
Landkey, Dev. 219
Lanercost, Cmb. 142
Lanerton, Cmb. 142
Lann Guorboe, LLD 144
Lannmaes, Wal. 165
lannteliau, LLD 130
Lanrekaythin, Cmb. 142
Lan Sent, Brt. 205
Lantokai, Som. 219
Lapwing (The), Drb., Shr. 62
Laughter Tor, Dev. 148
Lavernock, Gla. 154
Leather Tor, Dev. 148
Lech meneich, LLD 146
Lein, Brt. 146
Lether Hall, Gla. 147
Lew, Dev. 151
licat laguernnuc, LLD 154
Lindifferon, Fif. 82
Lindum, RB 149
Lin Garan, Sco. 102, 149
Lios na mBuachaillí, Irl. 34
Lisarcors, Wal. 8, 66, 150
Lis-ros, Brt. 203
Llainmoidir, Pem. 146
Llam Lleidr, Crn. 141
Llam y Lladron, Mer. 141
llam yr bwch, Wal. 26
Llamyrewig, Mtg. 142
Llam y Trwsgl, Crn. 142
Llancarfan, Gla. 41
Llandinabo, Hre. 144
Llandygai, Crn. 219
Llanfihangel Helygen, Rad. 128
Llangors, Bre. 66
Llangwm, Mon. 64
llan gynin ai waison, Crm. 124
Llangynwyd, Gla. 143
llann adneu, Wal. 144
llethir y brin, Wal. 147
Lliw, Wal. 151
Lluestcornicyll, Crd. 62
Lluest-hen, Crd. 129
Lluestygïach, Crd. 59
Llwyd(i)arth, Wal. 102
Llwytgoed, Wal. 153
Llyfni, Wal. 148
Loch an Chaca, Irl. 36

INDEX OF NON-CORNISH PLACE-NAMES

Locus Maponi, RB 151
Loeniou, Brt. 153
Logodec, Brt. 152
Loin, Brt. 153
Lostanenez, Brt. 154
lost ir inis, LLD 7, 154
Lost Valley, Sco. 48
Louargniec, Brt. 154
Loyn garth, Wal. 153
luch cinahi, LLD 152
luch edilbiu, LLD 152
luch i crecion, LLD 7
Luh-guiuuan, Brt. 152
luhinn maur, LLD 153
Lydeard, Som. 102
Lydiard, Wlt. 102

Machynlleth, Mtg. 156
Macoer, Brt. 156
Maen y Bugail, Agl. 34
Maesmawr, Mtg. 165
Maesmor, Mer. 165
Maes y Groes, Fli. 165
Magis, RB 155
Magoaerou, Brt. 156
-magus, Gau. 156
Mahalon, Brt. 156
main brith (i), LLD 32, 161
mainti, LLD 159
mair a chirig, Crm. 124
mais mail lochou, LLD 165
Marchat Rannac, Brt. 157
Marogilum, Gau. 139
Maroialus, Gau. 138
marulinniou, LLD 149
Matle, LLD 145
Maysycros, Crm. 165
Mean Nevez, Brt. 163
Medgarth, Wal. 158
Mediobogdum, RB 158
Mediolanum, Gau. 158
Mediomatrici, Gau. 158
Medionemetum, RB 158, 172
Medon, Brt. 158
Meiarth, Wal. 158, 159
Meidrim, Crm. 159
Meiros, Wal. 159
Melchennec, Brt. 160
Melchonnec, Brt. 160
Mellor, Lnc., Drb. 30, 167
Melrose, Rox. 200, 201
Mendy, Brt. 159
Menechi, LLD 163

Ménez-Klujeau, Brt. 61
Mezahibou, Brt. 165
Midglenn, Irl. 158
Midros, Irl. 159
Millinneuez, Brt. 160
Minihi, Brt. 163
Mochdre, Wal. 167
Moelfre, Wal. 30, 167
Moelnant, Wal. 23
Moel-y-Gest, Mer. 55
Moiaroc, Brt. 168
Moillewyk, Den. 96
Morchard, Dev. xv, 67
Morhoc'hed (Ar-), Brt. 169
Moridunum, RB 165
Morionoc, Brt. 169
Motrev, Brt. 167
Moult (Le), Brt. 168
Mynid du, LLD 163
Mynydd Epynt, Bre. 132

Nancollet, Brt. 63
nan pedecon, LLD 170
Nantbucelis, Wal. 34
Nant Caruguan, Gla. 7, 41, 170
nant cenou, LLD 50
nant duuin, LLD 88
Nant Ffrancon, Crn. 100
Nantgarw, Gla. 41
nant hi rotguidou, LLD 199
nant i buch, LLD 26
nant lechou, LLD 146
Nantmoel, Wal. 23
Nant Mynawyd, Mer. 162
Nantogilum, Gau. 138
Nantoi(a)lus, Gau. 138
nant pedecou, LLD 170
nant tauel, LLD 216
nant trineint, LLD 233
Nant y Lladron, Mer. 141
nant yr eguic, LLD 96
Nant yr Iar, Crn. 138
Neath/Nedd, Gla. 172
Neizbran, Brt. 172
Nemet, Brt. 172
Nemet(o)-, Gau. 172
Nine Stones, Eng. 172
Nouarnec (Le), Brt. 134
Nymet, Dev. 172
Nympsfield, Glo. 172

odencolc, Dev. 36, 173

INDEX OF NON-CORNISH PLACE-NAMES

ol i gabr, LLD 102
Orchard, Dor. 8

Paimbœuf, Brt. 32
Pant y bara, Wal. 17
Pantymenyn, Crm. 5
Pant (y) Meriog, Den. 7
Parc-an-Hinquin-Bihan, Brt. 129
Parret, Dor./Som., Glo./Wor. 176
Paunsett, Wlt. 181
Peart, Som. 183
Peffery, Ros. 184
Pembro, Brt. 32
Pembroke, Pem. 32
Pem-prat, Brt. 193
Penally, Pem. 5
Penalun, Pem. 5
Pénan, Brt. 182
Penard, Brt. 180
Penbedw, Wal. 19
Penboc'h, Brt. 26, 179
Penbwch, Gla. 179
Pencoed, Wal. xv, 181
Pencoyd, Hre. 181
Pencryc(h)gel, Wal. 74
Penfao, Brt. 96
Penfeunteuniou, Brt. 181
Penffynnon, Wal. 181
Penfro, Pem. 32
Penge, Sur. xv, 181
Penguilly, Brt. 180
Penhador, Brt. 35
Penharth, Brt. 180
Penharth, Gla. 180
Penharz, Brt. 180
Penhesgyn, Agl. 131
Penhoat, Brt. xv, 6, 181
Penhors, Brt. 66
Peniarth, Wal. 180
Penketh, Lnc. 181
Penllyn, Wal. 182
Pennal, Mer. 5
Pennant, Wal. 182
pennant ir caru, Wal. 41
Pennard, Som. 181
penncelli guennuc, LLD 180
penncelli gulible, LLD 180
Pennclecir, LLD 60
Penn hischin, Brt. 131
penn i celli, LLD 6, 180
penn i cgelli, LLD 180
Pennichen, Wal. 173
penn i prisc, LLD 194

penn Luhin latron, LLD 141, 153
Penn ohen, Brt. 173, 177, 179
Penntirh, Wal. 222
Penquit, Dev. 181
Pen-rhos, Wal. 183
Penrhyn, Wal. 183
Penros, Brt. 183
Penrose, Dev. 183
Penterc'h, Brt. 222
Pentraeth, Agl. 223
Pentrez, Brt. 223
Pentrich, Drb. xvi, 222
Pentridge, Dor. xvi, 222
Pentyrch, Gla., etc. xvi, 45, 222
Penyard, Hre. 181
Penyberth, Crn. 183
Pen-y-bont, Wal. 180
Penyfeidr, Pem. 28
Pen y Ffynnon, Wal. 181
Pen y Gadair, Wal. 35
Pen y Gadlas, Fli. 35
Penygelli, Mtg. 180
Pen y Gors, Pem. 66
Pen y Pil, Gla., Mon. 185
Perret, Brt. 196
Perros, Brt. 183
Perta, Gau. 183
Perth, Sco. 183
Perth Gelyn, Crn. 183
Petrucorii, Gau. 64
Pillaton Hall, Stf. 185
Pingewood, Brk. 181
Plo-Harnoc, Brt. 134
Poguisma, LLD 155
Pol-bleiz, Brt. 22
Polcasek, Brt. 42
Polgauer, Cmb. 102
Polhearn Farm, Dev. 134
Poltross Burn, Cmb. 236
Pont-ar-Ddulais, Gla. 9
Pontbren, Crm. 193
Pont-Christ, Brt. 190
Ponthou, Brt. 190
pont meiniauc, LLD 159
Ponugual, Wal. 114
Porthbinawid, Wal. 162
porthcassec, LLD 42
Porth-y-garan, Agl. 102
Porthyrysgraff, Agl. 205
Porz Hili, Brt. 131
Porz mousek, Brt. 169
Poul-an-Ouidi, Brt. 134
Poulbroc'h, Brt. 32

INDEX OF NON-CORNISH PLACE-NAMES

Poulhoas, Brt. 111
Poul-lazron, Brt. 141
Poull Brein, Brt. 31
Poulpri, Brt. 195
Prat ar c'houm, Brt. 64
Prawle Point, Dev. 113
Priddy, Som. 194
pridpull (i), LLD 195
Primrose Hill, Eng. 31
prysc katleu, Wal. 194
pul ir deruen, Wal. 80
pull lifan, LLD 146
pul retinoc, Wal. 196
Pumryd, Wal. 176
Pwllheli, Crn. 131
Pwll-y-chwil, Crd. 241

Quaerosan, Brt. 203
Quaking Bridge, Brk. 32
Quantock Hills, Som. 37
Quenmarch, Brt. 45
Quignénec, Brt. 50
Quillivic, Brt. 47
Quimper, Brt. 48

Rac Ynys, Wal. 195
Radenec, Brt. 196
Raguénès, Brt. 94, 195
ráith brain, Irl. 29
Rangof, Brt. 106
Rann-louuinid, Brt. 154
Ran Penpont, Brt. 180
Raswraget, Cmb. 124, 201, 202
Rattenuc, Brt. 196
Redgand, Brt. 37
Redmain, Cmb. 162, 198
Redynure, Crn. 196
Rethcand, Brt. 37
Rhiw Hiriaeth, Mtg. 133
Rhonynys, Agl. 203
Rhos, Den., Pem. 200
Rhosili, Gla. 200, 201
Rhydafallen, Wal. 13, 198
Rhyd Cryw, Mer. 70
Rhyd ddofn, Wal. 88
Rhyd Faen, Wal. 162, 198
Rhyd Fraith, Mon. 7
Rhyd Gau, Wal. 58
Rhyd Halog, Wal. 126, 197
Rhyd Hir, Wal. 197
Rhyd Loyw, Wal. 105
Rhyd Lydan, Wal. 145, 198
Rhyd Sych, Bre. 207

Rhyd Wen, Wal. 198
Rhyd y Bontbren, Gla. 193
Rhyd Ychen, Wal. 173
Rhydyfallen, Wal. 13
Rhyd y Llech, Gla. 146
Rhyd y Meirch, Wal. 157
Rhyd y Re, Wal. 198
rit ar trodi, LLD 99
ritec, LLD 196
rit i cambren, LLD 36
rit ir euic, LLD 96
rit litan, LLD 145
Riu, LLD 197
Riubrein, LLD 197
Riu Morgant, Wal. 196
Roches d'Argenton, Brt. 11
Rodwydd Forlas, Wal. 199
Roose, Lnc. 202
Ros, Som. 202
Roscaroc, Brt. 201
Roscoff, Brt. 106, 202
Rosiou an Fauh, Brt. 201
ros ir eithin, LLD 92, 201
Rosnonen, Brt. 174
Ross, Ntb. 202
Rossam (ad), Wml. 201
Rossen Clough, Che. 203
Rossendale, Lnc. 203
Rossington, YoW. 203
Rossmore, Dor. 202
Ross-on-Wye, Hre. 202
rosulgen, LLD 201
Roudouderc'h, Brt. 222
Rozart, Brt. 10
Rudfoss, Brt. 204
rudglann, Wal. 204
rudpull, LLD 204
Run-, Brt. 204
Runyou, Brt. 204
Ryt Uorlas, Wal. 199
ryt y cerr, LLD 55

Saethon, Crn. 209
Salia, Gau. 127
Scaouet, Brt. 206
Séac'h-Ségal, Brt. 207
Sechenent, Cmb. 207
Sechnant, Wal. 207
Selsey, Ssx. 204
Serïor, Den. 140
sichnant (ir), LLD 207
sich pull (hi), LLD 207
Sirior, Den. 140

INDEX OF NON-CORNISH PLACE-NAMES

Siviec, Brt. 209
Sixpenny Close, Eng. 85
Sparnacus, Gau. 210
*Sparnomagos, Gau. 210
Spethot, Brt. 211
Squiber Nevez, Brt. 206
Squiffiec, Brt. 206
Squiviec, Brt. 206
Stockwood, Glo. 58
Stradey, Crm. 212
Stubbing(s) Wood, Eng. 58
Stubwood, Brk. 58
Sychnant, Wal. xv, 207

Talar Gerwin, Crn. 215
Tal-ar-iz-ar-sornigou, Brt. 137
Talar Rett, Brt. 215
Taldrogh, Den. 235
tal ir brinn, LLD 31, 214
tal ir cecyn, LLD 7, 45
tal ir fos, LLD 7, 214
Tallentire, Cmb. 214
Tal-y-bont, Wal. 190
*Tannoialon, Gau. 139
taranpull, LLD 215
Tenby, Pem. 84
Terley, Dev. 229
Threepenny Close, Eng. 85
tig-guocobauc, Ntt. 78, 107
Tindale, Cmb. 139
Tinduff, Brt. 217
Tintern, Mon. 84
Tintinhull, Som. 84
tir hiernin, LLD 217
tir retoc, LLD 217
tnou guinn ('r), LLD 218
Tnou-guydel, Brt. 123
Tnou melin, Brt. 160
Tnou Mern, Brt. 218
tnou mur, LLD 218
Tolchet, Som. 219
tollcoit, LLD 219
Tolmaen, Brt. 220
Tolven, Brt. 220
Tonfannau, Mer. 218
Torbant, Crn. 221
torr ir allt, LLD 221
Toularhoat, Brt. 220
Toulgoet, Brt. 219
Toulhoat, Brt. 220
Toull Ifern, Brt. 136
Trawsfynydd, Mer. 89
Trawsgoed, Wal. 89

Traymill, Dev. 229
Treable, Dev. 225
Treales, Lnc. 151
Treberfedd, Wal. 183
treb guidauc, Wal. 122, 227
Trebick, Dev. 21, 225
Tredown, Dev. 229
Tref carn, LLD 40
tref eithinauc, LLD 92
Treffuortre, Brt. 110
Trefhelygen, Crd. 128
Tref Iulitt, Brt. 227
Trefles, Brt. 151
Treflys, Crm. 151
Tref redinauc, Wal. 196, 226
Tref ret, LLD 226
Trefriw, Wal. 197
Tréguier, Brt. 64
Trehill, Dev. 229
Trelana, Dev. 229
Treleigh, Dev. 229
Trelleck, Mon. 233
Trellick, Dev. 225
Trem canus, LLD 227
Trem carn, LLD 227
Trem Gyllicg, LLD 227
Tremlech, Wal. 146, 227
Trem y crucou, Wal. 227
Tretire, Hre. 197
Trevenn, Dev. 229
Treveri, Gau. 233
Trevidel, Brt. 123
Trewsbury, Glo. 89
Tricorii, Gau. 64
Tridour, Brt. 233
trinanto, Gau. 233
Trostre, Mon. 89
Troustrie, Fif. 89
Trunch, Nrf. 235
Trusley, Drb. 89
Trussenhayes, Wlt. 89
Trussmore, Glo. 89
Tryfan, Crn. 233
trylec bechan, LLD 233
Tulketh, Lnc. 219
Twelmin, Wml. 83
Twmbarlwm, Mon. 17
Tŷ-croes, Wal. 72
Tŷ-fry, Agl. 30
Tyllbrys, Mtg. 220
Tyllgoed, Wal. 219
Tyn-y-ton, Wal. 221
Tŷ'r Bwci, Crm. 33

INDEX OF NON-CORNISH PLACE-NAMES

Tŷ'rfeidr, Pem. 28
Tywyn, Wal. 222

uernemetis, Gau. 172
Ushant, Brt. 238
Uxisama, Gau. 238

Ventry, Irl. 120, 223
Vornometum, RB 172
Vertamocori, Gau. 64
Vieille (La), Brt. 123

Waunarlwydd, Gla. 11
Wele Conws, Crn. 117
Wenferð, Wor. 100
Wheelock, Che. 241

Whyle (The), Hre. 241
Winford, Som. 100
win monid, Brt. 120, 163
Woolpit, Sfk., Sur. 22
Wrac'h (Ar), Brt. 123
Wylfa Head, Agl. 113
Wynford, Dor. 100

Yetts o' Muckart, Per. 12
Ynysoedd y Moelrhoniaid, Agl. 203
Yondercott, Dev. 233
Yonderlake, Dev. 233
Ysgeifiog, Fli. 206
Ystradau, Crm. 212
Ystumcoed, Crd. 213
ystum Guy, LLD 213

INDEX OF REJECTED ELEMENTS

('No evidence' means that no evidence has been found for
the occurrence of the word in place-names.)

'Gover': J. E. B. Gover, *The Place-Names of Cornwall* (1948: typescript at the Royal Institution of Cornwall, Truro); 'Nance': R. Morton Nance, *A New Cornish–English Dictionary* (St Ives, 1938), and *A Guide to Cornish Place-Names* (5th edition, Marazion, 1967).

a-barth 'on the side of' (Nance): no evidence
a-bell 'far off' (Nance): no evidence (but see **pell**)
aber 'mouth, estuary' (Gover): no evidence
adar 'outside, yonder' (Nance): no evidence
*ayr 'air' (Nance): no evidence

*bagas 'bush' (Nance): no evidence
*balek 'jutting' (Gover): a doubtful word
*bank 'bank' (Nance): no evidence
*bans 'height' (Nance): no such word, see ***pans** and **ban**
*barlys 'barley' (Nance): no evidence
*bath 'boar' (Nance): no evidence
bern 'heap, rick' (Nance): no evidence
*bissoc 'fork' (Gover): a doubtful word
bos 'bush' (Nance): no evidence

*cabul 'tangle' (Nance): no evidence
*kalar 'dung, mud' (Gover): a doubtful word
cals 'heap' (Nance): no evidence
*canabyer 'hemp-field' (Nance): no evidence
*carker 'prison' (Gover): no such word
*cardhen 'thicket' (Nance): no such word, see **kerden**
*keger 'hemlock' (Nance): no evidence
*kenak 'worm' (Nance): no evidence (cf. ***keun**)
*kenethel 'tribe' (Nance): no evidence
*kergh 'oats' (Nance): no evidence
*keres 'cherries' (Nance): no evidence
*kewargh 'hemp' (Nance): no evidence
*kewny 'moss' (Nance): no evidence
*kydel 'stake-net' (Nance): no evidence
*kylyn 'creek' (Nance): a doubtful word
*clask 'pile, heap' (Gover): a doubtful word
*cok 'boat' (Nance): no evidence
*cod 'bag, pouch' (Gover): a doubtful word
*col 'peak' (Gover): a doubtful word
*cop 'summit' (Nance): no evidence
*cort 'court' (Nance): no evidence

INDEX OF REJECTED ELEMENTS

*cot 'short' (Nance): no evidence
*cothyn 'cavity' (Gover): no such word
*crogla 'gibbet' (Nance): probably no such word
*chyvylas 'beast-house' (Nance): no evidence

da 'good' (Nance): no evidence (but see **daek**)
*devetty 'sheep-cote' (Nance): no evidence
*dyfyth 'wilderness' (Nance): no evidence
*dol 'dale' (Nance): probably no such word
*dorge 'earthen hedge' (Nance): no evidence
drok 'bad' (Nance): no evidence

efan 'wide' (Nance): no evidence
*elyl 'plot' (Nance): probably no such word
eneval 'animal' (Nance): no evidence
*erlewys 'prepared' (Nance): no such word
*ewhyas 'horseman' (Nance): no evidence

*fynny 'coarse grass' (Nance): no evidence
*fon 'hay' (Nance): no evidence

*glawjy 'rain shelter' (Nance): no evidence
*glyp 'wet' (Nance): no evidence
*gonys 'working' (Nance): no evidence
*gora 'hay' (Nance): no evidence
*gorch 'defence work' (Gover): probably no such word
*goskes 'shelter' (Nance): no evidence, probably no such word
*graghell 'heap' (Nance): no evidence
*grunjy 'grange' (Nance): no evidence

*gwastas 'level' (Nance): no evidence
*gwavwels 'winter pasture' (Nance): a doubtful word
*gwedrot 'wether' (Gover): a doubtful word
*gwelen 'rod' (Nance): no evidence
*gwyls 'wild' (Nance): no evidence
*gwyndon 'lay land' (Nance): no evidence
*gwresen 'warm, fertile ground' (Nance): no evidence; probably no such word

*haun 'ravine, hollow' (Nance): no such word
holan 'salt' (Nance): no evidence

*lab 'barn' or 'strip of land' (Gover): a doubtful word
lagen 'pool' (Nance): no evidence
*laid 'mud' (Gover): a doubtful word
*laun 'open working, cleared space' (Nance): no evidence
*lawns 'sward' (Nance): probably no such word
*lejek 'heifer' (Nance): no evidence

334

INDEX OF REJECTED ELEMENTS

*lerion 'sea water' (Gover): a doubtful word
*long 'ship' (Gover): probably no such word

*maner 'manor' (Nance): no evidence
*marred 'cultivated' (Gover): probably no such word (see ***kevar** ?)
*maw 'boy' (Gover): no evidence
*mer 'steward' (Nance): no evidence
*mew 'den' (Nance): probably no such word
*my 'field, plain' (Nance): probably no such word
*mig 'bog' (Gover): a doubtful word
myl 'beast' (Nance): no evidence
*myldyr 'mile' (Nance): no evidence
*mon 'manure' (Nance): no evidence

*nos 'mark' (Nance): no evidence

olas 'hearth' (Nance): no evidence
*olcan 'metal' (Nance): no evidence
*or 'boundary' (Nance): no evidence, probably no such word

*penwyth 'end' (Nance): no evidence, probably no such word
*pesky 'to feed' (Nance): no evidence
*pila 'finch' (Gover): probably no such word
*pon 'light dust' (Nance): no evidence
*porvyn 'rush-bog' (Nance): no evidence

*resk 'marsh' (Gover): no such word
ryal 'noble' (Nance): no evidence
*roper 'ropemaker' (Nance): no evidence
rounsan 'ass' (Nance, Gover): no evidence (Goenrounsen contains a personal name)

*sagh 'stagnant' (Nance): no evidence
*saya 'sift' (Nance): no evidence, probably no such word
*sarn 'causeway' (Nance): no evidence
*saworys 'well-savoured' (Nance): no evidence, probably no such word
*skew 'shelter' (Nance, Gover): no such word, see **scawen**
*skidi 'fallow land' (Gover): probably no such word
*sclew 'shelter' (Nance): no such word
*sols 'demesne land' (Nance): no evidence, probably no such word

*teghyjy 'hermitage' (Nance): no such word
*ter 'clear, pure' (Nance): no such word
*tewas 'sand' (Nance): no evidence
*tyeges 'farmer's wife' (Nance): no evidence (Forty Acres, fld., is English, ironic)
*tolgh 'hillock' (Nance): no such word
*tulf 'hump' (Gover): no such word

335

INDEXES OF WELSH AND BRETON COGNATES

Note that the Welsh and Breton forms given are not necessarily the exact equivalents of the Cornish head-forms. (E.g. **benyges** is ppp. but Welsh *bendigo* and Breton *binnigañ* are v.ns.)

Breton head-forms follow those of GIB, and Welsh ones follow GPC where available, otherwise Geir.Mawr.

WELSH COGNATES

We.	Co.	We.	Co.
a(c)	ha(g)	bara	bara
adwy	*aswy	barcud	bargos
aethnen	aidlen	bas	*bas
afal	aval	baw	*boðour
afallen	auallen	bedwen	bedewen
afon	auon	bedd	beth
agalen	*ygolen	bendigo	benyges
agos	ogas	berwr	beler
angen	anken	bery(f)	*bery
alan	*elen	beu-	bowyn
alarch	elerhc	beudy	boudzhi
alaw	*alaw	biw (MlWe.)	*byu
allt	als	blaen	blyn
anial	*enyal	blaidd	bleit
annedd	anneth	?blen	blyn
annel	antell	blwch	blogh
ar	*ar, war	bod	*bod
arab	*gortharap	?bogail	begel
aradr	ardar	?bôn	ben
ardd	*arð	both	*both
argel	*argel	braen	*breyn
arglwydd	arluth	braich	bregh
arian(t)	arghans	brân	bran
arwystl	*arwystel	bras	bras
asgwrn	ascorn	bre	*bre
?aweddwr	*aweth	brennig	brenigan
awel	awel	?briallu	breilu
-awr (MlWe.)	-yer	brith	*bryth
		briw	brew
bach	bagh	bro	bro
bagl	bagyl	broch	brogh
ban	ban	bron	bron
bana(d)l	banathel	brwyn	bronnen
bannog	*bannek	brych	*brygh
bar	bar	bryn	*bren

WELSH AND BRETON COGNATES

We.	Co.	We.	Co.
buarth	*buorth	cenau	*kenow
buddyn	budin	cennin	kenin
bugail	bugel	cerddin	kerden
bu(w)ch	bugh	?cern	*kernan
bwci	bucka	cesail	*casel
bwch	bogh	cest	*kest
bwlch	*bolgh	cethr	kenter
bwr	bor	ceunant	*kew-nans
*bych	*bich	ci	ky
bychan	byghan	cil	*kyl
byr	ber	clafdy	*clav-jy
		clais	*cleis
cach	*cagh	clawdd	cleath
cad	cas	cleg(y)r	*cleger
cadair	cadar	clep	clap
cadlys	*cad-lys	clud	*clus
cadwydd	*caswyth	clun	*clun
cae	kee	cnau	*cnow
caer	*ker	cnwch	*cnegh
?cafall	*kevyl	coch	cough
?cagl	*cagh	coed	cos
cain	keyn	coeg	cuic
calch	kalx	cog	*cok
caled	cales	coll	*coll
calen	*ygolen	colli	kellys
cal(y)	kal	collwydd	*collwyth
callestr	cellester	côr	*cor²
cam	cam	cordd	*cor¹
camas	*camas	cor(dd)lan	*cor-lann
can	can	cored	*cores
canel	*canel	corff	corf
cant	cans	corn	corn
câp	capa	?cornchwiglen	kodna huilan
capel	chapel	cors	*cors
car	*car	cosgor(dd)	coscor
carn	carn	crac	crak²
carreg	karrek	craidd	cres
carw	carow	craig	*crak¹
caseg	cassec	crau	krow
castell	castell	crawn	*creun
cath	cath	crib	krib
cau	*kew	crin	*cryn
cawn	*keun	croeniog	cronek
cawsai	*cawns	croes	crous
cefn	keyn	croesbren	crouspren
?ceffyl	*kevyl	crofft	*croft
cegin	*kegen	crogen	crogen
ceibr	keber	cromlech	*cromlegh
ceiliog	kullyek	crug	cruc
cêl	*kel	crwm	crom
celyn	kelin	crwys	crous
celli	kelli	crych	*crygh

338

WELSH AND BRETON COGNATES

We.	Co.	We.	Co.
crychydd	cherhit	draw	*treu
?cryd	*crys	dros	dres
crŷn	*cren	drum	*drum
cryw	*crew	drys	dreys
cudyn	cudin	du	du
cul	cul	duw	dev
?curnen	*kernan	dwfn	down
cwm	*comm	dŵr	dour
cwning	kynin	dwrn	dorn
cyfamwg	*kevammok	Dy-	*to-
cyfar	*kevar	dyfnant	*downans
cyfranc	keverang	dyfrgi	doferghi
cyff	*kyf	dyn	den
cylch	*kylgh		
cymer	*kemer	ebol	ebel
Cymro	*Kembro	-edig	-esyk
cymwys	compes	edn	ethen
?cymyn	*kemyn	efwr	*evor
cynaeaf-	*kyniaf-vod	eglwys	eglos
cyndy	*cun-jy	eidion	odion
cynnud	kunys	eiddew	idhio
cynnull	cuntell	eithin	eithin
cyrnig	*kernyk	elain	*elen
		elestr	elester
chwaer	wuir	elin	elin
chwarae	guary	elor	geler
chwe	whe	-en	-en
chwedl	whelth	erw	erw
chwefr-	*whevrer	eryr	er[1]
?Chwefror	*whevrer	esgair	*esker
chwil	hwilen	esgob	epscop
chwŷl	wheyl	ewig	*ewyk
chwyth	*wheth	ewythr	euiter
daear	dor	ffa	fav
dafad	daves	ffau	*fow
dail	deyl	ffawydd	*faw
dall	dall	fferyll	*feryl
dan	dan	ffin	fyn
danadl	lynas	ffod(i)og	fodic
dâr	dar	ffog	fok
dau	dew	ffordd	forth
dawns	*dons	ffos	fos
deifio	*dywys	ffranc	Frank
derw	dar	ffrwd	frot
deuddeg	dewthek	ffwrn	forn
di-	dy-	ffynnon	fenten
diawl	dyowl		
diles	*dy-les	gadu	gesys
din	*dyn	gaeaf	goyf
dinas	*dynas	gaeafod	*gwavos
drain	dreyn	gafl	*gawl

WELSH AND BRETON COGNATES

We.	Co.	We.	Co.
gafr	gaver	gwaun	goon
gaing	*genn	gweilgi	*gwailgi
gar	*gar	gwely	guely
garan	garan	gwell	guella
garth	*garth	gwellt	gwels
garw	garow	gwenith	gwaneth
gefail	gofail	gwennol	guennol
gêl	ghel	gwern	guern
genau	ganow	gweryd	gweras
gïach	kio	gwest	gvest
gïât	yet	gweu	gwîa
glan	glan	gwig	*gwyk
glas	glas	gwlad	gulas
?glesin	*glasen	gwledig	*gwlesyk
glo	glow[2]	gwlych	*gwleghe
gloyw	*glow[1]	gŵr	gour
glyn	*glynn	gwrach	gruah
go-	*go-	gwraig	gwrek
gobant	*gobans	?gwrdd-dref	*gor-dre
godech	*godegh	gwrm	*gorm
godref	*godre	?gwrtharab	*gortharap
gof	gof	gwybed	guibeden
gofer	gover	gŵydd 'goose'	goth[2]
gogledd	*gogleth	gwŷdd 'trees'	gwyth[1]
gogof	googoo	?Gwyddel	*gwythel
golau	golow	?gwyddwal	*gwythel
gor-	*gor-	gŵyl	gol
?gorwedd	*gorweth	gwylfa	guillua
?gorwydd	*gorweth	gw(y)mon	gubman
gosgor(dd)	coscor	gwyn	guyn
gradell	*radgel	gwynt	guyns
graean	grow	gwyrdd	guirt
gras	gras	gwŷs	guis
gre	*gre	gŵyth	goth[1]
grelyn	grelin	gylfin	geluin
grofft	*croft		
gronyn	gronen	haearn	horn
grug	*gruk	haf	haf
gwaedu	*gosa	hafar	*havar
gwaelod	goles	hafod	*havos
?gwäell	guel	hagr	hager
gwag	gvak	?haidd	*heth
gwaith	guyth[2]	hâl	hal
gwal	*gwal	hanner	hanter
gwaladr	*gwalader	haul	houl
gwallt	gols	hawdd	hueth
gwan	guan	heb	hep
gwanwyn	guaintoin	heli	hyly
gwarae	guary	hely	*helgh
gwarthaf	guartha	helygen	heligen
gwartheg	guarthek	hen	hen
gwas	guas	hendref	*hendre

WELSH AND BRETON COGNATES

We.	Co.	We.	Co.
hendy	hensy	llwm	*lom
henffordd	*hen-forth	llwnc	*lonk
hesgen	heschen	llwyd	los
-hin (OWe.)	*hin	llwyn	*lon
hir	hyr	llydan	ledan
hobi	*hoby	(cyn)llyfan	lovan
hwch	hoch	llyfn	leven
hwnt	hans	llyffant	*lefant
hwrdd	horþ	llygod	logaz
hwyad	hos	llyn	lyn
?hydd	*heth	llys	*lys
hynt	-hins-	llysiau	les
		llywarn	lowarn
iäen	yeyn	llywiawdr	lêuiader
iâr	yar	llywydd	leuuit
iarll	*yarl		
iau	ieu	*ma	*ma
Iau	yow	maen	men
-ig	-yk	maendy	*meyn-dy
ing	yn	maenglawdd	*men-gleth
-iol	*-yel	maen hir	*men hyr
ir	êr²	maes	mes
is	*is¹	magu	maga
isel	ysel	magwyr	*magoer
ithr (OWe.)	ynter	malwod	*melwhes
iwrch	yorch	mam	mam
		march	margh
llaeth	leth	marchnad	marghas
llain	*leyn	marchog	marrek
llam	lam	mawr	meur
llan	*lann	meddyg	methek
llannerch	lanherch	meidr	bounder
llawen	lowen	*mei(dd)-	*með
llech	*legh	meillionen	melhyonen
llefrith	leuerid	meinog	*meynek
lleidr	lader	melin	melin¹
llestr	lester	melin wynt	*melyn wyns
llethr	*lether	melog	*melek
lleyg	lek	melyn	melyn²
lliain	lyen	merthyr	*merther
lliw	lyw	mesglyn	mesclen
llo	lugh	milgi	mylgy
lloc (MlWe.)	*lok	min	myn
llosg	losc	moch	mogh
llosgi	leskys	moel	*moyl
llost	lost	mollt	mols
llu	lu	môr	mor
lludw	lusew	moreb	*morrep
luird (OWe.)	lowarth	morfa	*morva
llurig	*luryk	morhwch	morhoch
llus	*lus	moydir	bounder
llwch	*loch	mŵn	*mon²

WELSH AND BRETON COGNATES

We.	Co.	We.	Co.
mws	mosek	plwyf	plu
mwyalch	moelh	poeth	*poth
mwyar	moyr-	pont	pons
mwynglawdd	*mon-gleth	porchell	porhel
mynach	manach	porth	porth
mynawyd	*menawes	post	post
mynydd	meneth	pren	pren
myrion (MlWe.)	*moryon	pridd	pry
		prins	pryns
nadd	*nath	pryf	prif
nant	nans	prys(g)	*prys(k)
naw	naw	pur	pur
nerth	nerth	pwdr	podar
nesaf	nessa	pwll	pol
newydd	nowyth	pys	pêz
noeth	noth		
nyfed	*neved	rhag-	rag
nyth	nyth	*rhed	*red
		rhedyn	reden
		rhiw	*rew
odyn	*oden	rhodwydd	*rodwyth
oen	on	-rhon	ruen
ofnog	ownek	rhos	*ros
o vaes (MlWe.)	aves	*rhosan	*rosan
-og	-ek	rhudd	ruth
ogof	googoo	rhwd	*rud
-ol	*-yel	rhyd	rid
onnen	onnen	rhyn	*rynn
padell	padel	Saeson	Zowzon
pant	*pans	saeth	seth
parc	park	safn	sawn
pau	pow	Sais	*Seys
pawr	*peur	sant	sans
pedwar	peswar	sawdd	*soð
pefr	*pever	seren	steren
pêl	*pel	serth	serth
pell	pell	siambr	chammbour
pellen	pellen	sofl	zowl
pen	pen	swrn	sorn
*penardd	*pen-arth	sych	segh
*pen(i)arth	*pen-arth	syfi	sevi
penrhyn	*pen-ryn		
pentir	*pen-tyr	tafod	taves
perfedd	perveth	tagell	*tagell
perth	*perth	tail	teil
pibell	*pybell	tâl	tal
pig	*pyk	?talar	*talar
pîn	pin-	tam	tam
pistyll	*pystyll	?taradr	*talar
pit	pyt	taran	taran
plas	plas	tarw	tarow

WELSH AND BRETON COGNATES

We.	Co.	We.	Co.
tawel	*tawel	uch	ugh
teg	teg	uchel	ughel
tenau	tanow	*udd	*yuf
tenewyn	tenewen	uffern	yfarn
tin	tyn	ugain(t)	ugens
tir	tyr	us	usion
tomen	tûban		
ton	ton	wast	wast
top	top	wicwr	guicgur
tor	tor	wrth	worth
torth	*torthell	ŵy	oye
traeth	trait		
trai	trig	ychen	oghan
trap	*trap	ŷd	eys
*traw	*treu	ymenyn	amanen
traws	dres	ymyl	*amal
tre(f)	tre	ynial	*enyal
tri	try¹	ynys	enys
trigfa	trygva	y(r)	an
?trochi	*trokya	ysbyddad	spethas
troed	tros	ysgaw	scawen
trum	*drum	ysglyf	scoul
trwch	trogh	*ysgod	schus
trwyn	tron	ysgraff	skath
try-	*try-²	ysgrif	scrife
tu	tu	ysgubor	skyber
tud	tus	ysgwfl	scoul
twll	toll	ysgwr	scorren
twrch	torch	ysgyfarn	scovern
twym	tum	ystabl	*stable
tŷ	chy	ystaen	stean
tyfu	tevys	ystafell	steuel
tyno	*tnou	ystrad	*stras
tywarch	*towargh	(y)stryd	*stret
tywyll	tewl	ystum	*stum
tywyn	towan	ywen	hiuin

BRETON COGNATES

Br.	Co.	Br.	Co.
alarc'h	elerhc	?arabad	*gortharap
amanenn	amanen	arar	ardar
amoug	*kevammok	arc'hant	arghans
an	an	aruuistl (OBr.)	*arwystel
anken	anken	askorn	ascorn
annez	anneth	auo(u)n (MlBr.)	auon
ant	nans	a-vaez	aves
antell	antell	aval	aval
aod	als	avalenn	auallen
aonik	ownek	avel	awel
?aos	*aweth	awrec	*havrek

WELSH AND BRETON COGNATES

Br.	Co.	Br.	Co.
bac'h	**bagh**	?c'hwiliañ	**wheyl**
banal	**banathel**		
bann	**ban**	dall	**dall**
bara	**bara**	dan	**dan**
barged	**bargos**	dañs	***dons**
barr	**bar**	dañvad	**daves**
bas	***bas**	daou	**dew**
begel	**begel**	daouzek	**dewthek**
beler	**beler**	dar (OBr.)	**dar**
berr	**ber**	De-	***to-**
bevin	**bowyn**	deil	**deyl**
bez	**beth**	den	**den**
bezv	**bedewen**	?derch (MlBr.)	***derch**
bihan	**byghan**	derv	**dar**
bili	***byly**	devet	***dywys**
binnigañ	**benyges**	di-	**dy-**
bioù	***byu**	diaoul	**dyowl**
blein	**blyn**	dibr	**diber**
bleiz	**bleit**	?dillo	***dyllo**
blouch (MlBr.)	**blogh**	din (OBr.)	***dyn**
bouc'h	**bogh**	diner	**dyner**
bougeo	**vooga**	don	**down**
boulc'h	***bolgh**	dorn	**dorn**
bour-	**bor**	dorojoù	**daras**
boutig (OBr.)	**boudzhi**	douar	**dor**
bran	**bran**	doue	**dev**
bras	**bras**	dour	**dour**
bre	***bre**	dourgi	**doferghi**
brec'h 'pox'	***brygh**	drein	**dreyn**
brec'h 'arm'	**bregh**	dreist	**dres**
brein	***breyn**	-dreu (MlBr.)	***treu**
brennig	**brenigan**	drez	**dreys**
brizh	***bryth**	du	**du**
bro	**bro**		
broc'h	**brogh**	ebeul	**ebel**
broenn	**bronnen**	ed	**eys**
bronn	**bron**	eflenn	**aidlen**
brug	***gruk**	ejen	**odion**
brulu	**breilu**	-ek	**-ek**
bugel	**bugel**	elestr	**elester**
buoc'h	**bugh**	enez	**enys**
buorth (OBr.)	***buorth**	enk	**yn**
		-enn	**-en**
chacc (MlBr.)	***chas**	eontr	**euiter**
chaoser	***cawns**	er(er)	**er**[1]
chapel	**chapel**	erv	**erw**
c'hoar	**wuir**	esker	***esker**
c'hoari	**guary**	eskob	**epscop**
c'hwec'h	**whe**	ethin (OBr.)	**eithin**
?C'hwevrer	***whevrer**	-etic (OBr.)	**-esyk**
c'hwezh	***wheth**	etre	**ynter**
c'hwil	**hwilen**	?evlec'h	**elaw**

WELSH AND BRETON COGNATES

Br.	Co.	Br.	Co.
evn	ethen	goulou	golow
evor	*evor	goumon	gubman
ezhomm	ethom	gour	gour
ezlen (MlBr.)	aidlen	gour-gourrin (MlBr.)	*gor-*hin
faou	*faw	?gourvez	*gorweth
fav	fav	gov	gof
fennañ	*fenna	govel	gofail
feunteun	fenten	gras	gras
fin	fyn	gre	*gre
finvez	fynweth	greunenn	gronen
forn	forn	grouan	grow
foz	fos	gwadañ	*gosa
?frank	Frank	gwak	gvak
fri	*fry	gwan	guan
froud	frot	gwaz 'goose'	goth²
fubu	guibeden	gwaz 'servant'	guas
		gwazh	goth¹
gaol	*gawl	gwazhell	*gothel
gaonac'h	gawna	gweañ	gwîa
gar	*gar	gueid- (OBr.)	guyth²
garan	garan	gwele	guely
garv	garow	gwellañ	guella
garzh	*garth	gwenn	guyn
gavr	gaver	gwennili	guennol
gelaouenn	ghel	gwent	guyns
geler	geler	gwer	guirt
genn	*genn	gueretreou (OBr.)	gweras
genou	ganow	gwern	guern
geot	gwels	guest- (OBr.)	gvest
geun	goon	gwez	gwyth¹
gioc'h	kio	guic (OBr.)	*gwyk
glad	gulas	?-guydel (MlBr.)	*gwythel
glann	glan	guygourr (MlBr.)	guicgur
glaou	glow²	gwinizh	gwaneth
glas	glas	gwiz	guis
glasten	glastan	guolt (OBr.)	gols
glazenn	*glasen	gwrac'h	gruah
glec'h	*gwleghe	gwreg	gwrek
glenn	*glynn		
gloedic (MlBr.)	*gwlesyk	ha(g)	ha(g)
-gloeu (OBr.)	*glow¹	hakr	hager
goañv	goyf	hal	hal
goast (MlBr.)	wast	halegenn	heligen
golbinoc (OBr.)	geluin	hanter	hanter
gored	*cores	hañv	haf
gortoz	gortos	havreg	*havrek
gou-goubant	*go-*gobans	?heiz	*heth
gouel	gol		
goueled	goles		
gouer	gover		

WELSH AND BRETON COGNATES

Br.	Co.	Br.	Co.
?heizez	*heth	karreg	karrek
helc'hiñ	*helgh	karv	carow
hen	hen	kastell	castell
hent	-hins-	kazeg	cassec
heol	houl	kazel	*casel
hep	hep	kazh	cath
hesk	heschen	keal	whelth
(qe-)-hezl	whelth	kebr	keber
higolenn	*ygolen	kef	*kyf
hili	hyly	kegin	*kegen
?hinkin	*henkyn	kein	keyn
hir	hyr	kelc'h	*kylgh
hoc'h	hoch	kelenn	kelin
hogenn	*hogen	kelvez	*collwyth
hogos	ogas	kember	*kemer
hont	hans	?kemenañ	*kemyn
houad	hos	?kemenn	*kemyn
houarn	horn	quemiada (MlBr.)	cummyas
huel	ughel	kenkiz	*kenkith
hwibed	guibeden	kentr	kenter
		keñver	*kevar
-ic (OBr.)	-yk	kêr	*ker
-idik	-esyk	quercheiz (MlBr.)	cherhit
ifern	yfarn	?kern	*kernan
iliav	idhio	?kernigell	kodna huilan
ilin	elin	kerzhinenn	kerden
iliz	eglos	kest	*kest
is (OBr.)	*is[1]	keuneud	kunys
islonk	*yslonk	kev	*kew
ivin	hiuin	ki	ky
izel	ysel	kignen	kenin
		kil	*kyl
jard(r)in	jarden	kilhog	kullyek
jentil	gentyl	*killi	kelli
		klañvdi	*clav-jy
kab	capa	kleger	*cleger
kador	cadar	kleuz	cleath
kae	kee	klud	*clus
?kagal	*cagh	koad	cos
kailhastr	cellester	coll (OBr.)	*coll
kalc'h	kal	kollet	kellys
kalet	cales	komm	*comm
kambr	chammbour	kompez	compes
kamm	cam	contulet (OBr.)	cuntell
-cann (OBr.)	can	*cor (OBr.)	*cor[1]
kanol	*canel	korf	corf
kant	cans	korn	corn
kaoc'h	*cagh	korz	*cors
karn	carn	koskor	coscor
karr	*car	coufranc (MlBr.)	keverang
karrbont	*car-bons		
karrdi	*car-jy		

WELSH AND BRETON COGNATES

Br.	Co.	Br.	Co.
koulin	**kynin**	linad	**lynas**
koulomer	***colomyer**	liorzh	**lowarth**
kozh	**coth**	liv	**lyw**
cragg (MlBr.)	***crak**[1]	livrizh	**leuerid**
?krak	**crak**[2]	loc'h	***loch**
krann	***crann**	logod	**logaz**
kraoñ	***cnow**	Lok-	***lok**
kraou	**krow**	lonk	***lonk**
krec'h 'hill'	***cnegh**	losk	**lose**
krec'h 'crinkled'	***crygh**	lost	**lost**
kreiz	**cres**	louan	**lovan**
kren	***cren**	louarn	**lowarn**
?kreun	***creun**	louet	**los**
krib	**krib**	louzoù	**les**
krin	***cryn**	-lu (OBr.)	**lu**
?crit (OBr.)	***crys**	ludu	**lusew**
?kriz	***crys**	lus	***lus**
kroaz	**crous**		
croaz pren		*ma	*ma
(MlBr.)	**crouspren**	maen	**men**
?kroc'henek	**cronek**	maen-hir	***men hyr**
croes (MlBr.)	**crous**	maez	**mes**
krogen	**crogen**	magañ	**maga**
kromm	**crom**	malan	**manal**
krommlec'h	***cromlegh**	mamm	**mam**
krug	**cruc**	manac'h	**manach**
kudenn	**cudin**	maouez	**mowes**
cul (OBr.)	**cul**	maout	**mols**
?kumun	***kemyn**	marc'h	**margh**
kutuilh	**cuntell**	marc'had	**marghas**
cuuranc (MlBr.)	**keverang**	marc'heg	**marrek**
		-med (OBr.)	***með**
laer	**lader**	meder	**midzhar**
laerez	***ladres**	meinek	***meynek**
laezh	**leth**	melchonenn	**melhyonen**
laezhdi	***lety**	melc'houed	***melwhes**
lamm	**lam**	melek	***melek**
*lann	***lann**	melen	**melyn**[2]
laouen	**lowen**	menez	**meneth**
le (OBr.)	**le**	mengleuz	***men-gleth**
lec'h	***legh**	merien	***moryon**
ledan	**ledan**	merzher	***merther**
lenn	**lyn**	meskl	**mesclen**
leskiñ	**leskys**	meur	**meur**
lestr	**lester**	mezeg	**methek**
leue	**lugh**	milin	**melin**[1]
levenez	**lowen**	miliner	***melynder**
leviader (?)	**lêuiader**	minaoued	***menawes**
lez	***lys**	minic'hi	***meneghy**
lien	**lyen**	moal	***moyl**
lik	**lek**	moan	**mon**[1]
limn (OBr.)	**leven**	moc'h	**mogh**

WELSH AND BRETON COGNATES

Br.	Co.	Br.	Co.
moger	*magoer	poull	pol
mor	mor	poull-rod	*pol ros
morhoc'h	morhoch	poul-pry	
mostoer (MlBr.)	*mynster	(MlBr.)	pwl prî
moualc'h	moelh	prad	pras
mouar	moyr-	prenn	pren
?moudenn	mÿdzhovan	preñv	prif
mougev	vooga	pri	pry
mouz	mosek	priñs	pryns
munut	munys	puñs	*puth
		pur	pur
nav	naw		
?naoz	*aweth	radell	*radgel
neizh	nyth	raden	reden
nerzh	nerth	rak	rag
nesañ	nessa	red	*red
nevez	nowyth	reunig	ruen
noazh	noth	-rinn (OBr.)	*rynn
		rit (OBr.)	rid
oan	on	*ros	*ros
oc'hen	oghan	*rosan	*rosan
ode	*aswy	roudouz	*rodwyth
onnenn	onnen	run	*run
ourgouilh(o)us		ruz	ruth
(MlBr.)	*orguilus		
ouzh	worth	saezh	seth
		sant	sans
?padell	padel	sa(o)n	sawn
park	park	Saozon	Zowzon
pell	pell	sec'h	segh
pellenn	pellen	sec'hor	sichor
penn	pen	segal	sÿgal
pennek	*pennek	serzh	serth
*penn-tir	*pen-tyr	sivi	sevi
permed (OBr.)	perveth	skaf	skath
peulvan	*peulvan	skav	scawen
peur	*peur	skeud	schus
pevar	peswar	skiber	skyber
pezel	padel	skouarn	scovern
pig	*pyk	skoul	scoul
?pil	pyl	skourr	scorren
pilad	*pylas	skrivañ	scrife
pin	pin-	soul	zowl
piz	pêz	spern	spern
plas	plas	spezad	spethas
ploue	plu	staen	stean
poazh	*poth	?stag	*stak
pont	pons	sterenn	steren
porc'hell	porhel	straal (OBr.)	strail
porzh	porth	stronk	*stronk
post	post	stumm	*stum
pou (OBr.)	pow		

WELSH AND BRETON COGNATES

Br.	Co.	Br.	Co.
tagell	*tagell	treizh	*treth
tal	tal	trev	tre
?talar	*talar	tri	try[1]
tamm	tam	tri-	*try-[2]
tanav	tanow	trionenn	*tryonenn
taol-vaen	*tol-ven	troad	tros
taouarc'h	*towargh	troc'h	trogh
taran	taran	tu	tu
tarv	tarow	tud	tus
?tec'h	*godegh	tuell(enn)	*tewel
teil	teil		
teñval	tewl	(a) uch (MlBr.)	ugh
teod	taves	ugent	ugens
teskoù	*tescow	uhel	ughel
tevenn	towan	?(a) us	*ussa
ti	chy	uzien	usion
tir	tyr		
tomm	tum	vi	oye
tonnen	ton		
tor	tor	war	war
torzhell	*torthell		
toull	toll	Yaou	yow
tourc'h	torch	yar	yar
touskan	*tusk	yen	yeyn
traezh	trait	-(y)er	-yer
traoñ	*tnou	yev	ieu
trap	*trap	-(i)ol (OBr.)	*-yel
tre (OBr.)	trig	yourc'h	yorch

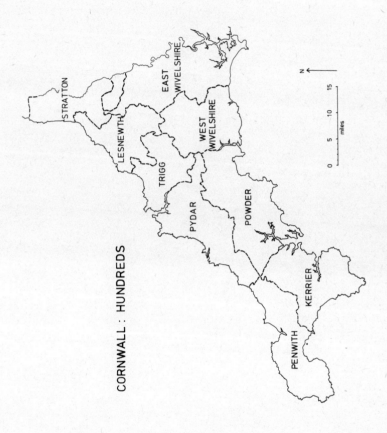

CORNWALL:
TRE as B2

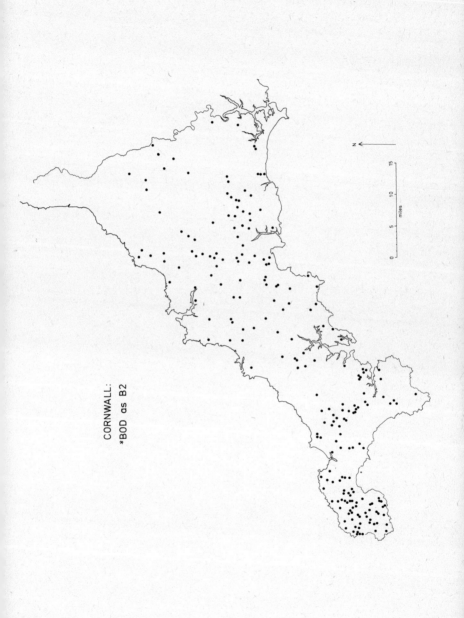

CORNWALL:
*BOD as B2

CORNWALL:
*KER as B2